The Analysis of Variance

A WILEY PUBLICATION IN MATHEMATICAL STATISTICS

The Analysis of Variance

HENRY SCHEFFÉ
Professor of Statistics
University of California, Berkeley

Wiley Classics Library Edition Published 1999

A Wiley-Interscience Publication
JOHN WILEY & SONS, INC.
New York • Chichester • Weinheim • Brisbane • Singapore • Toronto

This text is printed on acid-free paper. ♾

Published simultaneously in Canada.

Wiley Classics Library Edition published 1999.

Library of Congress Cataloging in Publication Data:

Library of Congress Catalog Card Number: 59-14994
ISBN 0-471-75834-5
ISBN 0-471-34505-9 (Classics Edition)

Printed in the United States of America

10 9 8 7 6 5 4 3 2 1

To

Maud Susan Sherwood

Preface

In this book I have tried to elucidate in a unified way what appears to me at present to be the basic theory of the analysis of variance. This necessitates considering several different mathematical models for the subject. The theory of Part I, namely that for fixed-effects models with independent observations of equal variance, I judge to be jelled into a fairly permanent form, but the theory of Part II, namely that under other models, I expect will undergo considerable extension and revision. Perhaps this presentation will help stimulate the needed growth. What I feel most apologetic about is the little I have to offer the reader on the unbalanced cases of the random-effects models and mixed models. These cannot be generally avoided in planning biological experiments, especially in genetics, the situation being unlike that in physical science. This gap in the theory I have not been able to fill.

The mathematical background necessary for the reader to understand this book is a course in calculus at some time in the past, and at least occasional use of some mathematical notation in the present. Very little of the calculus is actually employed, but the reader who never had it would be unlikely to have developed sufficient ease in the necessary language of mathematics. Most of the derivations in the book are of an algebraic nature. To facilitate the derivations in Chs. 1, 2, and 6, vector and matrix methods are extensively employed. The exposition of the needed vector and matrix algebra in Apps. I and II should make the book self-contained for the reader with the minimal mathematical background indicated above. The reader not at home with matrix notation should write out in longhand without this notation some of the first equations he encounters in this notation. Then soon he will reach the stage where matrix formulations are not only easier to look at and to write, but also to think in.

My decision to use matrix notation may be further justified in the following way. It is well known that one unifying and insightful way of regarding the analysis of variance is from the geometrical viewpoint: it may be viewed as a method of resolving the vector of observations into vectors lying in certain specified spaces corresponding to different sources of variation in the observations, and to each of which a meaningful interpretation can be given. For understanding the geometry of such resolutions and the geometrical interpretation of the statistics used to test whether the magnitudes of some of the component vectors associated with different sources are significant, the concept of orthogonality of vectors and spaces is indispensable. The easiest way of defining, applying, and manipulating this geometric concept is, I believe, through the use of matrix notation.

The statistical background necessary for the reader is knowledge equivalent to that aimed at in a sound year course in statistics stressing the concepts of elementary probablility, confidence intervals, and the power (or operating characteristics) of tests, and including use of the t-, χ^2, and F-distributions.

This book contains 117 problems at the ends of the chapters and appendices, of which 38 require numerical computations with "real" data. The variety of applications in these 38 problems should give some idea of the broad applicability of the analysis of variance, even though the problems were chosen only because they furnish suitable examples of the methods described in the text, and with no conscious attempt at inclusion of many substantive fields. The importance of carrying through a considerable amount of numerical work is greater here than it is in learning most branches of statistics. Indeed, some practitioners of the analysis of variance would regard the computational techniques as the most important part of the subject, and consider as perverted my emphasis on the choice of mathematical models. I realize that many practitioners have developed reliable intuitive and verbal paths to the correct analysis in given situations without defining the model, but I find it easier to follow the path to which I am constrained by the choice of model; the approach of choosing the model and then making the analysis dictated by it seems to me also to be simpler to teach, as well as more appropriate for a book on the theory of the subject.

The book is intended as a text for a one-semester or two-quarter course at the senior or graduate level, and for self-study. At Berkeley in a semester graduate course meeting for three lecture hours and two laboratory sessions per week the material in the book is covered except for Chs. 5, 8, and 9, which are included with other topics in a course on the design of experiments for which this course is prerequisite. In future we will expect the student before starting this course to have acquired a knowledge of matrix algebra at least equal to that obtainable from an elementary course, or to have worked his way through

Apps. I and II of this book. For a shorter course Ch. 6 and parts of Ch. 7 might also be omitted.

The following topics are omitted, since one purpose of the book is to serve as a text for a course in the analysis of variance, and these omissions are usually covered in other courses in statistics departments: the multivariate generalization of the exclusively univariate theory developed here, sequential methods in the analysis of variance, and nonparametric theory, except for the permutation tests based on the F-statistics for the Latin-square and incomplete-blocks designs. For the same reason the design of experiments is touched on only incidentally, and the theories of confounding, fractional replication, response-surface exploration, and the more complicated experimental designs receive no mention. However, the omission of the decision-theory approach to the subject is mainly for another reason: Except possibly for one problem (experiments designed for choosing the best of a set of treatments, see sec. 3.7), this approach seems to me to have yielded as yet no important new useful methods in this, perhaps the most widely used, branch of statistics, where typically many possible decisions from a set of data are considered.*

I earnestly hope this book will be suitable for self-study, the route which many users of statistical methods have had to follow because the subject is still not available, or not encouraged, in many college and university programs of training for scientific and engineering professions. For the reader wishing to master the subject in this way the above remark about the importance of some numerical computation is especially pertinent, and I urge him to work most of the 38 problems involving data. If he cannot find access to a desk calculating machine he will generally have to calculate directly from the definitions of the various sums of squares and not, because of the consequent loss of significant figures, from the computing formulas given in the book and intended for use on machines, as explained at the end of sec. 3.1. As for the remaining problems, the reader without benefit of a teacher should not feel discouraged if he cannot solve them all, for they vary in mathematical difficulty from being easy for all to being easy only for professional mathematicians. The lone reader may also find it helpful, when the argument is geometrical or permits a geometrical interpretation, to sketch figures, like Fig. 2.9.1, which suggest the n-dimensional geometrical relationships in a two-dimensional representation.

I am also hopeful that the reader will not experience the abundance

* However, recent work by Kiefer and Wolfowitz (1958) opens the possibility that a game-theoretic approach to the problems of optimum experimental design may yield new solutions which are useful and computable.

of footnotes as a pedantic obstacle but will realize they are an efficient device for writing simultaneously for two classes of readers with different degrees of scholarly interest: The footnotes concern (i) literature citations and other indications of the history of the subject, and (ii) finer points of the theory. Most readers will wish to read all or none, and will quickly decide to which class they belong. A few footnotes are of a different nature; special attention is called to them in the body of the text.

HENRY SCHEFFÉ

Princeton, N. J.
November 1958

Acknowledgments

Dr. Jean-Pierre Imhof wrote up lecture notes for the course on this subject which I gave in Berkeley in the spring of 1954. Although this book drastically revises and extends those notes, it never would have gotten off the ground without that springboard. Dr. Imhof also helped me with the problems at the ends of the chapters and appendices: He originated the 12 problems numbered 2.3, 2.7, 2.8, 4.4, 4.15, 9.1, I.1, II.1, II.2, II.9, II.10, III.1, and he helped prepare and tried out the 14 problems numbered 3.1, 4.2, 4.3, 4.5, 4.6, 4.7, 4.8, 5.1, 5.4, 5.7, 6.1, 6.3, 9.2, I.2.

Many helpful suggestions were made by Professor Werner Gautschi, Dr. Imhof, and Dr. Mervin E. Muller, who read the manuscript.

Professor E. S. Pearson has added much to the value of the book by his permission to reproduce the excellent tables and charts from *Biometrika*.

Dr. James Pacheres of Hughes Aircraft Company kindly accelerated the computation of the table of 10 per cent points of the Studentized range to permit its inclusion in this book.

When I wrote Mrs. Maxine Merrington for permission to publish the result of rounding off to three significant figures the five-figure *F*-table computed by her and Miss Catherine M. Thompson, she graciously offered to verify in my table, from a more exact table in her files, the entries where the last two figures rounded off were 50; she not only did this but checked my whole table.

Data that furnish clean-cut illustrations of statistical methods are generally not easy to find, and, in "borrowing" data from books and papers by fellow statisticians who cited other primary sources, I have credited the statistician, who saw the illustrative value of the data, rather than the original worker, who gathered them for their substantive value.

While reading over the manuscript as a whole before sending it to the publisher, I was struck by the realization of how many of the major (secs. 3.5 and 8.1) and minor pieces of my research that had gone into it and into my papers had grown out of problems suggested to me by Mr. Cuthbert Daniel, and by the thought that I had never acknowledged in print my debt to him.

Mrs. Julia Rubalcava contributed her unusual skill at mathematical typing.

The Office of Naval Research supported my work on the book during the summers of 1954, 1956, and 1958, and also that of Dr. Imhof and Mrs. Rubalcava. My work during the academic year 1958–59, which has included rounding out the problem sections, and will include some final revision and the many tasks connected with a book in press, is being supported by the National Science Foundation and the Office of Ordnance Research.

It turned out that I was impelled—mainly by the constructive needling of Dr. Mervin E. Muller, but also by my own perfectionist tendencies—to spend on polishing this book most of my work-time during the academic year 1958–59, during which I was supported by the National Science Foundation and the Office of Ordnance Research. My work of proof-reading and indexing during the summer of 1959 was supported by the Office of Naval Research.

H. S.

Contents

PART I. THE ANALYSIS OF VARIANCE IN THE CASE OF MODELS WITH FIXED EFFECTS AND INDEPENDENT OBSERVATIONS OF EQUAL VARIANCE

PART II. THE ANALYSIS OF VARIANCE IN THE CASE OF OTHER MODELS

PART I

The Analysis of Variance in the Case of Models with Fixed Effects and Independent Observations of Equal Variance

CHAPTER 1

Point Estimation

1.1. INTRODUCTION

The following rough definition of our subject may serve tentatively: The analysis of variance[1] is a statistical technique for analyzing measurements depending on several kinds of effects operating simultaneously, to decide which kinds of effects are important and to estimate the effects. The measurements or observations may be in an experimental science like genetics or a nonexperimental one like astronomy. A theory of analyzing measurements naturally has implications about how the experiment should be planned or the observations should be taken, i.e., *experimental design*. Historically, the present technique of analysis of variance has been developed mainly in connection with problems of agricultural experimentation.

An agricultural experiment of a relatively simple structure to which the analysis of variance would be applicable would be the following: In each of three localities four varieties of tomatoes are grown in tanks containing chemical solutions. Two different chemical solutions, which we shall call "treatments," are used, with different proportions of the chemicals. For each treatment in each locality there is a mixing tank from which the fluid is pumped to all the tanks on this treatment, connected "in parallel:" We do not want a "series" connection, where the outflow from one tank is the inflow to another, because this would *confound* the effects of the varieties in these two tanks with the effects (if any) of order in the "series" connection. The tanks are arranged outdoors with the same orientation,

[1] The analysis of variance, as commonly understood and practiced today, has been developed chiefly by R. A. Fisher (1918, 1925, 1935), who introduced the terms *variance* and *analysis of variance* into statistics. The latter term would seem more appropriate for the random-effects models (Ch. 7), and these may constitute the path by which Fisher himself originally approached the subject. For some historical background see Scheffé (1956*b*). Names followed by dates in parentheses refer to the author index and bibliography at the end of the book.

so that the plants in one tank will not appreciably shade those in another, etc. For each treatment in the three localities the chemicals are renewed according to the same specifications. Each variety is grown in a separate tank, with the same number of plants in each. The *yield* of each tank is the weight of ripe tomatoes produced. (Later we shall speak about an observed yield and a "true" or expected yield.)

The yield from a tank may depend on the variety, the chemical treatment, and the locality. In particular, it will depend on *interactions* among these factors, a useful concept of the analysis of variance that we will develop later (sec. 4.1). The sort of questions for which our theory offers answers is the following: Are the varieties different in yield when averaged over the two treatments and three localities? Do the yields demonstrate differential effects of the varieties for different localities? How can we quantitatively express the differences with a given degree of confidence? Etc.

The developments in the remainder of this chapter and in Ch. 2 are of a generality that may somewhat dismay a reader expecting to find results whose usefulness is clearly visible to him. Such a reader may be encouraged to work through these two chapters by the following remarks: Beginning with Ch. 3, he will find that most of the material in the rest of the book is of a more obvious usefulness. The general developments of Chs. 1 and 2 furnish the foundation not only for obtaining these results, but also for carrying out the analysis of variance in cases he may encounter that do not fall under any of those treated specifically and in detail in the rest of the book.

1.2. MATHEMATICAL MODELS

Suppose that we have n observations or measurements. In the mathematical models employed in this book it is assumed that these observations are values taken on by n random variables[2] $y_1, y_2, \cdots, y_n$, which are constituted of linear combinations of p unknown quantities $\beta_1, \beta_2, \cdots, \beta_p$ plus errors $e_1, e_2, \cdots, e_n$,

$$y_i = x_{1i}\beta_1 + x_{2i}\beta_2 + \cdots + x_{pi}\beta_p + e_i \qquad (i = 1, 2, \cdots, n), \tag{1.2.1}$$

where[3] the $\{x_{ji}\}$ are known constant coefficients. (The reader unfamiliar with the brace notation { } should read the footnote.[4]) The

[2] We will generally use the same symbols for random variables and for their observed values in this book. (Exceptions occur in secs. 2.10 and 9.3.)

[3] It might seem more natural to permute the subscripts on the x's in (1.2.1), but the present notation is standard. It would seem appropriate in situations where x_{ji} is the value assumed by an "independent" variable x_j in the ith observation; see sec. 6.1.

[4] The brace notation denotes the *set* of quantities indicated: In this case $\{x_{ji}\}$ means the set consisting of the np quantities x_{ji} with $j = 1, 2, \cdots, p$; $i = 1, 2, \cdots, n$.

$\{\beta_j\}$ are more or less idealized formulations of some aspects of interest to the investigator in the phenomena underlying the observations. The purpose of the analysis of variance is to make inferences about the $\{e_i\}$ and some of the $\{\beta_j\}$, the inferences to be valid regardless of the values of the other $\{\beta_j\}$, if any, which we may be more desirous of "eliminating" than "assessing."

A minimal assumption which is always made on the random variables $\{e_i\}$ is that their expected values are zero:

$$E(e_i) = 0 \qquad (i = 1, 2, \cdots, n). \tag{1.2.2}$$

We shall also usually assume that

$$E(e_i e_j) = \sigma^2 \delta_{ij}, \tag{1.2.3}$$

where σ^2 is an unknown constant and δ_{ij} is 0 or 1, according as $i \neq j$ or $i = j$, respectively. This is equivalent to saying that the random variables $\{e_i\}$ are uncorrelated (i.e., have zero coefficients of correlation) and have equal variance σ^2.

We may now make our definition in sec. 1.1 gradually more precise: The *analysis of variance* is a body of statistical methods of analyzing measurements assumed to be of the structure (1.2.1), where the coefficients $\{x_{ji}\}$ are integers usually 0 or 1. In order to clarify[5] this it is necessary to consider not only the values of the $\{x_{ji}\}$ but also their origin in the real situation being investigated: In the analysis of variance the $\{x_{ji}\}$ are the values of "counter variables" or "indicator variables" which refer to the presence or absence of the *effects* $\{\beta_j\}$ in the conditions under which the observations are taken: x_{ji} is the number of times β_j occurs in the ith observation, and this is usually[6] 0 or 1. If the $\{x_{ji}\}$ are values taken on in the observations not by counter variables but by continuous variables like t = time, T = temperature, t^2, e^{-t}, tT, t^0, etc. (these are called independent or concomitant variables, and the observations $\{y_i\}$ are then said to be on a dependent variable y; see sec. 6.1), we say we have a case of *regression analysis*. If there are some $\{x_{ji}\}$ of both kinds, we have an *analysis of covariance*. More natural and meaningful but equivalent definitions to distinguish among the three kinds of analysis, all of which fall under the general theory of Chs. 1 and 2, will be formulated in Ch. 6 after the reader has become accustomed to thinking in terms of the factors that are varied in an experiment or series of observations.

[5] These definitions and those of sec. 6.1 grew out of helpful discussions I had with Professor William Kruskal and Dr. Mervin Muller.

[6] For an example where some $x_{ji} = -1$ see Scheffé (1952), sec. 7; where some $x_{ji} = 2$, see Kempthorne (1952), sec. 6.8.

Up to now we have not specified the nature of the unknown effects $\{\beta_j\}$: They may be either unknown constants, which we then call parameters, or unobservable random variables subject to further assumptions about their distribution involving other unknown parameters. We shall call a model in which all the $\{\beta_j\}$ are unknown constants a *fixed-effects model.*[7] It often happens that one of the $\{\beta_j\}$ is a constant which occurs with every observation with coefficient 1 so that, for this j, $x_{ji} = 1$ for all i. We may call such a β_j an *additive constant* (in applications it is usually a "general mean" in some sense). A model in which all the $\{\beta_j\}$ are random variables, except possibly for one which is an additive constant, is called a *random-effects model.* Intermediate cases, where at least one β_j is a random variable and at least one is a constant not an additive constant, are called *mixed models.*

Examples: We wish to illustrate the notation now, but it is not convenient to use typical analysis-of-variance examples at this point because they would introduce other complications better postponed.

1. Consider the problem of fitting a polynomial of degree three, $y = a_0+a_1x+a_2x^2+a_3x^3$, to a set of observed values (x_i, y_i), $i = 1, \cdots, n$, assuming that y_i is a random variable, x_i is not, and the expected value of y_i is the ordinate on the cubic curve at $x = x_i$:

$$E(y_i) = a_0 + a_1x_i + a_2x_i^2 + a_3x_i^3.$$

We have in this case $p = 4$, $\beta_j = a_{j-1}$ $(j = 1, \cdots, 4)$. We note that the regression, in this case $a_0+a_1x+a_2x^2+a_3x^3$, need not be linear in the "independent" variable x, but only in the unknown parameters.

2. Another problem might be to fit a trigonometric polynomial to some periodic data with known period (which by change of the time scale we could make 2π):

$$\begin{aligned} E(y_i) = a_0 &+ a_1 \cos t_i + b_1 \sin t_i \\ &+ a_2 \cos 2t_i + b_2 \sin 2t_i \\ &+ a_3 \cos 3t_i + b_3 \sin 3t_i. \end{aligned}$$

Here the observation y_i is made at time t_i and the $\{\beta_j\}$ are the seven a's and b's.

These examples indicate that our models include a great variety of situations.

The development of the general theory in Chs. 1, 2, and 6 is greatly facilitated[8] by the use of vector and matrix algebra. The author hopes he has given a sufficient introduction to this in Apps. I and II. We define

[7] Fixed-effects models are also called Model I, and random-effects models, Model II, following Eisenhart (1947).

[8] See Preface.

the vectors (vectors and matrices will always be printed in **boldface** type)

$$\mathbf{y}^{n\times 1} = \begin{pmatrix} y_1 \\ y_2 \\ \cdot \\ \cdot \\ \cdot \\ y_n \end{pmatrix}, \qquad \boldsymbol{\beta}^{p\times 1} = \begin{pmatrix} \beta_1 \\ \beta_2 \\ \cdot \\ \cdot \\ \cdot \\ \beta_p \end{pmatrix}, \qquad \mathbf{e}^{n\times 1} = \begin{pmatrix} e_1 \\ e_2 \\ \cdot \\ \cdot \\ \cdot \\ e_n \end{pmatrix},$$

and the matrix

$$\mathbf{X}^{p\times n} = \begin{pmatrix} x_{11} & x_{12} & \cdots & x_{1n} \\ x_{21} & x_{22} & \cdots & x_{2n} \\ \cdot & \cdot & \cdots & \cdot \\ x_{p1} & x_{p2} & \cdots & x_{pn} \end{pmatrix},$$

where superscripts $r\times s$ on a matrix indicate that the matrix has r rows and s columns. When there is no risk of ambiguity we drop the superscripts. The set of equations (1.2.1) then takes the simple form

$$\mathbf{y} = \mathbf{X}'\boldsymbol{\beta} + \mathbf{e}, \tag{1.2.4}$$

where $\mathbf{X}'$ denotes the transpose of $\mathbf{X}$.

Matrix Random Variables

Definition: Given a matrix $\mathbf{V}^{r\times s}$ of jointly distributed random variables $\{v_{ij}\}$ with finite expectations,

$$\mathbf{V} = \begin{pmatrix} v_{11} & v_{12} & \cdots & v_{1s} \\ v_{21} & v_{22} & \cdots & v_{2s} \\ \cdot & \cdot & \cdots & \cdot \\ v_{r1} & v_{r2} & \cdots & v_{rs} \end{pmatrix},$$

we define the *expected value of the matrix* $\mathbf{V}$ to be the matrix

$$E(\mathbf{V}) = \begin{pmatrix} E(v_{11}) & E(v_{12}) & \cdots & E(v_{1s}) \\ E(v_{21}) & E(v_{22}) & \cdots & E(v_{2s}) \\ \cdot & \cdot & \cdots & \cdot \\ E(v_{r1}) & E(v_{r2}) & \cdots & E(v_{rs}) \end{pmatrix}. \tag{1.2.5}$$

This definition enables us to write the conditions (1.2.2) and (1.2.3) in the condensed matrix form

(1.2.6) $$E(\mathbf{e}) = \mathbf{0}, \qquad E(\mathbf{ee}') = \sigma^2\mathbf{I},$$

where $\mathbf{0}$ is the $n\times 1$ zero matrix and $\mathbf{I}$ is the $n\times n$ identity matrix.

Lemma: If $\mathbf{A}^{q\times r}$ and $\mathbf{B}^{s\times t}$ are matrices of constants and $\mathbf{V}^{r\times s}$ is a matrix of random variables we have the relation

(1.2.7) $$E(\mathbf{AVB}) = \mathbf{A}\, E(\mathbf{V})\mathbf{B}.$$

Proof: In the proof only the linear operator property of the operator E on ordinary random variables is utilized, i.e., $E(ax+by) = a\,E(x)+b\,E(y)$, if a and b are constants, and x and y are random variables.

Covariance Matrices

Consider a vector $\mathbf{v} = (v_1, \cdots, v_n)'$ of jointly distributed random variables all having finite variance. We call the matrix

(1.2.8) $$\boldsymbol{\Sigma}_v = (\text{Cov}\,(v_i, v_j)),$$

whose i,j element is the covariance of v_i and v_j, the *covariance matrix* of $\mathbf{v}$. Write $\mu_i = E(v_i)$, so $\text{Cov}\,(v_i, v_j) = E[(v_i-\mu_i)(v_j-\mu_j)]$. Then by (1.2.5) we may write

(1.2.9) $$\boldsymbol{\Sigma}_v = E[(\mathbf{v}-\boldsymbol{\mu})(\mathbf{v}-\boldsymbol{\mu})'],$$

where $\boldsymbol{\mu} = E(\mathbf{v})$.

We shall make frequent use of the following property: For a linear transformation

$$\mathbf{w}^{m\times 1} = \mathbf{A}^{m\times n}\mathbf{v}^{n\times 1}$$

from n random variables $v_1, \cdots, v_n$ to m random variables $w_1, \cdots, w_m$, with matrix $\mathbf{A}$, the covariance matrix of $\mathbf{w}$ is given by

(1.2.10) $$\boldsymbol{\Sigma}_w = \mathbf{A}\boldsymbol{\Sigma}_v\mathbf{A}'.$$

Proof: $\boldsymbol{\Sigma}_w = E([\mathbf{w} - E(\mathbf{w})][\mathbf{w} - E(\mathbf{w})]') = E(\mathbf{A}[\mathbf{v} - E(\mathbf{v})][\mathbf{v} - E(\mathbf{v})]'\mathbf{A}')$ $= \mathbf{A}\, E([\mathbf{v} - E(\mathbf{v})][\mathbf{v} - E(\mathbf{v})]')\mathbf{A}' = \mathbf{A}\boldsymbol{\Sigma}_v\mathbf{A}'$.

1.3. LEAST-SQUARES ESTIMATES AND NORMAL EQUATIONS

We use the symbol Ω throughout this book to denote a set of fundamental or underlying assumptions. Here we consider the following ones already introduced in sec. 1.2:

$$\Omega:\quad \mathbf{y}^{n\times 1} = \mathbf{X}'\boldsymbol{\beta}^{p\times 1} + \mathbf{e}^{n\times 1}, \qquad E(\mathbf{e}) = \mathbf{0}, \qquad E(\mathbf{ee}') = \sigma^2\mathbf{I},$$

which may be written even more briefly as

$$\Omega: \quad E(\mathbf{y}) = \mathbf{X}'\boldsymbol{\beta}, \qquad \boldsymbol{\Sigma}_y = \sigma^2\mathbf{I}.$$

Suppose that $b_1, \cdots, b_p$ denote quantities which we might consider using as estimates of $\beta_1, \cdots, \beta_p$. The $\beta_1, \cdots, \beta_p$ are fixed unknown constants, whereas the $b_1, \cdots, b_p$ will be quantities that we vary freely in deciding which are the "best" values in some sense. For any $\mathbf{b} = (b_1, \cdots, b_p)'$ we form

$$\mathscr{S}(\mathbf{y}, \mathbf{b}) = \sum_{i=1}^{n} \left(y_i - \sum_{j=1}^{p} x_{ji} b_j \right)^2. \tag{1.3.1}$$

This might be interpreted as $\Sigma_1^n \hat{e}_i^2$, where $\hat{e}_i$ denotes the estimate of the error e_i in the observation y_i in (1.2.1) if $\boldsymbol{\beta}$ is estimated by $\mathbf{b}$. It can be regarded as a measure of how well the model with $\boldsymbol{\beta}$ estimated by $\mathbf{b}$ fits the observations; the smaller $\mathscr{S}$ is, the better the fit. In matrix notation we may write $\mathscr{S} = (\mathbf{y}-\mathbf{X}'\mathbf{b})'(\mathbf{y}-\mathbf{X}'\mathbf{b})$ or, if the length of a vector $\mathbf{v}$ is denoted by $\|\mathbf{v}\|$, as

$$\mathscr{S}(\mathbf{y}, \mathbf{b}) = \|\mathbf{y}-\mathbf{X}'\mathbf{b}\|^2. \tag{1.3.2}$$

Definition: A set of functions[9] of $\mathbf{y}$ (i.e., a set of *statistics*),

$$\hat{\beta}_1 = \hat{\beta}_1(\mathbf{y}), \cdots, \hat{\beta}_p = \hat{\beta}_p(\mathbf{y}),$$

such that the values $b_j = \hat{\beta}_j$ ($j = 1, \cdots, p$) minimize $\mathscr{S}(\mathbf{y}, \mathbf{b})$, is called a set of *LS* (*least-squares*)[10] *estimates* of the $\{\beta_j\}$.

Normal Equations

We shall see that LS estimates always exist, but need not be unique. Later it will be seen that any set of LS estimates satisfies the conditions (the reader whose calculus is rusty should read the footnote[11])

$$\partial\mathscr{S}(\mathbf{y}, \mathbf{b})/\partial b_\nu = 0 \qquad (\nu = 1, \cdots, p).$$

[9] For the mathematically advanced reader we remark that here and elsewhere we mean *measurable* functions. If the $\{\hat{\beta}_j\}$ are unique they turn out to be linear functions of the $\{y_i\}$; if they are not unique there are infinitely many which are linear functions, and it might be convenient (for example in considering their covariance matrix) to restrict them to linear functions.

[10] The LS (least-squares) method of estimation was invented independently, and published in books on astronomical problems, by Gauss (1809) and Legendre (1806).

[11] The notation $\partial\mathscr{S}(\mathbf{y}, \mathbf{b})/\partial b_\nu$ denotes the partial derivative of $\mathscr{S}(\mathbf{y}, \mathbf{b})$ with respect to b_ν, meaning an ordinary derivative with respect to b_ν when the other $\{b_j\}$ are held constant. All the partial derivatives needed in this book can be written down at once by the following rule: We shall always want the partial derivative with respect to some variable θ of a function $\mathscr{S}$ which is a sum of squares of expressions each of which is linear in θ (i.e., of the form $A+B\theta$, where A and B do not depend on θ). This partial derivative is equal to twice the sum of products of those expressions (not their squares!) containing θ by the coefficients of θ in those expressions.

These give

$$\partial\mathscr{S}/\partial b_\nu = -2\sum_{i=1}^{n}\left(y_i - \sum_{j=1}^{p} x_{ji}b_j\right)x_{\nu i} = 0 \qquad (\nu = 1, \cdots, p),$$

or

$$\sum_{i=1}^{n}\sum_{j=1}^{p} x_{\nu i}x_{ji}b_j = \sum_{i=1}^{n} x_{\nu i}y_i \qquad (\nu = 1, \cdots, p). \tag{1.3.3}$$

These equations may now be written in matrix form

$$\mathbf{XX'b} = \mathbf{Xy},$$

or, with $\mathbf{S} = \mathbf{XX'}$,

$$\mathbf{Sb} = \mathbf{Xy}. \tag{1.3.4}$$

These are the *normal equations*. We use the symbol $\hat{\hat{\boldsymbol\beta}}$ for any solution $\mathbf{b}$ of them, reserving the symbol $\hat{\boldsymbol\beta}$ for LS estimates exclusively. However, we are going to show that every solution of the normal equations is a set of LS estimates, and every set of LS estimates satisfies the normal equations, so that thereafter it will be justified to write simply $\hat{\boldsymbol\beta}$ instead of $\hat{\hat{\boldsymbol\beta}}$. In setting up and solving the normal equations in practice, we do not distinguish in the notation between $\mathbf{b}$ and $\boldsymbol\beta$, setting $\partial\mathscr{S}(\mathbf{y}, \boldsymbol\beta)/\partial\beta_\nu = 0$ $(\nu = 1, \cdots, p)$, solving for $\boldsymbol\beta$, and then denoting the solution $\boldsymbol\beta$ by $\hat{\boldsymbol\beta}$. We hope that it will not be confusing after this explanation.

Geometrical Interpretation

We are now going to prove the existence of the LS estimates and their equivalence to the solutions of the normal equations. For this purpose we use results from vector algebra which are derived in App. I.

In the n-dimensional space V_n we introduce the vector of means $\boldsymbol\eta = E(\mathbf{y})$; so under Ω

$$\boldsymbol\eta^{n\times 1} = \mathbf{X'}\boldsymbol\beta, \tag{1.3.5}$$

which we also write as (this is obvious by the interpretation of matrix multiplication above (II.7) of App. II)

$$\boldsymbol\eta = \beta_1\boldsymbol\xi_1 + \beta_2\boldsymbol\xi_2 + \cdots + \beta_p\boldsymbol\xi_p,$$

where the vector $\boldsymbol\xi_j^{n\times 1}$ is the jth column of $\mathbf{X'}$.

Let r denote the rank of $\mathbf{X}$, and V_r the r-dimensional vector space spanned (see App. I) by the vectors $\boldsymbol\xi_1, \cdots, \boldsymbol\xi_p$. Then a vector $\mathbf{z}^{n\times 1}$ lies in V_r if and only if there exist coefficients $b_1, \cdots, b_p$ such that $\mathbf{z} = b_1\boldsymbol\xi_1 + \cdots + b_p\boldsymbol\xi_p$. In particular, $\boldsymbol\eta \in V_r$ under Ω.

Let $\mathbf{z} = \mathbf{X'b}$, where we think of varying $\mathbf{b}$. Then from Theorem 2 of App. I it follows that $\mathscr{S}(\mathbf{y}, \mathbf{b}) = \|\mathbf{y}-\mathbf{z}\|^2$ has a minimum which is attained when and only when $\mathbf{z}$ is the vector $\hat{\boldsymbol{\eta}}$ defined to be the projection of $\mathbf{y}$ on V_r. Since $\hat{\boldsymbol{\eta}}$ is a vector in V_r, it can be written as a linear combination of $\boldsymbol{\xi}_1, \cdots, \boldsymbol{\xi}_p$; i.e., there exist $b_1, \cdots, b_p$ such that

$$\hat{\boldsymbol{\eta}} = b_1\boldsymbol{\xi}_1 + \cdots + b_p\boldsymbol{\xi}_p; \tag{1.3.6}$$

here $\hat{\boldsymbol{\eta}}$ is unique but the $\{b_j\}$ in general are not. Since $\hat{\boldsymbol{\eta}}$ is a function of $\mathbf{y}$ only, and not of unknown parameters, the $\{b_j\}$ in (1.3.6) may also be taken to be functions of $\mathbf{y}$ only, and they are then LS estimates—whose existence we have now demonstrated. Furthermore, any $\{b_1, \cdots, b_p\}$ which are functions of $\mathbf{y}$ only will be a set of LS estimates if and only if $\mathbf{X}'\mathbf{b} = \hat{\boldsymbol{\eta}}$, which is true if and only if each of the following statements holds (each statement is true if and only if the following one is true; the symbol $\perp$ denotes "is orthogonal to"),

$$\begin{aligned} &\mathbf{X}'\mathbf{b} = \hat{\boldsymbol{\eta}}, \\ &\mathbf{y} - \mathbf{X}'\mathbf{b} \perp V_r, \\ &\mathbf{y} - \mathbf{X}'\mathbf{b} \perp \boldsymbol{\xi}_j \qquad (j = 1, \cdots, p), \\ &\boldsymbol{\xi}_j'(\mathbf{y} - \mathbf{X}'\mathbf{b}) = 0 \qquad (j = 1, \cdots, p), \\ &\mathbf{X}(\mathbf{y} - \mathbf{X}'\mathbf{b}) = \mathbf{0}, \end{aligned} \tag{1.3.7}$$

$$\mathbf{X}\mathbf{X}'\mathbf{b} = \mathbf{X}\mathbf{y}. \tag{1.3.8}$$

Here (1.3.7) follows from Lemma 8 of App. I, and (1.3.8) states that $\mathbf{b}$ satisfies the normal equations. We have now proved that LS estimates $\hat{\beta}_1, \cdots, \hat{\beta}_p$ always exist, that any set of LS estimates satisfies the normal equations, and that any solution $\hat{\hat{\beta}}_1, \cdots, \hat{\hat{\beta}}_p$ of the normal equations, which is a function of $\mathbf{y}$ only, is a set of LS estimates.

Thus, there is no reason for using the symbol $\hat{\hat{\boldsymbol{\beta}}}$ any more, and $\hat{\boldsymbol{\beta}}$ will denote a solution of the normal equations as well as a set of LS estimates. The situation can be easily visualized as in Fig. 1.3.1.

Notation: We use the symbol $\mathscr{S}_\Omega$ for the minimum value of $\mathscr{S}(\mathbf{y}, \mathbf{b})$,

$$\mathscr{S}_\Omega = \mathscr{S}(\mathbf{y}, \hat{\boldsymbol{\beta}}), \tag{1.3.9}$$

where $\hat{\boldsymbol{\beta}}$ is any set of LS estimates, or any solution of the normal equations. We shall call $\mathscr{S}_\Omega$ the *error sum of squares*, because in sec. 1.6 we shall see that it provides an estimate of the error variance σ^2. Although $\hat{\boldsymbol{\beta}}$ is not in general unique, $\mathscr{S}(\mathbf{y}, \hat{\boldsymbol{\beta}})$ is. A very useful expression for $\mathscr{S}_\Omega$ is

$$\mathscr{S}_\Omega = \sum_{i=1}^{n} y_i^2 - \sum_{\nu=1}^{p} \hat{\beta}_\nu r_\nu, \tag{1.3.10}$$

where $\{\hat{\beta}_1, \cdots, \hat{\beta}_p\}$ is any set of LS estimates, and r_ν is the right member

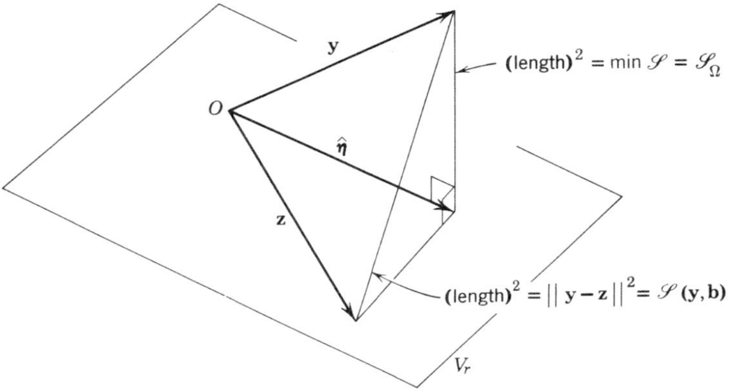

FIG. 1.3.1

of the νth normal equation (1.3.3). This expression may be derived as follows:

$$\mathscr{S}_\Omega = (\mathbf{y}-\mathbf{X}'\hat{\boldsymbol{\beta}})'(\mathbf{y}-\mathbf{X}'\hat{\boldsymbol{\beta}}) = \mathbf{y}'\mathbf{y} - \hat{\boldsymbol{\beta}}'(\mathbf{X}\mathbf{y}) + \hat{\boldsymbol{\beta}}'(\mathbf{X}\mathbf{X}'\hat{\boldsymbol{\beta}}-\mathbf{X}\mathbf{y}),$$

where we have used the fact that $\mathbf{y}'\mathbf{X}'\hat{\boldsymbol{\beta}}$ equals its transpose because it is a 1×1 matrix. This reduces to (1.3.10) because $\hat{\boldsymbol{\beta}}$ satisfies the normal equations $\mathbf{X}\mathbf{X}'\boldsymbol{\beta} = \mathbf{X}\mathbf{y}$.

Case where $\hat{\boldsymbol{\beta}}$ is Unique

The case where the $p\times n$ matrix $\mathbf{X}$ is of rank p is often called the *case of maximal rank*, or the *case of full rank*, because usually $p < n$. If rank $\mathbf{X} = p$ then (1.3.4) has a unique solution (and only then). In Theorem 7 of App. II we prove that rank $\mathbf{S}$ = rank $\mathbf{X}$, and hence in this case $\mathbf{S}$ is nonsingular. Thus $\mathbf{S}^{-1}$ exists, and the solution is given uniquely by

$$\hat{\boldsymbol{\beta}} = \mathbf{S}^{-1}\mathbf{X}\mathbf{y}. \tag{1.3.11}$$

Applying (1.2.10), we then obtain for the covariance matrix of $\hat{\boldsymbol{\beta}}$

$$\boldsymbol{\Sigma}_{\hat{\beta}} = (\mathbf{S}^{-1}\mathbf{X})\,\boldsymbol{\Sigma}_y(\mathbf{S}^{-1}\mathbf{X})'.$$

But $\mathbf{S}^{-1}$ is symmetric since $\mathbf{S}$ is, hence

$$\boldsymbol{\Sigma}_{\hat{\beta}} = \sigma^2\mathbf{S}^{-1}\mathbf{X}\mathbf{X}'\mathbf{S}^{-1},$$

and so finally

$$\boldsymbol{\Sigma}_{\hat{\beta}} = \sigma^2\mathbf{S}^{-1}. \tag{1.3.12}$$

Remarks: The case in which rank $\mathbf{X} = p$ sometimes occurs in practice. (It occurs usually in regression theory but not usually in the analysis of variance.) One does not then need the matrix $\mathbf{S}^{-1}$ in order to solve the normal equations. However, it is generally a good idea to compute $\mathbf{S}^{-1}$ first and then $\hat{\boldsymbol{\beta}}$ from (1.3.11) since it is almost always desirable subsequently to get also the covariance matrix $\boldsymbol{\Sigma}_{\hat{\beta}}$. The lack of uniqueness of the LS estimates $\{\hat{\beta}_j\}$ in the case where rank $\mathbf{X} < p$ is related to a similar nonuniqueness of the parameter values $\{\beta_j\}$; this is discussed further at the end of sec. 1.4. In connection with the result (1.3.10) involving the right members of the normal equations, and the result (1.3.12) if we think of $\mathbf{S}$ denoting the matrix of coefficients of the left members of the normal equations, it is of course essential that the normal equations be in exactly the form (1.3.3), where the νth equation is obtained by dividing by -2 the equation $\partial\mathscr{S}/\partial\beta_\nu = 0$ and transposing to the right member the known term resulting from the differentiation.

1.4. ESTIMABLE FUNCTIONS. THE GAUSS–MARKOFF THEOREM

The useful concept of *estimable functions*[12] is formulated in the following two definitions.

Definition: A *parametric function* is defined to be a linear function of the unknown parameters $\{\beta_1, \cdots, \beta_p\}$ with known constant coefficients $\{c_1, \cdots, c_p\}$,

$$\psi = \sum_{j=1}^{p} c_j\beta_j. \tag{1.4.1}$$

We introduce the vector $\mathbf{c}^{p\times 1} = (c_1, \cdots, c_p)'$; then we can write $\psi = \mathbf{c}'\boldsymbol{\beta}$.

Definition: A parametric function ψ is called an *estimable function* if it has an unbiased linear estimate, in other words, if there exists a vector of constant coefficients $\mathbf{a}^{n\times 1}$ such that

$$E(\mathbf{a}'\mathbf{y}) = \psi, \tag{1.4.2}$$

identically in $\boldsymbol{\beta}$ (i.e., no matter what the true values of the unknown parameters $\{\beta_j\}$).

Theorem 1: The parametric function $\psi = \mathbf{c}'\boldsymbol{\beta}$ is estimable if and only if $\mathbf{c}'$ is a linear combination of the rows of $\mathbf{X}'$, i.e., if and only if there exists a vector $\mathbf{a}^{n\times 1}$ such that

$$\mathbf{c}' = \mathbf{a}'\mathbf{X}'.$$

[12] Due to R. C. Bose (1944).

Proof: $\psi = \mathbf{c}'\boldsymbol{\beta}$ is estimable if and only if there exists $\mathbf{a}^{n\times 1}$ such that (1.4.2) is satisfied. But $E(\mathbf{a}'\mathbf{y}) = \mathbf{a}'E(\mathbf{y}) = \mathbf{a}'\mathbf{X}'\boldsymbol{\beta}$, and the condition $\mathbf{a}'\mathbf{X}'\boldsymbol{\beta} = \mathbf{c}'\boldsymbol{\beta}$ is satisfied identically in $\boldsymbol{\beta}$ if and only if $\mathbf{a}'\mathbf{X}' = \mathbf{c}'$.

We note that in nonmatrix notation the totality of estimable functions is $\{\Sigma_{i=1}^{n} a_i\eta_i\}$ where $\eta_i = E(y_i) = \Sigma_{j=1}^{p} x_{ji}\beta_j$, and $\{a_1, \cdots, a_n\}$ is an arbitrary set of n known constants.

For the proof[13] of the main theorem of this section we shall use the

Lemma: If $\psi = \mathbf{c}'\boldsymbol{\beta}$ is estimable, and if V_r is the space spanned by the columns of $\mathbf{X}'$, there exists a unique linear unbiased estimate of ψ, say $\mathbf{a}^{*\prime}\mathbf{y}$, with $\mathbf{a}^* \in V_r$. If $\mathbf{a}'\mathbf{y}$ is any unbiased linear estimate of ψ, then $\mathbf{a}^*$ is the projection of $\mathbf{a}$ on V_r.

Proof: Since ψ is estimable there exists an $\mathbf{a}^{n\times 1}$ for which $E(\mathbf{a}'\mathbf{y}) = \psi$. Let $\mathbf{a} = \mathbf{a}^* + \mathbf{b}$, where $\mathbf{a}^* \in V_r$, $\mathbf{b} \perp V_r$. Then

$$\psi = E(\mathbf{a}'\mathbf{y}) = E(\mathbf{a}^{*\prime}\mathbf{y}) + E(\mathbf{b}'\mathbf{y}) = E(\mathbf{a}^{*\prime}\mathbf{y}),$$

since $E(\mathbf{b}'\mathbf{y}) = \mathbf{b}'\mathbf{X}'\boldsymbol{\beta}$ and $\mathbf{b}'\mathbf{X}' = \mathbf{0}$ by orthogonality of $\mathbf{b}$ to the columns of $\mathbf{X}'$. Thus $\mathbf{a}^{*\prime}\mathbf{y}$ is an unbiased linear estimate of ψ with $\mathbf{a}^* \in V_r$. Suppose that the same is true of $\boldsymbol{\alpha}'\mathbf{y}$. Then we have identically in $\boldsymbol{\beta}$

$$\mathbf{0} = E(\mathbf{a}^{*\prime}\mathbf{y}) - E(\boldsymbol{\alpha}'\mathbf{y}) = (\mathbf{a}^* - \boldsymbol{\alpha})'\mathbf{X}'\boldsymbol{\beta},$$

so $(\mathbf{a}^* - \boldsymbol{\alpha})'\mathbf{X}' = \mathbf{0}$. Thus $\mathbf{a}^* - \boldsymbol{\alpha} \perp V_r$ and $\in V_r$, and hence $= \mathbf{0}$. This proves the uniqueness of $\mathbf{a}^{*\prime}\mathbf{y}$. The earlier part of the proof shows that for *any* unbiased estimate $\mathbf{a}'\mathbf{y}$, $\mathbf{a}^*$ is the projection of $\mathbf{a}$ on V_r.

Theorem 2 (Gauss–Markoff Theorem): Under the assumptions Ω: $E(\mathbf{y}) = \mathbf{X}'\boldsymbol{\beta}$, $\boldsymbol{\Sigma}_y = \sigma^2\mathbf{I}$, every estimable function $\psi = \mathbf{c}'\boldsymbol{\beta}$ has a unique unbiased linear estimate $\hat{\psi}$ which has minimum variance in the class of all unbiased linear estimates. The estimate $\hat{\psi}$ may be obtained from $\psi = \Sigma_{j=1}^{p} c_j\beta_j$ by replacing the $\{\beta_j\}$ by any set of LS estimates $\{\hat{\beta}_1, \cdots, \hat{\beta}_p\}$.

Proof: Let $\mathbf{a}^{*\prime}\mathbf{y}$ be the unbiased linear estimate of ψ with $\mathbf{a}^* \in V_r$, whose existence and uniqueness is given by the lemma, and let $\mathbf{a}'\mathbf{y}$ be any unbiased linear estimate of ψ. Then $\mathbf{a}^*$ is the projection of $\mathbf{a}$ on V_r, by the lemma, and

$$\|\mathbf{a}\|^2 = \|\mathbf{a}^*\|^2 + \|\mathbf{a} - \mathbf{a}^*\|^2.$$

By (1.2.10) with $m = 1$,

$$\begin{aligned} \text{Var}\,(\mathbf{a}'\mathbf{y}) &= \mathbf{a}'\boldsymbol{\Sigma}_y\mathbf{a} = \sigma^2\|\mathbf{a}\|^2 = \sigma^2\|\mathbf{a}^*\|^2 + \sigma^2\|\mathbf{a} - \mathbf{a}^*\|^2 \\ &= \text{Var}\,(\mathbf{a}^{*\prime}\mathbf{y}) + \sigma^2\|\mathbf{a} - \mathbf{a}^*\|^2. \end{aligned}$$

Thus $\text{Var}\,(\mathbf{a}'\mathbf{y}) \geqq \text{Var}\,(\mathbf{a}^{*\prime}\mathbf{y})$ with equality only if $\mathbf{a} = \mathbf{a}^*$. Hence $\mathbf{a}^{*\prime}\mathbf{y}$ is the unique unbiased linear estimate of ψ with minimum variance.

[13] This method of proof of the Gauss–Markoff theorem was suggested to me by Professor Werner Gautschi.

It remains to prove that $\mathbf{a}^{*\prime}\mathbf{y} = \mathbf{c}'\hat{\boldsymbol{\beta}}$. Now $\mathbf{a}^{*\prime}(\mathbf{y}-\hat{\boldsymbol{\eta}}) = 0$, where $\hat{\boldsymbol{\eta}} = \mathbf{X}'\hat{\boldsymbol{\beta}}$ is the projection of $\mathbf{y}$ on V_r, since $\mathbf{a}^* \in V_r$ and $\mathbf{y}-\hat{\boldsymbol{\eta}} \perp V_r$. Also $\mathbf{c}' = \mathbf{a}^{*\prime}\mathbf{X}'$ since $\mathbf{c}'\boldsymbol{\beta} \equiv E(\mathbf{a}^*\mathbf{y}) \equiv \mathbf{a}^*\mathbf{X}'\boldsymbol{\beta}$ identically in $\boldsymbol{\beta}$. Hence $\mathbf{a}^{*\prime}\mathbf{y} = \mathbf{a}^{*\prime}\hat{\boldsymbol{\eta}} = \mathbf{a}^{*\prime}\mathbf{X}'\hat{\boldsymbol{\beta}} = \mathbf{c}'\hat{\boldsymbol{\beta}}$.

Definition: For any estimable function ψ, its unique minimum-variance unbiased linear estimate $\hat{\psi}$, whose existence and structure are given by Theorem 2, will be called the *LS estimate of* ψ.

We have previously employed the terminology of "LS estimates" for the LS estimates $\{\hat{\beta}_j\}$ of the $\{\beta_j\}$. It might be extended to calling $\Sigma_1^p c_j\hat{\beta}_j$ the LS estimate of any linear function $\Sigma_1^p c_j\beta_j$, if the $\{\hat{\beta}_j\}$ are any set of LS estimates of the $\{\beta_j\}$. It would then follow that the LS estimate of $\Sigma_1^p c_j\beta_j$ is unique if and only if $\Sigma_1^p c_j\beta_j$ is estimable. However, we shall be interested in LS estimates only of estimable functions and of the $\{\beta_j\}$.

Corollary 1: If $\{\psi_1, \cdots, \psi_q\}$ are estimable functions every linear combination $\psi = \Sigma_1^q h_i\psi_i$ is estimable and its LS estimate $\hat{\psi}$ is $\Sigma_1^q h_i\hat{\psi}_i$, where $\hat{\psi}_i$ is the LS estimate of ψ_i.

Proof: Since $\Sigma_1^q h_i\hat{\psi}_i$ is an unbiased linear estimate of ψ, ψ is estimable. Suppose that $\psi_i = \Sigma_{j=1}^p c_{ij}\beta_j$. Then $\psi = \Sigma_j(\Sigma_i h_i c_{ij})\beta_j$. Applying Theorem 2 to both ψ_i and ψ, we find their LS estimates to be $\hat{\psi}_i = \Sigma_j c_{ij}\hat{\beta}_j$ and $\hat{\psi} = \Sigma_j(\Sigma_i h_i c_{ij})\hat{\beta}_j$, where the $\{\hat{\beta}_j\}$ are any set of LS estimates of the $\{\beta_j\}$. Hence $\hat{\psi} = \Sigma_i h_i\hat{\psi}_i$.

Side Conditions on the Parameters and Estimates

If rank $\mathbf{X} < p$ then we have seen that the LS estimates $\{\hat{\beta}_1, \cdots, \hat{\beta}_p\}$ are not unique, since they are any set $\{b_1, \cdots, b_p\}$ of statistics satisfying

$$b_1\boldsymbol{\xi}_1 + \cdots + b_p\boldsymbol{\xi}_p = \hat{\boldsymbol{\eta}}, \tag{1.4.3}$$

where $\boldsymbol{\xi}_j$ is the jth column of $\mathbf{X}'$, and $\hat{\boldsymbol{\eta}}$ is the projection of $\mathbf{y}$ on V_r, the space spanned by the $\{\boldsymbol{\xi}_j\}$. A similar indeterminacy affects the parameters $\{\beta_1, \cdots, \beta_p\}$ through the relation

$$\beta_1\boldsymbol{\xi}_1 + \cdots + \beta_p\boldsymbol{\xi}_p = \boldsymbol{\eta}, \tag{1.4.4}$$

in the sense that different sets of values for the $\{\beta_j\}$ will give the same $\boldsymbol{\eta}$ and hence the same vector of observations $\mathbf{y} = \boldsymbol{\eta}+\mathbf{e}$. We note however that if $\mathbf{c}'\boldsymbol{\beta}$ is any estimable function it has the same value regardless of which $\boldsymbol{\beta}$ is used in (1.4.4), since by Theorem 1 there exists a constant vector $\mathbf{a}$ such that $\mathbf{c}' = \mathbf{a}'\mathbf{X}'$, and thus $\mathbf{c}'\boldsymbol{\beta} = \mathbf{a}'\boldsymbol{\eta}$ depends only on $\boldsymbol{\eta}$. If it is desired to eliminate these indeterminacies two courses are open:

(i). Consider a "reduced" problem with only r parameters $\{\beta_j\}$. This can be achieved by choosing r linearly independent vectors from the set $\{\boldsymbol{\xi}_1, \cdots, \boldsymbol{\xi}_p\}$ as a basis for V_r, as in the proof of Lemma 2 of App. I, and keeping only the r corresponding $\{\beta_j\}$. This gives us a new $n \times r$ matrix of coefficients instead of the old $\mathbf{X}'$ and the resulting "reduced" problem is a case of maximal rank, i.e., the rank of the new $\mathbf{X}$ equals the new number of $\{\beta_j\}$.

(ii). Put suitable side conditions on the p parameters $\{\beta_j\}$ and their estimates. Thus, we would achieve the same result as in (i) if we agreed that for the $p-r$ parameters $\{\beta_j\}$ there discarded we always take $\beta_j = 0$ and $\hat{\beta}_j = 0$. In most analysis-of-variance situations it is convenient to add linear restrictions of a more general form than this to produce the desired uniqueness. We therefore consider subjecting the $\{\beta_j\}$ to t $(t \geqq p-r)$ linear restrictions $\mathbf{H}'\boldsymbol{\beta} = \mathbf{0}$, where $\mathbf{H}'$ is a $t \times p$ matrix of known constants. It will usually be almost obvious that the restrictions adopted in practice make the $\{\beta_j\}$ unique in the sense that for every possible set $\{\beta_j\}$ in the original problem there will exist a unique set $\{\tilde{\beta}_j\}$ satisfying

$$(1.4.5) \qquad \mathbf{X}'\boldsymbol{\beta} = \mathbf{X}'\tilde{\boldsymbol{\beta}} \quad \text{and} \quad \mathbf{H}'\tilde{\boldsymbol{\beta}} = \mathbf{0}.$$

The first of these conditions says the $\{\tilde{\beta}_j\}$ give the same $\boldsymbol{\eta} = \mathbf{X}'\boldsymbol{\beta}$ as the $\{\beta_j\}$. The two conditions (1.4.5) will then make the $\{\tilde{\beta}_j\}$ uniquely determined functions of the $\{\beta_j\}$. We will prove below that these are estimable functions in the original problem, so that then *every* parametric function $\mathbf{c}'\tilde{\boldsymbol{\beta}}$ in the new problem is an estimable function in the old problem. We shall also show that there is then a unique set of LS estimates $\{\hat{\beta}_j\}$ which satisfy the side conditions $\mathbf{H}'\hat{\boldsymbol{\beta}} = \mathbf{0}$, i.e., a unique solution of the normal equations which satisfies the side conditions. In later parts of the book, when applying this theory of making the parameters and their estimates unique by subjecting them to appropriate side conditions, we shall omit the tildes $(\sim)$ from the $\{\tilde{\beta}_j\}$, but it will clarify the derivation of the theory to keep the distinction in the notation for the present. (If the non-mathematically inclined reader is willing to accept all this without proof he may skip the rest of this section.)

We will see in a moment that we shall have to consider the trivial estimable function $\mathbf{c}'\boldsymbol{\beta}$ all of whose coefficients are zero, which we shall write $\mathbf{0}'\boldsymbol{\beta}$. It is clear that, aside from $\mathbf{0}'\boldsymbol{\beta}$, an estimable function $\mathbf{c}'\boldsymbol{\beta}$ can take on every possible value k for suitable choice of $\boldsymbol{\beta}$ (for some ν such that $c_\nu \neq 0$ take $\beta_j = k\delta_{\nu j}/c_\nu$). Denote the t rows of $\mathbf{H}'$ by $\mathbf{h}_1', \cdots, \mathbf{h}_t'$. Evidently we cannot let any $\mathbf{h}_i'\boldsymbol{\beta}$ be estimable, unless it is $\mathbf{0}'\boldsymbol{\beta}$, since we are going to add the restriction $\mathbf{H}'\tilde{\boldsymbol{\beta}} = \mathbf{0}$, or all $\mathbf{h}_i'\tilde{\boldsymbol{\beta}} = 0$, and, if $\mathbf{h}_i'\boldsymbol{\beta}$ is estimable, $\mathbf{h}_i'\tilde{\boldsymbol{\beta}} = \mathbf{h}_i'\boldsymbol{\beta}$ by the remark made after (1.4.4), and can take on every value k. More generally, no linear combination of the $\{\mathbf{h}_i'\boldsymbol{\beta}\}$ must be an estimable function except $\mathbf{0}'\boldsymbol{\beta}$ (it is quite possible that $\mathbf{0}'\boldsymbol{\beta}$ may be a linear combination of the $\{\mathbf{h}_i'\boldsymbol{\beta}\}$ with coefficients not all 0, since we permit the $\{\mathbf{h}_i\}$ to be linearly dependent). By Theorem 1 we may state this as follows: No linear combination of the rows of $\mathbf{H}'$ except[14] $\mathbf{0}'$ can be a linear combination of the rows of $\mathbf{X}'$. On the other hand if the solution $\tilde{\boldsymbol{\beta}}$ of (1.4.5) is to be unique, the rank of the composite $(n+t)\times p$ matrix

$$\mathbf{G}' = \begin{pmatrix} \mathbf{X}' \\ \mathbf{H}' \end{pmatrix} \tag{1.4.6}$$

must be p: For, by use of partitioned matrices (see end of App. II), (1.4.5) may be written as $\mathbf{G}'\tilde{\boldsymbol{\beta}} = \boldsymbol{\eta}^*$, where

$$\boldsymbol{\eta}^* = \begin{pmatrix} \mathbf{X}'\boldsymbol{\beta} \\ \mathbf{0} \end{pmatrix} \tag{1.4.7}$$

is a vector with $n+t$ components, or

$$\tilde{\beta}_1\mathbf{g}_1 + \tilde{\beta}_2\mathbf{g}_2 + \cdots + \tilde{\beta}_p\mathbf{g}_p = \boldsymbol{\eta}^*, \tag{1.4.8}$$

where $\mathbf{g}_j$ is the jth column of $\mathbf{G}'$. By Lemma 3 of App. I, the coefficients $\{\tilde{\beta}_j\}$ in (1.4.8) will be unique if and only if the $\{\mathbf{g}_j\}$ are linearly independent, i.e., rank $\mathbf{G}' = p$.

That the two necessary conditions we have found on the matrix $\mathbf{H}'$ are also sufficient for our purpose will follow as Corollary 2 to the following theorem. The theorem is stated as a purely algebraic result. In its statistical interpretation, condition (b) of the theorem is equivalent to (b′): No linear combination of the rows of $\mathbf{H}'\boldsymbol{\beta}$ (i.e., no linear combination of the parametric functions we equate to zero in the side conditions) is an estimable function except $\mathbf{0}'\boldsymbol{\beta}$.

Theorem 3: Suppose that $\mathbf{X}'$ is $n\times p$, $\mathbf{H}'$ is $t\times p$, rank $\mathbf{X}' = r$ $(p > r,\ t \geq p - r)$, and V_r is the space spanned by the columns of $\mathbf{X}'$. Then the system

$$\mathbf{X}'\mathbf{b} = \mathbf{z}, \qquad \mathbf{H}'\mathbf{b} = \mathbf{0} \tag{1.4.9}$$

has a unique solution $\mathbf{b}^{p\times 1}$ for every $\mathbf{z}^{n\times 1} \in V_r$ if and only if the following two conditions are satisfied: (*a*) The rank of the composite matrix

$$\mathbf{G}' = \begin{pmatrix} \mathbf{X}' \\ \mathbf{H}' \end{pmatrix}$$

is p. (*b*) No linear combination of the rows of $\mathbf{H}'$ is a linear combination of the rows of $\mathbf{X}'$ except $\mathbf{0}'$.

Proof: Most of the proof consists of showing that there exists a solution $\mathbf{b}$ for every $\mathbf{z} \in V_r$ if and only if condition (*b*) is satisfied. It follows by the argument stated above in connection with (1.4.8) that if a solution $\mathbf{b}$ exists it is unique if and only if (*a*) is satisfied.

[14] We write $\mathbf{0}'$ because we are thinking here of $\mathbf{0}$ as the zero vector; however it would be perfectly correct to write $\mathbf{0}$ instead for the $t\times 1$ zero matrix.

Write (1.4.9) as $\mathbf{G}'\mathbf{b} = \mathbf{z}^*$, where $\mathbf{z}^*$ is the vector with $n+t$ components

$$\mathbf{z}^* = \begin{pmatrix} \mathbf{z}^{n\times 1} \\ \mathbf{0}^{t\times 1} \end{pmatrix},$$

or

$$b_1\mathbf{g}_1 + b_2\mathbf{g}_2 + \cdots + b_p\mathbf{g}_p = \mathbf{z}^*,$$

where $\mathbf{g}_j$ is the jth column of $\mathbf{G}'$. Then a solution $\mathbf{b}$ is seen to exist if and only if $\mathbf{z}^* \in W$, where W is the space of vectors of $n+t$ components spanned by the $\{\mathbf{g}_j\}$. By Theorem 3 of App. I, $\mathbf{z}^* \in W$ if and only if $\mathbf{u}'\mathbf{z}^* = 0$ for every $\mathbf{u} \perp W$. If we partition $\mathbf{u}$,

$$\mathbf{u} = \begin{pmatrix} \mathbf{v}^{n\times 1} \\ \mathbf{w}^{t\times 1} \end{pmatrix},$$

so that

$$\mathbf{u}'\mathbf{z}^* = (\mathbf{v}', \mathbf{w}')\begin{pmatrix} \mathbf{z} \\ \mathbf{0} \end{pmatrix} = \mathbf{v}'\mathbf{z},$$

we see that $\mathbf{u}'\mathbf{z}^* = 0$ if and only if $\mathbf{v}'\mathbf{z} = 0$. Also $\mathbf{u} \perp W$ if and only if $\mathbf{u}$ is orthogonal to the columns of $\mathbf{G}'$ which span W; hence if and only if $\mathbf{u}'\mathbf{G}' = \mathbf{0}'$,

$$(\mathbf{v}', \mathbf{w}')\begin{pmatrix} \mathbf{X}' \\ \mathbf{H}' \end{pmatrix} = \mathbf{0}',$$

or

$$\mathbf{v}'\mathbf{X}' + \mathbf{w}'\mathbf{H}' = \mathbf{0}'. \tag{1.4.10}$$

We now have that a solution $\mathbf{b}$ exists for a given $\mathbf{z} \in V_r$ if and only if $\mathbf{v}'\mathbf{z} = 0$ for every $\mathbf{v}^{n\times 1}$ and $\mathbf{w}^{t\times 1}$ satisfying (1.4.10).

Suppose first that condition (b) is satisfied and suppose that $\mathbf{v}$ and $\mathbf{w}$ satisfy (1.4.10). Then $\mathbf{v}'\mathbf{X}' = -\mathbf{w}'\mathbf{H}'$ is a linear combination of the rows of $\mathbf{X}'$ and also of the rows of $\mathbf{H}'$ and hence must be $\mathbf{0}'$ by (b). Then $\mathbf{v}'\mathbf{X}' = \mathbf{0}'$ implies that $\mathbf{v}$ is orthogonal to the columns of $\mathbf{X}'$, thus $\mathbf{v} \perp V_r$; therefore $\mathbf{v} \perp \mathbf{z}$ for every $\mathbf{z} \in V_r$, i.e., $\mathbf{v}'\mathbf{z} = 0$,. and hence there exists a solution $\mathbf{b}$ for every $\mathbf{z} \in V_r$. Suppose next that (b) is not satisfied, so that there exists a linear combination of the rows of $\mathbf{H}'$, say $-\mathbf{w}'\mathbf{H}'$ which is a linear combination of the rows of $\mathbf{X}'$, say $\mathbf{v}'\mathbf{X}'$, and is not $\mathbf{0}'$: $\mathbf{v}'\mathbf{X}' = -\mathbf{w}'\mathbf{H}' = \boldsymbol{\lambda}'$, say, where $\boldsymbol{\lambda}^{p\times 1} \neq \mathbf{0}$. Now take $\mathbf{z} = \mathbf{X}'\boldsymbol{\lambda}$, so $\mathbf{z} \in V_r$. Then $\mathbf{v}'\mathbf{z} = \mathbf{v}'\mathbf{X}'\boldsymbol{\lambda} = \boldsymbol{\lambda}'\boldsymbol{\lambda} \neq 0$, while $\mathbf{v}$ and $\mathbf{w}$ satisfy (1.4.10). Thus for this $\mathbf{z} \in V_r$ there is no solution $\mathbf{b}$.

By taking $\mathbf{b} = \tilde{\boldsymbol{\beta}}$ and $\mathbf{z} = \mathbf{X}'\boldsymbol{\beta}$ in Theorem 3 we get

Corollary 2: The system (1.4.5) has a unique solution $\hat{\boldsymbol{\beta}}$ for every $\boldsymbol{\beta}$ if and only if conditions (a) and (b) of Theorem 3 are satisfied.

Recalling that any $\{b_1, \cdots, b_p\}$ that satisfy (1.4.3) and are functions of

$\mathbf{y}$ only[15] constitute a set of LS estimates, and taking $\mathbf{z} = \hat{\boldsymbol{\eta}}$ in Theorem 3, we get

Corollary 3: If conditions (*a*) and (*b*) of Theorem 3 are satisfied, there exists a unique set of LS estimates $\{\hat{\beta}_1, \cdots, \hat{\beta}_p\}$ (i.e., a unique solution of the normal equations) for which $\mathbf{H}'\hat{\boldsymbol{\beta}} = \mathbf{0}$.

This says that we may subject the LS estimates to the same side conditions as the parameters. Finally we need the following result, which implies that every linear combination of the parameters $\{\tilde{\beta}_j\}$, the $\{\tilde{\beta}_j\}$ being subject to the side conditions, is estimable:

Theorem 4: If conditions (*a*) and (*b*) of Theorem 3 are satisfied, so that the $\{\tilde{\beta}_j\}$ are functions of the $\{\beta_j\}$ determined uniquely by (1.4.5), then the $\{\tilde{\beta}_j\}$ are estimable functions.

Proof: We will obtain an explicit formula for the $\{\tilde{\beta}_j\}$ in terms of the $\{\beta_j\}$: For any $\boldsymbol{\beta}$ let $\tilde{\boldsymbol{\beta}}$ be the unique solution of (1.4.5), so

$$\mathbf{G}'\tilde{\boldsymbol{\beta}} = \begin{pmatrix} \mathbf{X}'\boldsymbol{\beta} \\ \mathbf{0} \end{pmatrix}.$$

Multiply by $\mathbf{G}$ on the left to get

$$\mathbf{G}\mathbf{G}'\tilde{\boldsymbol{\beta}} = (\mathbf{X}, \mathbf{H}) \begin{pmatrix} \mathbf{X}'\boldsymbol{\beta} \\ \mathbf{0} \end{pmatrix} = \mathbf{X}\mathbf{X}'\boldsymbol{\beta}.$$

Now by Theorem 7 of App. II the rank of the $p \times p$ matrix $\mathbf{G}\mathbf{G}'$ is equal to rank $\mathbf{G} = p$; so $\mathbf{G}\mathbf{G}'$ has an inverse, and thus $\tilde{\boldsymbol{\beta}} = (\mathbf{G}\mathbf{G}')^{-1}\mathbf{X}\mathbf{X}'\boldsymbol{\beta}$, or

$$\tilde{\boldsymbol{\beta}} = (\mathbf{X}\mathbf{X}' + \mathbf{H}\mathbf{H}')^{-1}\mathbf{X}\mathbf{X}'\boldsymbol{\beta},$$

the promised formula. Since $E(\mathbf{y}) = \mathbf{X}'\boldsymbol{\beta}$, $\tilde{\boldsymbol{\beta}}$ has the unbiased estimate $(\mathbf{X}\mathbf{X}' + \mathbf{H}\mathbf{H}')^{-1}\mathbf{X}\mathbf{y}$.

1.5. REDUCTION OF THE CASE WHERE THE OBSERVATIONS HAVE KNOWN CORRELATIONS AND KNOWN RATIOS OF VARIANCES

We consider now the case where the covariance matrix $\boldsymbol{\Sigma}_y$ of the observations $\{y_i\}$ is not of the form $\sigma^2\mathbf{I}$ but $\boldsymbol{\Sigma}_y$ is known except for a scalar factor, i.e., $\boldsymbol{\Sigma}_y = \theta\mathbf{B}$, where θ is an unknown positive constant and $\mathbf{B}^{n\times n}$ is a known constant matrix; $\mathbf{B}$ is necessarily symmetric and positive indefinite, and we shall assume furthermore that it is nonsingular (see App. V). This is equivalent to knowing the correlation coefficients of all pairs of observations y_i and the ratios of their variances.

[15] That these $\{b_j\}$ are functions of $\mathbf{y}$ only, and in fact *linear* functions, follows from their being the unique solution of the linear system $\mathbf{X}\mathbf{X}'\mathbf{b} = \mathbf{X}\mathbf{y}$, $\mathbf{H}'\mathbf{b} = \mathbf{0}$.

Our underlying assumptions are now

$$(1.5.1)\qquad \Omega:\quad E(\mathbf{y}) = \mathbf{X}'\boldsymbol{\beta},\qquad \boldsymbol{\Sigma}_y = \theta\mathbf{B},\qquad |\mathbf{B}| \neq 0,\qquad \text{rank } \mathbf{X}' = r.$$

This case may be reduced to that previously considered, where $\boldsymbol{\Sigma}_y = \sigma^2\mathbf{I}$, by appealing to Lemma 11′ and the discussion following it in App. II, which says there exists a nonsingular $\mathbf{P}^{n\times n}$ such that $\mathbf{P}'\mathbf{B}\mathbf{P} = \mathbf{I}$. Let $\tilde{\mathbf{y}} = \mathbf{P}'\mathbf{y}$. Then

$$E(\tilde{\mathbf{y}}) = \mathbf{P}'\, E(\mathbf{y}) = \mathbf{P}'\mathbf{X}'\boldsymbol{\beta} = \tilde{\mathbf{X}}'\boldsymbol{\beta},$$

where $\tilde{\mathbf{X}}' = \mathbf{P}'\mathbf{X}'$; so rank $\tilde{\mathbf{X}}'$ = rank $\mathbf{X}' = r$, and

$$\boldsymbol{\Sigma}_{\tilde{y}} = \mathbf{P}'\boldsymbol{\Sigma}_y\mathbf{P} = \theta\mathbf{P}'\mathbf{B}\mathbf{P} = \sigma^2\mathbf{I},$$

where $\sigma^2 = \theta$. We may thus write (1.5.1) as

$$\Omega:\quad E(\tilde{\mathbf{y}}) = \tilde{\mathbf{X}}'\boldsymbol{\beta},\qquad \boldsymbol{\Sigma}_{\tilde{y}} = \sigma^2\mathbf{I},\qquad \text{rank } \tilde{\mathbf{X}}' = r,$$

which is the case previously considered.

In applications the transformed "observations" $\{\tilde{y}_i\}$ are tedious to calculate and one usually prefers to work with the actual observations $\{y_i\}$. The LS estimates of the parameters $\{\beta_j\}$ may then be found by minimizing the following sum of squares involving the $\{y_i\}$ and $\{\beta_j\}$

$$(1.5.2)\qquad \mathscr{S}(\mathbf{y}, \boldsymbol{\beta}) = (\mathbf{y} - \mathbf{X}'\boldsymbol{\beta})'\, \mathbf{B}^{-1}(\mathbf{y} - \mathbf{X}'\boldsymbol{\beta}).$$

To see this we note that in the transformed problem, which falls under our previous theory, the $\{\hat{\beta}_j\}$ are found by minimizing

$$(1.5.3)\qquad \tilde{\mathscr{S}}(\tilde{\mathbf{y}}, \boldsymbol{\beta}) = (\tilde{\mathbf{y}} - \tilde{\mathbf{X}}'\boldsymbol{\beta})'(\tilde{\mathbf{y}} - \tilde{\mathbf{X}}'\boldsymbol{\beta}).$$

Now $\tilde{\mathbf{y}}-\tilde{\mathbf{X}}'\boldsymbol{\beta} = \mathbf{P}'(\mathbf{y}-\mathbf{X}'\boldsymbol{\beta})$, and substituting this in (1.5.3) and using $\mathbf{P}\mathbf{P}' = \mathbf{B}^{-1}$, we get that $\tilde{\mathscr{S}}(\tilde{\mathbf{y}}, \boldsymbol{\beta})$ equals the $\mathscr{S}(\mathbf{y}, \boldsymbol{\beta})$ defined by (1.5.2).

Besides the $\{\beta_j\}$ the model (1.5.1) contains the unknown parameter θ. In sec. 1.6 it will be shown that an unbiased estimate of σ^2 is $\tilde{\mathscr{S}}(\tilde{\mathbf{y}}, \hat{\boldsymbol{\beta}})/(n-r)$, where $\hat{\boldsymbol{\beta}}$ is any set of LS estimates. It follows that an unbiased estimate of the parameter θ is $\mathscr{S}(\mathbf{y}, \hat{\boldsymbol{\beta}})/(n-r)$, where $\mathscr{S}(\mathbf{y}, \hat{\boldsymbol{\beta}})$ is formed by replacing $\boldsymbol{\beta}$ by $\hat{\boldsymbol{\beta}}$ in (1.5.2).

The sum of squares (1.5.2) which is minimized to calculate the LS estimates may be called the "weighted sum of squares": In the particular case where the observations are uncorrelated, $\mathbf{B}$ is a diagonal matrix, and if we then write the ith diagonal element of $\mathbf{B}$ as w_i^{-1}, the $\{w_i\}$ are inversely proportional to the variances of the observations $\{y_i\}$, and (1.5.2) becomes

$$\mathscr{S}(\mathbf{y}, \boldsymbol{\beta}) = \sum_i w_i\Big(y_i - \sum_j x_{ji}\beta_j\Big)^2.$$

The case $\boldsymbol{\Sigma}_y = \sigma^2\mathbf{I}$ is the special case where the weights $\{w_i\}$ are all equal.

Sometimes in applications we may have some doubt about the correct weights, and we may then find some comfort in the fact that the method of least squares used with incorrect weights still leads to unbiased estimates; however, our calculations of the variances[16] of the estimates will be invalidated by incorrect weights. More generally, it is true that the use of any positive definite matrix $\mathbf{B}$ whatever (not just a correct one of the form $\theta^{-1}\mathbf{\Sigma}_y$) in (1.5.2) leads to unbiased estimates of estimable functions if the LS estimates of the $\{\beta_j\}$ are calculated by minimizing (1.5.2). We shall prove this only for the case where $\mathbf{X}'$ is of rank p:

Let $\mathbf{P}$ be defined as above for the $\mathbf{B}$ actually used, and again transform to $\tilde{\mathbf{y}}$ and $\tilde{\mathbf{X}}'$ as above. In the transformed problem the normal equations are $\tilde{\mathbf{X}}\tilde{\mathbf{X}}'\boldsymbol{\beta} = \tilde{\mathbf{X}}\tilde{\mathbf{y}}$, and the solution, which we will denote by $\hat{\boldsymbol{\beta}}^*$, is

$$\hat{\boldsymbol{\beta}}^* = (\tilde{\mathbf{X}}\tilde{\mathbf{X}}')^{-1}\tilde{\mathbf{X}}\tilde{\mathbf{y}}.$$

But this solution will be the same as that found from minimizing (1.5.2). Since

$$\hat{\boldsymbol{\beta}}^* = (\tilde{\mathbf{X}}\tilde{\mathbf{X}}')^{-1}\tilde{\mathbf{X}}\mathbf{P}'\mathbf{y},$$

therefore

$$E(\hat{\boldsymbol{\beta}}^*) = (\tilde{\mathbf{X}}\tilde{\mathbf{X}}')^{-1}\tilde{\mathbf{X}}\mathbf{P}'\mathbf{X}'\boldsymbol{\beta}.$$

Substituting $\mathbf{P}'\mathbf{X}' = \tilde{\mathbf{X}}'$ in this expression, we get $E(\hat{\boldsymbol{\beta}}^*) = \boldsymbol{\beta}$.

1.6. THE CANONICAL FORM OF THE UNDERLYING ASSUMPTIONS Ω. THE MEAN SQUARE FOR ERROR

Let us introduce in the sample space V_n of the observation vector $\mathbf{y}^{n\times 1}$ the orthonormal basis $\{\boldsymbol{\rho}_1, \boldsymbol{\rho}_2, \cdots, \boldsymbol{\rho}_n\}$, where $\boldsymbol{\rho}_i = (\delta_{i1}, \delta_{i2}, \cdots, \delta_{in})'$ (this is the basis R of the example after Theorem 1 of App. I), so $\mathbf{y} = \Sigma_1^n y_i\boldsymbol{\rho}_i$. Let us also introduce an orthonormal basis $\{\boldsymbol{\alpha}_1, \cdots, \boldsymbol{\alpha}_r\}$ for V_r, the space spanned by the columns of $\mathbf{X}'$, and complete it to an orthonormal basis $\{\boldsymbol{\alpha}_1, \cdots, \boldsymbol{\alpha}_r, \boldsymbol{\alpha}_{r+1}, \cdots, \boldsymbol{\alpha}_n\}$ for V_n; this is always possible (Lemmas 6 and 7 of App. I). Write

$$\mathbf{y} = \sum_{i=1}^{n} z_i\boldsymbol{\alpha}_i, \tag{1.6.1}$$

where $\{z_i\}$ are the coordinates of $\mathbf{y}$ relative to the new basis, and hence $z_i = \boldsymbol{\alpha}_i'\mathbf{y}$, as we see by multiplying (1.6.1) by $\boldsymbol{\alpha}_\nu'$. This relation between the coordinates $\{z_i\}$ and $\{y_i\}$ may be written $\mathbf{z} = \mathbf{P}\mathbf{y}$, where $\mathbf{P}^{n\times n}$ is the orthogonal matrix whose ith row is $\boldsymbol{\alpha}_i'$. Let $\zeta_i = E(z_i)$, so $\zeta_i = E(\boldsymbol{\alpha}_i'\mathbf{y}) = \boldsymbol{\alpha}_i'\boldsymbol{\eta}$. It follows that for all values of the parameters, $\zeta_i = 0$ for $i > r$

[16] Bounds on the bias of the estimated covariance matrix are derived for some cases in Watson (1955).

since $\boldsymbol{\eta} \in V_r \perp \boldsymbol{\alpha}_i$ for $i > r$. Furthermore we have for the covariance matrix of the transformed "observations" $\{z_i\}$,

$$\boldsymbol{\Sigma}_z = \mathbf{P}\boldsymbol{\Sigma}_y\mathbf{P}' = \sigma^2\mathbf{P}\mathbf{P}' = \sigma^2\mathbf{I}.$$

We have now shown that by a suitable orthogonal transformation (not depending on unknown parameters) we can always reduce the Ω-assumptions to the *canonical form*

$$\Omega: \begin{cases} \mathbf{z} = (z_1, \cdots, z_n)', \\ E(z_i) = \zeta_i \qquad (i = 1, \cdots, r), \\ E(z_i) = 0 \qquad (i = r+1, \cdots, n), \\ \boldsymbol{\Sigma}_z = \sigma^2\mathbf{I}, \end{cases}$$

where $\zeta_1, \cdots, \zeta_r$, and σ^2 are unknown parameters, and the $\{z_i\}$ are a known transformation of the observations.

Since we do not actually use the canonical form in analyzing data, we will never need to calculate the transformation matrix $\mathbf{P}$ explicitly (although it could be done by calculating its rows $\{\boldsymbol{\alpha}_i'\}$ by the Schmidt process of Lemma 6 of App. I). However, the canonical form is very useful for the derivation of distribution theory, for example:

An Unbiased Estimate of σ^2

The error sum of squares $\mathscr{S}_\Omega$ introduced at the end of sec. 1.3, namely

$$\mathscr{S}_\Omega = \sum_{i=1}^{n} \left(y_i - \sum_{j=1}^{p} x_{ji}\hat{\beta}_j\right)^2, \tag{1.6.2}$$

where $\{\hat{\beta}_j\}$ is any set of LS estimates, may be written $\mathscr{S}_\Omega = \|\mathbf{y}-\hat{\boldsymbol{\eta}}\|^2$, where $\hat{\boldsymbol{\eta}}$ is the projection of $\mathbf{y}$ on V_r. But $\mathbf{y} = \Sigma_1^n z_i\boldsymbol{\alpha}_i$, and $\hat{\boldsymbol{\eta}} = \Sigma_1^r z_i\boldsymbol{\alpha}_i$, where $\{\boldsymbol{\alpha}_1, \cdots, \boldsymbol{\alpha}_n\}$ is the above basis for the canonical form, and so $\mathscr{S}_\Omega = \|\Sigma_{r+1}^n z_i\boldsymbol{\alpha}_i\|^2$, or

$$\mathscr{S}_\Omega = \sum_{i=r+1}^{n} z_i^2. \tag{1.6.3}$$

Now for $i > r$, $E(z_i) = 0$, which implies that $E(z_i^2) = \text{Var}(z_i) = \sigma^2$. Hence from (1.6.3), $E(\mathscr{S}_\Omega) = (n-r)\sigma^2$. If we define

$$s^2 = \mathscr{S}_\Omega/(n-r), \tag{1.6.4}$$

we have $E(s^2) = \sigma^2$, that is, s^2 is an unbiased estimate of σ^2. The quantity s^2 is called the *mean square for error* (later written also as MS_e) and it is said to have *$n-r$ degrees of freedom.* In general the number of degrees of freedom of a quadratic form in the observations is defined to be its rank (i.e., the rank of the symmetric matrix of the quadratic form), and we see from (1.6.3) that the rank of $\mathscr{S}_\Omega$ is $n-r$.

This result for estimating σ^2 is a supplement to the Gauss–Markoff theorem of great practical importance, since in applications we want some idea of the accuracy of our unbiased point estimates. If $\psi = \mathbf{c}'\boldsymbol{\beta}$ is an estimable function, then by the theorem there exists a unique linear combination of the observations, $\hat{\psi} = \mathbf{a}^{*\prime}\mathbf{y}$ which is the optimum estimate of ψ. Then the variance of the estimate $\hat{\psi}$ is $\sigma^2_{\hat{\psi}} = \mathbf{a}^{*\prime}\mathbf{a}^{*}\sigma^2$, and this may then be estimated by $\hat{\sigma}^2_{\hat{\psi}} = \mathbf{a}^{*\prime}\mathbf{a}^{*}s^2$. This estimate of the variance is evidently unbiased, and has been shown to have other optimum properties.[17] The expected value of the mean square for error s^2 in the case where the observations $\{y_i\}$ have unequal variances $\{\sigma_i^2\}$, but s^2 is calculated in the above way as though the $\{\sigma_i^2\}$ were equal, is given by the rule at the beginning of sec. 10.4.

Estimation and Error Spaces

Consider the set of all *linear forms* $\Sigma_1^n a_i y_i = \mathbf{a}'\mathbf{y}$ in the observations. The coefficients $\{a_i\}$ are assumed to be known constants (i.e., they do not depend on unknown parameters); we may call $\mathbf{a}$ the *coefficient vector* of the linear form $\mathbf{a}'\mathbf{y}$. We see there is a one-to-one correspondence between the totality of linear forms $\mathbf{a}'\mathbf{y}$ and the totality of vectors $\mathbf{a} \in V_n$, and that addition of linear forms or multiplication of a linear form by a constant corresponds to the same operation on the coefficient vectors. It is convenient to speak of *spaces of linear forms spanned by a given set of linear forms, independence of linear forms, orthogonality of forms and of spaces,* etc., the terms being defined by use of the corresponding properties of the coefficient vectors of the forms.

The canonical variables $\{z_1, \cdots, z_n\}$ are linear forms in the observations $\{y_i\}$, and they may be used to define two interesting orthogonal spaces of linear forms, namely the space spanned by $\{z_1, \cdots, z_r\}$, called the *estimation space*, and that spanned by $\{z_{r+1}, \cdots, z_n\}$, called the *error space*.[18] Since $z_i = \boldsymbol{\alpha}_i'\mathbf{y}$, we see that the forms $\{z_1, \cdots, z_n\}$ constitute an orthonormal basis for the n-dimensional space of forms (because their coefficient vectors constitute an orthonormal basis for V_n), and so the two spaces are orthogonal.

The reason for calling the latter the error space is that the error sum of squares $\mathscr{S}_\Omega$ involves only the set $\{z_{r+1}, \cdots, z_n\}$. It is easily shown that a linear form $\mathbf{a}'\mathbf{y}$ is in the error space if and only if its expected value is identically zero in the parameters: The relation $\mathbf{z} = \mathbf{P}\mathbf{y}$ may be inverted, $\mathbf{y} = \mathbf{P}'\mathbf{z}$ since $\mathbf{P}'\mathbf{P} = \mathbf{I}$, hence $\mathbf{a}'\mathbf{y} = \mathbf{b}'\mathbf{z}$, where $\mathbf{b} = \mathbf{P}\mathbf{a}$, and so $E(\mathbf{a}'\mathbf{y}) = \mathbf{b}'\boldsymbol{\zeta} = \Sigma_1^r b_i\zeta_i = 0$ if and only if $b_1 = b_2 = \cdots = b_r = 0$, i.e., if and only if $\mathbf{a}'\mathbf{y} = \Sigma_{r+1}^n b_i z_i$. The former space is called the estimation space

[17] By P. L. Hsu (1938*b*).

[18] By R. C. Bose (1944).

because if ψ is any estimable function and $\hat{\psi}$ is its LS estimate then the linear form $\hat{\psi}$ is a linear combination of $\{z_1, \cdots, z_r\}$ only, i.e., $\hat{\psi}$ is in the estimation space. To see this note that the columns of $\mathbf{P}'$ are $\{\boldsymbol{\alpha}_1, \cdots, \boldsymbol{\alpha}_n\}$, the orthonormal basis for V_n used in deriving the canonical form. If ψ is estimable, by the Gauss–Markoff theorem its LS estimate $\hat{\psi}$ is of the form $\mathbf{a}^{*\prime}\mathbf{y}$ with $\mathbf{a}^* \in V_r$, i.e., $\mathbf{a}^* \perp \boldsymbol{\alpha}_j$ for $j > r$. Now $\hat{\psi} = \mathbf{a}^{*\prime}\mathbf{y} = \mathbf{c}'\mathbf{z}$, where $\mathbf{c}' = \mathbf{a}^{*\prime}\mathbf{P}'$ is a row matrix whose jth element is $c_j = \mathbf{a}^{*\prime}\boldsymbol{\alpha}_j$, so $c_j = 0$ for $j > r$. Hence $\hat{\psi} = \Sigma_1^r c_j z_j$.

Although the linear forms $\{z_1, \cdots, z_n\}$ depend on the choice of the basis $\{\boldsymbol{\alpha}_1, \cdots, \boldsymbol{\alpha}_n\}$, it is clear that the estimation and error spaces do not, since the first is the space of all $\hat{\psi}$, and the second is the space of all $\mathbf{a}'\mathbf{y}$ for which $E(\mathbf{a}'\mathbf{y}) = 0$.

PROBLEMS

1.1. First- and second-degree polynomials are fitted by LS to n points (x_i, y_i), $i = 1, \cdots, n$. Let ω and Ω denote the assumptions*

$$\omega: \quad y_i = \alpha + \beta x_i + e_i, \qquad E(e_i) = 0, \qquad E(e_i e_{i'}) = \sigma^2 \delta_{ii'},$$
$$\Omega: \quad y_i = \alpha + \beta x_i + \gamma x_i^2 + e_i, \qquad E(e_i) = 0, \qquad E(e_i e_{i'}) = \sigma^2 \delta_{ii'}.$$

Find by differentiation the normal equations for the estimates of α and β under ω, and of α, β, and γ under Ω. Solve the former explicitly and indicate the solution of the latter by using determinants. Save the results of Problems 1.1, 1.2, and 1.3 for later use in Ch. 2.

1.2. In Problem 1.1 find the variances and covariance of the estimates of α and β under ω. Show that if we write $\delta + \beta(x_i - \bar{x})$ in place of $\alpha + \beta x_i$ in ω, then under ω, $\hat{\delta} = \bar{y}$ and Cov $(\hat{\delta}, \hat{\beta}) = 0$.

1.3. In Problem 1.1 express Var $(\hat{\gamma})$ under Ω by using determinants.

1.4. Prove the following lemma: If $\mathbf{y} = (y_1, \cdots, y_n)'$, $E(\mathbf{y}) = \boldsymbol{\eta}$, $\mathbf{e} = \mathbf{y} - \boldsymbol{\eta}$, and $Q(\mathbf{y})$ is a quadratic form in $\mathbf{y}$, then $E(Q(\mathbf{y})) = Q(\boldsymbol{\eta}) + E(Q(\mathbf{e}))$. Note that $Q(\boldsymbol{\eta})$ may be evaluated by replacing the $\{y_i\}$ by their expectations in $Q(\mathbf{y})$, and that $E(Q(\mathbf{e}))$ is the value of $E(Q(\mathbf{y}))$ when $\boldsymbol{\eta} = \mathbf{0}$.

1.5. Prove the following result, of importance in the theory of the design of experiments: Under Ω: $E(\mathbf{y}) = \Sigma_1^p \beta_j \boldsymbol{\xi}_j$ and $\boldsymbol{\Sigma}_y = \sigma^2 \mathbf{I}$, if $\boldsymbol{\xi}_\nu = \boldsymbol{\xi}_\nu^* + \boldsymbol{\xi}_\nu^\perp$, where $\boldsymbol{\xi}_\nu^*$ is the projection of $\boldsymbol{\xi}_\nu$ on the space spanned by the other $\{\boldsymbol{\xi}_j\}$, and if $\boldsymbol{\xi}_\nu^\perp \neq \mathbf{0}$, then β_ν is estimable, and the variance of its LS estimate $\hat{\beta}_\nu$ is $\|\boldsymbol{\xi}_\nu^\perp\|^{-2}\sigma^2$. [*Hint:* Assume $\nu = 1$ and take the vector $\boldsymbol{\alpha}_1$ of the canonical form of sec. 1.6 in the direction of $\boldsymbol{\xi}_\nu^\perp$.]

* It is convenient to denote the underlying assumptions by ω and Ω rather than Ω_1 and Ω_2 for later use in Ch. 2.

CHAPTER 2

Construction of Confidence Ellipsoids and Tests in the General Case Under Normal Theory

2.1. UNDERLYING ASSUMPTIONS Ω AND DISTRIBUTION OF POINT ESTIMATES UNDER Ω

The theory of this chapter, like that of Ch. 1, is general, in that it is not restricted to the case where the elements of $\mathbf{X}'$ have integer values: Even though this case is the one of primary interest for this book, we do not now restrict ourselves to it because it would not simplify the derivation of the results in these two chapters. In addition to the underlying assumptions already made that $E(\mathbf{y}) = \mathbf{X}'\boldsymbol{\beta}$ and $\boldsymbol{\Sigma}_y = \sigma^2\mathbf{I}$, we now assume further that the observations $\{y_i\}$ have a joint normal distribution.[1] This further assumption permits us to derive (i) confidence intervals for the values of estimable functions of the parameters, whose point estimation was treated in Ch. 1, and, more generally, confidence sets for the simultaneous estimation of more than one estimable function, and (ii) tests of certain kinds of hypotheses about the parameter values, and the power of these tests. The effects of departures from these underlying assumptions on the statistical inferences derived from them will be discussed in Ch. 10. They may be written

$$\Omega: \quad \mathbf{y}^{n\times 1} \quad \text{is} \quad N(\mathbf{X}'\boldsymbol{\beta}^{p\times 1}, \sigma^2\mathbf{I}), \qquad \text{rank } \mathbf{X}'^{n\times p} = r.$$

The small number of results from multivariate theory needed in this chapter may be found in App. V.

[1] This together with the assumption $\boldsymbol{\Sigma}_y = \sigma^2\mathbf{I}$ implies that the observations $\{y_i\}$ are statistically independent. The more general assumptions about $\boldsymbol{\Sigma}_y$ made in sec. 1.5 could evidently be made here also, and could be followed by the same transformation which reduces to the present assumptions.

Let $\psi_1, \psi_2, \cdots, \psi_q$ be any set of q estimable functions,

$$(2.1.1) \qquad \psi_i = \sum_{j=1}^{p} c_{ij}\beta_j,$$

where the $\{c_{ij}\}$ are known constant coefficients, and let $\hat{\psi}_1, \hat{\psi}_2, \cdots, \hat{\psi}_q$ be their LS estimates: By the Gauss–Markoff theorem (sec. 1.4) these are uniquely determined linear functions of the observations,

$$(2.1.2) \qquad \hat{\psi}_i = \sum_{j=1}^{n} a_{ij}y_j \qquad (i = 1, \cdots, q).$$

We may think of determining the coefficients $\{a_{ij}\}$ in two different ways: (i) Let $\{\hat{\beta}_j\}$ be any solution of the normal equations: The $\{\beta_j\}$ can always be taken as linear functions of the $\{y_i\}$, and may then be substituted into $\hat{\psi}_i = \Sigma_{j=1}^{p} c_{ij}\hat{\beta}_j$. (ii) Since any estimable function is a linear combination of the expected values of the observations $\{y_i\}$, $\psi_i = \mathbf{b}_i'\boldsymbol{\eta}$, where each $\mathbf{b}_i$ is a vector of constant coefficients and $\boldsymbol{\eta} = E(\mathbf{y})$. Then, by the proof of the Gauss–Markoff theorem in sec. 1.4, $\hat{\psi}_i = \mathbf{a}_i'\mathbf{y}$, where $\mathbf{a}_i$ is the projection of $\mathbf{b}_i$ on V_r, the space spanned by the columns of $\mathbf{X}'$ (and $\mathbf{a}_i$ could be calculated by constructing an orthogonal basis for V_r through application of the Schmidt process of Lemma 6, App. I, to the columns of $\mathbf{X}'$). The equations (2.1.1) and (2.1.2) may be written in matrix form

$$(2.1.3) \qquad \boldsymbol{\psi} = \mathbf{C}\boldsymbol{\beta},$$

$$(2.1.4) \qquad \hat{\boldsymbol{\psi}} = \mathbf{A}\mathbf{y},$$

where $\boldsymbol{\psi}^{q\times 1} = (\psi_1, \cdots, \psi_q)$, $\hat{\boldsymbol{\psi}}^{q\times 1} = (\hat{\psi}_1, \cdots, \hat{\psi}_q)$, $\mathbf{C}^{q\times p} = (c_{ij})$, and $\mathbf{A} = (a_{ij})$. The covariance matrix of the estimates $\{\hat{\psi}_i\}$ is then $\boldsymbol{\Sigma}_{\hat{\psi}} = \sigma^2\mathbf{A}\mathbf{A}'$, and an unbiased estimate of σ^2 is the mean square for error

$$s^2 = \mathscr{S}_\Omega/(n-r)$$

considered in sec. 1.6.

The joint distribution of the estimates $\{\hat{\psi}_i\}$ and the error sum of squares $\mathscr{S}_\Omega$ is given by the following

Theorem: Under the above Ω-assumptions $\hat{\boldsymbol{\psi}}$ is $N(\boldsymbol{\psi}, \boldsymbol{\Sigma}_{\hat{\psi}})$ and statistically independent of $\mathscr{S}_\Omega/\sigma^2$, which has a chi-square distribution with $n-r$ d.f.

Proof: That $\hat{\boldsymbol{\psi}}$ has a multivariate normal distribution follows from the same for $\mathbf{y}$ and the linear relation (2.1.4), as shown in App. V. From the Gauss–Markoff theorem we know that $E(\hat{\boldsymbol{\psi}}) = \boldsymbol{\psi}$. To establish the rest of the theorem we use the canonical form of sec. 1.6: If $\{z_1, \cdots, z_n\}$ are the canonical variables, then $\mathbf{z} = \mathbf{P}\mathbf{y}$, where $\mathbf{P}'\mathbf{P} = \mathbf{I}$, so $\mathbf{z}$ is $N(\boldsymbol{\zeta}, \sigma^2\mathbf{I})$, and $\zeta_i = 0$ for $i > r$. We found at the end of sec. 1.6 that $\hat{\boldsymbol{\psi}}$ is a function only of the set $\{z_1, \cdots, z_r\}$, and $\mathscr{S}_\Omega$ only of the set $\{z_{r+1}, \cdots, z_n\}$, and since the two sets are statistically independent so are $\hat{\psi}$ and $\mathcal{S}_\Omega$. Finally, $\mathcal{S}_\Omega/\sigma^2 = \Sigma_{r+1}^n(z_i/\sigma)^2$, and since the $\{z_i/\sigma\}$ for $i > r$ are independently $N(0, 1)$, therefore $\mathcal{S}_\Omega/\sigma^2$ is chi-square with $n-r$ d.f.

2.2. NOTATION FOR CERTAIN TABLED DISTRIBUTIONS

In this section we adopt a notation for the per cent points of the F-distribution, which we will hereafter encounter frequently, and for the per cent points of some other distributions for which we will have more occasional use. We will refer the reader to tables for the per cent points needed, and also to existing tables of the corresponding cumulative distribution functions. By the *upper* α *point* (upper 100 α per cent point) of a random variable, or of its distribution, we mean a value whose probability of being exceeded by the random variable is α, i.e., z_α is the upper α point of the random variable z if $\Pr\{z > z_\alpha\} = \alpha$. By the *cumulative distribution function* of a random variable we mean that function of x which gives for every value x the probability that the random variable does not exceed x, i.e., the cumulative distribution function of the random variable z is $\Pr\{z \leq x\}$.

We write χ_ν^2 to denote a chi-square variable with ν d.f. (degrees of freedom), and $\chi_{\alpha;\nu}^2$ for its upper α point, so $\Pr\{\chi_\nu^2 > \chi_{\alpha;\nu}^2\} = \alpha$. We write F_{ν_1,ν_2} for an F-variable with ν_1 and ν_2 d.f., and $F_{\alpha;\nu_1,\nu_2}$ for its upper α point; t_ν for a t-variable with ν d.f., and $t_{\alpha;\nu}$ for its upper α point. The "central" variables χ_ν^2, F_{ν_1,ν_2}, t_ν are the special cases of the respective noncentral variables $\chi'^2_{\nu,\delta}, F'_{\nu_1,\nu_2;\delta}$, $t'_{\nu,\delta}$ defined in App. IV, when the noncentrality parameter $\delta = 0$.

For $\alpha = 0.005, 0.01, 0.025, 0.05, 0.10$ the values of $F_{\alpha;\nu_1,\nu_2}$ are given in the F-table at the end of this book. If interpolation is necessary in ν_1 or ν_2 one uses linear interpolation not in ν_1 and ν_2 but their reciprocals: The tables are arranged to facilitate linear interpolation in $120/\nu_1$ and $120/\nu_2$. For the usual F-tests only the upper α points are needed (for small values of α), but for certain two-sided confidence intervals (for example in Ch. 8) a lower α point is also needed. For the lower α point of F_{ν_1,ν_2} we may write $F_{1-\alpha;\nu_1,\nu_2}$, and from the definition of F_{ν_1,ν_2} it follows that (note the numbers of d.f. get reversed):

$$F_{1-\alpha;\nu_1,\nu_2} = 1/F_{\alpha;\nu_2,\nu_1}.$$

Values of $\chi_{\alpha;\nu}^2$ and $t_{\alpha;\nu}$ may also be obtained from the table of $F_{\alpha;\nu_1,\nu_2}$, by the relations

$$\chi_{\alpha;\nu}^2 = \nu F_{\alpha;\nu,\infty},$$

$$t_{\alpha/2;\nu} = (F_{\alpha;1,\nu})^{1/2}.$$

(Note that the upper α point of F corresponds to the "two-tailed" α point of t.)

The cumulative distribution function of χ^2_ν is tabled in the *Biometrika Tables for Statisticians* edited by E. S. Pearson and Hartley (1954), Table 7; 1 minus the cumulative distribution function is there called the "probability integral." The probability integral of F_{ν_1,ν_2} may be found from Karl Pearson's (1934) *Tables of the Incomplete Beta Function*; $\Pr\{F_{\nu_1,\nu_2} > F_0\} = I_{x_0}(\frac{1}{2}\nu_1, \frac{1}{2}\nu_2)$, where $x_0 = \nu_1 F_0/(\nu_2 + \nu_1 F_0)$, and $I_x(p, q)$ is Karl Pearson's notation for the incomplete-beta function he tabled.[2] The cumulative distribution function of t_ν is tabled in the *Biometrika Tables for Statisticians*, Table 9, and is there called the "probability integral."

Some tables and charts for the noncentral distributions are described in sec. 2.8.

In the next chapter we will apply the distribution of the Studentized range $q_{k,\nu}$ and its upper α point, $q_{\alpha;k,\nu}$. The distribution of $q_{k,\nu}$ is defined as follows: Let $x_1, \cdots, x_k$ be independently $N(\mu_x, \sigma_x^2)$, let R be the range of the $\{x_i\}$, i.e.,

$$R = \max_i x_i - \min_i x_i,$$

and let s_x^2 be an independent mean-square estimate of σ_x^2 with ν d.f., i.e., $\nu s_x^2/\sigma_x^2$ is χ^2_ν and statistically independent of R. Then the distribution of the *Studentized range* $q_{k,\nu}$ is that of R/s_x. Its upper α point is tabled for $\alpha = 0.01$, 0.05, and 0.10 in the Studentized range table at the end of the book.

2.3. CONFIDENCE ELLIPSOIDS AND CONFIDENCE INTERVALS FOR ESTIMABLE FUNCTIONS

Confidence sets are generalizations of the familiar notion of confidence intervals: Suppose that $\{y_1, \cdots, y_n\}$ are observations whose distribution is completely determined by the unknown values of parameters $\{\theta_1, \cdots, \theta_m\}$, and that $\{\psi_1, \cdots, \psi_q\}$ are specified functions of the parameters (which are presumably of especial interest to us for some application). Denote the three points with coordinates $\{y_1, \cdots, y_n\}$, $\{\theta_1, \cdots, \theta_m\}$, $\{\psi_1, \cdots, \psi_q\}$ respectively by $\mathbf{y}$, $\boldsymbol{\theta}$, $\boldsymbol{\psi}$, so that $\boldsymbol{\psi}$ is a point, determined by the value of $\boldsymbol{\theta}$, in a q-dimensional ψ-space. Suppose for every possible $\mathbf{y}$ in the sample space a region[3] $R(\mathbf{y})$ in the q-dimensional ψ-space is determined. Then, if the region $R(\mathbf{y})$ has the property that the probability

[2] Which is the incomplete-beta function usually defined, divided by $B(p, q) = I_1(p, q)$.

[3] Point set.

that it cover the true point $\boldsymbol{\psi}$ is a preassigned constant $1-\alpha$, no matter what the unknown true parameter point $\boldsymbol{\theta}$ is, we say that $R(\mathbf{y})$ is a *confidence set* for $\boldsymbol{\psi}$ with *confidence coefficient* $1-\alpha$. The frequency interpretation is that in the succession of different situations where a statistician employs confidence sets with confidence coefficients $1-\alpha$, in the long run a proportion $1-\alpha$ of his sets will cover the true $\boldsymbol{\psi}$'s being estimated by the confidence sets (in general there are not only different $\mathbf{y}$'s in the different situations but different n's, m's, $\boldsymbol{\theta}$'s, distributions, and $\boldsymbol{\psi}$-functions of interest). A *confidence interval* is the special case where $q = 1$ and $R(\mathbf{y})$ is an interval in the one-dimensional ψ-space.

Let $\psi_1, \psi_2, \cdots, \psi_q$ denote q estimable functions. In this section we will obtain a confidence set in the form of an ellipsoid[4] (App. III) for the point $(\psi_1, \cdots, \psi_q)$ in a q-dimensional ψ-space. The most important application is perhaps the S-method of multiple comparison (secs. 3.4, 3.5).

We may assume that $\psi_1, \cdots, \psi_q$ are linearly independent; that is, if $\boldsymbol{\psi}^{q\times 1} = \mathbf{C}^{q\times p}\boldsymbol{\beta}^{p\times 1}$, the rows of $\mathbf{C}$ are linearly independent. If this were not the case we could find m $(m < q)$ linearly independent ψ_i such that the rest are linear combinations of these. Suppose that the $\{\psi_i\}$ were renumbered so that $\{\psi_{m+1}, \cdots, \psi_q\}$ are linear combinations (with known coefficients) of $\{\psi_1, \cdots, \psi_m\}$, the latter set being linearly independent. Then for every position of the point $(\psi_1, \cdots, \psi_m)$ the point $(\psi_1, \cdots, \psi_q)$ is uniquely determined, and so if we have a confidence set for the former point we also have one for the latter.

We use now the notations (2.1.1) to (2.1.4) of sec. 2.1. We see that rank $\mathbf{C} = q$ since the $\{\psi_i\}$ are linearly independent. From the theorem in sec. 2.1 we know that $\hat{\boldsymbol{\psi}}$ is $N(\boldsymbol{\psi}, \sigma^2\mathbf{B})$, where

$$\mathbf{B} = \sigma^{-2}\boldsymbol{\Sigma}_{\hat{\psi}} = \mathbf{A}\mathbf{A}', \tag{2.3.1}$$

and $\hat{\boldsymbol{\psi}}$ is statistically independent of $\mathscr{S}_\Omega = \sigma^2\chi^2_{n-r}$. We shall show below that $\mathbf{B}$ is nonsingular. It will then follow (App. V) that

$$(\hat{\boldsymbol{\psi}}-\boldsymbol{\psi})'\,\mathbf{B}^{-1}(\hat{\boldsymbol{\psi}}-\boldsymbol{\psi}) \quad \text{is} \quad \sigma^2\chi^2_q \tag{2.3.2}$$

and independent of

$$\mathscr{S}_\Omega/\sigma^2, \quad \text{which is} \quad \chi^2_{n-r}, \tag{2.3.3}$$

and hence that

$$q^{-1}(\hat{\boldsymbol{\psi}}-\boldsymbol{\psi})'\,\mathbf{B}^{-1}(\hat{\boldsymbol{\psi}}-\boldsymbol{\psi})/s^2 \quad \text{is} \quad F_{q,n-r}, \tag{2.3.4}$$

where $s^2 = \mathscr{S}_\Omega/(n-r)$ is the mean square for error.

[4] The concept of confidence ellipsoids was introduced by Hotelling (1929, 1931). The general theory of confidence intervals was founded by Neyman (1937).

To prove $\mathbf{B}$ nonsingular we take expected values in $\hat{\boldsymbol{\psi}} = \mathbf{Ay}$ to find $\boldsymbol{\psi} = \mathbf{X}'\boldsymbol{\beta} = \mathbf{C}\boldsymbol{\beta}$ identically in $\boldsymbol{\beta}$; hence $\mathbf{C} = \mathbf{AX}'$, and so $q = \text{rank } \mathbf{C} = \text{rank } \mathbf{AX}' \leqq \text{rank } \mathbf{A}^{q\times n} \leqq q$, or rank $\mathbf{A} = q$. But by Theorem 7 of App. II applied to (2.3.1) we get rank $\mathbf{B}$ = rank $\mathbf{A}$, and so $\mathbf{B}^{q\times q}$ is nonsingular.

The desired confidence set falls out of (2.3.4): Under Ω the probability is $1-\alpha$ that the F-variable in (2.3.4) is $\leqq F_{\alpha;q,n-r}$, or that

$$(\boldsymbol{\psi}-\hat{\boldsymbol{\psi}})'\,\mathbf{B}^{-1}(\boldsymbol{\psi}-\hat{\boldsymbol{\psi}}) \leqq qs^2F_{\alpha;q,n-r}. \tag{2.3.5}$$

Inequality (2.3.5) determines an ellipsoid (see App. III) in the q-dimensional ψ-space with center at $(\hat{\psi}_1, \cdots, \hat{\psi}_q)$, and the probability that this random ellipsoid covers the true parameter point $(\psi_1, \cdots, \psi_q)$ is $1-\alpha$, no matter what the values of the unknown parameters $\beta_1, \cdots, \beta_p, \sigma^2$.

We may obtain a confidence interval for a single estimable function $\psi = \mathbf{c}'\boldsymbol{\beta}$ $(\mathbf{c} \neq \mathbf{0})$ by specializing the above calculation to $q = 1$. The resulting one-dimensional ellipsoid is the interval

$$b^{-1}(\hat{\psi}-\psi)^2 \leqq s^2F_{\alpha;1,n-r}, \tag{2.3.6}$$

where $\hat{\psi} = \mathbf{a}'\mathbf{y}$ is the LS estimate of ψ, and $b = \mathbf{a}'\mathbf{a}$. We estimate $\text{Var}(\hat{\boldsymbol{\psi}}) = \mathbf{a}'\mathbf{a}\sigma^2$ by

$$\hat{\sigma}_{\hat{\psi}}^2 = \mathbf{a}'\mathbf{a}s^2,$$

and so may write (2.3.6) as

$$\hat{\psi} - t_{\alpha/2;n-r}\hat{\sigma}_{\hat{\psi}} \leqq \psi \leqq \hat{\psi} + t_{\alpha/2;n-r}\hat{\sigma}_{\hat{\psi}}; \tag{2.3.7}$$

the probability that this random interval covers the unknown ψ is $1-\alpha$. The interval (2.3.7) could also be derived from the fact that

$$(\hat{\psi}-\psi)/\hat{\sigma}_{\hat{\psi}} \quad \text{is} \quad t_{n-r}. \tag{2.3.8}$$

One-sided confidence intervals for ψ are immediately obtained by using (2.3.8) in the relations $\Pr\{t_{n-r} \leqq t_{\alpha;n-r}\} = 1-\alpha$ or $\Pr\{t_{n-r} \geqq -t_{\alpha;n-r}\} = 1-\alpha$.

The reader is cautioned against using many "t-intervals" (2.3.7) calculated on the same data, each with, say, a 95 per cent confidence coefficient, and especially against using (2.3.7) on a ψ that has been selected because the data happen to give the estimate $\hat{\psi}$ a value large compared with $\hat{\sigma}_{\hat{\psi}}$, for he will then not know what "confidence" can be attached to his set of conclusions. A more correct method for such situations is given in secs. 3.4 and 3.5.

(δ, β) lies on the (boundary of the) ellipse. This problem may also be postponed until after sec. 3.6 and then solved as an application of the S-method of multiple comparison to the family $\{\delta + \beta(x-\bar{x})\}$, the estimation problem being the same as that for the two-dimensional space $\{c_1\delta + c_2\beta\}$ of estimable functions. By one of these two approaches, prove that the confidence band for the true line consists of all the points (x, y) satisfying

$$[y - \hat{\delta} - \hat{\beta}(x-\bar{x})]^2 \leq 2F_{\alpha;2,n-2}[n^{-1} + S_x^{-1}(x-\bar{x})^2]s^2.$$

Another band may be obtained by considering the result of Problem 2.10*b* for all values of x_0. Compare the shape and location of these bands and their interpretation.

2.13. If for each x (say in some interval) there is a random variable y whose distribution depends on x, so that $E(y)$ is a function of x, say $g(x)$, then $g(x)$ is called the *regression function*, or simply the *regression*, of y on x. Consider the hypothesis H that the regression is linear. The hypothesis H can only be tested under the underlying assumptions that the true regression is a quadratic, or a cubic, or some r-parameter function ($r < n$) including the two-parameter family of linear functions, *unless* for some x's the values of y are replicated (more than one observation taken), in which case no restriction need be made on the regression in the underlying assumptions. In the latter case derive the F-test of H by applying the method of sec. 2.5 with the following notation: Suppose n_i observations $\{y_{i1}, \cdots, y_{in_i}\}$ are made at x_i $(i = 1, \cdots, I)$, where $n_i > 1$ for some i, and $\Sigma_1^I n_i = n$. Let Ω and H be

Ω: The $\{y_{ij}\}$ are statistically independent and y_{ij} is $N(\eta_i, \sigma^2)$.

H: The $\{\eta_i\}$ satisfy $\eta_i = \alpha + \beta x_i$ for some unspecified constants α, β.

2.14. In a spectroscopic method for determining the per cent x of natural rubber content of vulcanizates the variable y used for determining x is $1 + \log_{10} r$, where r is the ratio of transmittances at two selected wavelengths. In order to establish a calibration curve, samples of vulcanizates with known x were tested with the following results.‡ (*a*) Plot the 24 data points. At each

x	0	20	40	60	80	100
	0.727	0.884	1.073	1.194	1.350	1.442
	0.721	0.880	1.050	1.184	1.291	1.369
y	0.742	0.885	1.045	1.205	1.291	1.458
	0.746	0.890	1.033	1.180	1.323	1.459

x plot also the average of the four observed y's and circle these. Fit a straight line by LS and plot it. Do the deviations of the circled points from the line appear to be large compared with the deviations of the original data points from the circled points? (*b*) Apply the test derived in Problem 2.13 to test the adequacy of linear regression in this problem. (*c*) Regardless of whether or not the result in (*b*) was significant, fit also a parabola by LS and plot the ordinates at the six values of x used. Visually, do you judge that there is a great

‡ From Table 4, p. 221 of "Determination of natural rubber in GR-S-natural rubber vulcanizates by infrared spectroscopy," by M. Tryon, E. Horowitz, and J. Mandel, *J. Research National Bureau of Standards*, Vol. 55, 1955. Reproduced with the kind permission of the authors and the editor.

improvement over the line? (*d*) Apply a test similar to (*b*) to test the adequacy of the parabolic regression. (*e*) Make a *t*-test of whether the coefficient of x^2 in (*c*) differs significantly from zero. (*f*) Considering all the previous results, would you use the straight line or the parabola as the calibration curve? [If polynomials are frequently fitted at equally spaced abscissas it is worth learning the use of orthogonal polynomials. See for example Ch. 16 of Anderson and Bancroft (1952).]

2.15. Suppose§ x denotes a point in a space of one or more dimensions, and the "true value" of the variable y at x is $\eta = g(x)$, in general an unknown function. A set of points $\{x_1, \cdots, x_n\}$ is chosen (the *design* of the experiment) and the value of y_i is observed for $\eta_i = g(x_i)$ at $x = x_i$ $(i = 1, \cdots, n)$. Let $\hat{g}(x)$ be an estimate of $g(x)$ formed by any method, for example fitting a polynomial by LS. The expected squared error $E[\hat{g}(x) - g(x)]^2$ then depends on x, on $\{x_i\}$, and on the method of estimation. Show that it may be decomposed into Var $(\hat{g}(x))$ plus a component $[E(\hat{g}(x)) - g(x)]^2$, which may be interpreted as the effect of bias due to inadequacy of the form of function chosen to approximate $g(x)$. In considering the efficiency of designs for predicting $g(x)$ the latter component is often forgotten although it is often the more important.

2.16. Denote the bias component $[E(\hat{g}(x)) - g(x)]^2$ in Problem 2.15 by $B(x)$. Show that among all functions $\hat{g}(x)$ of the type $\hat{g}(x) = \Sigma_{j=1}^p \tilde{\beta}_j h_j(x)$, where the $\{h_j(x)\}$ are given functions of x, and the $\{\tilde{\beta}_j\}$ are linear functions of the $\{y_i\}$, the $\hat{g}(x)$ formed from the $\{\tilde{\beta}_j\}$ which minimize $\Sigma_{i=1}^n [\Sigma_{j=1}^p \tilde{\beta}_j\, h_j(x_i) - y_i]^2$ also minimizes $\Sigma_{i=1}^n B(x_i)$.

2.17. Suppose the same parameters $\{\beta_1, \cdots, \beta_p\}$ are involved in K sets of measurements which are completely independent with equal variance σ^2 and with n_k in the kth set $(k = 1, \cdots, K)$. Let $\mathbf{y}_{(k)}$ denote the n_k-dimensional vector whose elements are the measurements in the kth set, and suppose $E(\mathbf{y}_{(k)}) = \mathbf{X}'_{(k)}\boldsymbol{\beta}$, where $\mathbf{X}'_{(k)}$ is $n_k \times p$. Prove that any set $\hat{\boldsymbol{\beta}}$ of LS estimates from the complete set of $\Sigma_k n_k$ measurements satisfies $(\Sigma_k \mathbf{W}_{(k)})\hat{\boldsymbol{\beta}} = \Sigma_k \mathbf{W}_{(k)}\hat{\boldsymbol{\beta}}_{(k)}$, where $\hat{\boldsymbol{\beta}}_{(k)}$ is any set of LS estimates from the kth set, and $\mathbf{W}_{(k)} = \mathbf{X}_{(k)}\mathbf{X}'_{(k)}$. [*Hint*: Partition the matrix $\mathbf{X}'$ and vector $\mathbf{y}$ of the combined set into those of the K sets.]

§ This problem was suggested by *A Basis for the Selection of a Response Surface Design* by G. E. P. Box and N. R. Draper, paper 66 presented at the 31st session of the International Statistical Institute in Brussels, Sept. 1958, Technical Report 23, Statistical Techniques Research Group, Princeton University.

CHAPTER 3

The One-Way Layout. Multiple Comparison

3.1. THE ONE-WAY LAYOUT

In this chapter we shall define the simplest case in which the analysis of variance is applied, namely that of the one-way layout. We shall use this to illustrate some of the general theory of estimation and testing of Chs. 1 and 2. We shall also introduce some new concepts and methods concerning problems of multiple comparison (a kind of simultaneous interval estimation and multiple significance testing), first in connection with the one-way layout, but then generalizing them in such a way as to supplement the general F-test of Ch. 2 and to throw more light on it: We shall see that the general null hypothesis H considered in Ch. 2 is equivalent to the statement that all parametric functions in a certain class have the value zero, and whenever "the" F-test rejects H we may use one of the multiple-comparison methods to decide which of the parametric functions in the class differ from zero and by how much. We shall also see that the F-test may be regarded (i) as a preliminary method to find whether or not this more elaborate analysis will yield anything; or (ii) as looking only at that estimate of the functions in the class which differs most from zero in a certain sense, and deciding whether it differs significantly from zero.

The *one-way layout* (also called the *one-way classification*) refers to the comparison of the means of several (univariate) populations. Denote the means by $\beta_1, \beta_2, \cdots, \beta_I$. We shall assume that the I populations are normal with equal variance σ^2, and that we have independent random samples of sizes $J_1, J_2, \cdots, J_I$ from the respective populations. Denote the sample from the ith population by $(y_{i1}, y_{i2}, \cdots, y_{iJ_i})$. Then our underlying assumptions (effects of departures from these assumptions are treated in Ch. 10) are equivalent to

$$\Omega: \begin{cases} y_{ij} = \beta_i + e_{ij} & (i = 1, \cdots, I; \quad j = 1, \cdots, J_i), \\ \{e_{ij}\} \text{ are independently } N(0, \sigma^2). \end{cases}$$

We wish to test the hypothesis

$$H:\quad \beta_1 = \beta_2 = \cdots = \beta_I,$$

and we shall derive "the" F-test of H under Ω by applying the general theory of sec. 2.5.

Let $n = \Sigma_1^I J_i$. The vector of observations, written $\mathbf{y} = (y_1, y_2, \cdots, y_n)'$ in the general theory of Chs. 1 and 2, must now be thought of with a double-subscript notation for its components, $\mathbf{y} = (y_{11}, y_{12}, \cdots, y_{1J_1}, y_{21}, \cdots, y_{2J_2}, \cdots, y_{I1}, \cdots, y_{IJ_I})'$. From $E(y_{ij}) = \beta_i$ we see that the $n \times I$ matrix $\mathbf{X}'$ now has the following form: The first J_1 rows are all $\boldsymbol{\rho}_1'$, the next J_2 rows are all $\boldsymbol{\rho}_2'$, $\cdots$, the last J_I rows are all $\boldsymbol{\rho}_I'$, where $\boldsymbol{\rho}_i' = (\delta_{i1}, \delta_{i2}, \cdots, \delta_{iI})$ is the ith row of the identity matrix. Thus the rank r (the number of linearly independent rows) of $\mathbf{X}'$ equals I (the number of columns of $\mathbf{X}'$, denoted by p in the general notation of Ch. 1), and hence all parametric functions are estimable (a different parameterization for which this is not true is discussed in sec. 3.2).

The sum of squares $\mathscr{S}(\mathbf{y}, \boldsymbol{\beta})$ to be minimized under Ω and $\omega = \Omega \cap H$ is

$$\mathscr{S}(\mathbf{y}, \boldsymbol{\beta}) = \sum_{i=1}^{I} \sum_{j=1}^{J_i} (y_{ij} - \beta_i)^2. \tag{3.1.1}$$

The normal equations under Ω are obtained by equating to zero

$$\partial \mathscr{S}/\partial \beta_\nu = -2 \sum_{j=1}^{J_\nu} (y_{\nu j} - \beta_\nu)$$

for $\nu = 1, \cdots, I$. This gives the LS estimates $\hat{\beta}_\nu = \Sigma_{j=1}^{J_\nu} y_{\nu j}/J_\nu$. With the very convenient *dot notation*, which we shall employ extensively henceforth, this may be written

$$\hat{\beta}_\nu = y_{\nu.} \qquad (\nu = 1, \cdots, I),$$

where replacing a subscript by a dot means that the arithmetic average of the quantities to which the subscript is attached has been taken over all possible values of the subscript.

Minimizing (3.1.1) under ω is the same as minimizing

$$\mathscr{S}' = \sum_{i=1}^{I} \sum_{j=1}^{J_i} (y_{ij} - \beta)^2,$$

where β denotes the common (unknown) value of $\beta_1, \beta_2, \cdots, \beta_I$. There is then only one normal equation,

$$\partial \mathscr{S}'/\partial \beta = -2 \sum_{i=1}^{I} \sum_{j=1}^{J_i} (y_{ij} - \beta) = 0,$$

which gives us (the caret over β indicates that it is the LS estimate of the unknown β)

$$\hat{\beta} = \sum_{i=1}^{I} \sum_{j=1}^{J_i} y_{ij}/n = \sum_{i=1}^{I} J_i y_{i.}/n.$$

We shall denote the last member by[1] $\bar{y}$. Thus

$$\hat{\beta}_{\nu,\omega} = \bar{y} \qquad (\nu = 1, \cdots, I),$$

where $\hat{\beta}_{\nu,\omega}$, written above as $\hat{\beta}$, denotes the LS estimate of β_ν under ω.

For expressing SS's and identities it is convenient to use the η-notation of Ch. 2: We recall that $\boldsymbol{\eta}$ denotes $E(\mathbf{y})$, and that $\hat{\boldsymbol{\eta}}$ and $\hat{\boldsymbol{\eta}}_\omega$ are the LS estimates of $\boldsymbol{\eta}$ under Ω and ω, equal to the projections of $\mathbf{y}$ on V_r and V_{r-q}, the spaces in which $\boldsymbol{\eta}$ is constrained to lie by Ω and ω, respectively. The components of the vectors $\boldsymbol{\eta}$, $\hat{\boldsymbol{\eta}}$, and $\hat{\boldsymbol{\eta}}_\omega$ must now be indexed with the same double-subscript notation as the components of $\mathbf{y}$. Then the "i,j component" (component with subscripts i,j) of $\boldsymbol{\eta}$ is written η_{ij}, and since $\eta_{ij} = E(y_{ij})$ therefore

$$\eta_{ij} = \beta_i \text{ under } \Omega, \tag{3.1.2}$$

and the i,j component of $\hat{\boldsymbol{\eta}}$ is obtained by replacing $\{\beta_1, \cdots, \beta_I\}$ in (3.1.2) by their LS estimates under Ω, giving

$$\hat{\eta}_{ij} = y_{i.}.$$

Operating similarly under ω, we find that the i,j component of $\hat{\boldsymbol{\eta}}_\omega$ is

$$\hat{\eta}_{ij,\omega} = \bar{y}.$$

We obtain the numerator and denominator SS's of the test-statistic $\mathscr{F}$ from the general identities of sec. 2.9

$$SS_H = \|\hat{\boldsymbol{\eta}} - \hat{\boldsymbol{\eta}}_\omega\|^2, \qquad SS_e = \|\mathbf{y} - \hat{\boldsymbol{\eta}}\|^2,$$

which become

$$SS_H = \sum_i \sum_j (\hat{\eta}_{ij} - \hat{\eta}_{ij,\omega})^2 = \sum_i \sum_j (y_{i.} - \bar{y})^2 = \sum_i J_i (y_{i.} - \bar{y})^2,$$

$$SS_e = \sum_i \sum_j (y_{ij} - \hat{\eta}_{ij})^2 = \sum_i \sum_j (y_{ij} - y_{i.})^2.$$

These formulas suggest simple intuitive interpretations: SS_H is a weighted measure of the spread of the sample means from the I populations, and SS_e is a combined measure of the spread of the observations from within each of the I samples; for these reasons SS_H is also called the SS *between*

[1] We do not write $y_{..}$, as we would in case the J_i were assumed equal, because this would suggest the arithmetic or unweighted average of the $\{y_{i.}\}$ instead of their weighted average $\bar{y}$.

groups, and SS_e the SS *within groups*. For numerical calculation of these SS's formulas different from the above are used, which also follow from general identities of sec. 2.9. One of those identities, the third of (2.9.2), written $SS_H = \|\hat{\boldsymbol{\eta}}\|^2 - \|\hat{\boldsymbol{\eta}}_\omega\|^2$, becomes in this case

$$SS_H = \sum_i \sum_j \hat{\eta}_{ij}^2 - \sum_i \sum_j \hat{\eta}_{ij,\omega}^2,$$

or

$$SS_H = \sum_i J_i y_{i.}^2 - n\bar{y}^2. \tag{3.1.3}$$

The general identity $\mathscr{S}_\omega = \mathscr{S}_\Omega + SS_H$, or $\|\mathbf{y} - \hat{\boldsymbol{\eta}}_\omega\|^2 = SS_e + SS_H$, becomes

$$\sum_i \sum_j (y_{ij} - \bar{y})^2 = SS_e + SS_H. \tag{3.1.4}$$

The SS on the left is called the *total* SS *about the grand mean*. We shall write it $SS_{\text{"tot"}}$—the quotation marks to distinguish it from $SS_{\text{tot}} = \Sigma_i \Sigma_j y_{ij}^2$. We may then write (3.1.4) as

$$SS_{\text{"tot"}} = SS_{\text{within groups}} + SS_{\text{between groups}}. \tag{3.1.5}$$

Finally, the general identity $\mathscr{S}_\omega = \|\mathbf{y}\|^2 - \|\hat{\boldsymbol{\eta}}_\omega\|^2$ becomes

$$SS_{\text{"tot"}} = \sum_{i=1}^{I} \sum_{j=1}^{J_i} y_{ij}^2 - n\bar{y}^2. \tag{3.1.6}$$

The hypothesis H which we wish to test can be specified in various ways by equating to zero $I-1$ linearly independent estimable functions, for example

$$H: \quad \beta_2 - \beta_1 = 0,\ \beta_3 - \beta_1 = 0 \cdots,\ \beta_I - \beta_1 = 0;$$

hence the number of d.f. for SS_H is $q = I-1$. We have already noticed that $r = I$, hence the number of d.f. for SS_e is $n-r = n-I$. The statistic $\mathscr{F}$ is thus MS_H/MS_e, where

$$MS_H = SS_H/(I-1), \qquad MS_e = SS_e/(n-I),$$

and "the" F-test of H of sec. 2.5 consists in rejecting H at the α level of significance if and only if $\mathscr{F} > F_{\alpha;I-1,n-I}$. Under Ω, $\mathscr{F}$ has the distribution of a noncentral F-variable, namely $\mathscr{F}$ is $F'_{I-1,n-I;\delta}$, where the noncentrality parameter δ is determined by Rule 1 of sec. 2.6: If we write $\bar{\beta}$ for $E(\bar{y})$, so that

$$\bar{\beta} = \sum_{i=1}^{I} J_i \beta_i / n,$$

then

$$\sigma^2 \delta^2 = \sum_{i=1}^{I} J_i (\beta_i - \bar{\beta})^2.$$

Some of these results are usually summarized in a table, as in Table 3.1.1.

TABLE 3.1.1

ANALYSIS OF VARIANCE OF ONE-WAY LAYOUT

Source	SS	d.f.	MS	E(MS)
Between groups	$SS_H = \sum_i J_i(y_{i.}-\bar{y})^2$	$I-1$	$SS_H/(I-1)$	$\sigma^2+(I-1)^{-1}\sum_i J_i(\beta_i-\bar{\beta})^2$
Within groups	$SS_e = \sum_i\sum_j(y_{ij}-y_{i.})^2$	$n-I$	$SS_e/(n-I)$	σ^2
Total about grand mean	$SS_{\text{"tot"}} = \sum_i\sum_j(y_{ij}-\bar{y})^2$	$n-1$	—	—

Calculations: Although the subtotals $\Sigma_j y_{ij}^2$ are not needed for the F-test, but only their total (over i), it is a good idea to record the subtotals in case we wish to look at the quantities

$$s_i^2 = \left(\sum_j y_{ij}^2 - J_i y_{i.}^2\right)/(J_i - 1),$$

s_i^2 being the estimate of the variance of the ith population that we would use if we did not assume equal population variances. In this connection see also secs. 3.8 and 10.6. The quantity $SS_{\text{"tot"}}$ is calculated from (3.1.6), SS_H from (3.1.3), and $SS_e = SS_{\text{within groups}}$ from (3.1.5) by subtracting SS_H from $SS_{\text{"tot"}}$. In the summary of the analysis, it is strongly recommended that besides an analysis-of-variance table like Table 3.1.1, a table of the I sample means $\{y_{i.}\}$, the sample sizes $\{J_i\}$, and perhaps the p sample variances $\{s_i^2\}$ should also be provided.

The methods of calculation suggested throughout this book are adapted to desk calculating machines, and often require that a large number of significant figures be retained until the final stages of the calculation because of the loss of significant figures possible when a SS is calculated by subtraction of other SS's. Until a good feel is developed for the right number of figures to carry, it is better to carry too many in the intermediate stages rather than too few, since if after a subtraction too few figures are left, the whole calculation may have to be repeated. On the other hand, the final results should always be pared down to a sensible number of figures, ordinarily such that a unit change in the last figure retained in the result is of the order of magnitude of a fifth of the estimated standard deviation of the result. A statistician cooperating with a chemist or engineer may discredit himself by thoughtlessly offering a confidence interval of the form 7.32179 ± 0.05248 instead of 7.32 ± 0.05, or by giving seven-figure coefficients for a straight line fitted to three-figure data.

For slide-rule or long-hand calculations SS's must be calculated directly from formulas like those shown in Table 3.1.1 instead of by subtraction, as in (3.1.3), to avoid the necessity of carrying a large number of significant figures. For calculations with or without machines (and especially on digital computers with a fixed number of digits) it is often worth reducing the data by subtracting a suitable constant, for example, if the data range from 151.2 to 158.7, by subtracting 150; the result of this on the conclusions is always obvious.

3.2. AN ILLUSTRATION OF THE THEORY OF ESTIMABLE FUNCTIONS

Although the parameterization we chose above for the one-way layout is the most natural one for this case, that of the following form of Ω and H resembles more nearly the parameterizations used in other cases of the analysis of variance; at the same time it affords us a simple and interesting illustration of the theory of estimable functions of sec. 1.4:

$$\Omega: \begin{cases} y_{ij} = \mu + \alpha_i + e_{ij} & (i = 1, \cdots, I;\ j = 1, \cdots, J_i), \\ \{e_{ij}\} \text{ are independently } N(0, \sigma^2), \end{cases}$$

(3.2.1)

$$H: \quad \alpha_1 = \alpha_2 = \cdots = \alpha_I.$$

It is clear that we can without loss of generality assume[2] that $\Sigma_i \alpha_i = 0$, since we can write

$$\eta_{ij} = E(y_{ij}) = \mu + \alpha_i = (\mu + \alpha_.) + (\alpha_i - \alpha_.),$$

and take as new μ and α_i the quantities $\tilde{\mu} = \mu + \alpha_.$ and $\tilde{\alpha}_i = \alpha_i - \alpha_.$, then $\Sigma_i \tilde{\alpha}_i = 0$. We also see that for any given $\{\eta_{ij}\}$ of the assumed form $\eta_{ij} = \mu + \alpha_i$, the new parameters $\tilde{\mu}$ and $\{\tilde{\alpha}_i\}$ satisfying the side condition $\Sigma_i \tilde{\alpha}_i = 0$, unlike the former μ and $\{\alpha_i\}$, are unique, for we may write

$$\tilde{\mu} + \tilde{\alpha}_i = \eta_{i1},$$

and taking the average on i, find that

$$\tilde{\mu} = \eta_{.1}, \qquad \tilde{\alpha}_i = \eta_{i1} - \eta_{.1}, \tag{3.2.2}$$

i.e., $\hat{\mu}$ and $\{\hat{\alpha}_i\}$ are uniquely determined by $\{\eta_{ij}\}$. (The $\{\eta_{ij}\}$ as the expected values of the observations have a direct probability meaning, the parameters μ and $\{\alpha_i\}$ separately do not.) It now follows from the general theory in the subsection of sec. 1.4 titled "Side Conditions on the Parameters

[2] Sometimes it is more convenient to take $\Sigma_i J_i \alpha_i = 0$.

and Estimates" that the parametric function $\Sigma_i\alpha_i$ used in the side condition is nonestimable, that, while the solution of the normal equations for the LS estimates $\hat{\mu}$ and $\{\hat{\alpha}_i\}$ is not unique, there is a unique solution satisfying $\Sigma_i\hat{\alpha}_i = 0$, and that every parametric function of the new parameters $\tilde{\mu}$ and $\{\tilde{\alpha}_i\}$ is estimable. We shall now verify these statements directly in this special case. (Some readers may wish to skip the rest of this section.)

With the parameterization (3.2.1) the number of parameters $p = I+1$. The vector $\boldsymbol{\beta}$ of parameters in the general theory may be taken as $(\mu, \alpha_1, \cdots, \alpha_I)'$. The $n\times(I+1)$ matrix $\mathbf{X}'$ of coefficients may then be obtained from the $n\times I$ matrix of the previous parameterization by adjoining a column of 1's on the left. Since the $n\times I$ matrix has rank I, and the column of 1's is the sum of the columns of the $n\times I$ matrix, the rank of the $n\times(I+1)$ matrix $\mathbf{X}'$ is also I, i.e., $r = I$. This value for r may also be arrived at more intuitively by recalling that r is the dimension of the space V_r to which $\boldsymbol{\eta}$ is restricted by Ω, and hence regarding r as the number of "independent" parameters in the problem, and "counting constants": Calculating r as the total number of parameters less the number of linearly independent side conditions on the parameters, we get $r =$ (no. of μ's) + (no. of $\{\alpha_i\}$) $-$ (no. of linearly independent side conditions) $= 1+I-1 = I$. We shall usually use this method of determining r in later cases of the analysis of variance.

In the parameterization (3.2.1) without the side condition, what are the estimable functions? In the notation of the general theory of sec. 1.4 they are the totality of parametric functions of the form $\Sigma_i a_i E(y_i)$, where $\{a_i\}$ is any set of n (known) constants; in the present notation they are $\Sigma_i\Sigma_j a_{ij}\, E(y_{ij}) = \Sigma_i\Sigma_j a_{ij}(\mu+\alpha_i) = \Sigma_i c_i(\mu+\alpha_i)$, where $\{c_i = \Sigma_j a_{ij}\}$ is an arbitrary set of I constants. Thus the estimable functions are the totality of functions ψ of the form

$$\psi = c_1\alpha_1 + c_2\alpha_2 + \cdots + c_I\alpha_I + \left(\sum_{i=1}^{I} c_i\right)\mu.$$

Hence the parametric function $\Sigma_i\alpha_i$ used in the side condition is not estimable (since the coefficient of μ, namely 0, is not the sum of the coefficients of the $\{\alpha_i\}$). Neither is any of the parametric functions $\mu, \alpha_1, \cdots, \alpha_I$ estimable, nor is $\Sigma_i J_i\alpha_i$, which is sometimes used in a side condition. A linear function of the $\{\alpha_i\}$ will be estimable if and only if it is of the form[3] $\Sigma_1^I c_i\alpha_i$ with $\Sigma_1^I c_i = 0$. In particular the $I-1$ parametric functions $\alpha_i-\alpha_1$ $(i = 2, \cdots, I)$, which may be used to describe H, are estimable.

[3] This form of parametric function is called a *contrast* in the $\{\alpha_i\}$; see sec. 3.4.

The $I+1$ normal equations for the parameterization (3.2.1), obtained by equating to zero the partial derivatives of

$$\sum_{i=1}^{I} \sum_{j=1}^{J_i} (y_{ij} - \mu - \alpha_i)^2$$

with respect to μ and the $\{\alpha_i\}$, are

$$n\hat{\mu} + \sum_i J_i\hat{\alpha}_i = \sum_i J_i y_{i.}, \tag{3.2.3}$$

$$\hat{\mu} + \hat{\alpha}_i = y_{i.} \qquad (i = 1, \cdots, I). \tag{3.2.4}$$

The equations are dependent, since if we multiply (3.2.4) by J_i and add with respect to i we get (3.2.3). Hence (3.2.3) will be satisfied by any solution of (3.2.4). But (3.2.4) will be satisfied if we choose $\hat{\mu}$ arbitrarily (but let us however restrict it to a linear function of the $\{y_{ij}\}$) and then take $\hat{\alpha}_i = y_{i.} - \hat{\mu}$; in this way we get the general solution of the normal equations. If we impose the side condition $\Sigma_i\hat{\alpha}_i = 0$ on the solution of the normal equations the solution becomes unique, since taking the sum on i of (3.2.4) then gives

$$\hat{\mu} = \sum_{i=1}^{I} y_{i.}/I, \qquad \hat{\alpha}_i = y_{i.} - \sum_{\nu=1}^{I} y_{\nu.}/I, \tag{3.2.5}$$

that is, the solutions[4] $\hat{\mu}$ and $\{\hat{\alpha}_i\}$ satisfying $\Sigma_i\hat{\alpha}_i = 0$ are uniquely determined as functions of the $\{y_{ij}\}$ by (3.2.5). Finally, to see that any parametric function of $\tilde{\mu}$ and $\{\tilde{\alpha}_i\}$, say

$$\psi = c_0\tilde{\mu} + \sum_{i=1}^{I} c_i\tilde{\alpha}_i, \tag{3.2.6}$$

is estimable, we may use (3.2.2) to write $\psi = c_0\eta_{.1} + \Sigma_i c_i(\eta_{i1} - \eta_{.1})$ to see that $c_0 y_{.1} + \Sigma_i c_i(y_{i1} - y_{.1})$ is an unbiased estimate of ψ. The LS estimate $\hat{\psi}$ is obtained by replacing $\tilde{\mu}$ and $\{\tilde{\alpha}_i\}$ in (3.2.6) by their LS estimates (3.2.5).

3.3. AN EXAMPLE OF POWER CALCULATIONS

The chief technique offered by statistical theory for deciding on the number of observations needed in an experiment is the calculation of the power of a statistical test. We shall now illustrate power calculations for the case of the one-way layout; calculations for other experimental designs are similar. Some practical suggestions concerning power calculations are given at the end of this section.

[4] We remark that, had we used the side condition $\Sigma_i J_i\alpha_i = 0$, $\hat{\mu}$ would have turned out to be the weighted average $\bar{y}$ of the $\{y_{i.}\}$ instead of the unweighted average in (3.2.5).

Suppose that eight different kinds of alloy steel are prepared by varying the composition or the method of manufacture.[5] It is expected from experience with similar experimental steels that the tensile strengths will be of the order of 150,000 psi (pounds per square inch), and that for a certain way of preparing the specimens for the tensile-strength test (the test destroys the specimen) the standard deviation σ of duplicate specimens from the same batch of steel will be about 3000 psi. Suppose further that the null hypothesis of no difference of tensile strengths of the alloys is to be tested at the 5 per cent level, and that it is considered of sufficient economic importance to reject the null hypothesis to warrant requiring a high probability of rejection, say 0.9 or more, if in fact two of the alloys differ by 10,000 psi or more. The question is, how many specimens to prepare and test for each alloy.

It is intuitively evident that if none of the eight alloys plays a distinguished role (a contrary case would be one where we tested seven experimental alloys against a control—see Problem 3.2), we should take equal numbers J of specimens for each. The numerator SS of the F-test will be

$$SS_H = J\sum_i (y_{i.} - y_{..})^2,$$

where y_{ij} is the measured tensile strength of the jth specimen of the ith alloy ($i = 1, \cdots, I$. Later we shall put $I = 8$). Rule 1 of sec. 2.6 tells us that the noncentrality parameter satisfies

$$\sigma^2\delta^2 = J\sum_i (\beta_i - \beta_.)^2,$$

where β_i is the "true" tensile strength of the ith alloy. Since the power of the test increases with δ, we next ask what is the minimum δ subject to the condition that two of the $\{\beta_i\}$ differ by Δ or more, where $\Delta = 10,000$ psi in the present example. If we obtain the specified power of $\beta = 0.9$ for this minimum δ satisfying the condition, then the power will be at least as great for all $\{\beta_i\}$ satisfying the condition. The minimum $\sigma^2\delta^2$ is obtained if two of the β_i differ by Δ and the remaining $I-2$ equal the average of these two; this is clear to our geometric or mechanical intuitions (consider minimizing the moment of inertia of a set of I equal

[5] This example was suggested to me by Mr. Cuthbert Daniel: The assumptions of independence and equal variance are reasonable here, and so is the normality assumption. The last would not be true for tensile strength of concrete, where the distribution has a large lower tail because occasionally a specimen will rupture at a very low stress because of the configuration of the gravel in the concrete. One might say that the effect of grain size is much greater with concrete than steel. Mr. Daniel remarked that if the required number J of specimens per alloy were more than a few one should consider the advantages of taking roughly the same total number IJ of observations in a factorial design comprising more factors.

masses placed at $\{\beta_i\}$ on a β-axis, about a perpendicular axis through the center of mass, subject to the condition that at least two of the masses are a distance apart $\geq \Delta$), or it may be shown analytically. Then $|\beta_i - \beta_.|$ $= \frac{1}{2}\Delta$ for the two extreme β_i, and $= 0$ for the others; hence

$$\sigma^2\delta^2 = J\Delta^2/2,$$

$$\delta = (J/2)^{1/2}(\Delta/\sigma).$$

By the definition of ϕ in sec. 2.8, with $q = I-1$, $\phi = I^{-1/2}\delta$, or

$$\phi = (\tfrac{1}{2}J/I)^{1/2}(\Delta/\sigma). \tag{3.3.1}$$

Putting $I = 8$, $\Delta = 10{,}000$ psi, $\sigma = 3000$ psi, we have

$$\phi = \tfrac{5}{6}J^{1/2}. \tag{3.3.2}$$

We now turn to the Pearson–Hartley chart with $\nu_1 = 7$ and $\alpha = 0.05$. To get a first approximation to J, we read on the curve for $\nu_2 = \infty$ that for $\beta = 0.9$, $\phi = 1.50$. Substituting this in (3.3.2) and solving for J, we get $J = 3.24$. A first approximation for J obtained from the curve for $\nu_2 = \infty$ will always be too small, since for any finite J we should be on a curve to the right of this one, where the ϕ at $\beta = 0.9$ is higher, and the J therefore by (3.3.2) likewise higher. If we took $J = 4$, the number of d.f. for error would be $\nu_2 = I(J-1) = 24$; while, from (3.3.2), $\phi = \frac{5}{3}$. Interpolating visually between the curves for $\nu_2 = 20$ and 30, we find for this ν_2 and ϕ that $\beta = 0.87$. Similarly, we find for $J = 5$ that $\nu_2 = 32$, $\phi = 1.86$, hence $\beta = 0.95$. The smallest J that gives a $\beta \geqq 0.9$ is thus $J = 5$; however, we might reconsider whether $\beta = 0.87$ is not sufficiently close to the nominal value of 0.9, in which case we could use $J = 4$: Anyway, unless one hits very close to the nominal value, it is a good idea to contemplate the two nearest attainable values that straddle it (0.87 and 0.95 in the above example).

Now suppose the value $\sigma = 3000$ psi that we used were wrong and σ were actually 50 per cent more, $\sigma = 4500$ psi. If we substitute $\Delta/\sigma = 10/4.5$, $I = 8$, and $J = 4$ in (3.3.1) we find that $\phi = 1.11$, and again interpolating between the curves, for this ϕ and $\nu_2 = 24$, we get $\beta = 0.47$. We are now more likely to accept than reject the null hypothesis in a situation where we wanted a high probability of rejection! This example illustrates how distressingly sensitive the power calculations are to the value of σ, treated as unknown in the Ω-assumptions. In some cases, as in the present example, it may be reasonable to set Δ, or a similar measure of the spread of the $\{\beta_i\}$ such as $[\Sigma_i(\beta_i - \beta_.)^2]^{1/2}$, as a given multiple of σ in order to avoid this difficulty. Thus, if we decide that it is important to discover differences of the order of 2σ, we might stipulate that we want a high power for $\Delta/\sigma = 2$. In other problems, differences of the order of σ or $\frac{1}{2}\sigma$ may be worth trying to detect. The reader may verify by using the above method that to get a power of at least 0.9 for $\Delta \geqq 2\sigma$, the first approximation is $J = 9.0$, and that for $J = 10$, $\beta = 0.90$.

Against the alternatives employed in the preceding power calculations a more powerful test is possible, as indicated by an intuitive argument at the end of sec. 3.7, namely reject H if the maximum and minimum of the observed means $\{y_{i.}\}$ differ by more than $J^{-1/2}q_{\alpha;J,\nu}s$, where $q_{\alpha;J,\nu}$ is the upper α point of the Studentized range defined in sec. 2.2, s^2 is MS_e, and ν is its number of d.f. Tables or charts for the power of this test are not available. However if we use the J calculated to produce the desired power with the F-test we shall exceed this power with the Studentized range test against the alternatives considered.

In designing an experiment it is sometimes advisable to present a small double-entry table of the power, for a few different sample sizes and alternative hypotheses (alternative Δ/σ in the present case). It may be necessary to lower the specified value of the power, or to increase the distance of the alternative where this power is attained, in order to settle for feasible sample sizes. It is wise to make power calculations for a contemplated experiment before, rather than after, the experiment, when it may be found that the number of observations was too few to give even a 50 per cent chance ($\beta < 0.5$) of detecting differences of practical importance, or sometimes, at the other extreme, that it was wastefully large.

An alternative—or better, supplementary—way of calculating sample sizes in many situations is to ask whether the confidence intervals for quantities of interest will be sufficiently narrow to be of value, calculating as though the error MS would come out equal to the guessed value of σ^2. These intervals should usually be those of the S- or T-method (secs. 3.5–3.7). It may be necessary to take some preliminary data[6] to get an adequate estimate of σ^2, unless we are willing to specify certain quantities as multiples of σ (the quantities being analogous for this purpose to the above Δ in the case of tests), namely the (rms) average lengths of the confidence intervals.

[6] An exact theory for two-stage tests whose power does not depend on σ was given by Stein (1945). This theory has been extended to the multiple-comparison methods by Healy (1956). Some practically oriented people would hesitate to apply the Stein theory because only the information about σ from the first stage is utilized (the second stage might "indicate" a very different value). The claimed probabilities of error are of course correct, but these unconditional probabilities, relevant to the experience of the statistician with all his clients, may be less appropriate to a particular client than certain conditional probabilities (given the particular experiment of the client). This problem, needing further formulation, was pointed out to me by Professor J. L. Hodges Jr.

3.4. CONTRASTS. THE S-METHOD OF JUDGING ALL CONTRASTS

If the hypothesis is rejected in actual applications of the F-test for equality of means in the one-way layout, the resulting conclusion that the means $\beta_1, \beta_2, \cdots, \beta_I$ are not all equal would by itself usually be insufficient to satisfy the experimenter. Methods of making further inferences about the means are then desirable. A similar problem may arise after "the" F-test has been made for any Ω and H, as described in the general context of sec. 2.5. Simple answers to the question of what further inferences can be made about the means are offered by the methods of multiple comparison which we shall call the S-method and the T-method, considered in secs. 3.5–3.7.

In the case of the one-way layout the multiple comparison methods permit us to make statements similar to the following: Suppose that $I = 7$ and we relabel the true means in the order of the observed means so that $\hat{\beta}_1 \leqq \hat{\beta}_2 \leqq \cdots \leqq \hat{\beta}_7$. Then the methods applied to the data might permit us to state, for example:

β_7 is greater than β_4 (and hence $\beta_3, \beta_2, \beta_1$ if the sample sizes $\{J_i\}$ are equal). There is insufficient evidence that β_7 is greater than β_6 or β_5; if it is, the differences do not exceed $\beta_7 - \beta_6 \leqq 2.1$, $\beta_7 - \beta_5 \leqq 3.2$. β_1 and β_2 are less than β_4 (and hence $\beta_5, \beta_6, \beta_7$ if the $\{J_i\}$ are equal). It was noted after analysis of the data that, where the treatments 1, 2, 3, 4 use aluminum, the treatments 5, 6, 7 use copper. This suggests considering the difference of the averages

$$\psi = \tfrac{1}{3}(\beta_5 + \beta_6 + \beta_7) - \tfrac{1}{4}(\beta_1 + \beta_2 + \beta_3 + \beta_4).$$

This difference satisfies the inequality $8.4 \leqq \psi \leqq 9.5$. Inequalities are available for all other contrasts (see below) that may be of interest, including any suggested by the way the data fall out. The over-all confidence coefficient is 90 per cent for the totality of these statements about contrasts, the specific ones we made and the potential ones we didn't make.

Contrasts

Definition: A *contrast* among the parameters $\beta_1, \cdots, \beta_I$ is a linear function of the β_i, $\Sigma_{i=1}^I c_i\beta_i$, with known constant coefficients subject to the condition $\Sigma_{i=1}^I c_i = 0$.

For example, the difference of any two means, $\beta_i - \beta_{i'}$, is a contrast, as is the average of any subset of the $\{\beta_i\}$ minus the average of any other subset. (Other useful examples of contrasts are the parametric functions called interactions, introduced in Ch. 4.) The hypothesis H of sec. 3.1 is equivalent to the statement that all contrasts among the $\{\beta_i\}$ are zero.

In the one-way layout, a contrast

$$\psi = \sum_i c_i\beta_i \qquad \left(\sum_i c_i = 0\right)$$

From this relationship of the S-method to the F-test springs perhaps its chief usefulness: Whenever a hypothesis H is rejected by the F-test we can investigate the different estimable functions in L to find out which ones are responsible for rejecting H. Of course the S-method yields much more than this. Frequently it may even be regarded as the statistical technique of main interest, and the F-test as a preliminary technique to find if the other is worth trying.

This relation, showing that the S-method has in a certain sense the same sensitivity as the F-test may help to overcome objections to the S-method from people accustomed to making F-tests, followed in the case of rejection by the dubious practice[11] of calculating many confidence intervals from the same data, all using the upper $\alpha/2$ point of the t-distribution: they tend to object to the length of the S-intervals, although they do not usually complain about the insensitivity of the F-test. The writer suggests the practice of applying the S-method with $\alpha = 10$ per cent instead of making individual interval estimates suggested by inspection of the data and employing with these the upper $2\frac{1}{2}$ per cent (two-tailed 5 per cent) point of the t-distribution: A guaranteed 90 per cent confidence coefficient is preferable to a nominal 95 per cent one if, as usually, we have no idea how far the true value falls below 95 per cent.

The relationship may be expressed in another way, which gives further intuitive insight into the nature of the F-test: Let us denote by L' the set obtained by deleting from L the trivial ψ which is identically zero for all $\{\beta_j\}$. Now $\hat{\psi}$ is significantly different from zero if and only if

$$|\hat{\psi}| > S\hat{\sigma}_{\hat{\psi}}. \tag{3.5.6}$$

It follows that, for any $\psi \in L'$ and any constant multiple $\tilde{\psi} = k\psi$ with $k \neq 0$, $\hat{\tilde{\psi}}$ is significantly different from zero if and only if $\hat{\psi}$ is: For, (3.5.6) will be satisfied for $\hat{\tilde{\psi}}$ if and only if it is satisfied for $\hat{\psi}$, because $\hat{\tilde{\psi}} = k\hat{\psi}$, $\hat{\sigma}_{\hat{\tilde{\psi}}} = k\hat{\sigma}_{\hat{\psi}}$. To ascertain whether there is any $\psi \in L$ for which $\hat{\psi}$ is significantly different from zero we may thus confine our attention to the subset L'' of L defined as follows: L'' consists of all $\psi \in L$ for which the variance of the LS estimate $\hat{\psi}$ is $C\sigma^2$, where C is a positive constant that we chose arbitrarily and then hold fixed. We may restrict our attention to L'' because to every $\psi \in L'$ there corresponds a constant multiple $\tilde{\psi} = k\psi \in L''$; if $\text{Var}(\hat{\psi}) = A\sigma^2$ we may take $k = (C/A)^{1/2}$. We may call the $\tilde{\psi}$ in L'' thus corresponding to any $\psi \in L'$ the corresponding *normalized* estimable function. Let us consider the $\psi \in L''$ for which $\hat{\psi}$ is a maximum, and denote this $\hat{\psi}$ by $\hat{\psi}_{\max}$. Since for all $\psi \in L''$, $\hat{\sigma}^2_{\hat{\psi}} = Cs^2$,

[11] See Scheffé (1953), Tables 5 and 6, for some numerical results on the consequences of this practice.

it follows from (3.5.6) that there will be a $\psi \in L''$ for which $\hat{\psi}$ is s.d.f.z. (significantly different from zero) if and only if $\hat{\psi}_{\max}$ is s.d.f.z. We now have the following chain of implications, each statement being true if and only if the following is: (i) The F-test rejects H. (ii) For some $\psi \in L$, $\hat{\psi}$ is s.d.f.z. (iii) For some $\psi \in L''$, $\hat{\psi}$ is s.d.f.z. (iv) $\hat{\psi}_{\max}$ is s.d.f.z.

We might thus interpret the F-test as looking only at $\hat{\psi}_{\max}$, that estimate of a normalized estimable function which is maximum for the observations obtained, and rejecting H if and only if this is significantly different from zero by the S-criterion.

The above argument shows that the F-test is equivalent to rejecting H if $\hat{\psi}_{\max}$ is s.d.f.z. by the S-criterion, that is, if

$$\hat{\psi}_{\max} > S\hat{\sigma}_{\hat{\psi}_{\max}}.$$

This is equivalent to

$$\hat{\psi}^2_{\max} > S^2Cs^2,$$

or

$$q^{-1}\hat{\psi}^2_{\max}/\text{MS}_e > CF_{\alpha;q,n-r},$$

and the last form suggests that if we choose the normalization with $C = 1$, the numerator SS of "the" F-test of H may actually be $\hat{\psi}^2_{\max}$. This is indeed always the case, and may be proved easily by use of the canonical form of sec. 2.6. Since in the canonical form, H is the hypothesis that $\zeta_1 = \zeta_2 = \cdots = \zeta_q = 0$, it follows that the space L of estimable functions for which H states $\psi = 0$ for all $\psi \in L$ is the set of all ψ of the form $\psi = \Sigma_1^q b_i\zeta_i$, where the $\{b_i\}$ are constants. For such a ψ, $\hat{\psi} = \Sigma_1^q b_i z_i$, and $\text{Var}(\hat{\psi}) = \sigma^2\Sigma_1^q b_i^2$. The normalization with $\text{Var}(\hat{\psi}) = \sigma^2$ thus imposes the condition

$$\sum_1^q b_i^2 = 1. \tag{3.5.7}$$

For fixed $\{z_1, \cdots, z_q\}$, the maximum of $\hat{\psi} = \Sigma_1^q b_i z_i$ subject to (3.5.7) is $\hat{\psi}_{\max} = (\Sigma_1^q z_i^2)^{1/2}$, as can be seen geometrically by interpreting $\hat{\psi}$ as the projection of the fixed vector $(z_1, \cdots, z_q)'$ on the variable unit vector $(b_1, \cdots, b_q)'$, and noting that this projection is maximum when the latter vector has the same direction as the former, that is, when $b_i = \lambda z_i$, where λ is a positive constant, whose value may be determined from (3.5.7) to be $\lambda = (\Sigma_1^q z_i^2)^{-1/2}$. We now have

$$\hat{\psi}^2_{\max} = \left(\sum_1^q b_i z_i\right)^2 = \left(\sum_1^q \lambda z_i^2\right)^2 = \lambda^2\left(\sum_1^q z_i^2\right)^2 = \sum_1^q z_i^2,$$

and the last expression is the numerator SS of the F-test of H. An application of this interpretation of the numerator SS is made in sec. 4.4.

3.6. THE *T*-METHOD OF MULTIPLE COMPARISON

Whereas the *S*-method utilizes the *F*-distribution, the *T*-method utilizes the distribution of $q_{k,\nu}$, the Studentized range, defined at the end of sec. 2.2. The *T*-method can be used to make simultaneous confidence statements about contrasts among a set of parameters $\{\theta_1, \cdots, \theta_k\}$ in terms of unbiased estimates $\{\hat{\theta}_1, \cdots, \hat{\theta}_k\}$ and an estimate s^2 of error if certain restrictions are satisfied. One of these restrictions is that the $\{\hat{\theta}_i\}$ have equal variances; thus, if we wish to apply the method to the one-way layout of sec. 3.1, so that the $\{\theta_i\}$ are the means $\{\beta_i\}$, then the sample sizes $\{J_i\}$ must be equal. We shall state the method first for the special case where the $\{\hat{\theta}_i\}$ are statistically independent and the only contrasts considered are the $\frac{1}{2}k(k-1)$ differences $\{\theta_i - \theta_{i'}\}$; $i, i' = 1, \cdots, k$. The assumptions on the statistics $\{\hat{\theta}_i\}$ and s^2 are then the following:

(3.6.1) Ω: The $\{\hat{\theta}_i\}$ are statistically independent and $\hat{\theta}_i$ is $N(\theta_i, a^2\sigma^2)$, $i = 1, \cdots, k$, where a is a known positive constant. s^2 is an independent quadratic estimate of σ^2 with ν d.f., i.e., $\nu s^2/\sigma^2$ is χ^2_ν and statistically independent of the $\{\hat{\theta}_i\}$.

The *T*-method then rests on

Theorem 1: Under the assumptions (3.6.1) the probability is $1-\alpha$ that all the $\frac{1}{2}k(k-1)$ differences $\{\theta_i - \theta_{i'}\}$ simultaneously satisfy

$$\hat{\theta}_i - \hat{\theta}_{i'} - Ts \leqq \theta_i - \theta_{i'} \leqq \hat{\theta}_i - \hat{\theta}_{i'} + Ts, \tag{3.6.2}$$

where the constant T is

$$T = aq_{\alpha;k,\nu},$$

and $q_{\alpha;k,\nu}$ is the upper α point of the Studentized range $q_{k,\nu}$ defined at the end of sec. 2.2.

Proof: Let $u_i = \hat{\theta}_i - \theta_i$. Under Ω the $\{u_i\}$ are independently $N(0, a^2\sigma^2)$, and statistically independent of $(\nu a^2 s^2)/(a^2\sigma^2)$, which is χ^2_ν. Therefore, by the definition of sec. 2.2, the range of the $\{u_i\}$ divided by as is distributed like $q_{k,\nu}$, and so the probability is $1-\alpha$ that

$$(\max_i u_i - \min_i u_i)/(as) \leqq q_{\alpha;k,\nu},$$

or

$$\max u_i - \min u_i \leqq Ts.$$

The latter inequality is equivalent to

$$|u_i - u_{i'}| \leqq Ts \quad \text{for all } i, i',$$

or

$$-Ts \leqq (\theta_i - \theta_{i'}) - (\hat{\theta}_i - \hat{\theta}_{i'}) \leqq Ts,$$

which is the same as (3.6.2).

Extension of the T-Method to the Set of All Contrasts

Theorem 2: Under the assumptions (3.6.1) the probability is $1-\alpha$ that the values of all contrasts $\psi = \Sigma_{i=1}^k c_i\theta_i$ ($\Sigma_i c_i = 0$) simultaneously satisfy

$$\hat{\psi} - Ts\left(\tfrac{1}{2}\sum_{i=1}^k |c_i|\right) \leqq \psi \leqq \hat{\psi} + Ts\left(\tfrac{1}{2}\sum_{i=1}^k |c_i|\right), \tag{3.6.3}$$

where $\hat{\psi} = \Sigma_{i=1}^k c_i\hat{\theta}_i$, $T = aq_{\alpha;k,\nu}$, and $q_{\alpha;k,\nu}$ is defined in sec. 2.2.

We note that, if a contrast ψ is a difference $\theta_i - \theta_{i'}$ $(i \neq i')$, then $\frac{1}{2}\Sigma|c_i| = 1$, and in that case (3.6.3) is identical with (3.6.2). Thus if (3.6.3) is true for all contrasts ψ it implies (3.6.2) for all differences $\theta_i - \theta_{i'}$. Consequently the proof of Theorem 2 will be complete if we show that if (3.6.2) is true for all differences $\theta_i - \theta_{i'}$ then (3.6.3) is true for all contrasts ψ. But this is an immediate consequence of the following lemma, with $\{u_i\}$ corresponding to $\{\hat{\theta}_i - \theta_i\}$, and h to Ts:

Lemma: If $|u_i - u_{i'}| \leqq h$ for all $i, i' = 1, \cdots, k$, then

$$\left|\sum_{i=1}^k c_i u_i\right| \leqq h\left(\tfrac{1}{2}\sum_{i=1}^k |c_i|\right) \tag{3.6.4}$$

for all $\{c_i\}$ such that $\Sigma_{i=1}^k c_i = 0$.

Proof: If all $c_i = 0$, (3.6.4) is trivially true, so suppose that not all $c_i = 0$. Let P be the set of subscripts i for which $c_i > 0$, and N the set for which $c_i < 0$. Then

$$0 = \sum_{i=1}^k c_i = \sum_{i \in P} c_i + \sum_{i \in N} c_i.$$

Let $g = \Sigma_{i\in P}\, c_i$; so $\Sigma_{i'\in N}(-c_{i'}) = g$, also. Furthermore,

$$\tfrac{1}{2}\sum_{i=1}^k |c_i| = \tfrac{1}{2}\sum_{i\in P} c_i + \tfrac{1}{2}\sum_{i'\in N}(-c_{i'}) = g.$$

If we multiply and divide the right member of

$$\sum_{i=1}^k c_i u_i = \sum_{i\in P} c_i u_i - \sum_{i'\in N}(-c_{i'})u_{i'}$$

by g, we get

$$\begin{aligned}
\sum_i c_i u_i &= g^{-1}\left\{\sum_{i\in P} c_i u_i \sum_{i'\in N}(-c_{i'}) - \sum_{i\in P} c_i \sum_{i'\in N}(-c_{i'})u_{i'}\right\} \\
&= g^{-1}\left\{\sum_{i\in P}\sum_{i'\in N}[c_i(-c_{i'})u_i - c_i(-c_{i'})u_{i'}]\right\} \\
&= g^{-1}\left\{\sum_{i\in P}\sum_{i'\in N} c_i(-c_{i'})(u_i - u_{i'})\right\}.
\end{aligned}$$

Since c_i and $-c_{i'}$ are positive, the absolute value of the terms in the last sum is $c_i(-c_{i'})|u_i-u_{i'}|$, and hence each term is in absolute value $\leqq c_i(-c_{i'})h$. Thus

$$\left|\sum_i c_i u_i\right| \leqq g^{-1}h \sum_{i\in P} c_i \sum_{i'\in N}(-c_{i'}) = gh.$$

But gh is the right member of (3.6.4).

Extension to the Case where the $\{\hat{\theta}_i\}$ have Equal Variances and Equal Covariances

Our last extension will be to a case where the $\{\hat{\theta}_i\}$ need not be independent. The assumptions are

$$\text{(3.6.5)} \quad \Omega': \begin{cases} \{\hat{\theta}_i\} \text{ are jointly normal and statistically independent of } s^2. \\ \nu s^2/\sigma^2 \text{ is } \chi^2_\nu. \quad E(\hat{\theta}_i) = \theta_i \ (i = 1, \cdots, k). \quad \text{Var}(\hat{\theta}_i) = a^2\sigma^2 \\ \text{for all } i. \quad \text{Cov}(\hat{\theta}_i, \hat{\theta}_{i'}) = b\sigma^2 \text{ for all } i, i' \text{ with } i \neq i'. \\ a \text{ and } b \text{ are known constants with}^{12} \ -a^2 \leqq (k-1)b \leqq 0. \end{cases}$$

Theorem 3: Under the assumptions (3.6.5) the probability is $1-\alpha$ that all contrasts $\psi = \Sigma_{i=1}^k c_i\theta_i$ simultaneously satisfy (3.6.3) with $\hat{\psi} = \Sigma_{i=1}^k c_i\hat{\theta}_i$,

$$T = (a^2 - b)^{1/2} q_{\alpha;k,\nu},$$

and $q_{\alpha;k,\nu}$ defined in sec. 2.2.

Proof: We adjoin to the random variables $\{\hat{\theta}_i\}$ a new random variable x which is $N(0, \sigma_x^2)$, statistically independent of the $\{\hat{\theta}_i\}$, and with σ_x^2 determined so that, if $\tilde{\theta}_i = \hat{\theta}_i + x$, the $\{\tilde{\theta}_i\}$ are statistically independent: This is possible since the $\{\tilde{\theta}_i\}$ will be jointly normal and

$$\begin{aligned} \text{Cov}(\tilde{\theta}_i, \tilde{\theta}_{i'}) &= E[(\hat{\theta}_i-\theta_i+x)(\hat{\theta}_{i'}-\theta_{i'}+x)] = \text{Cov}(\hat{\theta}_i, \hat{\theta}_{i'}) + \text{Var}(x) \\ &= b\sigma^2 + \sigma_x^2, \end{aligned}$$

which will be zero if $\sigma_x^2 = -b\sigma^2$. Then

$$\text{Var}(\tilde{\theta}_i) = \text{Var}(\hat{\theta}_i) + \text{Var}(x) = (a^2 - b)\sigma^2,$$

and if we apply Theorem 2 to the statistically independent $\{\tilde{\theta}_i\}$ we get Theorem 3.

3.7. COMPARISON OF THE *S*- AND *T*-METHODS. OTHER MULTIPLE-COMPARISON METHODS

Since the T-method applies only when the space L of estimable functions introduced in sec. 3.5 is a class of contrasts in parameters $\{\theta_i\}$, and when

[12] That $b \leqq 0$ we need in the proof. That $b \geqq -a^2/(k-1)$ may be proved from the fact that the covariance matrix of the $\{\hat{\theta}_i\}$ must be positive indefinite. In applications b is usually 0 or $-a^2/(k-1)$.

certain other conditions (equal variance of the estimates $\hat{\theta}_i$, etc.) are satisfied, we shall have to restrict the comparison of the S- and T-methods to this case. If there are k parameters $\{\theta_i\}$, as in sec. 3.6, then the dimension of L, denoted by q in sec. 3.5, is $q = k-1$. The S-method then gives simultaneous confidence intervals for all possible contrasts $\psi \in L$, while the T-method was originally designed to give intervals for the differences $\{\theta_i - \theta_{i'}\}$ only (Theorem 1 of sec. 3.6). Therefore, we might expect the T-method to give shorter intervals for these differences. The situation is generally reversed when we apply the T-method to more complicated contrasts.

We illustrate the relative efficiency of the two methods by considering the ratio R of the squared[13] lengths of the intervals given by the S- and T-methods, respectively. We take the case where $k = 6$, $\nu = \infty$, $\alpha = 0.05$ in Theorem 2 of sec. 3.6; in the S-method we then have to use $q = k-1$. Suppose that we consider contrasts which are differences between averages of the $\{\theta_i\}$. We shall write $[n_1, n_2]$ for a contrast which is a difference between an average of n_1 of the $\{\theta_i\}$ and an average of n_2 of the $\{\theta_i\}$; this is usually the most important type of contrast. For example,

$$\tfrac{1}{2}(\theta_1+\theta_4) - \tfrac{1}{4}(\theta_2+\theta_3+\theta_5+\theta_6) \quad \text{is a [2, 4] contrast,}$$
$$\tfrac{1}{3}(\theta_2+\theta_3+\theta_5) - \theta_4 \quad \text{is a [3, 1] contrast.}$$

The relative efficiency of the two methods for this case is shown in Table 3.7.1. For example, the entry $R = 0.68$ in the table for [2, 2] means that only about 68 per cent as many observations are needed by the S-method as compared with the T-method for the same accuracy on a contrast of the type [2, 2]. The contrasts for linear, quadratic, etc., effects in Table 3.7.1 are those used in fitting orthogonal polynomials when the values $\{\theta_i\}$ correspond to equal steps of an independent variable.[14]

Later (sec. 4.1) we shall introduce another important class of contrasts called interactions; for these also the S-method seems to be generally better. The table shows that the T-method is preferable when one is interested only in contrasts of the type $\theta_i - \theta_{i'}$, while the S-method generally gives shorter intervals for more complicated contrasts.

We remark that little is known about the operating characteristic of the T-method: The T-method would correspond to a test based on the Studentized range of the $\{\hat{\theta}_i\}$ in the same way that the S-method corresponds to the F-test, each method finding at least one contrast significantly

[13] The reason for considering the squared length rather than the length is that in general for a particular method of estimation, this, or its expected value, tends to be inversely proportional to the number n of observations, at least for large n.

[14] Fisher and Yates (1943), Table XXIII.

TABLE* 3.7.1

RELATIVE EFFICIENCY OF S- AND T-METHODS

R = ratio of squared lengths of intervals,
$k = 6, \nu = \infty, \alpha = 0.05$

Type of Contrast	$1/R$	R
[1, 1]	0.73	
[1, 2]	0.98	
[1, 3]		0.91
[1, 4]		0.85
[1, 5]		0.82
[2, 2]		0.68
[2, 3]		0.57
[2, 4]		0.51
[3, 3]		0.45
Linear		0.59
Quadratic		0.57
Cubic		0.48

* Reproduced with the kind permission of the editor from p. 93 of "A method for judging all contrasts in the analysis of variance" by H. Scheffé, *Biometrika*, Vol. 40 (1953).

different from zero if and only if the corresponding test rejects the hypothesis that all contrasts are zero. The power of the corresponding test is well known for the S- but not for the T-method. Furthermore, formulas are available for the probabilities of two kinds of error with the S-method, somewhat analogous to the two kinds in the theory of testing hypotheses, the first kind connected with statements that the estimates of zero contrasts are significantly different from zero, the second with the opposite statements for nonzero contrasts.[15] The T-method is of more limited applicability since it is available only for the case of equal variances of the $\{\hat{\theta}_i\}$. A further advantage of the S-method is that it is known to be insensitive to violations of the assumptions of normality and of equality of variance (sec. 10.6).

[15] See Scheffé (1953), which contains also further comparisons of the two methods besides Table 3.7.1.

Other Multiple-Comparison Methods

In this subsection we consider briefly several other methods of multiple comparison.[16] One of these is based on the Studentized augmented range, and another on the Studentized maximum modulus, which we now define: Suppose that $\{x_1, \cdots, x_k\}$ are independently $N(0, \sigma_x^2)$, and $\nu s_x^2/\sigma_x^2$ is χ_ν^2 and independent of the $\{x_i\}$, as at the end of sec. 2.2. The maximum modulus of the $\{x_i\}$ is

$$M = \max_i |x_i|,$$

the *Studentized maximum modulus* is

$$m_{k,\nu} = M/s_x,$$

and its upper α point will be denoted by $m_{\alpha;k,\nu}$. This has been tabled for $\alpha = 0.05$ by Pillai and Ramachandran (1954). The augmented range R' of the $\{x_i\}$ is the maximum of M and R, where R is the range of the $\{x_i\}$, the *Studentized augmented range* is

$$q'_{k,\nu} = R'/s_x,$$

and its upper α point will be denoted by $q'_{\alpha;k,\nu}$. This has not been tabled, but for $k > 2$ and $\alpha \leqq 0.05$ it has been shown[17] to differ from the corresponding upper α point $q_{\alpha;k,\nu}$ of the Studentized range (sec. 2.2) by an amount that is practically negligible. A relation we shall exploit is that the augmented range of the sample $x_1, \cdots, x_k$ is the ordinary range of the set of $k+1$ random variables $x_0, x_1, \cdots, x_k$, where the "random variable" x_0 is defined to be identically zero.

Under the assumptions (3.6.1) we might wish to consider the simultaneous estimation of *all* linear functions $\{\psi = \Sigma_1^k c_i\theta_i\}$, not just the contrasts as before. Then Theorem 2 of sec. 3.6, which is the basis of the T-method, remains true[18] if $q_{\alpha;k,\nu}$ is replaced by $q'_{\alpha;k,\nu}$, and $\frac{1}{2}\Sigma_1^k|c_i|$ is replaced by g_ψ, where g_ψ is defined to be the larger of (i) the sum of the

[16] I have not included the multiple-comparison methods of D. B. Duncan because I have been unable to understand their justification. Duncan was one of the earliest workers in this field. His rules for deciding for all pairs of means whether the means differ (and if so, in which direction) do not have the rapidly increasing insensitivity with increasing q or k that may disappoint users of the S- and T-methods, the insensitivity of these being related to the increasing lengths of the intervals for contrasts. The interested reader is referred to an expository paper, Duncan (1955), and subsequent papers, Kramer (1956) and Duncan (1957). The paper, Duncan (1952), giving the justification originally advanced for the methods, contains errors, e.g., the F in the basic distribution (2) cannot be *central* F as claimed.

[17] Tukey (1953), Ch. 14.

[18] This result, and the device of proving it by writing a linear function of k quantities as a contrast in the k quantities and zero, is due to Tukey (1953), Ch. 10.

positive $\{c_i\}$, and (ii) the negative of the sum of the negative $\{c_i\}$. This may be proved by the device of defining $\hat{\theta}_0 \equiv \theta_0 \equiv 0$ and writing ψ formally as a contrast in $\{\theta_0, \theta_1, \cdots, \theta_k\}$, namely $\psi = \Sigma_0^k c_i \theta_i$, where $c_0 = -\Sigma_1^k c_i$. Letting $u_i = \hat{\theta}_i - \theta_i$, we see that the set $\{u_0, u_1, \cdots, u_k, a^2 s^2\}$ is distributed like $\{x_0, x_1, \cdots, x_k, s_x^2\}$ above with $\sigma_x = a\sigma$. The probability is thus $1-\alpha$ that

$$\max_{i=0,\cdots,k} u_i - \min_{i=0,\cdots,k} u_i \leqq T's, \tag{3.7.1}$$

where $T' = aq'_{\alpha;k,\nu}$, and, proceeding as in the proof of Theorem 1 of sec. 3.6, we find that (3.7.1) is equivalent to $|u_i - u_{i'}| \leqq T's$ for i, $i' = 0, 1, \cdots, k$. Theorem 2 with $q_{\alpha;k,\nu}$ and $\frac{1}{2}\Sigma_1^k |c_i|$ replaced by $q'_{\alpha;k,\nu}$ and g_ψ now follows if we use the lemma below Theorem 2 with $\{u_1, \cdots, u_k\}$ replaced by $\{u_0, \cdots, u_k\}$, and note that, with the sets N and P of the proof of the lemma,

$$\sum_0^k |c_i| = \sum_1^k |c_i| + |c_0| = \sum_1^k |c_i| + \Big|\sum_1^k c_i\Big| = \sum_{i\in P} c_i - \sum_{i\in N} c_i + \Big|\sum_{i\in P} c_i + \sum_{i\in N} c_i\Big|,$$

which is easily seen to be twice the g_ψ defined above. The former proof of Theorem 3 for dependent $\{\hat{\theta}_1, \cdots, \hat{\theta}_k\}$ may be used to show that if the assumptions (3.6.5) are satisfied, the totality of linear functions $\{\psi = \Sigma_1^k c_i \theta_i\}$ satisfies Theorem 3 with the inequality (3.6.3) modified by replacing $T = aq_{\alpha;k,\nu}$ by $T' = aq'_{\alpha;k,\nu}$, and $\frac{1}{2}\Sigma_1^k |c_i|$ by g_ψ.

Since for $k > 2$ and $\alpha \leqq 0.05$, $q'_{\alpha;k,\nu}$ is practically equal to $q_{\alpha;k,\nu}$, we use $q_{\alpha;k,\nu}$ instead, and having "paid" for the intervals for the contrasts in Theorems 2 and 3, we get "free" the intervals for all the other linear functions.

In some situations we might be more interested in getting simultaneous interval estimates for the true means $\{\theta_1, \cdots, \theta_k\}$ themselves rather than for contrasts or other linear functions; this might be the case if the $\{\theta_i\}$ were true ordinates in a regression problem where we do not wish to assume that the regression function is of a known form, or if the $\{\theta_i\}$ are means of groups or cells in an analysis of variance and we are mainly interested directly in these means. If the estimates $\{\hat{\theta}_i\}$ satisfy the assumptions (3.6.1), and if we let $u_i = \hat{\theta}_i - \theta_i$, and proceed as in the proof of Theorem 1 of sec. 3.6, we find that the probability is $1-\alpha$ that all the $\{\theta_i\}$ simultaneously satisfy

$$\hat{\theta}_i - am_{\alpha;k,\nu}s \leqq \theta_i \leqq \hat{\theta}_i + am_{\alpha;k,\nu}s,$$

where $m_{\alpha;k,\nu}$ is the upper α point of the Studentized maximum modulus defined above. The method, being designed for this specific purpose, gives shorter intervals for the $\{\theta_i\}$ than those yielded by the last extension of the T-method, or by the S-method with $q = k$.

If an experiment is designed specifically to investigate certain estimable functions $\{\psi_1, \psi_2, \cdots, \psi_p\}$, the principle of Problem 2.5 may be applied to obtain simultaneous inferences about the $\{\psi_i\}$ as follows[19]: Before inspecting the data, specify besides the functions $\{\psi_i\}$ a set of corresponding positive numbers $\{\alpha_i\}$, whose sum is of the order of magnitude of the α to which the error probability is usually controlled. For each ψ_i form a confidence interval by using the t-distribution with probability α_i in the tails; thus with equal tails the interval will be of the form (2.3.7) with the symbols ψ, $\hat{\psi}$, α replaced by ψ_i, $\hat{\psi}_i$, α_i. The probability that all p of these intervals simultaneously cover the true values is then $\geqq 1 - \Sigma_1^p \alpha_i$. (The number p of this paragraph and the next is not intended to be the same as the number of parameters $\{\beta_j\}$: we have begun to run out of suitable symbols.)

The S-method is a tool well adapted to "data-snooping," i.e., in this case, making inferences about estimable functions that are suggested by the configuration of the data, as opposed to functions that constitute a relatively small set specifically formulated before the experiment is performed, as discussed in the last paragraph. A method that permits the data-snooping valid under the S-method while avoiding the wide intervals it gives for specific functions which an experiment is designed to study, consists in allocating beforehand the chosen total error probability α into components $\{\alpha_1, \cdots, \alpha_p\}$ for the specific functions to be covered by t-intervals, and a (usually larger) component α_0 for the S-method, so that $\Sigma_0^p \alpha_i = \alpha$. The probability of all intervals (the t- and the S-intervals) simultaneously covering is then $\geqq 1 - \alpha$.

These methods involving a valid use of several simultaneous intervals based on t will usually require per cent points not listed in the common t-tables. The following approximation[20] for the upper α point of t_ν is then very satisfactory:

$$t_{\alpha;\nu} \doteq z_\alpha + (4\nu)^{-1}(z_\alpha^3 + z_\alpha),$$

where z_α denotes the upper α point of $N(0, 1)$ and is widely available for any α of practical interest.

A rather different type of multiple-comparison problem arises with experiments whose only purpose is to select the best of a set of treatments. The simplest case would be one in which we could obtain N independent measurements of equal known variance on each of a set $\{\theta_i\}$ of true means,

[19] The greater the correlations of the estimates $\{\hat{\psi}_1, \cdots, \hat{\psi}_p\}$, the less efficient this method may be expected to be. An efficient method as discussed in connection with (3.7.2) exists in principle, but the necessary constants determining the widths of the intervals are in general difficult to compute. Similarly there exists in principle a more efficient method than the mixed t- and S-methods defined in the next paragraph.

[20] Due to Peiser (1943).

and we wish to decide which is the largest θ_i. This would be difficult if the largest θ_i differed by a small amount from the next largest, and there would then be little harm in deciding wrongly that this is the largest. The problem may be formulated statistically, that we want the probability of correctly choosing the largest θ_i to be $\geqq$ a specified value P if the largest differs from the next largest by an amount $\geqq$ a specified value δ. Tables are available[21] giving the N required for given P and δ. This kind of solution of the problem could be useful in agricultural trials and other situations where sequential experimentation is not feasible. If the observations can be obtained sequentially a really satisfactory method would stop taking observations on each θ_i as soon as the measurements gave sufficient evidence that it is not among the highest of the $\{\theta_i\}$, and would continue taking measurements on the rest. Such a procedure has been suggested[22] but has not yet been worked out in detail. Sequential methods are also necessitated if the variance of the measurements is unknown, but of those that have been developed so far none is of the desirable type just mentioned, as they continue taking measurements on all the $\{\theta_i\}$ until a decision is reached. The interested reader is referred to a paper by Bechhofer (1958) which includes references to previous work on this type of problem.

More General Approach to Methods of Multiple Comparison

Mastery of the rest of this section (the less mathematical reader attempting it should read the footnote[23]) is not necessary for understanding subsequent parts of this book. The multiple-comparison methods that we have considered, except those for selecting the largest mean, are all included in the following more general description: Suppose we have the Ω-assumptions of sec. 2.1. Let M be a set of estimable functions of primary interest in the problem, for which we want a multiple-comparison method. The set M may be finite, like the set of $\frac{1}{2}k(k-1)$ differences in the first statement of the T-method (Theorem 1 of sec. 3.6), or infinite, like the space L of the S-method (sec. 3.5). The set M will span some space of estimable functions; call the space $L(M)$, and its dimension q, and let $\{\psi_1, \cdots, \psi_q\}$ be a basis for $L(M)$. Now suppose we have a method of multiple comparison which gives for each $\psi \in M$ an interval

$$(3.7.2) \qquad \hat{\psi} - h_\psi s \leqq \psi \leqq \hat{\psi} + h_\psi s,$$

[21] Bechhofer (1954).

[22] Stein (1948).

[23] The *intersection* of two or more sets of prints is the set of points that are contained in all the sets. A set of points is *convex* if for every pair of points in the set the line segment joining them is contained in the set.

where h_ψ is a constant depending on the coefficients $\{c_i\}$ in $\psi = \Sigma_1^q c_i \psi_i$ but not on the unknown $\{\psi_i\}$; $\hat{\psi} = \Sigma_1^q c_i \hat{\psi}_i$, where $\hat{\psi}$ and $\{\hat{\psi}_i\}$ are respectively the LS estimates of ψ and $\{\psi_i\}$; s^2 is the error mean square; and the probability that the inequalities (3.7.2) are simultaneously true for all $\psi \in M$ is $1-\alpha$.

The inequality (3.7.2), which may be written

$$\left| \sum_{i=1}^{q} c_i(\psi_i - \hat{\psi}_i) \right| \leq h_\psi s,$$

may be interpreted geometrically to mean that the point $(\psi_1, \cdots, \psi_q)$ lies in a strip of the q-dimensional space between two parallel planes orthogonal to the vector $\mathbf{c} = (c_1, \cdots, c_q)'$, the point $(\hat{\psi}_1, \cdots, \hat{\psi}_q)$ being midway between the planes. The intersection of these strips for all $\psi \in M$ determines a certain convex set $\mathscr{C}$, and (3.7.2) holds for all $\psi \in M$ if and only if the point $(\psi_1, \cdots, \psi_q)$ lies in $\mathscr{C}$. One may thus approach the problem of multiple comparison by starting with a convex confidence set $\mathscr{C}$ instead of a set M of estimable functions. From any convex confidence set $\mathscr{C}$ one can then derive simultaneous confidence intervals for the infinite set of all estimable functions in the space $L(M)$ in a way similar to our derivation of the S-method, by starting with the relation that the point $(\psi_1, \cdots, \psi_q)$ lies in $\mathscr{C}$ if and only if it lies between every parallel pair of planes of support of $\mathscr{C}$. This approach could be used to define and study optimum properties of methods of multiple comparison.

In the S-method $\mathscr{C}$ is the confidence ellipsoid of sec. 2.3; in the T-method, where $q = k-1$, $\mathscr{C}$ is the polyhedron whose $\frac{1}{2}k(k-1)$ pairs of parallel faces satisfy the equations obtained by expressing Δ_j and $\hat{\Delta}_j$ in terms of $\{\psi_i\}$ and $\{\hat{\psi}_i\}$ respectively in the equations $|\Delta_j - \hat{\Delta}_j| = Ts$ $(j = 1, \cdots, \frac{1}{2}k(k-1)$, where $\{\Delta_j\}$ are the $\frac{1}{2}k(k-1)$ differences, and $\{\hat{\Delta}_j\}$ their LS estimates. Suppose we take the basis $\{\psi_1, \cdots, \psi_q\}$ to be the $\{\zeta_1, \cdots, \zeta_q\}$ of the canonical form of the hypothesis H: $\psi = 0$ for all $\psi \in L(M)$, and write $\mathscr{C}_T(\hat{\psi}_1, \cdots, \hat{\psi}_q; s)$ for the convex polyhedron of the T-method. The corresponding convex set $\mathscr{C}$ of the S-method associated with H is then a sphere $\mathscr{C}_S(\hat{\psi}_1, \cdots, \hat{\psi}_q; s)$. For both $\mathscr{C}_T(\hat{\psi}_1, \cdots, \hat{\psi}_q; s)$ and $\mathscr{C}_S(\hat{\psi}_1, \cdots, \hat{\psi}_q; s)$, the point $(\hat{\psi}_1, \cdots, \hat{\psi}_q)$ is a center of symmetry and s is a scale factor.

In the discussion of sec. 2.10 we remarked that the F-test of H is optimum if for any fixed σ and $c > 0$ we have a "uniform" interest in rejecting H if the true parameter point $(\psi_1, \cdots, \psi_q)$ is anywhere on the sphere $\mathscr{C}_S(0, \cdots, 0; c)$. In situations where one is interested mainly in the $\frac{1}{2}k(k-1)$ differences and has an equal interest in these, one might decide that, for any σ and $c > 0$, one has a "uniform" interest in rejecting

H if the true parameter point $(\psi_1, \cdots, \psi_q)$ is anywhere on the polyhedron $\mathscr{C}_T(0, \cdots, 0; c)$. It is intuitively evident that in this case the Studentized range test of H associated with the T-method is better than the F-test associated with the S-method, since in the corresponding multiple-comparison methods the T-method gives shorter intervals for the differences than the S-method.

3.8. COMPARISON OF VARIANCES

Although the Ω-assumptions for the analyses of variance considered in this book always include the assumption of equal variances within the "cells," it is not recommended that a preliminary test of this assumption ordinarily be made before applying the analysis of variance, for reasons to be explained in Ch. 10. Sometimes however there is a direct interest in comparing the variances of several populations. These populations might be the "cells" in a one-way or higher layout. The standard test[24] for homogeneity of variance is extremely sensitive to nonnormality (Ch. 10). We shall consider an approximate test based on the analysis of variance of the logarithms of the sample variances:[25] the problem is thus transformed to the comparison of means, and, as indicated in Ch. 10 the analysis of variance is fairly insensitive to the shape of the distributions of the estimated means.

Suppose that s^2 denotes the sample variance of a random sample of n from a population with variance σ^2, so that if the sample is $\{x_1, x_2, \cdots, x_n\}$,

$$s^2 = \sum_{i=1}^{n} (x_i - x_.)^2/(n-1).$$

Then $E(s^2) = \sigma^2$, and it may be shown from the lemma at the end of sec. 7.6 that

$$\text{Var}(s^2) = \sigma^4\left(\frac{2}{n-1} + \frac{\gamma_2}{n}\right), \tag{3.8.1}$$

where γ_2, a measure of "kurtosis" (discussed in paragraphs 2, 3, 4 of sec. 10.1), is defined as

$$\gamma_2 = \sigma^{-4}\mu_4 - 3, \tag{3.8.2}$$

[24] Bartlett's (1937) test is a modification of the likelihood-ratio test proposed by Neyman and Pearson (1931); it is described, with tables, in Pearson and Hartley (1954), pp. 57ff.

[25] I benefited from a conversation with Dr. G. E. P. Box and Professor J. W. Tukey about this test.

μ_4 is the fourth central moment of the population,[26] and so for a normal population $\gamma_2 = 0$. Let[27]

$$y = \log s^2,$$

the logarithms in this section being all to the base e. The usual approximate formulas for the mean and variance of a function of a random variable (the reader not familiar with these should read the footnote[28]) then give

$$E(y) \sim \log \sigma^2,$$

$$\text{Var}\,(y) \sim \frac{2}{n-1} + \frac{\gamma_2}{n}.$$

We consider the case where the populations whose variances are to be compared, and from each of which we have a sample of two or more observations, fall into I sets for which we are willing to assume that populations in the same set all have the same variance. Thus if there are J_i populations in the ith set and s_{ij}^2 is the sample variance for the sample from the jth population in the ith set, we include under the Ω-assumptions

$$E(s_{ij}^2) = \sigma_i^2 \qquad (j = 1, 2, \cdots, J_i).$$

We wish to test the hypothesis

$$H: \quad \sigma_1^2 = \sigma_2^2 = \cdots = \sigma_I^2.$$

Let

$$y_{ij} = \log s_{ij}^2.$$

Then under Ω

$$E(y_{ij}) \sim \eta_i,$$

where

$$\eta_i = \log \sigma_i^2,$$

and

$$\text{Var}\,(y_{ij}) \sim \frac{2}{n_{ij}-1} + \frac{\gamma_{2,ij}}{n_{ij}},$$

where n_{ij} is the size of the sample from which s_{ij}^2 is calculated, and $\gamma_{2,ij}$ is the kurtosis measure (3.8.2) for the corresponding population. We now add to Ω the assumption that the kurtosis measure $\gamma_{2,ij}$ has the same value γ_2 for all the populations.

[26] We assume that the population is infinite and μ_4 is finite.

[27] The use of this transformation for applying the analysis of variance to compare variances was suggested by Bartlett and D. G. Kendall (1946); however, they treated it only for the case where the observations are normal.

[28] If $y = f(z)$, then $\mu_y \sim f(\mu_z)$ and $\sigma_y^2 \sim [f'(\mu_z)\sigma_z]^2$, where $\mu_z = E(z)$, $\sigma_z^2 = \text{Var}\,(z)$, etc. These formulas are obtained by approximating y by a linear function of z in the neighborhood of $z = \mu_z$.

The hypothesis H is equivalent to $\eta_1 = \eta_2 = \cdots = \eta_I$. We treat the case where not each of the I sets consists of a single population, i.e., not all $J_i = 1$, because the number of d.f. for the denominator of the approximate F-test to be obtained will be $\Sigma_i(J_i - 1)$; the case where all $J_i = 1$ or $\Sigma_i(J_i - 1)$ is very small is considered below. If the sample sizes $\{n_{ij}\}$ are all equal, then (to the above approximations) the $\{y_{ij}\}$ all have the same variance, and this falls under sec. 3.1, except that the $\{y_{ij}\}$ are not normal.[29] The test statistic is

$$\frac{\nu_e}{I-1}\,\frac{\sum_i J_i(y_{i.} - \bar{y})^2}{\sum_i\sum_j(y_{ij} - y_{i.})^2},$$

where

$$\bar{y} = \sum_i J_i y_{i.}/\sum_i J_i, \qquad \nu_e = \sum_i(J_i - 1),$$

and has under H (approximately) the F-distribution with $I-1$ and ν_e d.f.

If the sample sizes $\{n_{ij}\}$ are not all equal, write

$$\nu_{ij} = n_{ij} - 1$$

for the number of d.f. of s_{ij}^2. Then Var (y_{ij}) is approximately inversely proportional to ν_{ij}, since

$$\text{(3.8.3)} \qquad \text{Var}\,(y_{ij}) \sim \frac{2}{\nu_{ij}} + \frac{\gamma_2}{\nu_{ij} + 1} = \frac{1}{\nu_{ij}}\left[2 + \frac{\gamma_2}{1 + \nu_{ij}^{-1}}\right],$$

and the quantity in brackets does not vary much with ν_{ij}. If we treat it as a constant θ,

$$\text{Var}\,(y_{ij}) \sim \theta/\nu_{ij},$$

then an analysis of variance based on weighted least squares (sec. 1.5) is appropriate,[30] with weights $\{\nu_{ij}\}$ associated with the $\{y_{ij}\}$. The minimum of the weighted SS

$$\mathscr{S} = \sum_i\sum_j \nu_{ij}(y_{ij} - \eta_i)^2$$

under Ω is found to be

$$\mathscr{S}_\Omega = \sum_i\sum_j \nu_{ij}(y_{ij} - \hat{\eta}_i)^2,$$

where

$$\hat{\eta}_i = \sum_j \nu_{ij} y_{ij}/\nu_i,$$

and

$$\nu_i = \sum_j \nu_{ij},$$

[29] However, they can be expected to be more nearly normal than the $\{s_{ij}^2\}$.

[30] The test and its power can also be easily obtained by transforming to $\{u_{ij} = \sqrt{\nu_{ij}}\, y_{ij}\}$ and applying the theory for equal variance to the $\{u_{ij}\}$.

while under $\omega = H \cap \Omega$ the minimum is

$$\mathscr{S}_\omega = \sum_i \sum_j \nu_{ij}(y_{ij} - \hat{\eta})^2,$$

where

$$\hat{\eta} = \sum_i \sum_j \nu_{ij} y_{ij}/\nu = \sum_i \nu_i \hat{\eta}_i/\nu$$

and

$$\nu = \sum_i \nu_i = \sum_i \sum_j \nu_{ij}.$$

If we calculate the numerator SS as $\mathscr{S}_\omega - \mathscr{S}_\Omega$ we then find the F-statistic for testing H to be

$$\frac{\nu_e}{I-1} \frac{\sum_i \nu_i(\hat{\eta}_i - \hat{\eta})^2}{\sum_i \sum_j \nu_{ij}(y_{ij} - \hat{\eta}_i)^2}, \tag{3.8.4}$$

with $I - 1$ and ν_e d.f., where $\nu_e = \Sigma_i(J_i - 1)$. For numerical computation this may be rewritten as

$$\frac{\nu_e}{I-1} \frac{\sum_i \nu_i \hat{\eta}_i^2 - \nu\hat{\eta}^2}{\sum_i \sum_j \nu_{ij} y_{ij}^2 - \sum_i \nu_i \hat{\eta}_i^2}.$$

The power of this F-test can be calculated by noting that the statistic (3.8.4) has under Ω approximately the noncentral F-distribution with $I-1$ and ν_e d.f. and noncentrality parameter

$$\delta^2 = \sum_i \nu_i(\eta_i - \bar{\eta})^2/\theta,$$

where

$$\bar{\eta} = \sum_i \nu_i \eta_i/\nu.$$

The variance of the estimate $\hat{\eta}_i$ of $\eta_i = \log \sigma_i^2$ is θ/ν_i, and an estimate of θ is the denominator MS, namely

$$\hat{\theta} = \sum_i \sum_j \nu_{ij}(y_{ij} - \hat{\eta}_i)^2/\nu_e,$$

the quantity $\nu_e\hat{\theta}/\theta$ being approximately $\chi^2_{\nu_e}$. The S-method of multiple comparison can be applied to the $\{\eta_i\}$, but the only inferences about the $\{\sigma_i^2\}$ thus obtained that would usually be of any interest would be from the contrasts in the $\{\eta_i\}$ which are differences, a statement of the form

$$A \leqq \eta_i - \eta_j \leqq B \tag{3.8.5}$$

being equivalent to $e^A \leqq \sigma_i^2/\sigma_j^2 \leqq e^B$. The T-method is of course more efficient for statements of the form (3.8.5), when it is applicable, but this will be only in the case when all $\{\nu_i\}$ are equal.

The above method of testing H is not possible if all $J_i = 1$, and not sensitive if $\Sigma_i(J_i - 1)$, the number of d.f. for the denominator of the F-statistic, is very small. For this case we suggest[31] dividing enough of the samples into two or more smaller samples so that the above method can then be applied with a reasonable ν_e. This subdivision will be possible only for samples of four or more, and should be done by use of a table of random numbers. Thus, to test the equality of ten variances, each estimated from a sample of five, we might subdivide each sample of five into a sample of two[32] and a sample of three, so that in applying the previous method $I = 10$, $J_1 = J_2 = \cdots = J_{10} = 2$, $n_{ij} = 2$ or 3, $n_{i1} + n_{i2} = 5$.

PROBLEMS

3.1. Table A gives the birth weights (in pounds) of Poland China pigs in eight litters. (*a*) Construct the analysis-of-variance table. Test at the 0.10

TABLE* A

1	2	3	4	5	6	7	8
2.0	3.5	3.3	3.2	2.6	3.1	2.6	2.5
2.8	2.8	3.6	3.3	2.6	2.9	2.2	2.4
3.3	3.2	2.6	3.2	2.9	3.1	2.2	3.0
3.2	3.5	3.1	2.9	2.0	2.5	2.5	1.5
4.4	2.3	3.2	3.3	2.0		1.2	
3.6	2.4	3.3	2.5	2.1		1.2	
1.9	2.0	2.9	2.6				
3.3	1.6	3.4	2.8				
2.8		3.2					
1.1		3.2					

* From Table 10.16.1, p. 269 of *Statistical Methods* by George W. Snedecor, Iowa State College Press, Ames, fifth edition, 1956. Reproduced with the kind permission of the author and the publisher.

level the hypothesis of no difference between mean weights in the eight litters. (*b*) Suppose that litters 1, 3, 4 were sired by one boar, the other five by another boar. Is there a significant difference between the mean weights in those two

[31] If γ_2 in (3.8.3) were known, the unbiased estimates $\{y_{ij}\}$ of the $\{\eta_i\}$ would have known variances, and a chi-square test could then be made for the equality of the $\{\eta_i\}$. It is tempting to try to estimate γ_2 from each of the samples, which will generally be small samples, combine the estimates, and replace the γ_2 entering the chi-square test by the combined estimate. The trouble with this is the difficulty of finding a satisfactory estimate of γ_2 from small samples.

[32] Some empirical evidence which indicates that the proposed test behaves satisfactorily even with n_{ij} as small as 2 was obtained from sampling experiments with rectangular populations by Box (1953), p. 332.

groups? (*c*) Is there a significant difference between the mean weight for large litters (nos. 1, 2, 3, 4) and that for small litters (nos. 5, 6, 7, 8)? [*Hints:* In deciding whether to use a *t*-test or the *S*-method in (*b*) and (*c*), assume that the experiment was planned to investigate among other things the difference between sires, but that question (*c*) was an afterthought which occurred during examination of the data. See the discussion in sec. 3.7.]

3.2. In an experiment where several experimental treatments are compared with a control, it may be desirable to replicate the control more than the experimental treatments, since it enters into every difference investigated. This would be justified by the following model: Each of m experimental treatments is replicated t times and the control c times. Let y_{ij} be the jth observation on the ith experimental treatment ($i = 1, \cdots, m$; $j = 1, \cdots, t$) and y_{0j} be the jth observation on the control ($j = 1, \cdots, c$). Assume that $y_{ij} = \tau_i + e_{ij}$, where the $\{e_{ij}\}$ are independent with zero means and equal variance σ^2. The LS estimate of the difference $\theta_i = \tau_i - \tau_0$ is then $\hat{\theta}_i = y_{i.} - y_{0.}$ ($i = 1, \cdots, m$). Prove that for a fixed total number of observations, i.e., $c + mt =$ constant, Var ($\hat{\theta}_i$) is minimized if the numbers of replications are in the ratio $c/t = \sqrt{m}$.

3.3. Table B gives the data from a taste-testing experiment for comparing four brands of a product. For each of the 12 ordered pairs (i, j), each of 12 subjects (different for each ordered pair) was asked to state his preference for brand i over brand j, which he was given to taste in the order: first i, then j. He was asked to make one of seven statements, scored as follows: "I prefer i to j strongly (score 3), moderately (score 2), slightly (score 1)," "No difference (score 0)," "I prefer j to i slightly (score -1)," etc. It was feared that inhomo-

TABLE† B

Pair	Frequency of Scores Equal to							Total
(i, j)	-3	-2	-1	0	1	2	3	Score
(1, 2)				3	3	2	4	19
(2, 1)	6	3	1		1	1		-22
(1, 3)			3	2	2	4	1	10
(3, 1)		4	1	2	4		1	-2
(1, 4)					2	5	5	27
(4, 1)	2	4	3	1		2		-13
(2, 3)	3	4	3		1	1		-17
(3, 2)			1	1	1	5	4	22
(2, 4)	1	2			1	4	4	14
(4, 2)	2	2	2		2	4		-2
(3, 4)					4	1	7	27
(4, 3)	5	1	1		1	3	1	-8

† From p. 391 of "An analysis of variance for paired comparisons" by H. Scheffé, *J. Amer. Stat. Assoc.*, Vol 47 (1952). Reproduced with the kind permission of the editor.

geneity of variance would be caused by the scores jamming against one end of the scale when the total score on an ordered pair was very high or low. Use the method of sec. 3.8 to compare the sample variances of the scores for the four pairs with the total scores highest in absolute value with those of the remaining eight pairs. (The F-test reduces to a t-test since there are only two groups of sample variances, and it would be appropriate to use a one-sided test.)

3.4. The result‡ of this problem is applicable to the interpretation of two methods of multiple comparison, in each of which intervals are calculated for m functions of the parameters chosen in advance, (a) with over-all confidence coefficient $1-\alpha$, and (b) with individual confidence coefficients $1-m^{-1}\alpha$. For example, in calculating intervals for the $m = \frac{1}{2}k(k-1)$ differences of $\{\theta_1, \cdots, \theta_k\}$, ($a$) might use the T-method with over-all coefficient $1-\alpha$, and (b) might use m intervals based on the t-distribution with individual coefficients $1-m^{-1}\alpha$. Consider a series of N independent experiments, with m_i statements made in the ith experiment, by method (a) and also by method (b), and two ways of scoring, $S_1 = N^{-1}$ times the number of experiments with one or more wrong statements, and $S_2 = N^{-1}$ times the total number of wrong statements, so that higher scores are worse and S_2 is more severe than S_1 since $S_2 \geqq S_1$ if both are applied to the same method. Show that, for method (a), S_1 converges in probability to α, and similarly for method (b) and S_2. [*Hints:* For method (b), calculate $E(S_2) = \alpha$, Var $(S_2) < N^{-1}m^2$, and apply the result of Problem IV.3a.]

3.5. For k, ν, α as in Table 3.7.1, compare the squared lengths of simultaneous confidence intervals obtained for the 15 differences of the $\{\theta_i\}$ by methods (a) and (b) of Problem 3.4.

3.6. Consider the following two methods of multiple comparison with over-all confidence coefficient $\geqq 0.90$: (a) the S-method with $\alpha = 0.10$, and (b) a combination of the S- and T-methods, each with $\alpha = 0.05$, the shorter of the two intervals being employed. If a table like Table 3.7.1 is made for the relative efficiency of these two methods, with R denoting the squared length of the interval from (a) divided by that from (b), calculate that for the first line $1/R = 0.88$, for the second line $R = 0.85$, and for all other lines $R = 0.83$.

‡ Mentioned to me by Professor J. W. Tukey.

CHAPTER 4

The Complete Two-, Three-, and Higher-Way Layouts. Partitioning a Sum of Squares

4.1. THE TWO-WAY LAYOUT. INTERACTION

In this chapter we apply the general theory of Chs. 1 and 2 to the simplest experimental designs for investigating the effects of two or more factors, which we shall call complete layouts. The important concept of interaction[1] which then enters is discussed at some length in this section. Later in the chapter we shall consider the general problem of partitioning a sum of squares: While this could have been treated in Chs. 1 and 2, it seems pedagogically advantageous to precede and follow it by interesting examples not possible before this chapter.

Suppose that two factors A and B vary in an experiment or in observational material, for example, in a tank experiment such as described in sec. 1.1, different varieties (A) are planted in different locations (B), all with the same chemical treatment, or in an astronomical investigation observations are made on several kinds of stars (A) on different nights (B). If there are I varieties and J locations in the first example, these are called the I *levels* of factor A and the J *levels* of factor B, respectively. The levels may be based on a qualitative classification, like varieties, or quantitative, like the distance of a star.

In such two-factor experiments (or nonexperimental investigations) the observations can be arranged in *a two-way layout* or $I \times J$ table by letting the rows of the table correspond to the levels of factor A and the columns to the levels of B, and writing in the "i,j cell," that is, in the ith row and jth column, the observations on the "i,j treatment combination" where

[1] The concept of interaction in multifactor experiments, and the other basic concepts of orthogonality of design, and of the importance of randomization, we owe to R. A. Fisher (1925, 1935).

factor A is at the ith level and B at the jth. In a *complete* layout there is at least one observation for every cell. Certain incomplete layouts will be considered in the next chapter. If we assume that the observations in the i,j cell are a random sample from a population corresponding to the cell, we may speak of the mean and variance of this population as the "*true*" *cell mean* and "*true*" *cell variance.* All the concepts of this section are defined in terms of "true" cell means, which are also called the "*true*" *responses* to the treatment combinations. We shall denote the "true" mean of the i,j cell by η_{ij}; its LS estimate, if no further assumptions are made about the $\{\eta_{ij}\}$, will later be found to be the average of the observations in the i,j cell, and this average is called the *observed cell mean* or *observed response.* We shall for brevity omit the word "true" in connection with the various means in the subsequent part of this section, since all will be "true."

Suppose a set of weights $\{w_j\}$ is chosen to be associated with the levels of the factor B. For example, if I varieties of cotton were tested in an experiment in J locations in California, on the basis of which a single variety were to be selected for all of California, it might be reasonable to weight the J locations with weights $\{w_j\}$ proportional to the total acreages of cotton in the regions of which the J locations are supposed to be typical. The *mean for the* ith *level of* A is defined to be the weighted mean of the cell means $\{\eta_{ij}\}$ in the ith row, the weights $\{w_j\}$ depending on the columns but not on the row; it is thus the mean response to the ith level of A averaged over the levels of B. Of the weights $\{w_j\}$ it is understood that all $w_j \geqq 0$ and not all $w_j = 0$, and so without loss of generality we may assume that $\Sigma_j w_j = 1$; otherwise the $\{w_j\}$ are to be regarded as arbitrary but fixed. Then the mean for the ith level of A is

$$A_i = \sum_j w_j \eta_{ij};$$

it is also called the ith *row mean.* Likewise, if $\{v_i\}$ is an arbitrary set of numbers with all $v_i \geqq 0$ and $\Sigma_i v_i = 1$, the *mean for the* jth *level of B*, or the jth *column mean*, is defined to be

$$B_j = \sum_i v_i \eta_{ij}.$$

The *general mean* is defined to be the average of the column means $\{B_j\}$ weighted with the $\{w_j\}$, which is the same as the average of the row means $\{A_i\}$ weighted with the $\{v_i\}$, and will be denoted by μ,

$$\mu = \sum_j w_j B_j = \sum_i v_i A_i = \sum_i \sum_j v_i w_j \eta_{ij}.$$

The *main effect of the* ith *level of* A is defined as the excess of the mean for the ith level over the general mean,

$$\alpha_i = A_i - \mu.$$

We note that the $\{\alpha_i\}$ satisfy the condition

$$\sum_i v_i \alpha_i = 0. \tag{4.1.1}$$

Similarly the *main effect of the* jth *level of* B is defined as

$$\beta_j = B_j - \mu,$$

so that

$$\sum w_j \beta_j = 0. \tag{4.1.2}$$

The main effects α_i and β_j are also called the ith *row effect* and jth *column effect*. We emphasize that the main effects for one factor are averages over the levels of the other factor and hence in general depend on what levels of the other factor are present in the experiment (in the fixed-effects models considered in Part I of this book).

If we were to define a main effect of the ith level of A *specific* to the jth level of B, it would naturally be the excess of the cell mean η_{ij} over the mean for the jth column, namely

$$\eta_{ij} - B_j. \tag{4.1.3}$$

The main effect of the ith level of A defined before is actually the weighted average of (4.1.3) over the columns, $\alpha_i = A_i - \mu = \Sigma_j w_j(\eta_{ij} - B_j)$. The excess of (4.1.3) over this average is called the *interaction of the* ith *level of* A *with the* jth *level of* B,

$$\gamma_{ij} = \eta_{ij} - B_j - A_i + \mu. \tag{4.1.4}$$

We would come to the same result (4.1.4) if we started with the main effect of the jth level of B specific to the ith level of A; the interaction is symmetric, and we may call γ_{ij} the interaction of the ith level of A *and* the jth level of B. We note that the IJ interactions satisfy the conditions

$$\sum_i v_i \gamma_{ij} = 0 \text{ for all } j, \qquad \sum_j w_j \gamma_{ij} = 0 \text{ for all } i. \tag{4.1.5}$$

On substituting $B_j = \mu + \beta_j$ and $A_i = \mu + \alpha_i$ in (4.1.4), we get

$$\eta_{ij} = \mu + \alpha_i + \beta_j + \gamma_{ij}. \tag{4.1.6}$$

We remark that if a set of constants $\{\mu, \alpha_i, \beta_j, \gamma_{ij}\}$ satisfies (4.1.6) this is not sufficient for them to be the general mean, main effects, and interactions, unless the side conditions (4.1.1), (4.1.2), and (4.1.5) are satisfied, in which case they are uniquely determined by the $\{\eta_{ij}\}$, as may be shown by assuming that $\eta_{ij} = \mu' + \alpha_i' + \beta_j' + \gamma_{ij}'$, calculating μ, A_i, B_j, and γ_{ij} from their definitions in terms of the $\{\eta_{ij}\}$, assuming that the primed quantities satisfy the side conditions, and thus finding $\mu = \mu'$, $\alpha_i = \alpha_i'$, $\beta_j = \beta_j'$, and $\gamma_{ij} = \gamma_{ij}'$.

Theorem 1: If the interactions $\{\gamma_{ij}\}$ are all zero for some system of weights $\{v_i\}$ and $\{w_j\}$, then they are all zero for every system of weights. In that case every contrast in the main effects $\{\alpha_i\}$ or $\{\beta_j\}$ has a value that does not depend on the system of weights $\{v_i\}$ and $\{w_j\}$, and the same is true for contrasts in the means $\{A_i\}$ and $\{B_j\}$ for the levels of A and B, respectively.

Proof: Suppose that for a particular system of weights $\{v_i^0\}$ and $\{w_j^0\}$, the A_i, B_j, μ, α_i, β_j, γ_{ij} defined above are denoted by A_i^0, B_j^0, μ^0, α_i^0, β_j^0, γ_{ij}^0, and suppose that all $\gamma_{ij}^0 = 0$. Then from (4.1.6)

(4.1.7) $$\eta_{ij} = \mu^0 + \alpha_i^0 + \beta_j^0.$$

Now let $\{v_i\}$ and $\{w_j\}$ be any other system of weights, so that $v_i \geqq 0$, $\Sigma_i v_i = 1$, $w_j \geqq 0$, $\Sigma_j w_j = 1$. Then substituting (4.1.7) in $A_i = \Sigma_j w_j \eta_{ij}$, we get

(4.1.8) $$A_i = \mu^0 + \alpha_i^0 + \sum_{j'} w_{j'}\beta_{j'}^0,$$

and similarly,

$$B_j = \mu^0 + \beta_j^0 + \sum_{i'} v_{i'}\alpha_{i'}^0,$$

$$\mu = \mu^0 + \sum_{i'} v_{i'}\alpha_{i'}^0 + \sum_{j'} w_{j'}\beta_{j'}^0.$$

On substituting these three expressions and (4.1.7) into (4.1.4) we find that $\gamma_{ij} = 0$, proving the first part of the theorem.

Next, write (4.1.8) in the form $A_i = \alpha_i^0 + D$, where the constant D depends on the system of weights $\{w_j\}$ but not on i. Let $\psi = \Sigma_i c_i A_i$ be any contrast in the $\{A_i\}$, so that $\Sigma_i c_i = 0$. Then $\psi = \Sigma_i c_i(\alpha_i^0 + D) = \Sigma_i c_i \alpha_i^0$, which is independent of the weights $\{v_i\}$ and $\{w_j\}$. If $\psi = \Sigma_i c_i \alpha_i$ is a contrast in the $\{\alpha_i\}$, then by definition, $\alpha_i = A_i - \mu$, where μ depends on the weights; hence $\psi = \Sigma_i c_i(A_i - \mu) = \Sigma_i c_i A_i$, for which we have just derived the desired property of independence of the weights. Similarly we may derive it for contrasts in the $\{B_j\}$ and in the $\{\beta_j\}$.

In the rest of this book unless otherwise specified (*sec.* 4.4), *we shall use equal weights* $\{v_i\}$ and $\{w_j\}$ in the definitions of the general mean, main effects, and interactions, so

(4.1.9)

$$\mu = \eta_{..}, \quad \alpha_i = \eta_{i.} - \eta_{..}, \quad \beta_j = \eta_{.j} - \eta_{..}, \quad \gamma_{ij} = \eta_{ij} - \eta_{i.} - \eta_{.j} + \eta_{..},$$

and

$$(4.1.10) \qquad \alpha_{.} = 0, \quad \beta_{.} = 0, \quad \gamma_{i.} = 0 \text{ for all } i, \qquad \gamma_{.j} = 0 \text{ for all } j.$$

When we then speak of a *case of no interactions*, that is, all $\gamma_{ij} = 0$, it follows from the above theorem that anyone using different weights would agree it is a case of no interactions, and he would agree with our values for all contrasts in the main effects $\{\alpha_i\}$ and $\{\beta_j\}$ even though he calculates his $\{\alpha_i\}$ and $\{\beta_j\}$ with different weights. A case of no interactions is also called a case of *additivity* of effects: This may be formally defined as a case where there exist constants $\{a_i\}$, $\{b_j\}$, c such that $\eta_{ij} = a_i + b_j + c$ for all i,j; it then easily follows, as in the proof of the above theorem, that all $\gamma_{ij} = 0$.

The interpretation of an analysis of variance is much simpler when we decide (on statistical or other grounds) that there are no interactions. Our inferences about the main effects (and perhaps the general mean) then usually suffice to summarize the whole analysis. Thus, if we compare the first and second varieties in the example of varieties and locations, and infer that the contrast (main effect of the first variety) − (main effect of the second variety) is positive, then if we assume additivity, this implies that the first variety is better than the second by the same amount in all the locations of the experiment, but if we do not make the assumption of additivity, our conclusion must be that, *averaged over the J locations in the experiment*, the first variety is better than the second; however, it might happen that in a particular location in the experiment the second variety was better than the first. If one interprets an analysis of variance under the assumption of additivity, and if this assumption has been accepted only because the hypothesis of no interactions was not rejected by the appropriate F-test, it is wise to try to answer the question whether this test has a reasonable power to reject the hypothesis if the interactions are in fact large enough to invalidate seriously the interpretation based on this hypothesis.

It happens occasionally that the hypothesis of no interactions will be rejected by a statistical test but the hypotheses of zero main effects for both factors will be accepted. The correct conclusion is then *not* that no differences have been demonstrated: If there are (any nonzero) interactions there must be (nonzero) differences among the cell means. The conclusion should be that there are differences, but that when the effects of the levels of one factor are averaged over the levels of the other, no difference of these averaged effects has been demonstrated.

It is easy to verify that if additivity exists it is preserved under a linear transformation of the measurements and their means $\{\eta_{ij}\}$, but that it is in general destroyed by a nonlinear transformation. The problem of interest is, if there are interactions, does there exist a suitable transformation of the scale of measurement so that on the new scale there will be additivity? The practical problem is extremely complicated[2] by the fact that we do not know the interactions but have only estimates which we know are in error. In the rest of this section there is some discussion of the more academic problem of transformation if the interactions were known. While this has some mathematical interest and may enrich our intuition of interaction, the reader may wish to skip it because it is of little practical use; skipping this will not interfere with understanding the rest of the book.

Elimination of Known Interactions by Transformation of Scale of Measurement

We shall restrict the transformations $z = f(y)$ to be strictly increasing, that is, $f(y') > f(y'')$ for every y', y'' such that $y' > y''$. The reason for this is that we wish to preserve the rank order of the cell means $\{\eta_{ij}\}$ and the observations. It is much easier to treat the

Case where the Factors are Quantitative

By this we mean that the levels of A correspond to values $u = u_1, \cdots, u_I$ of a continuous variable u (like temperature, pressure, weight of fertilizer, etc.) and the levels of B to values $v = v_1, \cdots, v_J$ (the reader will of course not confuse these with the weights $\{v_i\}$ above!) of a continuous variable v, and there exists a (regression) function $\eta(u, v)$ such that $\eta_{ij} = \eta(u_i, v_j)$. The function $\eta(u, v)$ may be called *additive* if there exist functions $g(u)$ and $h(v)$ such that $\eta(u, v) = g(u) + h(v)$. The set $\{\eta_{ij}\}$ will then have zero interactions for every choice of $\{u_i\}$ and $\{v_j\}$.

In the following theorem η_u, η_v, and η_{uv} denote the partial derivatives $\partial/\partial u$, $\partial/\partial v$, and $\partial^2/\partial u\,\partial v$ of $\eta(u, v)$, respectively.

Theorem[3] 2: For a given function $\eta(u, v)$ there exist functions $f(\eta)$, $g(u)$, $h(v)$ such that

(4.1.11) $$f(\eta(u, v)) = g(u) + h(v)$$

and $f'(\eta) > 0$, if and only if

(4.1.12) $$\eta_{uv}/(\eta_u \eta_v) = w(\eta),$$

[2] A lesser difficulty is that if we assume the original observations normal with equal variance the same will generally not be true for the transformed observations.

[3] The proof of this theorem can be made rigorous by specifying that $\eta(u, v)$, $f(\eta)$, $g(u)$, and $h(v)$ are twice differentiable in appropriate regions and that $w(\eta)$ is integrable. The condition $f'(\eta) > 0$ may be relaxed to permit $f'(\eta) = 0$ at a finite number of points. In a more rigorous statement of the theorem, the identity (4.1.12) would be cleared of fractions to prevent zero denominators.

i.e., $\eta_{uv}/(\eta_u\eta_v)$ depends on u and v only through the intermediary of $\eta(u, v)$. Then the functions $f(\eta)$, $g(u)$, $h(v)$ may be determined as follows: $f(\eta)$ is given by

$$f(\eta) = c_1 \int \exp\left[-\int w(\eta)\, d\eta\right] d\eta + c_2, \tag{4.1.13}$$

where $\{c_i\}$ denote constants and $c_1 > 0$; $\eta_u f'(\eta)$ is a function of u only, say $\phi(u)$; and

$$g(u) = \int \phi(u)\, du + c_3; \tag{4.1.14}$$

$f(\eta) - g(u)$ is a function of v only, and

$$h(v) = f(\eta) - g(u). \tag{4.1.15}$$

Proof: Suppose first that $f(\eta)$, $g(u)$, $h(v)$ exist so that (4.1.11) is satisfied. Take $\partial/\partial u$ in (4.1.11) to get

$$f'(\eta)\eta_u = g'(u),$$

and $\partial/\partial v$ to get

$$f''(\eta)\eta_u\eta_v + f'(\eta)\eta_{uv} = 0,$$

or (4.1.12) with

$$w(\eta) = -f''(\eta)/f'(\eta). \tag{4.1.16}$$

Next suppose that (4.1.12) is satisfied. We shall determine $f(\eta)$ to satisfy (4.1.16) and show that the resulting $f(\eta(u, v))$ has the structure (4.1.11) with the functions f, g, h of the form described in the theorem. Integrating (4.1.16) gives

$$f'(\eta) = c_1 \exp\left[-\int w(\eta)\, d\eta\right], \tag{4.1.17}$$

and $f'(\eta)$ will be positive if we choose $c_1 > 0$. Integrating (4.1.17) gives (4.1.13). To see that $\eta_u f'(\eta)$ is a function of u only, take $\partial/\partial v$,

$$\frac{\partial}{\partial v}[\eta_u f'(\eta)] = \eta_{uv} f'(\eta) + \eta_u\eta_v f''(\eta) = \eta_{uv} f'(\eta) - \eta_u\eta_v w(\eta) f'(\eta) = 0$$

for all v. Now define $g(u)$ by (4.1.14). To see that $f(\eta) - g(u)$ is a function of v only, take $\partial/\partial u$,

$$\frac{\partial}{\partial u}[f(\eta) - g(u)] = f'(\eta)\eta_u - \phi(u) = 0$$

for all u. Finally, define $h(v)$ by (4.1.15).

Example: Suppose $\eta(u, v) = uv$. Then $\eta_u = v$, $\eta_v = u$, $\eta_{uv} = 1$, and so $\eta_{uv}/(\eta_u\eta_v) = 1/(uv) = 1/\eta$. Thus $\eta(u, v)$ satisfies (4.1.12) and a transformation to additivity exists. It is given by (4.1.13) with $w(\eta) = 1/\eta$, namely

$$f(\eta) = c_1 \int \exp\left[-\int \eta^{-1}\, d\eta\right] d\eta + c_2 = c_1 \log \eta + c_2.$$

If we now form $f(\eta(u, v))$, we get for it $c_1 \log u + c_1 \log v + c_2$ and now there is obviously additivity, with $g(u) = c_1 \log u + c_3$ and $h(v) = c_1 \log v + c_2 - c_3$; however these could also have been calculated from (4.1.14) and (4.1.15).

Case where the Factors are Qualitative

Let us say that two rows in a two-way layout, say

$$\begin{matrix} a_1 & a_2 & \cdots & a_J \\ b_1 & b_2 & \cdots & b_J \end{matrix}$$

are *consistently ordered* if the J differences $\{a_j - b_j\}$ are either all >0 or all $=0$, or all <0 and similarly define consistent ordering of two columns. It is easily seen that a necessary condition that interactions in an $I \times J$ layout be removable by transformation is that all pairs of rows and all pairs of columns be consistently ordered: For if the interactions are removed all pairs of rows and all pairs of columns in the transformed layout must be consistently ordered, since for any pair the differences are all equal, and hence the original layout must satisfy the same condition, which is not affected by transformation by a strictly increasing junction.

The condition is easy to verify numerically by first rearranging the columns so that the first row is in nondecreasing order, and then rearranging the rows so that the first column is in nondecreasing order; then in the rearranged layout the condition is equivalent to the following: All rows and all columns must be in nondecreasing order, and if in some row (or column) two elements are equal then the two columns (or rows) containing these elements must be equal. For example, starting with

$$\begin{matrix} 2 & 6 & 5 \\ 3 & 8 & 7 \\ 0 & 4 & 1, \end{matrix}$$

we first rearrange the columns to put the first row in nondecreasing order,

$$\begin{matrix} 2 & 5 & 6 \\ 3 & 7 & 8 \\ 0 & 1 & 4. \end{matrix}$$

If any other row were not in nondecreasing order the condition would already be violated. We now arrange the rows so that the first column is in increasing order,

$$\begin{matrix} 0 & 1 & 4 \\ 2 & 5 & 6 \\ 3 & 7 & 8. \end{matrix} \tag{4.1.18}$$

Since all rows and columns are now in strictly increasing order, the condition is satisfied.

The condition, however, is not sufficient for the interactions to be removable by transformation, as the present example[4] shows: We shall prove that if we assume that there exists a strictly increasing function $f(\eta)$ which removes the interactions in this example then we shall be led to a contradiction. For every subtable

$$\begin{matrix} a & b \\ c & d \end{matrix} \tag{4.1.19}$$

we must have $f(a) + f(d) = f(b) + f(c)$. If we apply this to the subtables

$$\begin{matrix} 0 & 4 \\ 2 & 6 \end{matrix} \quad \text{and} \quad \begin{matrix} 0 & 1 \\ 3 & 7 \end{matrix}$$

of (4.1.18), we get

$$f(0) + f(6) = f(4) + f(2),$$
$$f(0) + f(7) = f(1) + f(3).$$

Since $f(\eta)$ is strictly increasing, $f(4) > f(3)$ and $f(2) > f(1)$; hence $f(0) + f(6) > f(0) + f(7)$, or $f(6) > f(7)$. But $f(7) > f(6)$, and this is the contradiction.

In the case where one factor is quantitative and the other qualitative, geometric conditions for the existence of a transformation which removes interactions are implied by the result of Problem 4.13, but, beyond a necessary condition analogous to that above of consistent ordering, their implications are not easy to see.

It may be shown[5] that the condition is sufficient in the case $I = J = 2$.

4.2. THE TWO-WAY LAYOUT WITH ONE OBSERVATION PER CELL

In this section we consider the two-way layout of sec. 4.1 in the case where there is just one observation in every cell, a case frequently occurring in practice. In order to get exact tests and confidence intervals concerning

[4] Constructed by Professor William H. Kruskal.

[5] We indicate the proof for the case where all four elements are different: The condition implies that the smallest element is in one corner and the largest in the opposite corner. One can bring the smallest into the a position of (4.1.19) and the largest into the d position by interchange of rows and columns, so $a < b < d$ and $a < c < d$. Define $f(\eta) = \eta$ for $\eta = a, b, c$ and $f(d) = b+c-a$. It is easily verified that $f(\eta)$ is strictly increasing.

the main effects it is generally necessary with the fixed-effects model[6] to assume that there are no interactions. (A test of the hypothesis of no interactions when there is one observation per cell, and a discussion of the effects of interactions on the inferences, is given in sec. 4.8.) This implies that the true cell means have the structure $\eta_{ij} = \mu+\alpha_i+\beta_j$, where $\alpha_. = \beta_. = 0$. If we add to this the assumptions of normality, of statistical independence, and of equality of the cell variances, and let y_{ij} denote the observation in the i,j cell, we have

$$\Omega: \begin{cases} y_{ij} = \mu + \alpha_i + \beta_j + e_{ij}, \\ \alpha_. = \beta_. = 0, \\ \{e_{ij}\} \text{ are independently } N(0, \sigma^2). \end{cases}$$

The hypotheses of chief interest are

$$H_A\text{: all } \alpha_i = 0 \quad \text{and} \quad H_B\text{: all } \beta_j = 0.$$

Hypothesis H_A says that the means $\{\mu+\alpha_i\}$ for the different levels of A are all equal, that is, that the different levels of A all have the same effect; similarly for H_B. Here and *throughout this book unless otherwise indicated* it will be understood that the ranges of subscripts i, j, etc., are respectively $i = 1, \cdots, I$; $j = 1, \cdots, J$; etc.

In practice, the statistical inferences based on the above model are not seriously invalidated by violation of the normality assumption, nor, if the numbers of observations in the cells are equal, as in the present case, by violation of the assumption of equality of the cell variances (Ch. 10). However, there are no such comforting considerations concerning violation of the assumption of statistical independence, except for experiments in which randomization has been incorporated into the experimental procedure, as indicated at the end of this section.

The matrix $\mathbf{X}'$ of the general assumption $E(\mathbf{y}) = \mathbf{X}'\boldsymbol{\beta}$ is exhibited inside the double lines of Table 4.2.1, the matrix being bordered with the vector $\mathbf{y}$ on the left and with $\boldsymbol{\beta}'$ at the top. The reader may find it instructive to prove directly that it has rank $I+J-1$, by showing that if we delete the first two columns the remaining $I+J-1$ are linearly independent, while the first is the sum of the last J, and the second is the sum of the last J minus the sum of the preceding $I-1$. That $I+J-1$ is the dimension of the space to which Ω restricts $\boldsymbol{\eta}$ may also be seen by "counting constants": the vector $\boldsymbol{\eta}$ is determined by $1+I+J$ parameters $\{\mu, \alpha_i, \beta_j\}$ subject to two (linearly independent) side conditions $\Sigma_i\alpha_i = 0$ and $\Sigma_j\beta_j = 0$.

[6] But not the random-effects model (Ch. 7) or the mixed model (Ch. 8).

TABLE 4.2.1

THE MATRIX $\mathbf{X}'$

y_{ij}	μ	α_1	α_2	$\cdots$	α_I	β_1	β_2	$\cdots$	β_J
	Coefficient in $E(y_{ij})$ of								
y_{11}	1	1	0	$\cdots$	0	1	0	$\cdots$	0
y_{12}	1	1	0	$\cdots$	0	0	1	$\cdots$	0
.	.	.	.	...	.	.	.	...	.
.	.	.	.	...	.	.	.	...	.
.	.	.	.	...	.	.	.	...	.
y_{1J}	1	1	0	$\cdots$	0	0	0	$\cdots$	1
y_{21}	1	0	1	$\cdots$	0	1	0	$\cdots$	0
y_{22}	1	0	1	$\cdots$	0	0	1	$\cdots$	0
.	.	.	.	...	.	.	.	...	.
.	.	.	.	...	.	.	.	...	.
.	.	.	.	...	.	.	.	...	.
y_{2J}	1	0	1	$\cdots$	0	0	0	$\cdots$	1
.	.	.	.	...	.	.	.	...	.
.	.	.	.	...	.	.	.	...	.
.	.	.	.	...	.	.	.	...	.
y_{I1}	1	0	0	$\cdots$	1	1	0	$\cdots$	0
y_{I2}	1	0	0	$\cdots$	1	0	1	$\cdots$	0
.	.	.	.	...	.	.	.	...	.
.	.	.	.	...	.	.	.	...	.
.	.	.	.	...	.	.	.	...	.
y_{IJ}	1	0	0	$\cdots$	1	0	0	$\cdots$	1

The SS to be minimized under Ω is

$$\mathscr{S} = \sum_i \sum_j (y_{ij} - \mu - \alpha_i - \beta_j)^2.$$

Equating to zero

$$\partial\mathscr{S}/\partial\mu = -2\sum_i \sum_j (y_{ij} - \mu - \alpha_i - \beta_j),$$

and using $\alpha. = \beta_. = 0$, we find

$$\hat{\mu} = y_{..}.$$

From

$$\partial\mathscr{S}/\partial\alpha_i = -2\sum(y_{ij} - \mu - \alpha_i - \beta_j) = 0,$$

we get $\hat{\mu} + \hat{\alpha}_i = y_{i.}$, hence

$$\hat{\alpha}_i = y_{i.} - y_{..};$$

and similarly

$$\hat{\beta}_j = y_{.j} - y_{..}.$$

The error SS, namely

$$\text{SS}_e = \sum_i \sum_j (y_{ij} - \hat{\mu} - \hat{\alpha}_i - \hat{\beta}_j)^2$$

thus becomes

$$\text{SS}_e = \sum_i \sum_j (y_{ij} - y_{i.} - y_{.j} + y_{..})^2, \tag{4.2.1}$$

which will later (sec. 4.3) be recognized as an "interaction SS;" it is also called the "residual SS." The number of d.f. for SS_e is $\nu_e = n-r = IJ-(I+J-1)$ or

$$\nu_e = (I-1)(J-1).$$

Under $\omega = \Omega \cap H_A$ we must minimize

$$\sum_i \sum_j (y_{ij} - \mu - \beta_j)^2$$

since $\alpha_i = 0$ under H_A. On equating to zero the partial derivatives we find that $\hat{\mu}_\omega$ and $\hat{\beta}_{j,\omega}$, the LS estimates of μ and β_j under ω, have the same values as under Ω, while of course $\hat{\alpha}_{i,\omega} = 0$. We remark here that the LS estimates of μ and β_j will not have the same value under ω as Ω in the case of inequality of the numbers of observations in the cells (sec. 4.4). If we take the SS in the numerator of $\mathscr{F}$ in the form $\text{SS}_H = \|\hat{\boldsymbol{\eta}} - \hat{\boldsymbol{\eta}}_\omega\|^2$, and write $\text{SS}_H = \text{SS}_A$, we get

$$\begin{aligned}\text{SS}_A &= \sum_i \sum_j (\hat{\eta}_{ij} - \hat{\eta}_{ij,\omega})^2 = \sum_i \sum_j (\hat{\mu} + \hat{\alpha}_i + \hat{\beta}_j - \hat{\mu}_\omega - \hat{\alpha}_{i,\omega} - \hat{\beta}_{j,\omega})^2 \\ &= \sum_i \sum_j \hat{\alpha}_i^2 = J \sum_i \hat{\alpha}_i^2,\end{aligned}$$

where we have used $\eta_{ij} = E(y_{ij}) = \eta + \alpha_i + \beta_j$, while $\hat{\eta}_{ij}$ and $\hat{\eta}_{ij,\omega}$ denote the LS estimates of η_{ij} under Ω and ω, respectively. Thus

$$\text{SS}_A = J \sum_i (y_{i.} - y_{..})^2 = J \sum_i y_{i.}^2 - IJy_{..}^2. \tag{4.2.2}$$

Since H_A states that $I-1$ linearly independent estimable functions are zero (see sec. 3.2), the number of d.f. for SS_A is $I-1$. Hence the F-test of H_A at significance level α consists in rejecting H_A if

$$\text{MS}_A / \text{MS}_e > F_{\alpha; I-1, \nu_e},$$

where $\text{MS}_A = \text{SS}_A/(I-1)$, $\text{MS}_e = \text{SS}_e/\nu_e$, and SS_A is given by (4.2.2),

SS_e by (4.2.1). In a similar way one would find that the F-test of H_B consists in rejecting H_B if

$$MS_B/MS_e > F_{\alpha;J-1,\nu_e},$$

where $MS_B = SS_B/(J-1)$, and

$$SS_B = I\sum_j(y_{.j} - y_{..})^2 = I\sum_j y_{.j}^2 - IJy_{..}^2. \tag{4.2.3}$$

The error SS is calculated by subtraction from

$$SS_e = SS_{\text{"tot"}} - SS_A - SS_B, \tag{4.2.4}$$

where

$$SS_{\text{"tot"}} = \sum_i\sum_j(y_{ij} - y_{..})^2 = \sum_i\sum_j y_{ij}^2 - IJy_{..}^2.$$

The identity (4.2.4) may be derived from the orthogonality relations to be stated below, or directly from the general identity $SS_e = \|\mathbf{y}\|^2 - \|\hat{\boldsymbol{\eta}}\|^2$, which gives

$$\begin{aligned} SS_e &= \sum_i\sum_j y_{ij}^2 - \sum_i\sum_j \hat{\eta}_{ij}^2 = \sum_i\sum_j y_{ij}^2 - \sum_i\sum_j(\hat{\mu} + \hat{\alpha}_i + \hat{\beta}_j)^2 \\ &= \sum_i\sum_j y_{ij}^2 - \sum_i\sum_j(\hat{\mu}^2 + \hat{\alpha}_i^2 + \hat{\beta}_j^2), \end{aligned}$$

the sums of cross-product terms in the last sum vanishing because of the side conditions $\hat{\alpha}_. = \hat{\beta}_. = 0$. Thus

$$SS_e = (\sum_i\sum_j y_{ij}^2 - IJy_{..}^2) - J\sum_i\hat{\alpha}_i^2 - I\sum_j\hat{\beta}_j^2,$$

or (4.2.4). The quantities SS_A and SS_B are called the SS's for main effects of A and B, respectively, or the SS's for rows and columns, respectively.

We summarize the results in Table 4.2.2. The last column is calculated in the usual way by Rule 2 of sec. 2.6; for example: The numerator MS of the statistic for testing H_A is $MS_A = J\sum_i\hat{\alpha}_i^2/(I-1)$. By Rule 2 we must replace the $\{y_{ij}\}$ involved in this expression by $E(y_{ij})$ under Ω, and add σ^2 to the result. But since $\hat{\alpha}_i$ is a linear function of the $\{y_{ij}\}$, this is equivalent to replacing $\hat{\alpha}_i$ by $E(\hat{\alpha}_i) = \alpha_i$ under Ω. Thus $E(MS_A) = \sigma^2 + J\Sigma_i\alpha_i^2/(I-1)$. The symbols σ_A^2 and σ_B^2 do not denote the variances of any random variable but are merely convenient abbreviations for the following functions of the parameters:

$$\sigma_A^2 = \sum_i\alpha_i^2/(I-1), \qquad \sigma_B^2 = \sum_j\beta_j^2/(J-1).$$

We note the hypotheses H_A and H_B may be expressed as H_A: $\sigma_A^2 = 0$ and H_B: $\sigma_B^2 = 0$.

TABLE 4.2.2

ANALYSIS OF VARIANCE FOR TWO-WAY LAYOUT WITH ONE OBSERVATION PER CELL

Source	SS	d.f.	MS	E(MS)
Rows	$SS_A = J\sum_i (y_{i.}-y_{..})^2$	$I-1$	$SS_A/(I-1)$	$\sigma^2+J\sigma_A^2$
Columns	$SS_B = I\sum_j (y_{.j}-y_{..})^2$	$J-1$	$SS_B/(J-1)$	$\sigma^2+I\sigma_B^2$
Residual	$SS_e = \sum_i\sum_j (y_{ij}-y_{i.}-y_{.j}+y_{..})^2$	$\nu_e = (I-1)(J-1)$	SS_e/ν_e	σ^2
Total	$SS_{\text{"tot"}} = \sum_i\sum_j (y_{ij}-y_{..})^2$	$IJ-1$	—	—

Calculations

For numerical calculations, before making a table like Table 4.2.2, a rectangular table of the observations should be made, bordered by the row means $\{y_{i.}\}$, the column means $\{y_{.j}\}$, and the general mean $y_{..}$, and this should be offered as part of the summary.[7] If an automatic electric calculating machine is used, the table of $\{y_{ij}\}$ can be bordered also with the row sums of squares $\Sigma_j y_{ij}^2$ and the column sums of squares $\Sigma_i y_{ij}^2$, these being obtained at the same time as the row and column sums needed for the row and column means; a check on the total SS, namely $\Sigma_i\Sigma_j y_{ij}^2$, is then obtained by summing it both ways; similarly $y_{..}$ can be checked. To construct the table like Table 4.2.2, SS_A is calculated from the last of the expressions (4.2.2), SS_B from the last of (4.2.3), and SS_e from (4.2.4). For a more thorough scrutiny of the data it is advisable to construct an $I\times J$ table in which the i,j entry is $\hat{\gamma}_{ij} = y_{ij}-y_{i.}-y_{.j}+y_{..}$. If there is a relatively large $\hat{\gamma}_{ij}$, it may suggest that the Ω-assumptions have somehow been violated, because of nonadditivity, or unequal variances of the observations, or a gross error in taking or recording the observation.

Contrasts among Main Effects

Let ψ be any linear function of the $\{\alpha_i\}$, $\psi = \Sigma_i c_i\alpha_i$. The set L of all such ψ is the same as the set of all contrasts among the true row means $\{A_i = \mu + \alpha_i = \eta_{i.}\}$, since $\Sigma_i c_i\alpha_i = \Sigma_i c_i'\eta_{i.}$, where $c_i' = c_i - c_.$, and hence $\Sigma_i c_i' = 0$; conversely, if $\Sigma_i c_i' = 0$, $\Sigma_i c_i'\eta_{i.} = \Sigma_i c_i'\alpha_i$. The LS estimate of

[7] The table is often bordered with totals instead of means, and the calculations are then also carried through with totals instead of means. I prefer the means because they are on the scale to which one becomes accustomed in handling the data. Since electric desk calculators usually have automatic division, even when they lack automatic multiplication, it is easy to get the mean whenever the total has been calculated, and then the total need not be copied off the machine.

ψ is $\hat{\psi} = \Sigma_i c_i \hat{\alpha}_i$. The estimates $\{\hat{\alpha}_i\}$, unlike the row means $\{y_{i.}\}$, are not independent; so it is easier to calculate Var $(\hat{\psi})$ from the formula $\hat{\psi} = \Sigma_i c_i' y_{i.}$, whence Var $(\hat{\psi}) = \Sigma_i c_i'^2 \sigma^2/J$. If the S-method (sec. 3.5) is applied to the set L of contrasts among the row means, then $q = I-1$ and $n-r = \nu_e$. The T-method (sec. 3.6) may also be used, and the discussion in sec. 3.7 of the relative advantages of the two methods applies without change. The S-method could also be applied with $q = I$ to the set of all linear functions of the row means; the extended T-method based on the augmented range (sec. 3.7) could also be applied to these. The column means could be treated similarly.

Orthogonality Relations

Consider the spaces spanned by the following four sets of linear forms in the observations $\{y_{ij}\}$:

Space	Spanned by	Dimension
$\mathscr{L}_\alpha$	$\hat{\alpha}_1, \cdots, \hat{\alpha}_I$	$I-1$
$\mathscr{L}_\beta$	$\hat{\beta}_1, \cdots, \hat{\beta}_J$	$J-1$
$\mathscr{L}_\mu$	$\hat{\mu}$	1
$\mathscr{L}_e$	$\{\hat{\gamma}_{ij} = y_{ij} - y_{i.} - y_{.j} + y_{..}\}$	$(I-1)(J-1)$

Two linear functions in different spaces are orthogonal, and hence, under the normality assumption in Ω, statistically independent. The orthogonality relations are proved as in the more complicated example we will encounter in sec. 4.5, by the "method of nested ω's" (sec. 2.9). It is convenient to define the direct sum of two linear spaces $\mathscr{L}_1$ and $\mathscr{L}_2$ as the set of all elements l_1+l_2 with $l_1 \in \mathscr{L}_1$ and $l_2 \in \mathscr{L}_2$, and to write it as $\mathscr{L}_1 \oplus \mathscr{L}_2$. Then the estimation space in the general theory of sec. 1.6 becomes in this case $\mathscr{L}_\alpha \oplus \mathscr{L}_\beta \oplus \mathscr{L}_\mu$, and the error space, $\mathscr{L}_e$.

The Randomized-Blocks Design

Suppose that I treatments are to be compared on certain experimental units, for example I varieties on certain plots, or I drugs on certain animals. It is often possible to make more precise comparisons of the treatments by grouping the experimental units into blocks of I units such that units within a block resemble each other more than units in different blocks; thus each block may consist of a row of I plots, or a set of I animals from the same litter. If there are J blocks we have a complete two-way layout, the factors being treatments at I levels and blocks at J levels. This two-way layout is called a *randomized blocks* design[8] if within each block the I treatments are assigned at random to the I experimental units in such a way that each of the $I!$ ways of assigning the treatments to the experimental units has the same probability of being adopted in the experiment, and the assignments in the different blocks are statistically independent. For achieving this a table of random permutations such as that in Cochran and Cox (1957), Ch. 15, is convenient.

There is nothing in the "normal-theory model" of the two-way layout we are now considering that reflects the increased accuracy possible by good blocking. Indeed, the present model is inappropriate to those randomized-blocks experiments where the "errors" are caused mainly by differences among the experimental units rather than measurement errors: We will find in Ch. 9 that these errors are correlated and that their variances may differ from block to block; the normality assumption of normal-theory models will in general be found to be less critical than the assumptions of independence and of equality of variance. Nevertheless we will see when we consider the more realistic model for randomized blocks in Ch. 9 that inferences about the treatment contrasts derived under the ill-fitting normal-theory model may still be regarded as a fair approximation if care is taken to incorporate the *randomization* described above into the assignment of treatments to experimental units.

Randomization

In any experiment where treatment combinations are assigned to experimental units this assignment should be done by randomization. The randomization may be achieved by coin tossing, card drawing, a table of random numbers or random permutations, etc. The randomizing may be subject to certain conditions imposed by the design, for example in the randomized-blocks design, each treatment must appear exactly once in each block of experimental units.

The intuitive justification for randomizing is that besides the factors being controlled in the experiment, such as A and B in the two-way layout, there may be other uncontrolled factors causing the experimental units to differ in their response to the treatment combinations, and one desires as far as possible to prevent their entering in a systematic way to alter the

[8] Logically this design and certain aspects of the three incomplete layouts considered in Ch. 5 do not belong in Part I of this book, but there are pedagogical advantages to my arrangement: (*i*) The estimates and the analysis-of-variance table except for the E(MS) column are derived from the general theory of the inappropriate model of Part I (distributions must later be reconsidered under the more realistic models, for which as yet we have no general theory). (*ii*) It is desirable that those basic designs and the homily on randomization be encountered as soon as possible; the logical position for randomized blocks, for example, would not be reached until Ch. 9.

apparent effects of the controlled factors. Thus, in the randomized-blocks design in agriculture, if the blocks consist of rows of I plots, the rows running in an East–West line, there might be a fertility gradient so that the plots are increasingly fertile from East to West in each block, and, if the varieties appear in the same order in each block, then, when there are no real differences among the varieties, their yields will tend to differ, increasing in the direction of the fertility gradient. In assigning treatment combinations to experimental units from groups of animals, people, sheets of material, etc., randomization safeguards the conclusions from systematic bias due to unconscious bias of the experimenter as well as other uncontrolled factors whose possible effects may or may not have been anticipated.

The logical reason for randomizing is that it is possible on the basis of a model reflecting the randomization to draw sound statistical inferences, the probability basis of the model being provided not by wishful thinking but by the actual process of randomization which is part of the experiment. It is fortunate that in situations where the randomization models are more appropriate, statistical inferences from the corresponding "normal-theory" models usually are fair approximations in the more realistic randomization models. However, to profit from this happy relationship of the two models *randomization must be incorporated into the experiment.*

4.3. THE TWO-WAY LAYOUT WITH EQUAL NUMBERS OF OBSERVATIONS IN THE CELLS

The number of observations in the i,j cell will be denoted by K_{ij}. To begin with, we shall assume about the $\{K_{ij}\}$ only that not all $K_{ij} = 0$ (else the number n of observations is zero); in a complete layout all K_{ij} are > 0, but some of the results we shall obtain here and in the sequel in sec. 4.4 will later be needed also for incomplete layouts. Beyond a certain point, the analysis for unequal $\{K_{ij}\}$ becomes much more complicated than for equal $\{K_{ij}\}$, and we shall then find it convenient to assume equal cell numbers, returning to the case of unequal numbers in the next section.

If y_{ijk} denotes the kth observation in the i,j cell, and D denotes the set of pairs $\{(i,j)\}$ which label the nonempty cells, then we assume that

$$\Omega: \begin{cases} y_{ijk} = \eta_{ij} + e_{ijk}, \\ \{e_{ijk}\} \text{ are independently } N(0, \sigma^2), \\ k = 1, \cdots, K_{ij}; \quad (i,j) \in D. \end{cases}$$

The hypotheses of chief interest concern the main effects and interactions; we shall formulate these below.

Under Ω we have to minimize

$$\mathscr{S} = \sum_{(i,j)\in D} \sum_{k=1}^{K_{ij}} (y_{ijk} - \eta_{ij})^2. \tag{4.3.1}$$

Only the $\{\eta_{ij}\}$ for the nonempty cells enter the observations, and they constitute the p parameters $\{\beta_j\}$ of the general theory of Chs. 1 and 2. Their LS estimates are

$$\hat{\eta}_{ij} = y_{ij.} \quad \text{for} \quad (i,j) \in D. \tag{4.3.2}$$

By considering the form of the matrix $\mathbf{X}'$ in this case, its rank r is easily seen to be equal to p. This can also be concluded from the fact that the normal equations have the unique solutions (4.3.2). The error SS, the minimum of (4.3.1), is

$$\mathrm{SS}_e = \sum_{(i,j)\in D} \sum_{k=1}^{K_{ij}} (y_{ijk} - y_{ij.})^2, \tag{4.3.3}$$

and its number of d.f. is $n - p$, where n is the number of observations, and p is the number of nonempty cells.

Since for this parameterization $r = p$, all linear functions of the p parameters $\{\eta_{ij}\}$ corresponding to the nonempty cells are estimable. If the layout is complete, by the Gauss–Markoff theorem (sec. 1.4), the LS estimates under Ω of all main effects and interactions, which are defined as certain linear combinations of the $\{\eta_{ij}\}$, are then obtained by replacing the η_{ij} in these linear combinations by (4.3.2). From (4.1.9), the LS estimates of the general mean, main effects, and interactions are then

(4.3.4)

$$\hat{\mu} = \hat{\eta}_{..}, \quad \hat{\alpha}_i = \hat{\eta}_{i.} - \hat{\eta}_{..}, \quad \hat{\beta}_j = \hat{\eta}_{.j} - \hat{\eta}_{..}, \quad \hat{\gamma}_{ij} = \hat{\eta}_{ij} - \hat{\eta}_{i.} - \hat{\eta}_{.j} + \hat{\eta}_{..},$$

the dot notation indicating *unweighted* averages of the observed cell means $\{\hat{\eta}_{ij}\}$. However, if there is even a single empty cell, then the general mean, main effects, and interactions are not estimable under Ω, since all these involve in their definitions the η_{ij} of the empty cell, which does not enter the observations.

The hypotheses that we usually wish to test are

$$H_A\colon \ \text{all } \alpha_i = 0,$$
$$H_B\colon \ \text{all } \beta_j = 0,$$
$$H_{AB}\colon \ \text{all } \gamma_{ij} = 0.$$

To simplify the tests and orthogonality relations *we assume in the remainder of this section that the cell numbers* $\{K_{ij}\}$ *are all equal to* $K > 1$.

The LS estimates (4.3.4) may now be written

$$\hat{\mu} = y_{...}, \quad \hat{\alpha}_i = y_{i..} - y_{...}, \quad \hat{\beta}_j = y_{.j.} - y_{...},$$
$$\hat{\gamma}_{ij} = y_{ij.} - y_{i..} - y_{.j.} + y_{...}.$$

The SS (4.3.1) may be written

$$\mathscr{S} = \sum_i \sum_j \sum_k (y_{ijk} - \mu - \alpha_i - \beta_j - \gamma_{ij})^2,$$

and if in $\mathscr{S}$ we substitute

$$y_{ijk} - \mu - \alpha_i - \beta_j - \gamma_{ij} = (y_{ijk} - \hat{\mu} - \hat{\alpha}_i - \hat{\beta}_j - \hat{\gamma}_{ij}) + (\hat{\mu} - \mu) + (\hat{\alpha}_i - \alpha_i)$$
$$+ (\hat{\beta}_j - \beta_j) + (\hat{\gamma}_{ij} - \gamma_{ij}),$$

and preserve the parentheses on squaring and summing over i, j, k, we find that the cross-product terms vanish because of the side conditions

$$\sum_i \alpha_i = 0, \quad \sum_j \beta_j = 0, \quad \sum_i \gamma_{ij} = 0 \quad \text{for all } j, \quad \sum_j \gamma_{ij} = 0 \quad \text{for all } i,$$

and the analogous side conditions on the $\{\hat{\alpha}_i\}$, $\{\hat{\beta}_j\}$, and $\{\hat{\gamma}_{ij}\}$, leaving

(4.3.5) $$\mathscr{S} = SS_e + IJK(\hat{\mu} - \mu)^2 + JK\sum_i (\hat{\alpha}_i - \alpha_i)^2 + IK\sum_j (\hat{\beta}_j - \beta_j)^2$$
$$+ K\sum_i \sum_j (\hat{\gamma}_{ij} - \gamma_{ij})^2.$$

It is clear from this expression that, except for the parameters which are zero under H_A, H_B, and H_{AB}, respectively, the LS estimates under these hypotheses are the same as under Ω. For example, under H_A, (4.3.5) becomes

(4.3.6) $$\mathscr{S} = SS_e + IJK(\hat{\mu} - \mu)^2 + JK\sum_i \hat{\alpha}_i^2$$
$$+ IK\sum_j (\hat{\beta}_j - \beta_j)^2 + K\sum_i \sum_j (\hat{\gamma}_{ij} - \gamma_{ij})^2.$$

This is obviously minimized by the values $\mu = \hat{\mu}$, $\beta_j = \hat{\beta}_j$, and $\gamma_{ij} = \hat{\gamma}_{ij}$, and thus the minimum value of $\mathscr{S}$ under H_A is

(4.3.7) $$\mathscr{S}_{\omega_A} = SS_e + JK\sum_i \hat{\alpha}_i^2.$$

It can be seen similarly that under any hypothesis H_α which specifies nothing about $\{\mu, \beta_j, \gamma_{ij}\}$ but only about the $\{\alpha_i\}$ (i.e., H_α states that the $\{\alpha_i\}$ satisfy given linear restrictions), the LS estimates of $\{\mu, \beta_j, \gamma_{ij}\}$ would still be the same. However, it would in general be necessary to use the Lagrange multiplier method to find the minimum of $JK\Sigma_i(\hat{\alpha}_i - \alpha_i)^2$ for varying $\{\alpha_i\}$ subject to H_α; this plus SS_e would then be the minimum $\mathscr{S}$ under H_α.

For testing H_A the numerator SS of $\mathscr{F}$, calculated as $\mathscr{S}_{\omega_A}-\mathscr{S}_{\Omega}$ (sec. 2.5), is thus from (4.3.7) evidently

$$SS_A = JK\sum_i \hat{\alpha}_i^2.$$

Similarly, we obtain for the numerator SS's for testing H_B and H_{AB},

$$SS_B = IK\sum_j \hat{\beta}_j^2,$$
$$SS_{AB} = K\sum_i\sum_j \hat{\gamma}_{ij}^2;$$

the denominator SS is of course SS_e in each case. The number of d.f. for SS_A is $I-1$, since this is the number of linearly independent estimable conditions imposed by H_A; similarly the number of d.f. for SS_B is $J-1$. The number of d.f. for SS_{AB} is $(I-1)(J-1)$: The number of estimable restrictions imposed by H_{AB}: all $\gamma_{ij}=0$, is IJ. However, if we think of arranging the $\{\gamma_{ij}\}$ in an $I\times J$ table, we see that, if all $\gamma_{ij}=0$ in the subtable obtained by deleting the last row and column, then $\gamma_{ij}=0$ in the whole table, since the row sums must be zero, and the columns sums, also. This suggests that the number of linearly independent restrictions imposed by H_{AB} is $(I-1)(J-1)$. A more rigorous argument is the following: Consider the dimension of the subspace V_{r-q} to which H_{AB} constrains $\boldsymbol{\eta}$. This subspace has the same dimension as the V_r (different r) of the Ω of sec. 4.2, if we assume zero interactions, and this dimension was found to be $I+J-1$. Thus $r-q=I+J-1$, and since $r=IJ$, therefore $q=(I-1)(J-1)$.

If we apply Rule 2 of sec. 2.6 to calculate the expected values of MS_A, MS_B, MS_{AB}, the resulting formulas suggest that we introduce the notations

$$\sigma_A^2 = (I-1)^{-1}\sum_i \alpha_i^2, \tag{4.3.7a}$$
$$\sigma_B^2 = (J-1)^{-1}\sum_j \beta_j^2,$$
$$\sigma_{AB}^2 = (I-1)^{-1}(J-1)^{-1}\sum_i\sum_j \gamma_{ij}^2.$$

Table 4.3.1 summarizes the results thus far. The entries in the mean-square column, which are not indicated, are the SS's divided by their numbers of d.f.

If the hypothesis H_A or H_B is rejected, one can use the S-method to find which contrasts are responsible for rejecting it. The T-method is also applicable to the main effects, as in any layout with equal cell numbers. If H_{AB} is rejected, one would ordinarily not explore the interactions further statistically, although, as with any F-test in fixed-effects models,

TABLE 4.3.1

ANALYSIS OF VARIANCE OF THE TWO-WAY LAYOUT WITH K OBSERVATIONS PER CELL

Source	SS	d.f.	MS	E(MS)
A main effects	$SS_A = JK\sum_i (y_{i..}-y_{...})^2$	$I-1$		$\sigma^2+JK\sigma_A^2$
B main effects	$SS_B = IK\sum_j (y_{.j.}-y_{...})^2$	$J-1$		$\sigma^2+IK\sigma_B^2$
AB interactions	$SS_{AB} = K\sum_i\sum_j (y_{ij.}-y_{i..}-y_{.j.}+y_{...})^2$	$(I-1)(J-1)$		$\sigma^2+K\sigma_{AB}^2$
Error	$SS_e = \sum_i\sum_j\sum_k (y_{ijk}-y_{ij.})^2$	$IJ(K-1)$		σ^2
"Total"	$SS_{\text{"tot"}} = \sum_i\sum_j\sum_k (y_{ijk}-y_{...})^2$	$IJK-1$	—	—

this could be done by the S-method, for one could then find interactions, or linear combinations of interactions, which are significantly different from zero by the S-criterion. The q (sec. 3.5) for the S-method applied to the whole space of interactions spanned by the $\{\gamma_{ij}\}$ would be the number of d.f. for SS_{AB}, namely $(I-1)(J-1)$. The T-method would not apply since the covariances of the $\{\hat{\gamma}_{ij}\}$ are not equal. Either method could be applied to contrasts among the cell means (q for S then being $IJ-1$). The S-method could also be applied to the set of all linear functions of the row means ($q = I$), or the set of all linear functions of the column means ($q = J$), or the set of all linear functions of the cell means ($q = IJ$); the extended T-method based on the augmented range (sec. 3.7) could also be applied to these sets. The multiple-comparison method based on the maximum modulus (sec. 3.7) could be applied to get simultaneous confidence statements about all the cell means.

Although the tests for main effects are valid regardless of the true values of the interactions, our interpretation of the results of the analysis will depend on whether or not we are willing actually to accept H_{AB}; this was discussed in sec. 4.1.

Calculations

The data should be arranged in an $I\times J$ table with the K observations on η_{ij} in the i,j cell of the table. The observed cell means $\{y_{ij.}\}$ should also be shown in these cells, or in a separate $I\times J$ table. The table with the cell means should be bordered by the row and column means $\{y_{i..}\}$ and $\{y_{.j.}\}$. The SS's for main effects are calculated in the now familiar way according to

$$SS_A = JK\sum_i y_{i..}^2 - C, \qquad SS_B = IK\sum_j y_{.j.}^2 - C,$$

where $C = IJKy^2_{\cdot\cdot\cdot}$. It is convenient to calculate a SS for "cell means about the grand mean,"

$$SS_{\text{"cells"}} = K\sum_i\sum_j y^2_{ij\cdot} - C.$$

Then the interaction SS may be obtained by subtraction,

$$SS_{AB} = SS_{\text{"cells"}} - SS_A - SS_B. \tag{4.3.8}$$

This formula is analogous to that for the residual SS in the two-way layout with one observation per cell. The total SS about the grand mean is calculated as usual from

$$SS_{\text{"tot"}} = \sum_i\sum_j\sum_k y^2_{ijk} - C,$$

and the error SS may then be formed by subtraction,

$$SS_e = SS_{\text{"tot"}} - SS_{\text{"cells"}}.$$

Orthogonality Relations

The orthogonality relations will now be derived by the "method of nested ω's" (sec. 2.9). The chain of hypotheses is shown in the first column of Table 4.3.2. The i,j,k component of $\hat{\boldsymbol{\eta}}_\omega$, where $\hat{\boldsymbol{\eta}}_\omega$ is the projection of $\mathbf{y}$ on the subspace to which ω constrains $\boldsymbol{\eta} = E(\mathbf{y})$, is denoted by $\hat{\eta}_{ijk,\omega}$ and listed in the second column of the table. The differences for the successive ω's are listed in the last column. From the general theory

TABLE 4.3.2

THE NESTED ω'S

Hypothesis ω	$\hat{\eta}_{ijk,\omega}$	Difference
Ω	$\hat{\mu}+\hat{\alpha}_i+\hat{\beta}_j+\hat{\gamma}_{ij}$	
		$\hat{\gamma}_{ij} = y_{ij\cdot}-y_{i\cdot\cdot}-y_{\cdot j\cdot}+y_{\cdot\cdot\cdot}$
$\omega_1 = \Omega \cap H_{AB}$	$\hat{\mu}+\hat{\alpha}_i+\hat{\beta}_j$	
		$\hat{\alpha}_i = y_{i\cdot\cdot}-y_{\cdot\cdot\cdot}$
$\omega_2 = \omega_1 \cap H_A$	$\hat{\mu}+\hat{\beta}_j$	
		$\hat{\beta}_j = y_{\cdot j\cdot}-y_{\cdot\cdot\cdot}$
$\omega_3 = \omega_2 \cap H_B$	$\hat{\mu}$	

of sec. 2.9 it follows that the five vectors listed in the first column of Table 4.3.3 lie in five mutually orthogonal spaces of the dimensions listed in the second column, and that the five sets of linear forms in the observations listed in the third column span five orthogonal spaces of linear forms of dimensions listed in the second column. These relations do not depend on the normality assumption. Under the normality assumption the five

SS's in the last column have independent noncentral chi-square distributions with the numbers of d.f. shown in the second column and noncentrality parameters given by Rule 1 of sec. 2.6.

TABLE 4.3.3

THE FIVE ORTHOGONAL SPACES

Vector $\mathbf{v}$	d.f.	Component v_{ijk}	$SS = \sum_i \sum_j \sum_k v_{ijk}^2$
$\mathbf{y}-\hat{\boldsymbol{\eta}}$	$IJ(K-1)$	$y_{ijk}-y_{ij.}$	SS_e
$\hat{\boldsymbol{\eta}}-\hat{\boldsymbol{\eta}}_{\omega_1}$	$(I-1)(J-1)$	$\hat{\gamma}_{ij}$	SS_{AB}
$\hat{\boldsymbol{\eta}}_{\omega_1}-\hat{\boldsymbol{\eta}}_{\omega_2}$	$I-1$	$\hat{\alpha}_i$	SS_A
$\hat{\boldsymbol{\eta}}_{\omega_2}-\hat{\boldsymbol{\eta}}_{\omega_3}$	$J-1$	$\hat{\beta}_j$	SS_B
$\hat{\boldsymbol{\eta}}_{\omega_3}$	1	$\hat{\mu} = y_{...}$	$IJKy_{...}^2$

The general identity

$$(4.3.9) \quad \|\mathbf{y}\|^2 = \|\mathbf{y}-\hat{\boldsymbol{\eta}}\|^2 + \|\hat{\boldsymbol{\eta}}-\hat{\boldsymbol{\eta}}_{\omega_1}\|^2 + \|\hat{\boldsymbol{\eta}}_{\omega_1}-\hat{\boldsymbol{\eta}}_{\omega_2}\|^2 + \|\hat{\boldsymbol{\eta}}_{\omega_2}-\hat{\boldsymbol{\eta}}_{\omega_3}\|^2 + \|\hat{\boldsymbol{\eta}}_{\omega_3}\|^2$$

states here that

$$(4.3.10) \qquad SS_{\text{tot}} = SS_e + SS_{AB} + SS_A + SS_B + IJKy_{...}^2.$$

Other identities could be obtained by using any m ($m < 5$) successive terms in the right of (4.3.9); for example, the identity (4.3.8) used above is that obtained by saying that the sum of the last four terms equals $\|\hat{\boldsymbol{\eta}}\|^2$, and writing $\hat{\eta}_{ijk} = y_{ij.}$.

4.4. THE TWO-WAY LAYOUT WITH UNEQUAL NUMBERS OF OBSERVATIONS IN THE CELLS

We continue now with the case of unequal numbers in the two-way layout, which we carried through in sec. 4.3 up to the point where we assumed all $K_{ij} = K$. We shall find that the tests for main effects and the associated multiple comparisons by the S-method remain relatively simple in the case of unequal numbers, but that the test for interactions is more difficult to compute, requiring the solution of m linear equations in m unknowns, where m is one less than the minimum of I and J.

We mention that a quick approximate method of analysis is described in sec. 10.6.

In this section we shall drop our usual convention of defining the general mean, main effects, and interactions only with equal weights $\{v_i\}$ *and* $\{w_j\}$, *as discussed in sec.* 4.1. The statistical inferences are just as easy to formulate for an arbitrary system of weights as for equal weights, and at one point there is a slight computational advantage in selecting a system other than of equal weights.

We take the Ω-assumptions as at the beginning of sec. 4.3, where we found $\mathscr{S}_\Omega$ (there written SS_e) to be

$$\mathscr{S}_\Omega = \sum_{(i,j)\in D} \sum_{k=1}^{K_{ij}} (y_{ijk} - y_{ij.})^2 \tag{4.4.1}$$

with $n - p$ d.f., where n is the number of observations, and p is the number of nonempty cells.

Test for Interactions

The hypothesis H_{AB} of no interactions, or of additivity, is

$$H_{AB}: \quad E(y_{ijk}) = \mu + \alpha_i + \beta_j,$$

where, as noted in Theorem 1 of sec. 4.1, the question of whether H_{AB} is true or false does not depend on the weights $\{v_i\}$ and $\{w_j\}$ used to define the general mean μ and main effects $\{\alpha_i\}$ and $\{\beta_j\}$. Throughout this section we will reserve the symbol ω for

$$\omega = \Omega \cap H_{AB}.$$

The effect under ω of choosing the weights $\{v_i\}$ and $\{w_j\}$ is only to impose the side conditions $\Sigma_i v_i \alpha_i = 0$ and $\Sigma_j w_j \beta_j = 0$ (see sec. 4.1). We may derive the normal equations under ω first and then choose convenient side conditions later.

Under ω we must minimize[9]

$$\sum_i \sum_j \sum_k (y_{ijk} - \mu - \alpha_i - \beta_j)^2.$$

Equating to zero $\partial/\partial\mu$, $\partial/\partial\alpha_i$, and $\partial/\partial\beta_j$ of this expression, we get

$$n\hat{\mu}_\omega + \sum_i G_i \hat{\alpha}_{i,\omega} + \sum_j H_j \hat{\beta}_{j,\omega} = \sum_i \sum_j \sum_k y_{ijk}, \tag{4.4.2}$$

$$G_i \hat{\mu}_\omega + G_i \hat{\alpha}_{i,\omega} + \sum_j K_{ij} \hat{\beta}_{j,\omega} = g_i \qquad (i = 1, \cdots, I), \tag{4.4.3}$$

$$H_j \hat{\mu}_\omega + \sum_i K_{ij} \hat{\alpha}_{i,\omega} + H_j \hat{\beta}_{j,\omega} = h_j \qquad (j = 1, \cdots, J), \tag{4.4.4}$$

where the subscript ω on the estimates indicates that they are LS estimates

[9] The pattern of nonempty cells in the two-way layout must satisfy a certain condition in order that the parameters μ, $\{\alpha_i\}$, $\{\beta_j\}$ be estimable under ω and the side conditions. I know of no simple way of formulating the condition; necessary and sufficient is the following: To every one of the IJ cells in the layout there corresponds a row of the matrix in Table 4.2.1, the i, j cell corresponding to the row following y_{ij} in the table. Consider the p-rowed submatrix whose rows correspond to the nonempty cells: Its rank must be $I+J-1$. A relevant criterion of connectedness is given by R. C. Bose (1949a) pp. 58–59.

under ω; $\{g_i\}$ and $\{h_j\}$ are respectively the row and column sums of the observations,

$$(4.4.5) \qquad g_i = \sum_j \sum_k y_{ijk}, \qquad h_j = \sum_i \sum_k y_{ijk};$$

and G_i and H_j, the row and column sums of the cell numbers,

$$(4.4.6) \qquad G_i = \sum_j K_{ij}, \qquad H_j = \sum_i K_{ij}.$$

The estimates will not be unique until the side conditions are chosen, but the various SS's involving them will nevertheless be unique.

From (4.4.4) we find

$$(4.4.7) \qquad \hat{\beta}_{j,\omega} = -\hat{\mu}_\omega + H_j^{-1}(h_j - \sum_{i'} K_{i'j}\hat{\alpha}_{i',\omega}),$$

and after substituting this into (4.4.3) we may write the result as

$$(4.4.8) \qquad \sum_{i'} a_{ii'}\hat{\alpha}_{i',\omega} = \mathscr{G}_i \qquad (i = 1, \cdots, I),$$

where

$$(4.4.9) \qquad \begin{aligned} a_{ii'} &= \delta_{ii'}G_i - \sum_j (K_{ij}K_{i'j}/H_j), \\ \mathscr{G}_i &= g_i - \sum_j (K_{ij}h_j/H_j). \end{aligned}$$

We remark that the elimination of the $\{\hat{\beta}_{j,\omega}\}$ resulted also in the elimination of $\hat{\mu}_\omega$. If we eliminated the $\{\hat{\alpha}_{i,\omega}\}$ instead of the $\{\hat{\beta}_{j,\omega}\}$ from the normal equations we would find

$$(4.4.10) \qquad \sum_{j'} b_{jj'}\hat{\beta}_{j',\omega} = \mathscr{H}_j \qquad (j = 1, \cdots, J),$$

where

$$(4.4.11) \qquad \begin{aligned} b_{jj'} &= \delta_{jj'}H_j - \sum_i (K_{ij}K_{ij'}/G_i), \\ \mathscr{H}_j &= h_j - \sum_i (K_{ij}g_i/G_i). \end{aligned}$$

Next we note that $\mathscr{S}_\omega$ can be expressed as

$$(4.4.12) \qquad \mathscr{S}_\omega = \sum_i \sum_j \sum_k y_{ijk}^2 - \sum_i \sum_j \sum_k y_{ijk}\hat{\mu}_\omega - \sum_i g_i\hat{\alpha}_{i,\omega} - \sum_j h_j\hat{\beta}_{j,\omega}.$$

This follows from a general result in the theory of least squares, namely (1.3.10) written for $\mathscr{S}_\omega$ instead of $\mathscr{S}_\Omega$, and for the LS estimates $\{\hat{\mu}_\omega, \hat{\alpha}_{i,\omega}, \hat{\beta}_{j,\omega}\}$ and the right members of the normal equations (4.4.2–4) under ω instead of Ω. We may eliminate the $\{\hat{\beta}_{j,\omega}\}$ from (4.4.12) by substituting (4.4.7), with the result

$$(4.4.13) \qquad \mathscr{S}_\omega = \sum_i \sum_j \sum_k y_{ijk}^2 - \sum_i \mathscr{G}_i\hat{\alpha}_{i,\omega} - \sum_j (h_j^2/H_j);$$

or we may similarly eliminate the $\{\hat{\alpha}_{i,\omega}\}$,

$$\mathscr{S}_{\omega} = \sum_i \sum_j \sum_k y_{ijk}^2 - \sum_j \mathscr{H}_j \hat{\beta}_{j,\omega} - \sum_i (g_i^2/G_i); \tag{4.4.14}$$

in both cases $\hat{\mu}_{\omega}$ drops out.

We make a final simplification in the calcu'ation of $\mathscr{S}_{\omega}$: We may choose the side conditions $\Sigma_i v_i \hat{\alpha}_{i,\omega} = 0$ and $\Sigma_j w_j \hat{\beta}_{j,\omega} = 0$ in the form $\hat{\alpha}_{I,\omega} = 0$ and $\hat{\beta}_{J,\omega} = 0$, corresponding to the weights $\{v_i = \delta_{iI}\}$ and $\{w_j = \delta_{jJ}\}$. Thus to calculate $\mathscr{S}_{\omega}$ we need solve only the following $I-1$ linear equations in $I-1$ unknowns $\{\hat{\alpha}_{i,\omega}\}$

$$\sum_{i'=1}^{I-1} a_{ii'} \hat{\alpha}_{i',\omega} = \mathscr{G}_i \qquad (\mathrm{i} = 1, \cdots, I-1), \tag{4.4.15}$$

where $\{a_{ii'}\}$ and $\{\mathscr{G}_i\}$ are defined by (4.4.9), (4.4.6), and (4.4.5), and then substitute these in (4.4.13) with $\hat{\alpha}_{I,\omega} = 0$. Or, we may solve the $J-1$ linear equation in $J-1$ unknowns $\{\hat{\beta}_{j,\omega}\}$

$$\sum_{j'=1}^{J-1} b_{jj'} \hat{\beta}_{j',\omega} = \mathscr{H}_j \qquad (j = 1, \cdots, J-1),$$

where the coefficients are defined by (4.4.11), and substitute in (4.4.14) with $\hat{\beta}_{J,\omega} = 0$. If $I \neq J$, we choose the method with the fewer number of unknowns.

The numerator SS of the statistic $\mathscr{F}$ for testing H_{AB} is $\mathscr{S}_{\omega} - \mathscr{S}_{\Omega}$, where $\mathscr{S}_{\Omega}$ is given by (4.4.1). Its number of d.f. may be calculated as follows: It is denoted by q in the general theory of Ch. 2, where ω constrains the vector of means $\boldsymbol{\eta}$ to an $(r-q)$-dimensional subspace of the r-dimensional space permitted by Ω. In the present case the parameters in $\boldsymbol{\eta}$ under ω are those of a two-way layout with additivity (sec. 4.2), so ω constrains $\boldsymbol{\eta}$ to a space of dimension $I+J-1$. Thus $r-q = I+J-1$, or $q = p-I-J+1$, since $r = p$. The statistic is thus

$$\frac{n-p}{p-I-J+1} \frac{\mathscr{S}_{\omega} - \mathscr{S}_{\Omega}}{\mathscr{S}_{\Omega}},$$

and has under H_{AB} the F-distribution with $p - I - J + 1$ and $n - p$ d.f.

Inferences about Main Effects, Assuming Additivity

The tests for and estimation of main effects depend on whether we incorporate the hypothesis of additivity H_{AB} into the underlying assumptions. If we do this, the underlying assumptions become the ω defined above. Suppose we wish to test

$$H_A: \text{ all } \alpha_i = 0$$

under ω. This is the hypothesis that all contrasts among the $\{\alpha_i\}$ are zero, and, as we remarked in sec. 4.1, under H_{AB} these contrasts are independent of the system of weights $\{v_i\}$ and $\{w_j\}$. The hypothesis

$$\omega_1 = \omega \cap H_A$$

is the same as the Ω for the one-way layout with J classes and H_j observations in the jth class, and so $\mathscr{S}_{\omega_1}$ is the error SS for the one-way layout, namely

$$(4.4.15a) \quad \mathscr{S}_{\omega_1} = \sum_i \sum_j \sum_k (y_{ijk} - H_j^{-1}h_j)^2 = \sum_i \sum_j \sum_k y_{ijk}^2 - \sum_j H_j^{-1}h_j^2,$$

since $H_j^{-1}h_j$ is the mean of all the observations in the jth class. To get the numbers of d.f. of the statistic $\mathscr{F}$ for testing H_A under ω we may reason as follows: We have already noted that ω constrains $\boldsymbol{\eta}$ to a space of dimension $I+J-1$; this now corresponds to the rank r of the general theory, where Ω corresponds to the present ω, so that the $n-r$ of the general theory becomes here $n-I-J+1$. The hypothesis H_A imposes $I-1$ linearly independent estimable restrictions. Thus the numbers of d.f. are $I-1$ and $n-I-J+1$, and the statistic

$$\frac{n-I-J+1}{I-1} \frac{\mathscr{S}_{\omega_1} - \mathscr{S}_{\omega}}{\mathscr{S}_{\omega}}$$

has under ω_1 the F-distribution with these numbers of d.f. The quantity $\mathscr{S}_{\omega}$ is calculated as described above.

If H_A is rejected we might consider continuing with the S-method of exploring the contrasts among the $\{\alpha_i\}$: If $\psi = \Sigma_i c_i \alpha_i$ is a contrast ($\Sigma c_i = 0$), its LS estimate $\Sigma_i c_i \hat{\alpha}_{i,\omega}$ may be written in terms of the solutions of (4.4.15), $\hat{\psi} = \Sigma_{i=1}^{I-1} c_i \hat{\alpha}_{i,\omega}$. The estimated standard deviation $\hat{\sigma}_{\hat{\psi}}$ needed for the S-method is rather tedious to compute: It is necessary to invert the $(I-1)$ by $(I-1)$ matrix $(a_{ii'})$ of the system (4.4.15). Denote the inverse matrix by $(a^{ii'})$. Then it may be shown that

$$\hat{\sigma}_{\hat{\psi}}^2 = s_{\omega}^2 \sum_{i=1}^{I-1} \sum_{i'=1}^{I-1} a^{ii'} c_i c_{i'},$$

where $s_{\omega}^2 = \mathscr{S}_{\omega}/(n-I-J+1)$. The constant needed for the S-method is $S = (I-1)F_{\alpha;I-1,n-I-J+1}$.

The test and estimation concerning the main effects $\{\beta_j\}$ are of course analogous.

Tests for Main Effects under Ω

Usually one would prefer to test and estimate the main effects without assuming additivity. Also, the computations are then much simpler. In this case the definition of the main effects depends on the system of weights $\{v_i\}$ and $\{w_j\}$. It is as easy to solve the present problem of testing and estimating the main effects under an arbitrary system of weights as under the usual system $\{v_i = 1/I\}$ and $\{w_j = 1/J\}$. In some applications, unequal weights might be desirable, as indicated in the example in sec. 4.1. As we remarked at the beginning of sec. 4.3, *it is now necessary to assume that there are no empty cells*, else the main effects are not estimable under Ω. Then $p = IJ$.

Suppose then that weights $\{v_i\}$ and $\{w_j\}$ are chosen subject to the condition that $\Sigma v_i = \Sigma w_j = 1$, and define the means for the ith level of A and the jth level of B as

$$A_i = \sum_j w_j \eta_{ij}, \qquad B_j = \sum_i v_i \eta_{ij}, \tag{4.4.16}$$

so that the main effects are

$$\alpha_i = A_i - \mu, \qquad \beta_j = B_j - \mu,$$

where

$$\mu = \sum_i \sum_j v_i w_j \eta_{ij}.$$

Then a contrast in the $\{\alpha_i\}$ has the same value as the corresponding contrast in the A_i, for, if $\psi = \Sigma_i c_i \alpha_i$ $(\Sigma c_i = 0)$, then $\psi = \Sigma_i c_i A_i$. From (4.4.16) we see that the LS estimate of A_i is

$$\hat{A}_i = \sum_j w_j \hat{\eta}_{ij} = \sum_j w_j y_{ij\cdot}. \tag{4.4.17}$$

Hence the LS estimate of ψ is $\hat{\psi} = \Sigma_i c_i \hat{A}_i$; similarly for contrasts in the $\{\beta_j\}$.

We shall now calculate SS_A, the numerator SS of the F-test of H_A by exploiting an interpretation of the F-test given in sec. 3.5, namely that

$$SS_A = \hat{\psi}^2_{\max},$$

where $\hat{\psi}_{\max}$, the maximum estimated normalized contrast in the main effects $\{\alpha_i\}$ of A is defined as follows: Let L be the space of all contrasts in the $\{\alpha_i\}$, and L'' the set of all normalized contrasts in L, i.e., the set of all ψ in L for which $\text{Var}(\hat{\psi}) = C\sigma^2$, and take $C = 1$. Then $\hat{\psi}_{\max}$ is the maximum $\hat{\psi}$ with ψ in L''.

If $\hat{\psi} = \Sigma_i c_i \hat{A}_i = \Sigma_i \Sigma_j c_i w_j y_{ij\cdot}$, then $\text{Var}(\hat{\psi}) = \Sigma_i \Sigma_j (c_i^2 w_j^2 / K_{ij})\sigma^2$; so the conditions for ψ to be in L'' are

$$\sum_i c_i = 0, \qquad \sum_i \sum_j (c_i^2 w_j^2 / K_{ij}) = 1. \tag{4.4.18}$$

The problem is to maximize $\Sigma_i c_i \hat{A}_i$ subject to (4.4.18); in this problem the $\{c_i\}$ are varied while the $\{\hat{A}_i\}$ are held fixed.

It is convenient to define numbers $\{W_i\}$ as

$$W_i = \left[\sum_j (w_j^2/K_{ij})\right]^{-1}. \tag{4.4.19}$$

Then the problem is to maximize $\Sigma_i c_i \hat{A}_i$ subject to

$$\sum_i c_i = 0, \qquad \sum_i (c_i^2/W_i) = 1. \tag{4.4.20}$$

This may be done by using Lagrange multipliers (see for example Courant (1950), Ch. III, sec. 6) λ_1 and λ_2, and equating to zero $\partial/\partial c_i$ of

$$\sum_i c_i \hat{A}_i + \lambda_1 \sum_i c_i + \lambda_2 \sum_i (c_i^2/W_i),$$

where $\{\lambda_i\}$ are treated as constants in taking $\partial/\partial c_i$, resulting in

$$\hat{A}_i + \lambda_1 + 2\lambda_2(c_i/W_i) = 0,$$
$$c_i = -W_i(\hat{A}_i + \lambda_1)/(2\lambda_2).$$

The condition $\Sigma c_i = 0$ gives

$$\sum_i W_i(\hat{A}_i + \lambda_1) = 0,$$

or $\lambda_1 = -\bar{A}$, where

$$\bar{A} = \sum_i W_i \hat{A}_i / \sum_i W_i.$$

We now have

$$c_i = -W_i(\hat{A}_i - \bar{A})/(2\lambda_2), \tag{4.4.21}$$

and substituting this into the other condition (4.4.20), we find

$$\sum_i W_i(\hat{A}_i - \bar{A})^2/(4\lambda_2^2) = 1,$$

$$4\lambda_2^2 = \sum_i W_i(\hat{A}_i - \bar{A})^2. \tag{4.4.22}$$

We finally get

$$\hat{\psi}_{\max}^2 = \left(\sum_i c_i \hat{A}_i\right)^2 = \left[\sum_i W_i(\hat{A}_i - \bar{A})^2\right]^2/(4\lambda_2^2) = \sum_i W_i(\hat{A}_i - \bar{A})^2,$$

with the aid of (4.4.21) and (4.4.22).

The numerator SS of the statistic is thus

$$SS_A = \sum_i W_i \hat{A}_i^2 - \left(\sum_i W_i\right)^{-1}\left(\sum_i W_i \hat{A}_i\right)^2,$$

where the $\{W_i\}$ are defined by (4.4.19) and the $\{\hat{A}_i\}$ by (4.4.17). In the case of equal weights, assumed elsewhere in this book, $w_j = 1/J$ in (4.4.19) and (4.4.17). The statistic MS_A/MS_e of the F-test of H_A under Ω has $I-1$ and $n-IJ$ d.f.

If H_A is rejected it is quite simple to follow up with the S-method, since, as we saw above, if $\psi = \Sigma_i c_i \alpha_i$ ($\Sigma c_i = 0$), then

$$\hat{\psi} = \sum_i c_i \hat{A}_i, \qquad \hat{\sigma}^2_{\hat{\psi}} = \sum_i \sum_j (c_i^2 w_j^2 / K_{ij}) s^2,$$

where $s^2 = \text{MS}_e$. The T-method is not applicable because the $\{\hat{A}_i\}$ have unequal variances.

The test of H_B and the S-method for contrasts among the $\{\beta_j\}$ are analogous.

Case of Proportional Frequencies

The normal equations (4.4.2–4) for the estimates $\{\hat{\mu}_\omega, \hat{\alpha}_{i,\omega}, \hat{\beta}_{j,\omega}\}$ under $\omega = \Omega \cap H_{AB}$ are easily solved in the particular *case of proportional frequencies*, defined to be the case where the cell numbers in any two rows (or columns) are proportional. It is easily shown that the numbers $\{K_{ij}\}$ in a two-way table have this property of proportionality if and only if they are related to the row totals $\{G_i\}$, column totals $\{H_j\}$, and the grand total n by the formula

$$K_{ij} = G_i H_j / n.$$

In this case

$$\sum_j K_{ij} \hat{\beta}_{j,\omega} = G_i \sum_j H_j \hat{\beta}_{j,\omega} / n,$$

and this will vanish for all i if we choose the side condition $\Sigma_j w_j \hat{\beta}_{j,\omega} = 0$ corresponding to weights $\{w_j = H_j/n\}$. Similarly

$$\sum_i K_{ij} \hat{\alpha}_{i,\omega} = H_j \sum_i G_i \hat{\alpha}_{i,\omega} / n$$

vanishes if we choose weights $\{v_i = G_i/n\}$. Then the normal equations (4.4.2–4) reduce to

$$\hat{\mu}_\omega = \sum_i \sum_j \sum_k y_{ijk} / n,$$
$$\hat{\alpha}_{i,\omega} = G_i^{-1} g_i - \hat{\mu}_\omega,$$
$$\hat{\beta}_{j,\omega} = H_j^{-1} h_j - \hat{\mu}_\omega.$$

It may be shown that in this case the three spaces of dimensions $I-1$, $J-1$, 1, spanned by the three sets of linear forms $\{\hat{\alpha}_{i,\omega}\}$, $\{\hat{\beta}_{j,\omega}\}$, $\{\hat{\mu}_\omega\}$ in the $\{y_{ijk}\}$ are mutually orthogonal, as in the case of equal cell numbers (which is a special case of proportional frequencies).

4.5. THE THREE-WAY LAYOUT

In this section we extend the concepts of main effects and interactions and the related statistical analysis to the case of three factors. Suppose then that there are three factors A, B, C affecting the observations, such as varieties, locations, and chemical treatments in the example of sec. 1.1, and we shall denote the number of levels for the factors by I, J, K, respectively. We let η_{ijk} denote the expected value of a measurement on the i,j,k "treatment combination," or in the i,j,k "cell," i.e., when A is at the ith level, B at the jth, C at the kth. We could carry over our two-factor concepts by considering a two-way table for each level of C; thus the interaction of the ith level of A and the jth level of B specific to the kth level of C would be

$$\eta_{ijk} - \eta_{i.k} - \eta_{.jk} + \eta_{..k}. \tag{4.5.1}$$

The average of these over the levels of C we shall call the interaction of the ith level of A with the jth level of B, and denote by

$$\alpha_{ij}^{AB} = \eta_{ij.} - \eta_{i..} - \eta_{.j.} + \eta_{...}, \tag{4.5.2}$$

the new symbolism, becoming more useful as the number of factors increases, showing in the superscripts the factors involved in the interaction or main effect, and in the subscripts the respective levels of these factors. The result is the same as though we applied our previous concepts to a two-way table obtained by averaging over the levels of C. From this two-way table, the main effects are then

$$\alpha_i^A = \eta_{i..} - \eta_{...},$$
$$\alpha_j^B = \eta_{.j.} - \eta_{...}.$$

If the main effect α_i^A were defined similarly from a two-way table of averages over the levels of B, the result would be the same; analogously for α_j^B. The corresponding definition of the main effect of the kth level of C is

$$\alpha_k^C = \eta_{..k} - \eta_{...}.$$

The general mean, defined from any of the three two-way tables of averages is

$$\mu = \eta_{...}.$$

The other two-factor interactions are of course defined similarly,

$$\alpha_{jk}^{BC} = \eta_{.jk} - \eta_{.j.} - \eta_{..k} + \eta_{...},$$
$$\alpha_{ik}^{AC} = \eta_{i.k} - \eta_{i..} - \eta_{..k} + \eta_{...}.$$

We can also define the mean for the ith level of A as $\eta_{i..} = \mu + \alpha_i^A$, etc. All these notions may then be regarded as familiar from the two-way layout.

We get something new when we consider how the $A \times B$ interactions (4.5.1) specific to the kth level of C differ for the different levels of C. These differential values of the $A \times B$ interaction may be expressed in terms of the specific value (4.5.1) minus the average value (4.5.2): There will be no dependence on the level of C if and only if all these differences are zero. The difference

$$\alpha_{ijk}^{ABC} = \eta_{ijk}-\eta_{ij.}-\eta_{i.k}-\eta_{.jk}+\eta_{i..}+\eta_{.j.}+\eta_{..k}-\eta_{...} \tag{4.5.3}$$

is called the *three-factor interaction* (or second-order interaction) between the ith level of A, the jth of B, and the kth of C. It is seen to be symmetrical in all three factors, that is, if we started with the differential effects for different levels of B of the $A \times C$ interactions, or the differential effects for different levels of A of the $B \times C$ interactions, we would reach the same result (4.5.3).

With these definitions we get

$$\eta_{ijk} = \mu+\alpha_i^A+\alpha_j^B+\alpha_k^C+\alpha_{ij}^{AB}+\alpha_{jk}^{BC}+\alpha_{ik}^{AC}+\alpha_{ijk}^{ABC},$$

where

$$\alpha_{.}^{A} = \alpha_{.}^{B} = \alpha_{.}^{C} = 0$$

$$\alpha_{i.}^{AB} = \alpha_{.j}^{AB} = \alpha_{j.}^{BC} = \alpha_{.k}^{BC} = \alpha_{i.}^{AC} = \alpha_{.k}^{AC} = 0,$$

$$\alpha_{ij.}^{ABC} = \alpha_{i.k}^{ABC} = \alpha_{.jk}^{ABC} = 0,$$

these conditions holding for all values of the subscripts i, j, k. *Additivity* is said to exist for the three factors if there exist constants a, $\{b_i\}$, $\{c_j\}$, $\{d_k\}$ such that $\eta_{ijk} = a+b_i+c_j+d_k$ for all i, j, k; it is easily shown that additivity exists if and only if all two-factor and all three-factor interactions are zero.

The development thus far could easily be extended to obtain definitions of *weighted* main effects, interactions, and general mean, with weights $\{u_i\}$, $\{v_j\}$, $\{w_k\}$, generalizing sec. 4.1. We also mention here that in "2^p" experiments, i.e., experiments with p factors each at two levels, it is conventional to define the main effects and interactions so that each has twice the value assigned by our definitions.

We shall treat the case where there are M ($M \geq 1$) observations on each of the treatment combinations (we sometimes say that there are M *replications*), denoting the mth observation on the i,j,k treatment combination by y_{ijkm}, and we shall assume that

$$\Omega: \begin{cases} y_{ijkm} = \eta_{ijk} + e_{ijkm}, \\ \{e_{ijkm}\} \text{ are independently } N(0, \sigma^2). \end{cases}$$

The LS estimates of the $\{\eta_{ijk}\}$ are found, by minimizing

$$\sum_i \sum_j \sum_k \sum_m (y_{ijkm} - \eta_{ijk})^2,$$

to be

$$\hat{\eta}_{ijk} = y_{ijk.}, \tag{4.5.4}$$

and it then follows from the Gauss–Markoff theorem that

$$(4.5.5)\quad \begin{aligned} \hat{\mu} &= y_{\cdot\cdot\cdot\cdot}, \\ \hat{\alpha}_i^A &= y_{i\cdot\cdot\cdot} - y_{\cdot\cdot\cdot\cdot}, \\ \hat{\alpha}_{ij}^{AB} &= y_{ij\cdot\cdot} - y_{i\cdot\cdot\cdot} - y_{\cdot j\cdot\cdot} + y_{\cdot\cdot\cdot\cdot}, \\ \hat{\alpha}_{ijk}^{ABC} &= y_{ijk\cdot} - y_{ij\cdot\cdot} - y_{i\cdot j\cdot} - y_{\cdot jk\cdot} + y_{i\cdot\cdot\cdot} + y_{\cdot j\cdot\cdot} + y_{\cdot\cdot k\cdot} - y_{\cdot\cdot\cdot\cdot}, \end{aligned}$$

etc., for the other main effects and interactions. The LS estimates under the various hypotheses usually tested may be found by the device used in (4.3.5), which makes the minimizing values obvious by inspection; for example, if the hypothesis concerns only the $B \times C$ interactions $\{\alpha_{jk}^{BC}\}$, the LS estimates of all the other interactions, of μ, and of the main effects are the same under the hypothesis as under Ω. This would not work with unequal cell numbers $\{M_{ijk}\}$. In that case the estimate (4.5.4) would still be correct, but the formulas (4.5.5) should be written with the dot notation applied to $\{\hat{\eta}_{ijk}\}$ instead of $\{y_{ijk\cdot}\}$, because these estimates are unweighted averages of the cell means $\hat{\eta}_{ijk}$, not averages weighted according to $\{M_{ijk}\}$, as the notation in (4.5.5) might suggest.

Orthogonality Relations

In the case of equal cell numbers we get the orthogonality relations indicated in Table 4.5.1 (these relations being, as always, independent of the normality assumption). The $IJKM$-dimensional space of linear forms in the observations $\{y_{ijkm}\}$ is decomposed into nine mutually orthogonal

TABLE 4.5.1

NINE ORTHOGONAL SPACES OF LINEAR FORMS

Space	Spanned by	Dimension
$\mathscr{L}_\mu$	$\hat{\mu}$	1
$\mathscr{L}_A$	$\hat{\alpha}_1^A, \cdots, \hat{\alpha}_I^A$	$I-1$
$\mathscr{L}_B$	$\hat{\alpha}_1^B, \cdots, \hat{\alpha}_J^B$	$J-1$
$\mathscr{L}_C$	$\hat{\alpha}_1^C, \cdots, \hat{\alpha}_K^C$	$K-1$
$\mathscr{L}_{AB}$	$\hat{\alpha}_{11}^{AB}, \cdots, \hat{\alpha}_{IJ}^{AB}$	$(I-1)(J-1)$
$\mathscr{L}_{BC}$	$\hat{\alpha}_{11}^{BC}, \cdots, \hat{\alpha}_{JK}^{BC}$	$(J-1)(K-1)$
$\mathscr{L}_{AC}$	$\hat{\alpha}_{11}^{AC}, \cdots, \hat{\alpha}_{IK}^{AC}$	$(I-1)(K-1)$
$\mathscr{L}_{ABC}$	$\hat{\alpha}_{111}^{ABC}, \cdots, \hat{\alpha}_{IJK}^{ABC}$	$(I-1)(J-1)(K-1)$
$\mathscr{L}_e$	$\{y_{ijkm} - y_{ijk\cdot}\}$	$IJK(M-1)$

spaces. If $M = 1$, the space $\mathscr{L}_e$ is empty. It is customary in this case to assume certain interactions zero in order to provide some d.f. for error;

this of course does not affect the orthogonality of the linear forms in the table.

Analysis-of-Variance Table

This is Table 4.5.2. To simplify the E(MS) column we have introduced the notations

$$\sigma_A^2 = (I-1)^{-1}\sum_i (\alpha_i^A)^2$$

$$\sigma_{AB}^2 = (I-1)^{-1}(J-1)^{-1}\sum_i\sum_j (\alpha_{ij}^{AB})^2,$$

$$\sigma_{ABC}^2 = (I-1)^{-1}(J-1)^{-1}(K-1)^{-1}\sum_i\sum_j\sum_k (\alpha_{ijk}^{ABC})^2,$$

with analogous definitions of σ_B^2, σ_C^2, σ_{BC}^2, σ_{AC}^2. Tables 4.5.1 and 4.5.2 are derived by now familiar methods. Under Ω the eight SS's (excluding the total SS) in Table 4.5.2 are independently distributed as σ^2 times noncentral chi-square with the indicated numbers of d.f. The noncentrality parameters are obtained by deleting the carets from the expressions in the second column, dividing by σ^2, and taking the square root, except for SS_e, for which $\delta = 0$.

TABLE 4.5.2

ANALYSIS OF VARIANCE OF THE THREE-WAY LAYOUT WITH M OBSERVATIONS PER CELL

Source	SS	d.f.	E(MS)
A main effects	$SS_A = JKM\sum_i(\hat{\alpha}_i^A)^2$	$I-1$	$\sigma^2+JKM\sigma_A^2$
B main effects	$SS_B = IKM\sum_j(\hat{\alpha}_j^B)^2$	$J-1$	$\sigma^2+IKM\sigma_B^2$
C main effects	$SS_C = IJM\sum_k(\hat{\alpha}_k^C)^2$	$K-1$	$\sigma^2+IJM\sigma_C^2$
AB interactions	$SS_{AB} = KM\sum_i\sum_j(\hat{\alpha}_{ij}^{AB})^2$	$(I-1)(J-1)$	$\sigma^2+KM\sigma_{AB}^2$
BC interactions	$SS_{BC} = IM\sum_j\sum_k(\hat{\alpha}_{jk}^{BC})^2$	$(J-1)(K-1)$	$\sigma^2+IM\sigma_{BC}^2$
AC interactions	$SS_{AC} = JM\sum_i\sum_k(\hat{\alpha}_{ik}^{AC})^2$	$(I-1)(K-1)$	$\sigma^2+JM\sigma_{AC}^2$
ABC interactions	$SS_{ABC} = M\sum_i\sum_j\sum_k(\hat{\alpha}_{ijk}^{ABC})^2$	$(I-1)(J-1)(K-1)$	$\sigma^2+M\sigma_{ABC}^2$
Error	$SS_e = \sum_i\sum_j\sum_k\sum_m(y_{ijkm}-y_{ijk.})^2$	$IJK(M-1)$	σ^2
Total about grand mean	$\sum_i\sum_j\sum_k\sum_m(y_{ijkm}-y_{....})^2$	$IJKM-1$	——

In the present fixed-effects model all the mean squares are tested against the MS_e. If $M = 1$, it is necessary in this model to add to the Ω assumptions the assumption that certain interactions are zero. If the $A \times B \times C$ interactions are assumed zero the SS_{ABC} of the above table becomes the SS_e; if we also assume the $B \times C$ interactions zero then $SS_{BC} + SS_{ABC}$ becomes the SS_e, etc., as explained further in sec. 4.6. If the SS_A is significant then one can follow up with the S-method for the contrasts in the main effects of A, etc.

Calculations

The SS's for main effects are computed directly, as usual. The SS's for two-factor interactions are best obtained with the help of three auxiliary two-way tables; thus for SS_{AB} we make a table of $\{y_{ij\cdot\cdot}\}$ and calculate from the identity

$$SS_{AB} = \left(KM\sum_i\sum_j y_{ij\cdot\cdot}^2 - IJKMy_{\cdot\cdot\cdot\cdot}^2\right) - SS_A - SS_B.$$

The SS for three-factor interactions is obtained as the difference between a SS for "cell means about the grand mean,"

$$M\sum_i\sum_j\sum_k y_{ijk\cdot}^2 - IJKMy_{\cdot\cdot\cdot\cdot}^2,$$

and the total of all the SS's on the lines above SS_{ABC} in Table 4.5.2. The error SS is finally obtained by subtracting, as usual, the SS for "cell means about the grand mean" from the total SS about the grand mean.

4.6. FORMAL ANALYSIS OF VARIANCE. PARTITION OF THE TOTAL SUM OF SQUARES

The results of sec. 4.5 can be formally generalized without difficulty. In the complete p-way layout there are a general mean, C_1^p sets of main effects, C_2^p sets of two-factor interactions, $\cdots$, 1 set of p-factor interactions,[10] where C_q^p denotes the binomial coefficient $p![q!(p-q)!]^{-1}$. If there is one observation per cell the total SS will be partitioned into 2^p component SS's associated with these effects, the SS for the p-factor interactions usually being assigned the role of the error SS; if there are M observations per cell there will be 2^p+1 component SS's, including a bona

[10] By successively taking differences as we did in obtaining (4.5.3) we may show that a q-factor interaction is the sum of 2^q terms: The first term is the true cell mean $\eta_{i,j,\text{etc.}}$ with subscripts (if any) not corresponding to the q factors replaced by dots, next there are C_1^q terms with minus signs, obtained by replacing the q subscripts by dots one a time, next C_2^q terms with plus signs, obtained by replacing the q subscripts by dots two at a time, next C_3^q terms with minus signs, etc.

fide error SS from within cells. Under the usual assumptions, including normality, statistical independence, and equality of the error variances σ^2, these SS's divided by σ^2 have independent noncentral chi-square distributions.

We illustrate the process on the three-way layout with more than one observation per cell already treated in sec. 4.5. In addition to the three factors A, B, and C, we introduce a fictitious "factor" D corresponding to m-fold replication of the observations in the cells of the three-way layout. (Actually, the true values of the main effects of D and of all interactions involving D are assumed to be zero.) We resolve the $IJKM$-dimensional space $\mathscr{L}$ of linear forms in the observations $\{y_{ijkm}\}$ into the direct sum of the 2^4 mutually orthogonal spaces indicated in Table 4.6.1. To each of

TABLE 4.6.1

SIXTEEN ORTHOGONAL SPACES OF LINEAR FORMS

Space	Spanned by	Dimension
$\mathscr{L}_\mu$	$y_{\cdots\cdot}$	1
$\mathscr{L}_A$	$y_{i\cdots}-y_{\cdots\cdot}$	$I-1$
$\mathscr{L}_B$	$y_{\cdot j\cdot\cdot}-y_{\cdots\cdot}$	$J-1$
$\mathscr{L}_C$	$y_{\cdot\cdot k\cdot}-y_{\cdots\cdot}$	$K-1$
$\mathscr{L}_D$	$y_{\cdot\cdot\cdot m}-y_{\cdots\cdot}$	$M-1$
$\mathscr{L}_{AB}$	$y_{ij\cdot\cdot}-y_{i\cdots}-y_{\cdot j\cdot\cdot}+y_{\cdots\cdot}$	$(I-1)(J-1)$
$\mathscr{L}_{AC}$	$y_{i\cdot k\cdot}-y_{i\cdots}-y_{\cdot\cdot k\cdot}+y_{\cdots\cdot}$	$(I-1)(K-1)$
$\mathscr{L}_{AD}$	$y_{i\cdot\cdot m}-y_{i\cdots}-y_{\cdot\cdot\cdot m}+y_{\cdots\cdot}$	$(I-1)(M-1)$
$\mathscr{L}_{BC}$	$y_{\cdot jk\cdot}-y_{\cdot j\cdot\cdot}-y_{\cdot\cdot k\cdot}+y_{\cdots\cdot}$	$(J-1)(K-1)$
$\mathscr{L}_{BD}$	$y_{\cdot j\cdot m}-y_{\cdot j\cdot\cdot}-y_{\cdot\cdot\cdot m}+y_{\cdots\cdot}$	$(J-1)(M-1)$
$\mathscr{L}_{CD}$	$y_{\cdot\cdot km}-y_{\cdot\cdot k\cdot}-y_{\cdot\cdot\cdot m}+y_{\cdots\cdot}$	$(K-1)(M-1)$
$\mathscr{L}_{ABC}$	$y_{ijk\cdot}-y_{ij\cdot\cdot}-y_{i\cdot k\cdot}-y_{\cdot jk\cdot}+y_{i\cdots}+y_{\cdot j\cdot\cdot}$ $+y_{\cdot\cdot k\cdot}-y_{\cdots\cdot}$	$(I-1)(J-1)(K-1)$
$\mathscr{L}_{ABD}$	$y_{ij\cdot m}-y_{ij\cdot\cdot}-y_{i\cdot\cdot m}-y_{\cdot j\cdot m}+y_{i\cdots}$ $+y_{\cdot j\cdot\cdot}+y_{\cdot\cdot\cdot m}-y_{\cdots\cdot}$	$(I-1)(J-1)(M-1)$
$\mathscr{L}_{ACD}$	$y_{i\cdot km}-y_{i\cdot k\cdot}-y_{i\cdot\cdot m}-y_{\cdot\cdot km}+y_{i\cdots}$ $+y_{\cdot\cdot k\cdot}+y_{\cdot\cdot\cdot m}-y_{\cdots\cdot}$	$(I-1)(K-1)(M-1)$
$\mathscr{L}_{BCD}$	$y_{\cdot jkm}-y_{\cdot jk\cdot}-y_{\cdot j\cdot m}-y_{\cdot\cdot km}+y_{\cdot j\cdot\cdot}$ $+y_{\cdot\cdot k\cdot}+y_{\cdot\cdot\cdot m}-y_{\cdots\cdot}$	$(J-1)(K-1)(M-1)$
$\mathscr{L}_{ABCD}$	$y_{ijkm}-y_{ijk\cdot}-y_{ij\cdot m}-y_{i\cdot km}-y_{\cdot jkm}+y_{ij\cdot\cdot}$ $+y_{i\cdot k\cdot}+y_{\cdot jk\cdot}+y_{i\cdot\cdot m}+y_{\cdot j\cdot m}+y_{\cdot\cdot km}$ $-y_{i\cdots}-y_{\cdot j\cdot\cdot}-y_{\cdot\cdot k\cdot}-y_{\cdot\cdot\cdot m}+y_{\cdots\cdot}$	$(I-1)(J-1)(K-1)(M-1)$

these 16 spaces there corresponds a sum of squares, namely the sum on i, j, k, m of the squares of the linear forms in the second column of the table, and

$$SS_{tot} = SS_\mu + SS_A + \cdots + SS_{BCD} + SS_{ABCD}.$$

If the Ω-assumptions are those of sec. 4.5 for the three-way layout with M observations per cell then the error SS is

$$SS_e = SS_D + SS_{AD} + SS_{BD} + SS_{CD} + SS_{ABD} + SS_{ACD} + SS_{BCD} + SS_{ABCD}.$$

If to Ω we added the assumption that all the three-factor interactions $\{\alpha_{ijk}^{ABC}\}$ were zero, we would "pool" SS_{ABC} into SS_e; if we also assumed all $\{\alpha_{ik}^{AC}\}$ zero, we would "pool" SS_{AC} into SS_e; etc.

Pooling Interaction SS's into the Error SS

A common but questionable practice is to pool into the error SS all interaction SS's found to be not significant at some chosen level.[11] This would be justified, we saw above, if the corresponding interactions were known to be absent—but in that case there would be no point in testing for their presence! The statistician is tempted to do this in order to get more "d.f. for error," in hopes of narrowing his confidence intervals and increasing the sensitivity of the tests where he substitutes the pooled error SS in the denominator of the test statistic. On the other hand, if the relevant interactions are wrongly assumed to be zero they tend to swell the expected value of the pooled mean square, and so there is a contrary effect of widening the confidence intervals and decreasing the sensitivity of the tests. Not very much is known[12] about the operating characteristics of these procedures, and it seems best to try to avoid such pooling by designing the experiment so that there will be a sufficient number of "d.f. for error" in the error SS selected in the design.

This problem should not be confused with that of choosing the proper denominator SS for the F-test in models other than the fixed-effects model, where we shall see later, for example, that the appropriate denominator SS for testing main effects may be not the error SS but a certain interaction SS. Pooling can also be considered in these other models, but then the

[11] Some statisticians use the rule of pooling an interaction SS into the error SS if and only if the statistic $\mathcal{F}$ for testing the interactions is not significant at the 5 per cent level and $\mathcal{F} < 2$, for example, Bennett and Franklin (1954); see also Paull (1950), where a similar rule is formulated for a different model.

[12] The problem has been studied by Bechhofer (1951). Results for other models have been obtained by Bozivich, Bancroft, and Hartley (1956); the effects of pooling in the situation treated by these authors are very different from those indicated above because pooling tends to shrink rather than swell the denominator MS of the test statistic.

pooling may tend in some situations to diminish (and in others, to swell) the denominator mean square, since the E(MS) for the SS pooled into the denominator may be always $\leqq$ the E(MS) for the "correct" denominator SS, unlike the above situation where the "correct" denominator SS is always the error SS, and this, of course, has the smallest E(MS) of any SS.

4.7. PARTITIONING A SUM OF SQUARES MORE GENERALLY

We have considered various ways of partitioning a sum of squares into other sums of squares, such that the component sums have independent noncentral chi-square distributions. Those cases might all be regarded as special cases of a more general method generated by repeated applications of Theorem 1 below.[13]

We consider the usual underlying hypothesis

$$\Omega: \quad \mathbf{y}^{n\times 1} \text{ is } N(\mathbf{X}'\boldsymbol{\beta}^{p\times 1}, \sigma^2\mathbf{I}^{n\times n}).$$

We also consider t linearly independent linear functions $\{g_i\}$ of the observations $\{y_1, \cdots, y_n\}$,

$$g_i = \sum_{j=1}^{n} a_{ij}y_j \qquad (i = 1, \cdots, t).$$

They span a t-dimensional space $\mathscr{L}$ of linear functions. In matrix notation, we write

$$\mathbf{g}^{t\times 1} = \mathbf{A}^{t\times n}\mathbf{y}^{n\times 1},$$

where the rows of $\mathbf{A}$ are the transposes of the coefficient vectors of the linear functions $\{g_i\}$.

Definition: *The sum of squares* $\mathrm{SS}_{\mathscr{L}}$ *associated with the space* $\mathscr{L}$ of linear functions is defined to be the quadratic form $\mathbf{g}'\mathbf{B}^{-1}\mathbf{g}$, where the matrix $\mathbf{B}^{t\times t}$ is given by the relation $\boldsymbol{\Sigma}_g = \sigma^2\mathbf{B}$, and $\boldsymbol{\Sigma}_g$ is the covariance matrix of the $\{g_i\}$.

In order to justify the definition, we note

(i) Since $g_1, \cdots, g_t$ are assumed to be linearly independent, rank $\mathbf{A} = t$. But $\boldsymbol{\Sigma}_g = \mathbf{A}\boldsymbol{\Sigma}_y\mathbf{A}' = \sigma^2\mathbf{A}\mathbf{A}'$, which shows that $\mathbf{B} = \mathbf{A}\mathbf{A}'$. From Theorem 7 of App. II, we deduce that rank $B = t$, i.e., $\mathbf{B}$ is nonsingular, and hence $\mathbf{B}^{-1}$ exists.

(ii) If $h_1, \cdots, h_t$ is another basis for $\mathscr{L}$, there exists a nonsingular matrix $\mathbf{C}^{t\times t}$ such that $\mathbf{h} = \mathbf{C}\mathbf{g}$. Then $\boldsymbol{\Sigma}_h = \mathbf{C}\boldsymbol{\Sigma}_g\mathbf{C}'$ and $\boldsymbol{\Sigma}_h^{-1} = \mathbf{C}'^{-1}\boldsymbol{\Sigma}_g^{-1}\mathbf{C}^{-1}$. Hence

$$\mathbf{h}'(\boldsymbol{\Sigma}_h/\sigma^2)^{-1}\mathbf{h} = \mathbf{g}'\mathbf{C}'\mathbf{C}'^{-1}(\boldsymbol{\Sigma}_g/\sigma^2)^{-1}\mathbf{C}^{-1}\mathbf{C}\mathbf{g} = \mathbf{g}'(\boldsymbol{\Sigma}_g/\sigma^2)^{-1}\mathbf{g}.$$

[13] The concepts of this section may be found in Bose (1949*a*).

This shows that $SS_{\mathscr{L}}$ does not depend on the basis chosen for the space $\mathscr{L}$.

The SS associated with $\mathscr{L}$ can also be expressed as a quadratic form in $\{y_1, \cdots, y_n\}$:

$$SS_{\mathscr{L}} = \mathbf{g}'(\mathbf{AA}')^{-1}\mathbf{g} = \mathbf{y}'\mathbf{Qy},$$

where the matrix $\mathbf{Q} = \mathbf{A}'(\mathbf{AA}')^{-1}\mathbf{A}$. In order to obtain the distribution of $SS_{\mathscr{L}}$, choose for a basis for $\mathscr{L}$, linear forms $\{z_1, \cdots, z_t\}$ such that Cov $(z_i, z_j) = \delta_{ij}\sigma^2$, as in the derivation of the canonical form in sec. 2.6. Since $SS_{\mathscr{L}}$ does not depend on the basis, we may express it relative to this basis where it takes the simple form

$$SS_{\mathscr{L}} = \mathbf{z}'(\mathbf{\Sigma}_z/\sigma^2)^{-1}\mathbf{z} = \sum_{i=1}^{t} z_i^2. \tag{4.7.1}$$

Hence, $SS_{\mathscr{L}}/\sigma^2$ has a noncentral chi-square distribution with t d.f. As usual, if δ is the noncentrality parameter, $\sigma^2\delta^2$ is obtained by substituting $E(g_i)$ for g_i in the definition of $SS_{\mathscr{L}}$.

Theorem 1: Consider a t-dimensional space $\mathscr{L}$ of linear functions of $\{y_1, \cdots, y_n\}$. Consider also m linear functions $f_i = \Sigma_{j=1}^n b_{ij}y_j$ $(i = 1, \cdots, m;\ m < t)$ belonging to $\mathscr{L}$. If the $\{f_i\}$ are orthogonal $(\Sigma_{k=1}^n b_{ik}b_{jk} = c_i\delta_{ij})$ and nonnull $(c_i \neq 0;\ i = 1, \cdots, m)$ then under Ω

$$\frac{f_1^2}{c_1}, \frac{f_2^2}{c_2}, \cdots, \frac{f_m^2}{c_m} \quad \text{and} \quad SS_{\mathscr{L}} - \sum_{i=1}^{m} \frac{f_i^2}{c_i}$$

are $m+1$ statistically independent quantities. After division by σ^2, they all have noncentral chi-square distributions with $1, 1, \cdots, 1$ and $t-m$ d.f., respectively. The noncentrality parameters are obtained by the usual rule (Rule 1 of sec. 2.6).

Proof: Let $z_i = c_i^{-1/2}f_i$, $i = 1, \cdots, m$. As in the derivation of the canonical form in sec. 2.6, adjoin to $\{z_1, \cdots, z_m\}$, $t-m$ linear functions $\{z_{m+1}, \cdots, z_t\}$ such that $\{z_1, \cdots, z_m, z_{m+1}, \cdots, z_t\}$ span $\mathscr{L}$ and are orthonormal [Cov $(z_i, z_j) = \sigma^2\delta_{ij}$]. Then from

$$SS_{\mathscr{L}} = \sum_1^t z_i^2, \quad \frac{f_1^2}{c_1} = z_1^2, \cdots, \quad \frac{f_m^2}{c_m} = z_m^2, \quad SS_{\mathscr{L}} - \sum_1^m \frac{f_i^2}{c_i} = \sum_{m+1}^t z_i^2,$$

the conclusion immediately follows.

Geometrical Interpretation

In order to obtain the geometrical interpretation of $SS_{\mathscr{L}}$ we establish the following

Theorem 2: For a t-dimensional space $\mathscr{L}$ of linear functions of $\{y_1, \cdots, y_n\}$, let V_t be the t-dimensional space of the coefficient vectors of the linear functions belonging to $\mathscr{L}$. Then $SS_{\mathscr{L}}$ is the squared length of the projection of the vector of observations $\mathbf{y}$ on the space V_t.

Proof: Choose orthonormal vectors $\{\boldsymbol{\alpha}_1^{n\times 1}, \cdots, \boldsymbol{\alpha}_t^{n\times 1}\}$ that span V_t. Define $z_i = \boldsymbol{\alpha}_i'\mathbf{y}$, so the linear forms $\{z_1, \cdots, z_t\}$ span $\mathscr{L}$. The squared length of the projection of $\mathbf{y}$ on V_t may be calculated as the sum of squares of the projections of $\mathbf{y}$ on the basis vectors $\{\boldsymbol{\alpha}_1, \cdots, \boldsymbol{\alpha}_t\}$, namely

$$\sum_1^t(\boldsymbol{\alpha}_i'\mathbf{y})^2 = \sum_1^t z_i^2 = \mathbf{z}'(\boldsymbol{\Sigma}_z/\sigma^2)^{-1}\mathbf{z} = SS_{\mathscr{L}}.$$

The geometrical interpretation of the partitioning theorem should now be pretty obvious.

Applications

For different kinds of examples where "single orthogonal d.f." are successively partitioned in a useful way out of a SS for treatment means we refer to the following:

(i) Fisher and Yates (1943), Table XXIII: Fitting of orthogonal polynomials when the means are the responses to equal steps in an independent variable (sec. 6.1. For a more detailed exposition see Anderson and Bancroft (1952), Ch. 16).

(ii) Loraine (1952): Problem of finding the break point in a case where a regression curve is known to follow a straight line up to a certain point and then curve away as the independent variable increases, again in the case of equal steps.

(iii) Cochran and Cox (1957), secs. 3.42–3.43. Various subdivisions of a treatment SS for different purposes.

In the next section we consider another example, in which we partition the usual error or residual SS in the two-way layout with one observation per cell.

4.8. INTERACTIONS IN THE TWO-WAY LAYOUT WITH ONE OBSERVATION PER CELL

In practice we are often concerned about the assumption of additivity for the two-way layout with one observation per cell (we made it in sec. 4.2). If y_{ij} is the observation in the i,j cell we consider first the general assumptions

$$\Omega: \begin{cases} y_{ij} = \mu + \alpha_i + \beta_j + \gamma_{ij} + e_{ij}, \\ \{e_{ij}\} \text{ are independently } N(0, \sigma^2), \\ \alpha_. = \beta_. = \gamma_{i.} = \gamma_{.j} = 0 \text{ for all } i,j. \end{cases}$$

The hypothesis we would like to test is

$$H: \quad \text{all } \gamma_{ij} = 0.$$

The usual F-test for interactions under Ω cannot be applied here, because there are no degrees of freedom left for an error SS. It is then natural to try to obtain a test of H under an Ω' which imposes some restrictions on the $\{\gamma_{ij}\}$. We shall motivate a test[14] of H by making the assumption that the $\{\gamma_{ij}\}$ are of the form

$$\gamma_{ij} = G\alpha_i\beta_j, \tag{4.8.1}$$

where G is a constant.

The restriction (4.8.1) may be made slightly more palatable by showing that if the interaction γ_{ij} for a cell is a function of the main effects α_i and β_j for that cell, and if this function is assumed to be a second-degree polynomial,

$$\gamma_{ij} = A + B\alpha_i + C\beta_j + D\alpha_i^2 + G\alpha_i\beta_j + H\beta_j^2, \tag{4.8.2}$$

then the polynomial is necessarily of the form (4.8.1). This is a consequence of the relations $\alpha_. = \beta_. = \gamma_{i.} = \gamma_{.j} = 0$: From (4.8.2) we calculate

$$\gamma_{i.} = A + B\alpha_i + D\alpha_i^2 + H\theta = 0, \qquad \text{where} \quad \theta = \sum_j \beta_j^2/J,$$

$$\gamma_{.j} = A + C\beta_j + D\phi + H\beta_j^2 = 0 \qquad \text{where} \quad \phi = \sum \alpha_i^2/I;$$

hence

$$B\alpha_i + D\alpha_i^2 = -A - H\theta,$$

$$C\beta_j + H\beta_j^2 = -A - D\phi.$$

Substitution in (4.8.2) gives

$$\gamma_{ij} = -A - H\theta - D\phi + G\alpha_i\beta_j. \tag{4.8.3}$$

But

$$\gamma_{i.} = -A - H\theta - D\phi = 0,$$

and so (4.8.3) reduces to (4.8.1).

The assumptions Ω' are then

$$\Omega': \quad \begin{cases} y_{ij} = \mu + \alpha_i + \beta_j + G\alpha_i\beta_j + e_{ij}, \\ \{e_{ij}\} \text{ are independently } N(0, \sigma^2), \\ \alpha_. = \beta_. = 0. \end{cases}$$

[14] The test is Tukey's (1949*a*), but the motivation is mine. The theorem and proof below are also due to Tukey.

We shall follow a heuristic approach up to (but not including) the theorem below. The derivation of LS estimates for the parameters μ, $\{\alpha_i\}$, $\{\beta_j\}$, G under Ω' would not be simple because the $\{E(y_{ij})\}$ are no longer linear functions of the parameters as assumed elsewhere in this book. But let us pretend for the moment that the $\{\alpha_i\}$ and $\{\beta_j\}$ are known. Under this fiction, we easily obtain a LS estimate $\tilde{G}$ of G, by minimizing

$$\mathscr{S} = \sum_i \sum_j (y_{ij} - \mu - \alpha_i - \beta_j - G\alpha_i\beta_j)^2.$$

The condition

$$\partial\mathscr{S}/\partial G = -2\sum_i \sum_j \alpha_i\beta_j(y_{ij} - \mu - \alpha_i - \beta_j - G\alpha_i\beta_j) = 0,$$

gives

$$\tilde{G} = \frac{\sum_i \sum_j \alpha_i\beta_j y_{ij}}{\sum_i \alpha_i^2 \sum_j \beta_j^2},$$

since

$$\sum_i \sum_j \alpha_i\beta_j = \sum_i \sum_j \alpha_i^2\beta_j = \sum_i \sum_j \alpha_i\beta_j^2 = 0.$$

We now consider a "SS for interactions" $\Sigma_i\Sigma_j\tilde{\gamma}_{ij}^2$, where $\tilde{\gamma}_{ij} = \tilde{G}\alpha_i\beta_j$,

$$\sum_i \sum_j \tilde{\gamma}_{ij}^2 = \tilde{G}^2 \sum_i \alpha_i^2 \sum_j \beta_j^2,$$

which can also be written

$$\sum_i \sum_j \tilde{\gamma}_{ij}^2 = \frac{\left[\sum_i \sum_j \alpha_i\beta_j y_{ij}\right]^2}{\sum_i \alpha_i^2 \sum_j \beta_j^2}.$$

Since we do not know the $\{\alpha_i\}$ and $\{\beta_j\}$ in this "SS for interactions" let us replace them by their estimates to obtain

$$SS_G = \frac{\left[\sum_i \sum_j \hat{\alpha}_i\hat{\beta}_j y_{ij}\right]^2}{\sum \hat{\alpha}_i^2 \sum \hat{\beta}_j^2},$$

where $\{\hat{\alpha}_i, \hat{\beta}_j\}$ are the LS estimates under Ω. We would like to use SS_G in a test of H, rejecting H for "large" SS_G. But can we find the distribution of SS_G? This is possible under $\omega = H \cap \Omega$. Even though SS_G is not a sum of squares of linear forms in the observations, but the quotient of a sixth-degree polynomial by a fourth-degree polynomial in the observations, it has under ω a chi-square distribution, as we shall now prove.

Theorem: Suppose that ω denotes the hypothesis

$$\omega: \begin{cases} y_{ij} = \mu + \alpha_i + \beta_j + e_{ij}, \\ \{e_{ij}\} \text{ are independently } N(0, \sigma^2), \\ \alpha_. = \beta_. = 0, \end{cases}$$

and we define

$$SS_G = \frac{\left[\sum_i \sum_j \hat{\alpha}_i \hat{\beta}_j y_{ij}\right]^2}{\sum_i \hat{\alpha}_i^2 \sum_j \hat{\beta}_j^2},$$

(4.8.4)
$$SS_{\text{int}} = \sum_i \sum_j (y_{ij} - y_{i.} - y_{.j} + y_{..})^2,$$

$$SS_{\text{res}} = SS_{\text{int}} - SS_G,$$

where

$$\hat{\alpha}_i = y_{i.} - y_{..}, \qquad \hat{\beta}_j = y_{.j} - y_{..}.$$

Then under ω, SS_G/σ^2 and SS_{res}/σ^2 are statistically independent and have chi-square distributions with one and $IJ-I-J$ d.f. respectively.

Proof: Let $\hat{\gamma}_{ij} = y_{ij} - y_{i.} - y_{.j} + y_{..}$. It is easy to verify that SS_G can be written

$$SS_G = \frac{\left[\sum_i \sum_j \hat{\alpha}_i \hat{\beta}_j \hat{\gamma}_{ij}\right]^2}{\sum_i \hat{\alpha}_i^2 \sum_j \hat{\beta}_j^2}.$$

Consider the $(I-1)(J-1)$-dimensional linear space $\mathscr{L}$ spanned by the linear forms $\{\hat{\gamma}_{ij}\}$. The SS associated with $\mathscr{L}$ is $SS_{\mathscr{L}} = SS_{\text{int}} = \Sigma_i \Sigma_j \hat{\gamma}_{ij}^2$, for this is the $\mathscr{L}_e$ of the subsection titled "orthogonality relations" in sec. 4.2, and the derivation of the orthogonality relations depends only on the covariance matrix assumed for the observations and not on their expectations. Consider also in $\mathscr{L}$ the linear form

(4.8.5)
$$f = \sum_i \sum_j a_i b_j \hat{\gamma}_{ij} = \sum_i \sum_j a_i b_j y_{ij},$$

where $\{a_i, b_j\}$ are constant coefficients subject to the conditions $\Sigma_i a_i = \Sigma_j b_j = 0$, $\Sigma_i a_i^2 > 0$, and $\Sigma_j b_j^2 > 0$. It follows from Theorem 1 of sec. 4.7 that

$$\frac{1}{\sigma^2} \frac{f^2}{\sum_i a_i^2 \sum_j b_j^2} \quad \text{and} \quad \frac{1}{\sigma^2}\left[SS_{\text{int}} - \frac{f^2}{\sum_i a_i^2 \sum_j b_j^2}\right]$$

are statistically independent and have chi-square distributions with one and $IJ-I-J$ d.f., respectively. Under ω we have $E(\hat{\gamma}_{ij}) = 0$, and hence both chi-square distributions are central.

We recall that the three sets of linear forms $\{\hat{\alpha}_i\}$, $\{\hat{\beta}_j\}$, and $\{\hat{\gamma}_{ij}\}$ span three mutually orthogonal spaces, and the three sets are thus statistically independent. Hence the conditional distribution of the $\{\hat{\gamma}_{ij}\}$, given the $\{\hat{\alpha}_i\}$ and $\{\hat{\beta}_j\}$, is identical with the unconditional distribution of the $\{\hat{\gamma}_{ij}\}$. Let us consider the $\{\hat{\alpha}_i\}$ and $\{\hat{\beta}_j\}$ fixed, and take $a_i = \hat{\alpha}_i$, $b_j = \hat{\beta}_j$ in (4.8.5). Since $\Sigma_i \hat{\alpha}_i^2 > 0$ and $\Sigma_j \hat{\beta}_j^2 > 0$ with probability one,[15] the joint conditional distribution under ω of SS_G/σ^2 and SS_{res}/σ^2, given the $\{\hat{\alpha}_i\}$ and $\{\hat{\beta}_j\}$, is that above of two statistically independent chi-square variables with one and $IJ-I-J$ d.f., respectively. Since this does not depend on the fixed values of the $\{\hat{\alpha}_i\}$ and $\{\hat{\beta}_j\}$, the unconditional distribution is the same as the conditional, and the theorem is proved.

A generalization of the theorem is given in Problem 4.19.

Application of the Theorem to Test Interactions

In analyzing a two-way layout with one observation per cell one may apply the preceding theorem to partition the usual error sum of squares into the two components SS_G and SS_{res}. The test for interactions, yielded by the theorem, consists in testing SS_G against SS_{res}. This is done with the statistic

$$(4.8.6) \qquad (IJ-I-J)SS_G/SS_{res},$$

which has under ω a central F-distribution with 1 and $IJ-I-J$ d.f.

The power of this test is unknown. The distribution of the test statistic under Ω or Ω' seems difficult to treat. Unfortunately we do not even know $E(SS_G)$ under Ω or Ω'. However, we may expect the test to be good against alternatives of the type (4.8.1).

Different tests for interactions in the two-way layout with one observation per cell are usually employed if one or both of the factors are quantitative (sec. 6.1) and the levels correspond to equal steps of a controlled variable: Suppose that factor A is qualitative and B quantitative. Then out of the interaction SS successive single "orthogonal d.f." may be partitioned for the interactions of A with the linear effect of B, with the quadratic effect of B, etc., as in Theorem 1 of sec. 4.7; see for example Davies (1956), secs. 8.3 and 8.4. If A is also quantitative, we may take single "orthogonal d.f." out of the interaction SS for linear $A \times$ linear B, linear $A \times$ quadratic B, quadratic $A \times$ linear B, etc.; see Problems 4.17 and 4.18. The residual SS after 3 d.f. are taken out in this way is then usually employed as the error SS.

The above test for interactions based on partitioning one d.f. out of the

[15] The random variables $\{\hat{\alpha}_2, \cdots, \hat{\alpha}_I, \hat{\beta}_2, \cdots, \hat{\beta}_J\}$ have a joint probability density and hence the probability that they are all zero is zero.

interaction SS in the two-way layout has been generalized[16] to other designs (see Problems 4.19 and 5.9).

Effect of Interactions on Inferences

The situation becomes complicated when the above test rejects H or if for other reasons we judge the interactions $\{\gamma_{ij}\}$ to be not negligible:[17] First, there is the usual difficulty about the practical interpretation of any conclusions which might be reached about the main effects in the presence of interactions. Second, there is the difficulty of reaching any conclusions about the main effects, because under Ω (or Ω') there is no unbiased estimate of σ^2 as is the case where some cells contain more than one observation. We shall consider now the effect of the $\{\gamma_{ij}\}$ on statistical inferences which would be valid if H were true.

The estimates of contrasts in the main effects which are the LS estimates under ω are easily seen to remain unbiased under Ω. (The only assumption needed for this is that $E(e_{ij}) = 0$ for all i,j.) For main effects for rows, if the contrast $\psi = \Sigma_i c_i \alpha_i$, where $\Sigma_i c_i = 0$, is estimated by $\hat{\psi} = \Sigma_i c_i \hat{y}_{i.}$, then under Ω, $E(\hat{\psi}) = \psi$, and $\sigma^2_{\hat{\psi}} = \sigma^2 \Sigma_i c_i^2 / J$. If we estimate σ^2 in the last equation by MS_{AB}, the quotient of (4.8.4) by $(I-1)(J-1)$, then this tends to overestimate the error in $\hat{\psi}$ since

$$E(\text{MS}_{AB}) = \sigma^2 + \sigma^2_{AB},$$

where

$$\sigma^2_{AB} = (I-1)^{-1}(J-1)^{-1} \sum_i \sum_j \gamma^2_{ij}.$$

This suggests that the intervals used to estimate the contrasts under ω are "too long" under Ω (by an unknown factor), and that if ψ is different from zero the probability of calling $\hat{\psi}$ not significantly different from zero is greater than that which would be calculated under ω.

We shall now treat the power under Ω of the usual test of H_A: all $\alpha_i = 0$, which consists in rejecting H_A if $\text{MS}_A/\text{MS}_{AB} > F_{\alpha;\nu_A,\nu_{AB}}$, where MS_A is the quotient of SS_A, the usual SS for row effects, by ν_A, and

$$\nu_A = I-1, \qquad \nu_{AB} = (I-1)(J-1).$$

Under Ω, SS_A/σ^2 and SS_{AB}/σ^2 have independent noncentral chi-square

[16] Tukey (1955).

[17] Tukey (1949*a*) suggests that if the test rejects H one should look for a transformation of the data which reduces (4.8.6) sufficiently so that it is not significant. He indicates a possible way of finding such a transformation by considering the contribution of each row to SS_G. The procedure seems to me hard to justify.

distributions with ν_A and ν_{AB} d.f., respectively, and noncentrality parameters δ_A and δ_{AB} determined by

$$\sigma^2\delta_A^2 = J\sum_i \alpha_i^2 = J\nu_A\sigma_A^2,$$

$$\sigma^2\delta_{AB}^2 = \sum_i\sum_j \gamma_{ij}^2 = \nu_{AB}\sigma_{AB}^2.$$

The test statistic MS_A/MS_{AB} thus has the distribution[18] of a doubly noncentral F-variable $F''_{\nu_A,\nu_{AB};\delta_A,\delta_{AB}}$ defined as the quotient of a $\nu_A^{-1}\chi'^2_{\nu_A,\delta_A}$ by an independent $\nu_{AB}^{-1}\chi'^2_{\nu_{AB},\delta_{AB}}$ (the noncentral chi-square variable $\chi'^2_{\nu,\delta}$ is defined in App. IV). The power of the test is thus

$$\text{(4.8.7)} \qquad \Pr\{F''_{\nu_A,\nu_{AB};\delta_A,\delta_{AB}} > F_{\alpha;\nu_A,\nu_{AB}}\}.$$

We may approximate (4.8.7) in terms of a central F-distribution, so that the approximation can be evaluated from Karl Pearson's *Tables of the Incomplete Beta Function* as explained in sec. 2.2. We first approximate $\chi'^2_{\nu,\delta}$ by $c\chi^2_{\tilde{\nu}}$ with c and $\tilde{\nu}$ determined by (IV.5) in App. IV. Substituting these approximations in the numerator and denominator of

$$\text{(4.8.8)} \qquad F''_{\nu_A,\nu_{AB};\delta_A,\delta_{AB}} = \frac{\nu_{AB}\chi'^2_{\nu_A,\delta_A}}{\nu_A\chi'^2_{\nu_{AB},\delta_{AB}}},$$

we find for it, after some simplification, the approximation

$$\frac{\sigma^2 + J\sigma_A^2}{\sigma^2 + \sigma_{AB}^2}\,\frac{\tilde{\nu}_{AB}\chi^2_{\tilde{\nu}_A}}{\tilde{\nu}_A\chi^2_{\tilde{\nu}_{AB}}},$$

where

$$\text{(4.8.9)} \qquad \tilde{\nu}_A = \nu_A\frac{(\sigma^2 + J\sigma_A^2)^2}{\sigma^2(\sigma^2 + 2J\sigma_A^2)}, \qquad \tilde{\nu}_{AB} = \nu_{AB}\frac{(\sigma^2 + \sigma_{AB}^2)^2}{\sigma^2(\sigma^2 + 2\sigma_{AB}^2)}.$$

Our central F approximation to (4.8.7) is thus

$$\text{(4.8.10)} \qquad \Pr\left\{F_{\tilde{\nu}_A,\tilde{\nu}_{AB}} > \frac{\sigma^2 + \sigma_{AB}^2}{\sigma^2 + J\sigma_A^2}F_{\alpha;\nu_A,\nu_{AB}}\right\},$$

where $\tilde{\nu}_A$ and $\tilde{\nu}_{AB}$ are given by (4.8.9).

Numerical calculation from (4.8.10) would show the power to be a rapidly decreasing function of σ_{AB}^2. That the power is a decreasing function of σ_{AB}^2 is already suggested by mere inspection of (4.8.10), since the right side of the inequality increases with σ_{AB}^2; however, this is not

[18] An expression for the probability density of the beta transform of F'' was given by Madow (1948) in the form of a doubly infinite series; its integration gives the cumulative distribution function of F'' as a doubly infinite series of incomplete-beta functions.

entirely conclusive since $\tilde{\nu}_{AB}$ on the left side also depends on σ^2_{AB}. A direct proof of this relation not based on approximations but on (4.8.8) is given in the appendix that follows this section. In particular, when H_A is true the probability of rejecting it, i.e., the probability of a type-*I* error, is a decreasing function of σ^2_{AB}, and hence the actual probability of rejecting H_A when $\sigma^2_{AB} > 0$ is always less than the nominal probability α, correct for $\sigma^2_{AB} = 0$. Since the validity of the *S*-method of multiple comparison depends only on the probability of a type-*I* error in the associated test of a hypothesis, it follows that if the *S*-method is used on the main effects $\{\alpha_i\}$ with over-all confidence coefficient $1-\alpha$, correct for $\sigma^2_{AB} = 0$, the actual probability of all statements being correct exceeds $1-\alpha$ for $\sigma^2_{AB} > 0$, and, more generally, is an increasing function of σ^2_{AB}.

Analogous results of course hold for inferences about the main effects for columns.

Appendix, Proving[19] that the Power of the Test of H_A is a Strictly Decreasing Function of σ^2_{AB}

If we substitute (4.8.8) in (4.8.7) and then apply (IV.1) of App. IV we find that the power is expressible as

$$(4.8.11) \qquad \Pr\left\{\frac{\nu_{AB}}{\nu_A}\,\frac{(x_1+\delta_A)^2+\left(\sum_2^{\nu_A} x_i^2\right)}{(y_1+\delta_{AB})^2+\left(\sum_2^{\nu_{AB}} y_i^2\right)} > F_{\alpha;\nu_A,\nu_{AB}}\right\},$$

where $\{x_1, \cdots, x_{\nu_A}, y_1, \cdots, y_{\nu_{AB}}\}$ are independently $N(0, 1)$. We now hold σ^2 and δ_A fixed. Since $\sigma^2_{AB} = \sigma^2\delta^2_{AB}/\nu_{AB}$, it suffices to show that (4.8.11) is a strictly decreasing function of $|\delta_{AB}|$. We may rewrite (4.8.11) as

$$(4.8.12) \qquad f(\delta_{AB}) = \Pr\{(y_1+\delta_{AB})^2 < z\},$$

where the random variable

$$z = \frac{\nu_{AB}\left[(x_1+\delta_A)^2+\sum_2^{\nu_A} x_i^2\right]}{\nu_A F_{\alpha;\nu_A,\nu_{AB}}} - \sum_2^{\nu_{AB}} y_i^2$$

is independent of y_1. If δ_1 and δ_2 denote any two values of δ_{AB} with $|\delta_1| < |\delta_2|$, we shall prove that for $f(\delta_{AB})$ defined by (4.8.12), $f(\delta_1) > f(\delta_2)$.

Now

$$f(\delta) = \int_0^\infty g_\delta(z)\,p(z)\,dz$$

where $p(z)$ is the probability density of z, and, for any positive number

[19] This proof was given to me by Professor W. H. Kruskal.

z', $g_\delta(z')$ denotes the conditional probability that $(y_1+\delta)^2 < z'$, given $z = z'$. However, this conditional probability must be the same as the unconditional probability, since y_1 and z are statistically independent. Thus $g_\delta(z')$ is the probability that the random variable y_1 fall in an interval of half length $\sqrt{z'}$ centered at $-\delta$, and since y_1 is $N(0, 1)$, this is a strictly decreasing[20] function of $|\delta|$. Thus $g_{\delta_1}(z') - g_{\delta_2}(z') > 0$ for all $z' > 0$. Hence

$$f(\delta_1) - f(\delta_2) = \int_0^\infty [g_{\delta_1}(z) - g_{\delta_2}(z)]p(z)\,dz > 0.$$

PROBLEMS

4.1. Show that for the two-way layout with factor A at two levels and B at J levels, and one observation per cell, the F-test of the hypothesis H_A is equivalent to a t-test based on the J differences $d_j = y_{1j} - y_{2j}$, and that hence in this case the Ω-assumptions can be considerably relaxed, so that we may take Ω: $y_{ij} = \mu + \alpha_i + \beta_j + e_{ij}$, $\alpha_. = 0$, $\beta_. = 0$, the J pairs (e_{1j}, e_{2j}) are independently bivariate normal with $E(e_{1j}) = E(e_{2j}) = 0$, $\text{Var}(e_{1j}) = \sigma_1^2$, $\text{Var}(e_{2j}) = \sigma_2^2$, $\text{Cov}(e_{1j}, e_{2j}) = \rho\sigma_1\sigma_2$.

4.2. The following data* show the effect of two soporific drugs on 10 patients: u_i is the increase in hours of sleep for the ith patient when using drug A, v_i the increase when using drug B. It is not assumed that the population variances

Patient	1	2	3	4	5	6	7	8	9	10
u_i	+0.7	−1.6	−0.2	−1.2	−1.0	+3.4	+3.7	+0.8	0.0	+2.0
v_i	+1.9	+0.8	+1.1	+0.1	−0.1	+4.4	+5.5	+1.6	+4.6	+3.4

σ_u^2 and σ_v^2 are equal. (*a*) Test at the 0.01 level of significance the hypothesis that there is no difference between the expected gain in sleep resulting from using one of the drugs rather than the other. [*Hint:* Consider differences $y_i = v_i - u_i$.] (*b*) State precisely the assumptions Ω and H underlying the analysis in (*a*). (*c*) Do the data provide conclusive evidence that the use of drug A produces an increase in sleep over the use of no drug? (*d*) Suppose we were confident that the variance σ_d^2 of $d = v - u$ is no more than 1.25, but would like to be 90 per cent sure of detecting a difference of 1-hour sleep increase between the two drugs. How many patients should we use in the experiment? (Adopt the 0.01 significance level.)

4.3. In order to investigate whether eight different fats are absorbed in different amounts by a doughnut mix during cooking, batches of 24 doughnuts were cooked on six different days in each of the eight fats. The results shown in Table A were obtained for grams of fat absorbed. (*a*) Make an analysis-of-variance table, including the column for $E(\text{MS})$. (*b*) Test at the 0.05 level of

[20] The proof that $f(\delta)$ defined by (4.8.12) is a strictly decreasing function of $|\delta|$ is valid for any independent random variables y_1 and z if y_1 has a density that is symmetric about $y_1 = 0$ and strictly decreasing for $y_1 > 0$, and $\Pr\{z > 0\} > 0$.

* From p. 20 of "The probable error of the mean" by Student, *Biometrika*, Vol. 6, 1908. Reproduced with the kind permission of the editor. The first example of Student's t-test in the literature was constructed from these data.

TABLE* A

Day	Fat No. 1	2	3	4	5	6	7	8
1	164	172	177	178	163	163	150	164
2	177	197	184	196	177	193	179	169
3	168	167	187	177	144	176	146	155
4	156	161	169	181	165	172	141	149
5	172	180	179	184	166	176	169	170
6	195	190	197	191	178	178	183	167

* Data from Example 18.3, p. 238, of *Statistical Theory in Research* by R. L. Anderson and T. A. Bancroft, McGraw-Hill, New York, 1952. Reproduced with the kind permission of the authors and the publisher.

significance the hypothesis that there is no difference among the (true) average amounts of fat absorbed when using the eight different fats. (*c*) State the assumptions Ω and H. (*d*) If the above hypothesis is rejected, does the S-method find any two fats that are absorbed in significantly different amounts? If not, find some contrast that is significantly different from zero. (*e*) Would the T-method find two fats that are absorbed in significantly different amounts?

4.4. (*a*) In the preceding problem, suppose that the eight means $\eta_i = \mu+\alpha_i$ fall into three groups: $\eta_5 = \eta_7 = \eta_8 = \eta$, say; $\eta_1 = \eta_2 = \eta_6 = \eta+12$; $\eta_3 = \eta_4 = \eta+22$. If σ^2 were equal to the mean square for error, what would then be the probability of rejecting the hypothesis of no difference among the eight means? (*b*) Since the fats 5, 7, and 8 appear to be the most economical, we plan to experiment further with them. How many observations should we take with each of them in order that the test at the 0.05 level of the hypothesis H: $\alpha_5 = \alpha_7 = \alpha_8$ should detect with probability $\geqq 0.8$ any difference $\geqq 10$ units between any two of the three fats, assuming that the value of σ^2 used in (*a*) will prove again to be acceptable?

4.5. Table B gives the yield of grain in grams per 16-foot row for each of eight varieties of oats in five replications in a randomized-block experiment.

TABLE* B

Variety	Block I	II	III	IV	V
1	296	357	340	331	348
2	402	390	431	340	320
3	437	334	426	320	296
4	303	319	310	260	242
5	469	405	442	487	394
6	345	342	358	300	308
7	324	339	357	352	220
8	488	374	401	338	320

* From Example 18.11, p. 245 of *Statistical Theory in Research* by R. L. Anderson and T. A. Bancroft, McGraw-Hill, New York, 1952. Reproduced with the kind permission of the authors and the publisher.

(*a*) Analyze as a two-way layout under the normal-theory model. Test at the 0.10 level for differences among the expected yields of the eight varieties. (*b*) Use the *T*-method to decide which pairs of varieties differ. Save your calculations for further use in Ch. 9.

4.6. A 24-hour cumulative sample of city gas is collected from a main and tested for calorific value each day from Monday through Saturday for nine weeks. The coded calorific values are shown in Table C. (*a*) The demand for gas

TABLE* C

Day	Week No. 1	2	3	4	5	6	7	8	9
Monday	5	1	−4	5	−13	−8	−2	−4	−10
Tuesday	3	6	−10	−2	−7	−2	−4	2	2
Wednesday	8	4	−14	−3	3	0	5	−11	−12
Thursday	8	10	−5	−1	4	−2	4	1	−12
Friday	4	−1	7	−5	5	−3	−7	−3	−6
Saturday	3	−9	3	−8	−6	0	−3	8	−1

* Data from Table IV, p. 20 of "Analysis of Variance" by B. A. Griffith, A. E. R. Westman, and B. H. Lloyd, *Industrial Quality Control*, Vol. 4, no. 6, May 1948. Reproduced with the kind permission of the authors and the editor.

depends to a considerable extent on the day of the week. Do the data show a significant day-to-day variation in calorific value? (Test at the 0.05 level.) (*b*) The calorific value may also vary from week to week with coal quality, etc. Test this. (The test can be invalidated by serial correlation between rows within columns—see sec. 10.5.) (*c*) How could we test (at the 0.05 level) the over-all hypothesis that there are neither day-to-day nor week-to-week variations? Does the conclusion of this test imply those arrived at in (*a*) and (*b*)? (*d*) Let y_{ij} be the coded observation on the *i*th day of the *j*th week. Assume that $E(y_{ij}) = \mu_i + \delta_j$. The engineer of the gas company desires to detect with probability 0.85 any pattern of the form $(\mu_1, \mu_2, \cdots, \mu_6) = (0, 4, 8, 12, 8, 4)$ or a permutation thereof. Assuming that σ^2 equals MS_e, are there enough weeks in the above experiment to satisfy this requirement?

4.7. Batches of ground meat from five different sources are charged consecutively into a rotary filling machine for packing into cans. The machine has six filling cylinders. Three filled cans are taken from each cylinder at random while each batch is being run. The coded weights of the filled cans are given in Table D. (*a*) Test at the 0.05 level of significance the hypothesis that the expected weight of filled can does not vary (*i*) from source to source and (*ii*) from cylinder to cylinder. (*b*) Test at the 0.05 level the hypothesis of no interactions. Save result for further use in Problem 4.11*c*. (*c*) Use the *T*-method to decide from which others source 3 is significantly different at the 0.95 level of confidence.

4.8. Table E gives the weight of hybrid female rats in a foster-nursing experiment with four types of rats. (The weights are litter averages in grams at 28 days. The within-litter variance was obviously negligible compared to the between-litter variance.) The factors in the two-way layout are the genotype of the foster mother and that of the litter. (*a*) Calculate the statistic $\mathscr{F}$ for testing interactions. (Further calculations are to be made without the assumption of

TABLE* D

Cylinder	Source 1	2	3	4	5
1	1, 1, 2	4, 3, 5	6, 3, 7	3, 1, 3	1, 3, 3
2	−1, 3, −1	−2, 1, 0	3, 1, 5	2, 0, 1	1, 0, 1
3	1, 1, 1	2, 0, 1	2, 4, 3	1, 3, 3	3, 3, 3
4	−2, 3, 0	−2, 0, 1	3, 3, 4	0, 0, 2	0, 1, 1
5	1, 1, −1	2, 1, 5	0, 1, 2	1, 0, −1	−2, 3, 1
6	0, 1, 1	0, 0, 3	3, 3, 4	3, 0, 2	3, 1, 2

* From Table IV, p. 21 of "Analysis of Variance" by B. A. Griffith, A. E. R. Westman, and B. H. Lloyd, *Industrial Quality Control*, Vol. 4, no. 6, May 1948. Reproduced with the kind permission of the authors and the editor. I have changed "batches" to "sources" since the former term suggests that the levels are those of a random-effects factor, and my mixed-model analysis in sec. 8.1 then requires that the number of levels be ≧ the number for the fixed-effects factor.

TABLE* E

Genotype of Litter	Genotype of Foster Mother *A*	*F*	*I*	*J*
A	61.5 68.2 64.0 65.0 59.7	55.0 42.0 60.2	52.5 61.8 49.5 52.7	42.0 54.0 61.0 48.2 39.6
F	60.3 51.7 49.3 48.0	50.8 64.7 61.7 64.0 62.0	56.5 59.0 47.2 53.0	51.3 40.5
I	37.0 36.3 68.0	56.3 69.8 67.0	39.7 46.0 61.3 55.3 55.7	50.0 43.8 54.5
J	59.0 57.4 54.0 47.0	59.5 52.8 56.0	45.2 57.0 61.4	44.8 51.5 53.0 42.0 54.0

* From Table B of the appendix to *The Inheritance of Maternal Influences on the Growth of the Rat* by D. W. Bailey, Ph.D. thesis, Univ. California, 1953. Reproduced with the kind permission of the author. I am grateful to Professor E. Dempster for giving me this example.

additivity.) (*b*) For each factor calculate the statistic $\mathscr{F}$ for testing main effects. (*c*) Let ψ denote the contrast which is the difference of the true main effects of the two types of foster mothers with the biggest difference of estimated main effects. Find a confidence interval for ψ by the *S*-method with $\alpha = 0.10$. Save your calculations for further use in Ch. 10.

4.9. In a 3^3 experiment (a q^p experiment involves p factors each at q levels) in a cannery on the drained weight of cherries (24 hours after canning) the factors studied were "fill" of the can, namely, the weight of raw cherries without syrup; "brix" of the syrup, a measure of the concentration of sugar in the syrup; and "condition" of the fruit, classified as *L*, *M*, *D*, i.e., light, medium, dark. In the Table F of drained weights the values for fill, brix, and drained weight are coded. Analyze the data, using the three-factor interaction MS for error.

TABLE F

	Brix								
	0			23			33		
Fill	Condition			Condition			Condition		
	L	*M*	*D*	*L*	*M*	*D*	*L*	*M*	*D*
0	55	95	169	55	69	163	49	88	153
1	200	232	223	183	215	207	148	200	245
2	233	285	291	236	259	278	233	223	259

4.10. An experiment to investigate the effects of certain factors on the insulation resulting from core-plate coatings on electrical steels was designed as a complete four-way layout with the following factors: four different coatings (levels of factor *A*) were used, two different curing temperatures of the coating (*B*), two different stress-relief annealing atmospheres (*C*), the panels of steel in the experiment being cut from four different coils of steel (*D*). The four different coils were selected by experts to represent when coated a range of insulation quality. Table G gives results of Franklin test (ASTM A-344-52)

TABLE* G

Level of *A*		1				2				3				4			
Level of *B*		1		2		1		2		1		2		1		2	
Level of *C*		1	2	1	2	1	2	1	2	1	2	1	2	1	2	1	2
Level of *D*	1	.25	.16	.30	.27	.41	.10	.13	.06	.44	.24	.22	.18	.43	.27	.26	.21
	2	.36	.02	.18	.03	.28	.04	.06	.03	.65	.08	.14	.36	.62	.03	.51	.03
	3	.36	.06	.44	.13	.33	.03	.19	.04	.42	.49	.17	.25	.47	.28	.21	.25
	4	.25	.10	.34	.04	.21	.01	.20	.01	.47	.14	.36	.19	.52	.07	.32	.38

* Provided by Mr. John D. Hromi and reproduced with the kind permission of the U.S. Steel Corporation.

measurements in amperes per square inch (at 500 psi pressure), so that smaller measurements indicate better insulation. (*a*) Make an analysis-of-variance table showing SS's and MS's for all main effects and interactions. (It is a good idea to look at the three- and -four-factor interaction MS's separately even if it is decided beforehand to pool the corresponding SS's for an error SS: a relatively large MS may suggest some interesting leads to follow.) (*b*) Using the pooled three- and four-factor interaction SS for error, which main effects and two-factor interactions do you judge to be important? (*c*) The four levels of factor A really constitute a 2×2 factorial experiment with two core-plate materials and two thicknesses of core plate: levels 1 and 2 of A are one material, 3 and 4 are the other, levels 1 and 3 are a lighter coating, and levels 2 and 4 are a heavier coating. Partition the SS for A into three "orthogonal d.f." for main effect of material, main effect of thickness, and interaction. What do you conclude from this? (In the actual experiment there was wisely included a fifth level of A corresponding to panels without any core-plate coating; we have omitted this because it caused an ABC interaction that would complicate the interpretation of this illustrative example.) (*d*) Considering further the four means (over levels of B, C, D) in the 2×2 experiment just discussed, we see that the effect of thickness is clearly negligible with the material of levels 3 and 4, but not obvious with the material of levels 1 and 2. To judge this, consider the estimated contrast which is the difference of the means for levels 1 and 2, and its estimated standard deviation. [*Hint:* The reader may verify whether he is proceeding correctly in the rather complicated calculations in (*a*) from the following partial results: The MS's for A, AB, ABC, $ABCD$ are respectively 0.1077, 0.0166, 0.0130, 0.0120.]

4.11. From the data of Problem 4.7 form a 6×5 table of cell sums; thus the entry in the first row and second column is $4+3+5=12$. (*a*) Pretend these are the observations in a two-way layout with one observation per cell, and apply the test of sec. 4.8 for interactions. (*b*) Repeat with the cubes of the cell entries in (*a*). (*c*) Explain why you would expect a small or large value of the F-statistic in each of (*a*) and (*b*).

4.12. If in an r-way layout with equal numbers in the cells one factor is present at only two levels, or if there are just two observations in each cell, some of the SS's can be calculated in an alternative way by using differences, for example: (*a*) In a two-way layout with the notation of sec. 4.3, suppose that $I=2$. Show

$$SS_A = \tfrac{1}{2}JK(y_{1..} - y_{2..})^2,$$

$$SS_{AB} = \tfrac{1}{2}K\sum_j (y_{1j.} - y_{2j.})^2 - SS_A.$$

(*b*) Suppose instead that $K=2$. Show that

$$SS_e = \tfrac{1}{2}\sum_i\sum_j (y_{ij1} - y_{ij2})^2.$$

4.13. Suppose that in a complete layout with equal numbers in the cells factor A is at I levels and the observed means for these levels are denoted by $\{\bar{y}_1, \cdots, \bar{y}_I\}$ (thus, in sec. 4.3, $\bar{y}_i = y_{i..}$), and suppose the I means are divided into two sets of I_1 and I_2 according to any method that does not depend on the outcome of the experiment, with $I_1+I_2=I$. Write the $\{\bar{y}_i\}$ in the first set as

$\{z_1, \cdots, z_{I_1}\}$, and those in the second set as $\{w_1, \cdots, w_{I_2}\}$. Show that the SS

$$\sum_{i=1}^{I} (\bar{y}_i - \bar{y}_.)^2,$$

which is the SS for main effects of A, except for a constant factor (JK in sec. 4.3), can be partitioned into SS's for each set about the mean of the set plus a SS between the means of the sets, namely,

$$\sum_{i=1}^{I_1} (z_i - z_.)^2 + \sum_{i=1}^{I_2} (w_i - w_.)^2 + I^{-1}I_1I_2(z_. - w_.)^2,$$

and argue that under the usual normal-theory assumptions these three SS's are statistically independent.

4.14. Obtain a result similar to that of Problem 4.13 for the case of three sets of means. [*Hint:* The generalization will be of the formula in Problem 4.13 obtained by leaving the last SS in the form $I_1(z_. - \bar{y}_.)^2 + I_2(w_. - \bar{y}_.)^2$.]

4.15. Verify the result of Problem 2.8 in the case of the hypothesis of zero row effects in the two-way layout with one observation per cell.

4.16. Table H lists the volumes in milliliters of loaves of bread made under controlled conditions from 100-gram batches of dough made with 17 different varieties of wheat flour and containing x milligrams of potassium bromate, for $x = 0, 1, 2, 3, 4$. (*a*) Analyze the data as a two-way layout. (*b*) Partition the SS for columns into single d.f. for linear, quadratic, cubic, and quartic

TABLE* H

	Loaf Volume for $x =$				
Variety	0	1	2	3	4
1	950	1075	1055	975	880
2	890	980	955	865	825
3	830	850	820	770	735
4	770	815	765	725	700
5	860	1040	1065	975	945
6	835	960	985	915	845
7	795	900	905	880	785
8	800	860	870	850	850
9	750	940	1000	960	960
10	885	1000	1015	960	895
11	895	935	965	950	920
12	685	835	870	875	880
13	615	665	650	680	660
14	885	910	890	835	785
15	985	1075	1070	1015	1005
16	710	750	740	725	720
17	785	845	865	825	820

* From p. 169 of *Methods of Statistical Analysis* by C. H. Goulden, John Wiley, New York, 1952, with the kind permission of the author and the publisher. Original data from Table II on p. 783 of "A comparison of hard red spring and hard red winter wheats" by R. K. Larmour, *Cereal Chemistry*, Vol. 18, 1941.

effects by using Table XXIII of Fisher and Yates (1943), and test the significance of each at the 0.05 level. (*c*) Fit a polynomial to the column means† as a function of x, of degree depending on your findings in (*b*). After calculating the equation of this polynomial, plot it and the column means on the same sheet of graph paper. [*Hint:* In calculating polynomial regression equations from points with equally spaced abscissas the explicit formulas for the first seven orthogonal polynomials given at the head of Table 47 of *Biometrika Tables for Statisticians* by Pearson and Hartley (1954) are useful; their Table 47 includes Fisher and Yates' Table XXIII.]

4.17. In a two-way layout where one factor is qualitative and one quantitative (sec. 6.1) the following results yield a useful partition of the interaction SS; see Davies (1956), sec 8.3, for an application. Suppose that $\{y_{ij}\}$ is a two-way layout of IJ variables. For a set of J variables $\{u_j\}$ let $\{L_r(u_1, \cdots, u_j) = \Sigma_j a_{rj} u_j\}$ be a set of R orthogonal linear functions ($R \leqq J$), and suppose that $a_{1j} = 1$, so that L_1 is the sum. If the R orthogonal functions $\{L_r\}$ are formed for each of the I rows of $\{y_{ij}\}$ we can make an $I \times R$ table of the resulting $\{L_{ir}\}$, where $L_{ir} = \Sigma_j a_{rj} y_{ij}$. Show that the $\{L_{ir}\}$ are orthogonal functions of the $\{y_{ij}\}$. In order to partition out of the interaction SS, SS's which are each a constant factor times $S_r = \Sigma_i (L_{ir} - L_{.r})^2$ for $r > 1$, we need to know that for $r > 1$ the linear form $L_{ir} - L_{.r}$ belongs to the "interaction space" $\mathscr{L}$ of forms spanned by $\{y_{ij} - y_{i.} - y_{.j} + y_{..}\}$: Prove this. [*Hint:* It is sufficient to show that $L_{ir} - L_{.r}$ is orthogonal to the $\{y_{i.}\}$ and the $\{y_{.j}\}$: Why?]

4.18. This problem is a continuation of Problem 4.17. If both factors are quantitative the following results are useful for partitioning the interaction SS; for an example see Davies (1956), sec. 8.4. Let $\{M_s\}$ denote a set of S orthogonal linear functions of I variables $\{v_i\}$, $M_s(v_1, \cdots, v_I) = \Sigma_i b_{si} v_i$, where $b_{1i} = 1$. By forming these orthogonal functions for each of the R rows of the above $I \times R$ table of $\{L_{ir}\}$, we get an $S \times R$ table of $\{M_{sr}\}$. Show that the same table is obtained if we operate on the $I \times J$ table of $\{y_{ij}\}$, first on columns with the $\{M_s\}$, and then on rows with the $\{L_r\}$. Prove that the $\{M_{sr}\}$ are orthogonal functions of the $\{y_{ij}\}$. Prove that for $s > 1$ and $r > 1$, M_{sr} belongs to the above "interaction space" $\mathscr{L}$.

4.19. Denote the assumption that $\mathbf{y}^{n \times 1}$ is $N(\mathbf{X}'_\omega \boldsymbol{\beta}, \sigma^2 \mathbf{I})$ by ω, $\mathbf{X}'_\omega \hat{\boldsymbol{\beta}}_\omega$ (where $\hat{\boldsymbol{\beta}}_\omega$ is any LS estimate of $\boldsymbol{\beta}$ under ω) by $\hat{\boldsymbol{\eta}}_\omega$, $\|\mathbf{y} - \hat{\boldsymbol{\eta}}_\omega\|^2$ by $\mathscr{S}_\omega$, and let $\mathbf{z}^{n \times 1} = f(\hat{\boldsymbol{\eta}}_\omega)$ be an arbitrary‡ function of $\hat{\boldsymbol{\eta}}_\omega$ (the function f is to be chosen before inspecting the outcome of the observation vector $\mathbf{y}$). Let $\hat{\boldsymbol{\zeta}}$ be the same linear function of $\mathbf{z}$ that $\hat{\boldsymbol{\eta}}_\omega$ is of $\mathbf{y}$; i.e., if $\hat{\boldsymbol{\beta}}_\omega = \mathbf{A}\mathbf{y}$, $\hat{\boldsymbol{\eta}}_\omega = \mathbf{X}'_\omega \mathbf{A}\mathbf{y} = \mathbf{B}\mathbf{y}$, say, then $\hat{\boldsymbol{\zeta}} = \mathbf{B}\mathbf{z}$. Suppose that rank $\mathbf{X}'_\omega = r_\omega$, so $\mathscr{S}_\omega/\sigma^2$ is $\chi^2_{\nu_\omega}$ with $\nu_\omega = n - r_\omega$ d.f. Define

$$SS_f = \|\mathbf{z} - \hat{\boldsymbol{\zeta}}\|^{-2} [\mathbf{z}'(\mathbf{y} - \hat{\boldsymbol{\eta}}_\omega)]^2.$$

Prove§ that, under ω, SS_f and $\mathscr{S}_\omega - SS_f$ are statistically independent, and, when

† A more detailed analysis can be made by calculating the linear, quadratic, cubic, and quartic contrasts separately for each variety and using the quartic × varieties interaction mean square for error; see for example Davies (1956), sec. 8.3; see also Problem 4.17.

‡ Borel-measurable, and such that $\|\mathbf{z} - \hat{\boldsymbol{\zeta}}\|$ defined below differs from zero with probability one.

§ The case where $\mathbf{z} = f(\hat{\boldsymbol{\eta}}_\omega)$ is defined by $z_i = a(\hat{\eta}_{\omega,i} - b)^2$, and a and b are constants, was stated by Tukey (1955).

divided by σ^2, have chi-square distributions with 1 and $\nu_\omega - 1$ d.f., respectively. We remark that the function $\|\mathbf{z}-\hat{\boldsymbol{\zeta}}\|^{-2}SS_f$ may be regarded as the square of the regression coefficient (Ch. 6) of the residual $\mathbf{y}-\hat{\boldsymbol{\eta}}_\omega$ under ω on $\mathbf{z}$. [*Hints:* $\mathbf{y}-\hat{\boldsymbol{\eta}}_\omega$ is independent of $\hat{\boldsymbol{\eta}}_\omega$ under ω. Consider the conditional distributions of SS_f and $\mathscr{S}_\omega - SS_f$, given $\hat{\boldsymbol{\eta}}_\omega$, so that conditionally $\mathbf{z}$ is a constant and $\mathbf{z}'(\mathbf{y}-\hat{\boldsymbol{\eta}}_\omega) = \mathbf{z}'(\mathbf{I}-\mathbf{B})\mathbf{y}$ is a linear form in $\mathbf{y}$, and proceed as in the proof of the theorem in sec. 4.8.]

4.20. Derive the theorem of sec. 4.8 by applying the result of Problem 4.19. [*Hint:* With double-subscript notation define $\mathbf{z} = f(\hat{\boldsymbol{\eta}}_\omega)$ by $z_{ij} = \hat{\eta}_{ij}^2$, where $\hat{\eta}_{ij} = \hat{\mu}+\hat{\alpha}_i+\hat{\beta}_j$.]

4.21. To see a geometrical interpretation of the existence of interactions between a qualitative factor and a quantitative factor, suppose factor A is qualitative and has I levels while the J levels of B correspond to the values $\{v_j\}$ of a continuous variable v. Let $\eta_i(v)$ denote the true response or regression function for the ith level of A, so that, in the notation introduced before the theorem near the end of sec. 4.1, $\eta_{ij} = \eta_i(v_j)$. Consider the graphs of the I functions $\{\eta_i(v)\}$ plotted against v, and show that there are no $A \times B$ interactions if and only if these graphs differ only by translations perpendicular to the v-axis.

4.22. Let $\eta = \eta(u, v)$ be the regression function introduced in sec. 4.1 in the case where both variables are quantitative, and consider the case where η is quadratic in u, v, so $\eta = A_0 + A_1 u + A_2 v + A_{11}u^2 + A_{22}v^2 + A_{12}uv$, where the A's are constants. (*a*) Prove that a transformation to additivity in u and v exists in the sense of sec. 4.1 if and only if $A_{12} = 0$, or $\eta = B(u-C)(v-D) + K$, or $\eta = (Eu+Fv+G)^2 + H$, where $E^2 + F^2 \neq 0$ and (u, v) is restricted to one of the half-planes where $Eu + Fv + G \geq 0$ or ≤ 0.

4.23. Table I gives the data from another experiment in the form of a complete four-way layout. The observations are the moisture content (in grams)

TABLE* I

Level of A	Level of B	Level of C 1, Level of D 1	Level of C 1, Level of D 2	Level of C 2, Level of D 1	Level of C 2, Level of D 2
1	1	8	5	8	4
	2	17	11	13	10
	3	22	16	20	15
2	1	7	3	10	5
	2	26	17	24	19
	3	34	32	34	29
3	1	10	5	9	4
	2	24	14†	24	16
	3	39	33	36	34

* Provided by Mr. Otto Dykstra Jr. and reproduced with the kind permission of the Research Center of the General Foods Corporation.

† Revised from 24.

of samples of a certain food product in the experimental stage. The levels of factor A are three kinds of salt, those of B correspond to the amounts of the salts (the same molar amounts for each salt), those of C are amounts of acid, and those of D are two different additives. Questions (*a*) and (*b*) are the same as in Problem 4.10. (*c*) The levels of factor B correspond to three equally spaced amounts of salt. Subdivide SS_B into two "orthogonal d.f" for linear and quadratic effects. What do you conclude?

4.24. Prove there always exists an orthogonal transformation from u, v in Problem 4.22 to u', v' such that η is additive in u' and v'.

CHAPTER 5

Some Incomplete Layouts: Latin Squares, Incomplete Blocks, and Nested Designs

5.1. LATIN SQUARES

Suppose an experiment is contemplated with p factors, the qth factor to assume I_q levels ($q = 1, \cdots, p$), so that there are $N = I_1 I_2 \cdots I_p$ cells or possible treatment combinations. We have seen that in the design we called a complete p-way layout there is at least one observation in each cell. Often this is impracticable or impossible and we must adopt a design in which there are no observations in some of the N cells; we shall call such a design an *incomplete p-way layout.* For the designs of any interest the subset of cells in which observations are taken is selected according to some pattern. Our first example will be the Latin square, where $p = 3$.

The Latin-square design is an incomplete three-way layout in which all three factors are at the same number m of levels, and observations are taken on only m^2 of the m^3 possible treatment combinations according to the pattern to be described below. The advantage over the complete three-way layout is of course that only $1/m$ times as many observations are needed. The chief disadvantage, we shall find later, is that *the analysis leans heavily on the assumption of additivity and can be misleading when interactions are actually present.* A more obvious limitation is that all three factors must be at the same number of levels.

Latin-square designs[1] originated in agricultural experimentation. Suppose five varieties of wheat were to be compared. To do this in a Latin-square experiment a rectangular field would be divided into 25 congruent rectangular plots arranged in five rows and five columns.

[1] As defined by Fisher (1926), including the randomization described later in this section.

Each variety would be planted in five plots, in such a way that it appears once in each row and once in each column. If the varieties are represented by the numbers 1, 2, 3, 4, 5 they might appear in the positions

$$\begin{array}{ccccc} 4 & 3 & 5 & 2 & 1 \\ 3 & 1 & 4 & 5 & 2 \\ 1 & 5 & 2 & 3 & 4 \\ 5 & 2 & 1 & 4 & 3 \\ 2 & 4 & 3 & 1 & 5. \end{array} \tag{5.1.1}$$

Such a square array of numbers (or other symbols) where each number appears once in each row and once in each column is called a Latin square. The mathematical models usually used assume that the "true" response in each plot is the sum of a row effect, a column effect, and a variety mean (i.e., variety effect plus general mean). This would be the case if the response were the sum of a variety mean and a fertility effect and (i) the fertility effect were a linear function $ax+by+c$ of Cartesian coordinates (x, y) in the plane of the field, or (ii) there were "waves of fertility" parallel to the rows of the layout; i.e., the fertility effect averaged over the plot were a function only of the row; or (iii) the same for columns, or[2] (iv) the fertility effect were a sum of (i), (ii), and (iii). However, this agricultural example would not conform to the normal-theory model of this section, which assumes independent errors with equal variances, for reasons similar to those mentioned for the randomized-blocks design (sec. 4.2). Nevertheless if the Latin square is chosen at random in a way we shall describe in a moment, if the number of varieties exceeds four,[3] and if the varieties do not interact with rows and columns, then we may expect that statistical inferences derived under the normal-theory model will be fair approximations to those valid under a more realistic randomization model (Ch. 9).

[2] An effect of the type (i) could be considered a sum of (ii) and (iii) for the special case where (ii) and (iii) are linear functions of the row number and column number, respectively.

[3] In the randomization theory (sec. 9.3) the statistic $\mathscr{F}$ has a discrete distribution generated by the actual randomization in the experiment. For $m \times m$ squares the number of values (not necessarily different) assumed by the statistic $\mathscr{F}$ is $N_m = (m-1)!S_m$, where S_m is the number of "standard squares," defined below, in the set from which we choose one at random. If we choose from all the standard squares that exist, so that S_m has its maximum value, then $S_2 = 1$, $S_3 = 1$, $S_4 = 4$, $S_5 = 56$, $S_6 = 9408$. (A general formula for S_m is not known.) Hence, for $m = 4$, $\mathscr{F}$ assumes 24 values. Under the hypothesis of no variety differences these are taken on with equal probability. The resulting discrete distribution would necessarily be wretchedly approximated in the tails by using the per cent points of the continuous F-distribution.

That Latin squares of all sizes exist is evident from the example

$$\begin{array}{ccccc} 1 & 2 & 3 & 4 & 5 \\ 2 & 3 & 4 & 5 & 1 \\ 3 & 4 & 5 & 1 & 2 \\ 4 & 5 & 1 & 2 & 3 \\ 5 & 1 & 2 & 3 & 4 \end{array} \tag{5.1.2}$$

which can be extended to any size. If columns of a Latin square are permuted the result is obviously again a Latin square; similarly for rows; also for numbers. In order to discuss the random choice of a Latin square we define the following: The totality of Latin squares obtainable from a single square by permutation of rows, columns, and numbers, is called a *transformation set*. A canonical form for any $m \times m$ Latin square, called a *standard square*, is the square obtained by permuting the columns so that the first row is $(1, 2, \cdots, m)$ and then the rows so that the first column is $(1, 2, \cdots, m)'$; this means that the first row will not be permuted. Thus, (5.1.2) is already a standard square, and the standard square corresponding to (5.1.1) is

$$\begin{array}{ccccc} 1 & 2 & 3 & 4 & 5 \\ 2 & 5 & 1 & 3 & 4 \\ 3 & 4 & 2 & 5 & 1 \\ 4 & 3 & 5 & 1 & 2 \\ 5 & 1 & 4 & 2 & 3. \end{array}$$

From a standard $m \times m$ square we may obtain $m!(m-1)!$ squares by making the $m!$ permutations of the columns and the $(m-1)!$ permutations of the rows which leave the first row in place. It may be argued that these $m!(m-1)!$ squares are all different and that hence the number of different squares with the same canonical form, i.e., standard square, is $m!(m-1)!$ From this it follows that the number of different squares in a transformation set is $m!(m-1)!$ times the number of standard squares in the set.

To choose a Latin square at random from a transformation set we may start with any square in the set and then randomize columns, rows, and numbers. (A convenient way of arranging a set of numbered objects—in this case, rows, or columns, or "numbers"—is to use the tables of random permutations in Cochran and Cox (1957), Ch. 15.) This gives any two different squares in the set the same probability of being selected.

In order to give *all* $m \times m$ Latin squares the same probability of being chosen we might choose a transformation set at random with probabilities proportional to the numbers of standard squares in the transformation sets. The same result would be achieved by selecting with equal probability from all the standard $m \times m$ squares and then randomizing columns, rows, and numbers (or just columns and rows excluding the first row). For more detailed instructions[4] and tables of standard squares the reader is referred to Fisher and Yates (1943), introduction to Tables XV and XVI.

We may consider using a Latin-square design in any three-factor experiment where each factor is at m levels: it consists in taking m^2 observations on the m^3 treatment combinations in such a way that each level of any factor appears exactly once with each level of each of the other factors in the m^2 treatment combinations observed. Thus if the rows in (5.1.1) are the levels of factor A, the columns are the levels of B, and the numbers are the levels of C (for example, the level of C observed with the second level of A and the fourth level of B is the fifth), then exactly the same experimental design would be pictured by the Latin square

5	4	2	1	3
2	5	1	3	4
1	3	4	5	2
3	2	5	4	1
4	1	3	2	5

in which the rows are the levels of A, the columns are the levels of C, and the numbers are the levels of B, or by the Latin square

3	5	2	1	4
2	4	1	5	3
4	3	5	2	1
5	1	3	4	2
1	2	4	3	5

[4] Statisticians would differ about whether, in agricultural applications where the rows and columns of the design are actual rows and columns of plots, one should use the square drawn at random if it happens to be a systematic square: A *systematic square* of size m is one of the $2m!$ squares which is of the type (5.1.2) or its mirror image, or differs from one of these only by a permutation of the numbers. It would evidently give the user biased results in the presence of certain kinds of fertility waves parallel to one of the diagonals. A similar question arises with the randomized-blocks design if the treatments occur in the same order in each block. The statistician (like

in which rows, columns, and numbers correspond to the levels of B, C, and A, respectively.[5]

In developing the "normal theory" of the analysis of variance of the Latin square we consider the probability relations, *given that a particular Latin square has been selected for the experiment.* If the Latin square is chosen at random, as it should be for reasons that will emerge later, then the probabilities and expectations calculated will be conditional, given that the particular square has been chosen.

We emphasize the symmetry of the design in our notation: We shall denote the observations by $\{y_{ijk}\}$, where y_{ijk} is the observation on the treatment combination where factor A is at the ith level, B at the jth, C at the kth, and the triples (i, j, k) take on only the m^2 values dictated by the particular Latin square selected for the experiment. We shall denote this set of m^2 values of (i, j, k) by D. The assumptions are then

$$\Omega: \begin{cases} y_{ijk} = \mu + \alpha_i^A + \alpha_j^B + \alpha_k^C + e_{ijk}, & (i, j, k) \in D, \\ \text{the } m^2 \text{ random variables } \{e_{ijk}\} \text{ are independently } N(0, \sigma^2), \\ \alpha_.^A = \alpha_.^B = \alpha_.^C = 0, \end{cases}$$

the notation $\{\alpha_i^A\}$ for the main effects of A, etc., being the same as in sec. 4.5 on the complete three-way layout, and the side conditions $\alpha_.^A = 0$, etc., implying as usual no loss of generality. The Ω-assumptions imply additivity or absence of interactions among the three factors. The hypothesis of zero main effects for factor A will be denoted by H_A: all $\alpha_i^A = 0$; similarly for H_B: all $\alpha_i^B = 0$, and H_C: all $\alpha_k^C = 0$.

Under Ω we minimize

$$\mathcal{S} = \sum_{(i,j,k)\in D} (y_{ijk} - \mu - \alpha_i^A - \alpha_j^B - \alpha_k^C)^2.$$

On equating to zero $\partial\mathcal{S}/\partial\mu$, we get after dividing by -2,

$$\sum_{(i,j,k)\in D} (y_{ijk} - \mu - \alpha_i^A - \alpha_j^B - \alpha_k^C) = 0.$$

---

me) who would discard a systematic design, and draw again, may hope that, since the resultant shrinkage in the large set of designs from which he draws with equal probability appears to be small, inferences about the treatments based on probability calculations for the larger set would not differ greatly from those that would be correctly based on probability calculations for the smaller set.

[5] The single three-dimensional pattern of a Latin-square design is described above by three different Latin squares, which are two-dimensional pictures. A single three-dimensional picture of the relationship is obtained by marking the m^2 treatment combinations used in the design in a cubic lattice of m^3 points representing the complete three-way layout.

When we sum over (i, j, k) in D we sum over all the observations, so that α_i^A occurs m times for each i, and hence

$$\sum_{(i,j,k)\in D} \alpha_i^A = \sum_i (m\alpha_i^A) = 0;$$

similarly for $\{\alpha_j^B\}$ and $\{\alpha_k^C\}$. Thus

$$\hat{\mu} = y_{...},$$

where $y_{...}$ is the arithmetic average of the m^2 observations $\{y_{ijk}\}$. From $\partial\mathscr{S}/\partial\alpha_k^C = 0$ we get

$$\sum_{(i,j)\in D_k} (y_{ijk} - \mu - \alpha_i^A - \alpha_j^B - \alpha_k^C) = 0, \tag{5.1.3}$$

where D_k is the set of m pairs (i, j) for which $(i, j, k) \in D$ and k has a fixed value. By the definition of the Latin square each i and each j occur exactly once with each k in the triples (i, j, k) that constitute D, and consequently D_k consists of m pairs (i, j), where i and j each take on once each of the values $\{1, 2, \cdots, m\}$. Hence

$$\sum_{(i,j)\in D_k} \alpha_i^A = \sum_{i=1}^{m} \alpha_i^A = 0, \qquad \sum_{(i,j)\in D_k} \alpha_j^B = \sum_{j=1}^{m} \alpha_j^B = 0, \tag{5.1.4}$$

and thus (5.1.3) reduces to

$$y_{..k} - \mu - \alpha_k^C = 0,$$

where $y_{..k}$ is the average of the m observations in which C is at level k. Hence

$$\hat{\alpha}_k^C = y_{..k} - y_{...},$$

and symmetrically,

$$\hat{\alpha}_i^A = y_{i..} - y_{...},$$
$$\hat{\alpha}_j^B = y_{.j.} - y_{...}.$$

If we calculate the error SS according to the general formula $SS_e = \|\mathbf{y}\|^2 - \|\hat{\boldsymbol{\eta}}\|^2$, we get

$$SS_e = \sum_{(i,j,k)\in D} y_{ijk}^2 - \sum_{(i,j,k)\in D} (\hat{\mu} + \hat{\alpha}_i^A + \hat{\alpha}_j^B + \hat{\alpha}_k^C)^2,$$

which may be written, after squaring, summing, and using the side conditions,

$$SS_e = SS_{\text{“tot”}} - SS_A - SS_B - SS_C,$$

where

$$SS_{\text{“tot”}} = \sum_{(i,j,k)\in D} y_{ijk}^2 - \mathscr{C},$$
$$SS_A = m\sum_i y_{i..}^2 - \mathscr{C},$$
$$SS_B = m\sum_j y_{.j.}^2 - \mathscr{C},$$

(5.1.5) $$\mathrm{SS}_C = m\sum_k y^2_{..k} - \mathscr{C},$$

and

$$\mathscr{C} = m^2 y^2_{...}.$$

Under Ω there are $1+3m$ parameters subject to three side conditions; so $r = 3m-2$, and thus the number of d.f. for SS_e is $n-r = m^2-3m+2$.

Let $\omega = \Omega \cap H_C$. The LS estimates under ω may be obtained by minimizing

(5.1.6) $$\sum_{(i,j,k)\in D} (y_{ijk} - \mu - \alpha_i^A - \alpha_j^B)^2,$$

or by obtaining an identity for $\mathscr{S}$ analogous to (4.3.5). By either method we would find that the LS estimates under ω are the same as under Ω for μ, $\{\alpha_i^A\}$, $\{\alpha_j^B\}$, while of course $\hat{\alpha}^C_{k,\omega} = 0$, and that the minimum of (5.1.6) is

$$\mathscr{S}_\omega = \mathscr{S}_\Omega + \mathrm{SS}_C,$$

where $\mathrm{SS}_C = m\Sigma_k(\hat{\alpha}_k^C)^2$ is calculated from (5.1.5). The number of d.f. for SS_C is $m-1$. The statistic $\mathscr{F}$ for testing H_C is thus $\mathrm{MS}_C/\mathrm{MS}_e$, where

$$\mathrm{MS}_C = \mathrm{SS}_C/(m-1), \qquad \mathrm{MS}_e = \mathrm{SS}_e/(m^2-3m+2).$$

Under ω, $\mathscr{F}$ is F_{m-1,m^2-3m+2}.

The hypotheses H_A and H_B are tested similarly. The analysis-of-variance table is Table 5.1.1. The expected mean squares under Ω shown in the table are calculated in the usual way and are written in terms of the symbols

$$\sigma_A^2 = \frac{\sum_i(\alpha_i^A)^2}{m-1}, \quad \sigma_B^2 = \frac{\sum_j(\alpha_j^B)^2}{m-1}, \quad \sigma_C^2 = \frac{\sum_k(\alpha_k^C)^2}{m-1}.$$

TABLE 5.1.1

ANALYSIS OF VARIANCE FOR THE $m \times m$ LATIN SQUARE

Source	Sum of Squares	d.f.	E(MS)
A	$\mathrm{SS}_A = m\sum_i(y_{i..}-y_{...})^2$	$m-1$	$\sigma^2+m\sigma_A^2$
B	$\mathrm{SS}_B = m\sum_j(y_{.j.}-y_{...})^2$	$m-1$	$\sigma^2+m\sigma_B^2$
C	$\mathrm{SS}_C = m\sum_k(y_{..k}-y_{...})^2$	$m-1$	$\sigma^2+m\sigma_C^2$
Residual	$\mathrm{SS}_e = \sum_{(i,j,k)\in D}(y_{ijk}-y_{i..}-y_{.j.}-y_{..k}+2y_{...})^2$	m^2-3m+2	σ^2
"Total"	$\mathrm{SS}_{\text{"tot"}} = \sum_{(i,j,k)\in D}(y_{ijk}-y_{...})^2$	m^2-1	——

The m^2-dimensional space of linear forms in the $\{y_{ijk}\}$ may be resolved into five mutually orthogonal spaces, namely the space of $y_{...}$, and the four spaces of linear forms whose squares occur in the first four rows of the second column of Table 5.1.1, the dimensions being respectively 1, $m-1$, $m-1$, $m-1$, and m^2-3m+2. This may be shown by applying the "method of nested ω's" (sec. 2.9) to the chain of hypotheses Ω, $\omega_1 = \Omega \cap H_C$, $\omega_2 = \omega_1 \cap H_B$, $\omega_3 = \omega_2 \cap H_A$. A consequence is the statistical independence under Ω of the five SS's consisting of $m^2y^2_{...}$ and the first four SS's in the table. The noncentrality parameters of their noncentral chi-square distributions may be found in the usual way from Rule 1 of sec. 2.6.

Numerical calculations for the Latin square are similar to those discussed for the two-way layout with one observation per cell (sec. 4.2), but require, in addition to the calculation of the row means $\{y_{i..}\}$ and column means $\{y_{.j.}\}$, the calculation of the m means $\{y_{..k}\}$ for the m levels of the third factor and SS_C from (5.1.5). Multiple comparison by the S- or T-method may be performed in an obvious way.

Effects of Interactions

In the complete p-way layout with equal numbers in the cells the estimate of any contrast in main effects calculated under the assumption that all interactions are zero is not biased by violation of this assumption (although the violation does in general bias the estimated variance of the contrast). We had an example of this in the two-way layout with one observation per cell in sec. 4.8, and it is easy to see in general: Suppose the contrast[6] is in the main effects of a factor indexed by i; then the estimated contrast is a linear function of the means $\{\bar{y}_i\}$, where $\bar{y}_i$ denotes the unweighted mean of all observations at the ith level of the factor. Since the observations entering $\bar{y}_i$ are summed over the subscripts of all the other factors, so are the interactions entering $E(\bar{y}_i)$, and they hence sum out by the side conditions on them. This nice property is not retained in the incomplete layouts, and in particular in the Latin square. Suppose we generalize the above Ω-assumptions to permit interactions, so that

$$(5.1.7)\quad y_{ijk} = \mu + \alpha_i^A + \alpha_j^B + \alpha_k^C + \alpha_{ij}^{AB} + \alpha_{jk}^{BC} + \alpha_{ik}^{AC} + \alpha_{ijk}^{ABC} + e_{ijk},$$

where the main effects $\{\alpha_i^A\}$, $\{\alpha_j^B\}$, $\{\alpha_k^C\}$ and interactions $\{\alpha_{ij}^{AB}\}$, $\{\alpha_{jk}^{BC}\}$, $\{\alpha_{ik}^{AC}\}$, $\{\alpha_{ijk}^{ABC}\}$ obey the usual side conditions

$$(5.1.8)\quad \alpha_.^A = \cdots = \alpha_{i.}^{AB} = \alpha_{.j}^{AB} = \cdots = \alpha_{ij.}^{ABC} = \alpha_{i.k}^{ABC} = \alpha_{.jk}^{ABC} = 0,$$

[6] The argument holds for any linear function, whose coefficients do not need to sum to zero. There is a similar result for the estimate of an interaction made by assuming all higher-order interactions zero.

for all i, j, k. We are now considering m^3 random variables $\{y_{ijk}\}$ of which only m^2 are observed in the actual experiment; y_{ijk} is the conceptual response that would be observed if a measurement were made on the i, j, k treatment combination. It is assumed that the m^3 random variables $\{e_{ijk}\}$ are independently $N(0, \sigma^2)$, or at least independently and identically distributed with zero means and variance σ^2.

To calculate $y_{..k}$, the mean of those m observations in the experiment where C is at the kth level, we sum (5.1.7) for $(i, j) \in D_k$, where the set D_k is defined below (5.1.3), and divide by m. But recalling (5.1.4) and noting

$$\sum_{(i,j)\in D_k} \alpha_{ik}^{AC} = \sum_{i=1}^{m} \alpha_{ik}^{AC} = 0, \qquad \sum_{(i,j)\in D_k} \alpha_{jk}^{BC} = \sum_{j=1}^{m} \alpha_{jk}^{BC} = 0,$$

we find that the expression obtained for $y_{..k}$ reduces to

$$y_{..k} = \mu + \alpha_k^C + \gamma_k + e_{..k}, \tag{5.1.9}$$

where

$$\gamma_k = m^{-1} \sum_{(i,j)\in D_k} (\alpha_{ij}^{AB} + \alpha_{ijk}^{ABC}). \tag{5.1.10}$$

It is clear that γ_k depends on the $A \times B$ and $A \times B \times C$ interactions and on the Latin square selected, but not on the $A \times C$ or $B \times C$ interactions. Since

$$E(y_{..k}) = \mu + \alpha_k^C + \gamma_k, \tag{5.1.11}$$

we see that $y_{..k}$ is a biased estimate of the true mean $\mu + \alpha_k^C$, and that if $\psi = \Sigma_k c_k \alpha_k^C$ is a contrast in the $\{\alpha_k^C\}$, then $\hat{\psi} = \Sigma_k c_k y_{..k}$ is a biased estimate of ψ since

$$E(\hat{\psi}) = \psi + \sum_k c_k \gamma_k. \tag{5.1.12}$$

The main effects for C are said to be confounded with the $A \times B$ and $A \times B \times C$ interactions (of course, the analogous property for the main effects of B or C is obtained by permuting the letters A, B, C in this statement). Thus in the above agricultural example, the main effects for varieties would be confounded with the row × column interactions and triple interactions, but not with the variety × row or the variety × column interactions.

Any bias of $\hat{\psi}$ will of course affect confidence intervals for ψ centered at $\hat{\psi}$. A complete determination of the biasing effect of interactions on the F-test of H_C would be complicated in the present "conditional" model, given the Latin square actually used. Some indication of the effect on the test can be obtained by considering the effect on the MS's. Since

MS_C depends only on the $\{y_{..k}\}$ it cannot depend on any interactions except the $A \times B$ and $A \times B \times C$ interactions insofar as they enter through the $\{\gamma_k\}$ in (5.1.9) and (5.1.10). From (5.1.9) we also find that

$$(5.1.13) \qquad E(\mathrm{MS}_C) = m(m-1)^{-1} \sum_{k=1}^{m} (\alpha_k^C + \gamma_k - \gamma_.)^2 + \sigma^2,$$

since

$$m(m-1)^{-1} \sum_{k=1}^{m} (e_{..k} - e_{...})^2$$

is identical with MS_C in the case where μ and all α's are zero, when $e_{ijk} = y_{ijk}$ and $E(\mathrm{MS}_C) = \sigma^2$. We see that if H_C is true, $E(\mathrm{MS}_C)$ will exceed σ^2 by the amount

$$m(m-1)^{-1} \sum_{k=1}^{m} (\gamma_k - \gamma_.)^2,$$

which could be large. On the other hand, if H_C is false, and the $\{\alpha_k^C\}$ are large they might be canceled by the $\{\gamma_k - \gamma_.\}$ in (5.1.13) to reduce $E(\mathrm{MS}_C)$ to σ^2. Thus the effect of the $A \times B$ and $A \times B \times C$ interactions on MS_C is complicated, depending on the Latin square used as well as on the values of these interactions, but the effect *may* be to inflate MS_C when H_C is true and to deflate it when H_C is false.

The effect of interactions on MS_e is even more complicated: While it is easy to see that MS_e does not depend on the general mean μ or the main effects, and that always $E(\mathrm{MS}_e) \geqq \sigma^2$ (this follows from Problem 1.4), the exact expression for $E(\mathrm{MS}_e)$ in the presence of interactions would be found to be messy and hard to interpret. To get some idea of the possible effect of the $A \times B$ interactions on MS_e we shall consider the simple case where these are the only interactions present, namely when

$$(5.1.14) \qquad y_{ijk} = \mu + \alpha_i^A + \alpha_j^B + \alpha_k^C + \alpha_{ij}^{AB} + e_{ijk}.$$

We easily calculate that

$$y_{ijk} - y_{i..} - y_{.j.} - y_{..k} + 2y_{...}$$
$$= (\alpha_{ij}^{AB} - \gamma_k) + (e_{ijk} - e_{i..} - e_{.j.} - e_{..k} + 2e_{...}),$$

where γ_k is given by (5.1.10) with all $\alpha_{ijk}^{ABC} = 0$, and hence

$$E(\mathrm{MS})_e = (m^2 - 3m + 2)^{-1} E(\mathrm{SS}_e)$$
$$= (m^2 - 3m + 2)^{-1} \sum_{(i,j,k)\in D} (\alpha_{ij}^{AB} - \gamma_k)^2 + \sigma^2$$

in this special case. We may verify from this the possibility that $E(\mathrm{MS}_e) = \sigma^2$ even when the $A \times B$ interactions are large, and hence that the interactions do not necessarily inflate $E(\mathrm{MS}_e)$ above σ^2; there *may* be a cancelation effect.

From the above discussion of the behavior of $E(\text{MS}_C)$ and $E(\text{MS}_e)$ we may conclude that if we are unlucky in selecting the Latin square in certain situations where the $A \times B$ interactions are large and H_C is true, we are likely to get significant values for the F-statistic for testing H_C, and in other situations where the $A \times B$ interactions are large and H_C is false, to get nonsignificant values.

Suppose now that the Latin square used was selected at random from a transformation set as described above (the transformation set might be chosen from a family of transformation sets in any way—possibly with probabilities proportional to the number of standard squares in the transformation sets, as mentioned above). Then (5.1.11) may be regarded as the conditional expected value of $y_{..k}$, given that the square defined by D has been selected. It will be shown in sec. 9.2 that under a "randomization model" based on the random selection of the Latin square the unconditional expected value of $y_{..k}$ is $\mu + \alpha_k^C$, from which it follows that unconditionally $y_{..k}$ and the estimate $\hat{\psi}$ defined above (5.1.12) are unbiased estimates.

The nature of the unbiasedness of $y_{..k}$ and $\hat{\psi}$ is somewhat subtle and needs clarification. In all our previous examples of an unbiased estimate $\hat{\psi}$, the estimate was unbiased for any particular design selected, whereas in the present case the estimate, given the Latin square actually used, has a known bias in the sense that it depends in a known way on the unknown interactions. It may help us in the case where the square is selected at random to rewrite (5.1.9) as

$$y_{..k} = \mu + \alpha_k^C + g_k + e_k, \tag{5.1.15}$$

where g_k is a random variable taking on values $\{\gamma_k\}$ depending on the square obtained, and e_k is a random variable with zero mean and variance σ^2/m representing the result $e_{..k}$ of the "technical errors" $\{e_{ijk}\}$. It will be shown in sec. 9.2 that $E(g_k) = 0$ and that g_k and e_k are statistically independent. We thus see from (5.1.15) that if we have randomized in selecting the Latin square, then γ_k, which appears as a bias from the "conditional" viewpoint, may be regarded as one of two components of random error, the other being e_k. Its role is similar to that of the effect of "unit errors" (this concept is developed in sec. 9.1) due to differences among experimental units when the units are selected at random. If the $A \times B$ interactions were known, so would be the value γ_k of g_k for the square used, and the estimate $y_{..k}$ of $\mu + \alpha_k^C$ should be adjusted accordingly. However, it is hard to imagine a situation where the means $\{\mu + \alpha_k^C\}$ are unknown and the $A \times B$ interactions are known. If the $A \times B$ interactions are unknown, we may inform the user just how they enter into the estimates, but this may be of as little use as telling him in the analogous situation where experimental units are selected at random, the experimental units whose unit errors enter the result and how they enter, when we do not know those unit errors. In either case we try to control the effect, which would be one of bias if the square or experimental units were not chosen at random, by performing an actual randomization before the data are taken and by basing the statistical inferences on the outcome of the experiment in view of this randomization. This is *not* one of those deplorable situations in which a statistician withholds pertinent information from his client and gives him a probability or expectation correct for the statistician's long-run experience with all his clients but wrong for the subclass to which this client's case would be restricted by the pertinent information.

Misbehavior of the "conditional" $E(\text{MS})$'s discussed above, namely that $E(\text{MS}_C)$ is large when H_C is true, or small when H_C is false, or that $E(\text{MS}_e)$ is not inflated by the $A \times B$ interactions may be regarded as unusual when the safeguard of randomization is used, in the light of the distribution of the MS's under randomization theory. (This optimism would be better founded if besides the $E(\text{MS})$ under randomization theory we knew more about the variation of the MS about its expectation.) The "unconditional" $E(\text{MS})$'s will be given in sec. 9.2. Exact tests will be considered in sec. 9.3.

Orthogonal Latin Squares

We mention these mainly to acquaint the reader with terminology which he may encounter elsewhere. Two Latin squares of size m are said to be *orthogonal* if when superimposed every one of the m^2 pairs of numbers k, k' $(k, k' = 1, 2, \cdots, m)$ occurs once. A set of Latin squares is called *orthogonal* if every pair of squares in the set is orthogonal. An example of three orthogonal Latin squares of size 4 is

1	2	3	4		1	2	3	4		1	2	3	4
2	1	4	3		3	4	1	2		4	3	2	1
3	4	1	2		4	3	2	1		2	1	4	3
4	3	2	1		2	1	4	3		3	4	1	2

If h orthogonal Latin squares of size m exist it is possible to use them to incorporate "orthogonally" $h+2$ factors each at m levels into an experiment with m^2 observations, by letting two of the factors correspond to rows and columns and the remaining h to the numbers in the h squares. This design would have the property that each level of any factor appears exactly once with each level of any other factor, and the experiment would be easy to analyze under the assumption of additivity and the usual normal-theory assumptions, the total SS about the grand mean with m^2-1 d.f. being resolved into the sum of $h+2$ SS's for main effects each with $m-1$ d.f. and a residual SS with $(m^2-1)-(h+2)(m-1)$ d.f. It follows that the number h of orthogonal Latin squares of size m must be less than m. In Fisher and Yates (1943) are exhibited sets of $m-1$ orthogonal Latin squares of side m for $m = 3, 4, 5, 7, 8, 9$. It has been proved[7] that no orthogonal squares of size 6 exist.

The term *Greco-Latin squares* is another name for a pair of orthogonal Latin squares, arising from a custom of replacing the numbers in the squares by Greek letters in one of the squares and Latin letters in the other; the term *hyper-Greco-Latin squares* refers to a set of more than two orthogonal squares. These designs of course suffer from the same disadvantages in the presence of interactions as the ordinary Latin square. They seem to be not much used[8] for the purpose of handling with only m^2 observations more than three factors each at m levels, but they are of incidental use in the construction of certain designs for a large number of factors each at two levels, and some other kinds of designs.[9]

Partially Replicated Latin Squares

This ingenious modification of the Latin-square design has been proposed to afford some information on the troublesome interactions usually assumed to be zero. It consists in taking m^2+m observations, with m^2 in an ordinary Latin square, and the additional m observations being duplicate measurements on certain treatment combinations. The duplicate measurements destroy the orthogonality of the design, but since they are incorporated in such a way that there is one duplication at each level of each factor, the resulting balance permits the calculations to be not drastically more complicated than for the usual case. An example is

1*	2	3	4	5
2	5*	1	3	4
3	1	4*	5	2
4	3	5	2*	1
5	4	2	1	3*,

[7] Tarry (1900), Bruck and Ryser (1949).

[8] An example of an application in which several Greco-Latin squares are used to handle four factors is given in Davies (1956), sec. 5.71.

[9] The reader is referred to Fisher and Yates (1943), introduction to Table XVI, on the use of orthogonal Latin squares, and to Mann (1949), Ch. 8, for an exposition of the number-theoretic approach to their construction.

the asterisks showing which treatment combinations are replicated. Rows may be permuted, also columns, also numbers. The design is of course impossible for the agricultural situation where the rows and columns are actual rows and columns of congruent plots. The m duplicate measurements yield an estimate of σ^2 which is unbiased, regardless of the interactions. However, the test then obtained for the interactions cannot be very sensitive: The denominator of the F-statistic for this test will have only m d.f. The point of view is the "conditional" one discussed above, where we saw that the interactions need not necessarily increase the "interaction" $E(\text{MS})$ beyond σ^2 with the ordinary Latin square; this possibility must exist also with the partially replicated square. Since all the measurements are taken on the same treatment combinations as in an ordinary Latin square it may be argued that the additional measurements in no way undo the confounding of the main effects with the two-factor interactions. As with the ordinary square this not only biases the estimates of the main effects but also the MS formed from these estimates to test the main effects, and possibly in the two unfortunate ways considered earlier. The reader is referred to the paper by Youden and Hunter (1955) for details of the analysis.

5.2. INCOMPLETE BLOCKS

In the randomized-blocks design (end of sec. 4.2) the size of the block of experimental units must be equal to the number of treatments to be compared. It is sometimes desirable or necessary to have the block size smaller than the number of treatments, as the following examples will illustrate:

(i) In a use test on several kinds of rubber heels the natural block consists of the two shoes of a subject.

(ii) In a taste-testing experiment on chocolate puddings, where a "block" could consist of the tasting of several brands by the same subject on the same occasion, the "error" would increase with the block size, and it would be advantageous to keep the block size down to three.

(iii) In comparing different makes of automobile tires the natural block consists of the four wheels of a car.

In the normal-theory models that we are using at present there is nothing that indicates why the error should increase with block size; this would be better reflected in a randomization model of the kind to be introduced later (sec. 9.1). In the above three examples it would usually be desirable to make our inferences about the effects of the treatments not in the particular blocks used in the experiment but in a conceptual population of blocks from which the blocks in the experiment might be regarded as a random sample. Since the treatment effects would then be regarded as constants, but the block effects as random variables, this would call for a mixed model, and will be considered later in this section. In any event we need to consider first the fixed-effects model under normal theory because in general only this automatically yields the estimates and SS's usually employed; their distribution may then be reconsidered under other models.

An *incomplete-blocks design*[10] is one in which the block size is smaller than the total number of treatments to be compared. We shall assume that each treatment is replicated the same number of times, that the blocks all have the same size, and that no treatment appears twice in the same block. Then, if (see footnote[11] concerning this notation)

$$I = \text{number of treatments},$$
$$J = \text{number of blocks},$$
$$r = \text{number of replications},$$
$$k = \text{block size},$$

we see that

$$rI = kJ, \tag{5.2.1}$$

since each side is the total number of observations. The analysis of the results will later be seen to be much simpler in the case of a *balanced incomplete-blocks design*, defined to be one where the number of blocks in which a particular pair of treatments occur together is the same for all pairs. An example in which seven treatments are compared in blocks of four is the following, where the numbers represent the treatments, and the columns represent the blocks:

(5.2.2)

3	1	1	1	2	1	2
5	4	2	2	3	3	4
6	6	5	3	4	4	5
7	7	7	6	7	5	6

The reader may verify[12] that each pair of treatments occurs in two different blocks.

[10] Invented by Yates (1936).

[11] The usual notation in tables of incomplete-block plans employs v or t in place of our I, b in place of our J, and E in place of our $\mathscr{E}$. I have adopted the standard notations k, r, and λ; in this section r does not denote rank $\mathbf{X}'$, as earlier.

[12] More insight into the structure of this design may be obtained by marking in a 7×7 two-way layout, where the i,j, cell corresponds to the ith treatment in the jth block, the cells in which observations are taken.

We shall let

$$\lambda_{ii'} = \text{number of blocks in which treatment } i \text{ occurs with treatment } i'. \tag{5.2.3}$$

Then for balanced incomplete blocks, all the $\lambda_{ii'}$ have the same value for $i \neq i'$, which we shall denote by λ, so that we may write

$$\lambda_{ii'} = \begin{Bmatrix} \lambda & \text{if } i \neq i' \\ r & \text{if } i = i' \end{Bmatrix} \quad \text{for balanced incomplete blocks.} \tag{5.2.4}$$

In this case we may deduce the value

$$\lambda = r(k-1)/(I-1), \tag{5.2.5}$$

by noting that any particular treatment occurs in r blocks, and considering the number of units on which the treatment does not appear in these r blocks: On the one hand it is $rk-r$, the total number of units in the r blocks minus the number on which the treatment appears; on the other hand it is $(I-1)\lambda$, the number of other treatments times the number of times the particular treatment appears with each. Equating $rk-r$ and $(I-1)\lambda$ gives (5.2.5).

An additional condition which must be satisfied by a balanced incomplete-blocks design is

$$I \leqq J, \tag{5.2.6}$$

which because of (5.2.1) is equivalent to $r \geqq k$. The proof[13] of (5.2.6) may be based on the nonsingularity of a symmetric $I \times I$ matrix $\mathbf{B}$ of the type

$$\mathbf{B} = \begin{pmatrix} r & \lambda & \lambda & \cdots & \lambda \\ \lambda & r & \lambda & \cdots & \lambda \\ \cdot & \cdot & \cdot & \cdots & \cdot \\ \lambda & \lambda & \lambda & \cdots & r \end{pmatrix}. \tag{5.2.6a}$$

The determinant $|\mathbf{B}|$ may be evaluated as

$$|\mathbf{B}| = [r + (I-1)\lambda](r-\lambda)^{I-1}$$

by the result of Problem II.4, and thus $\mathbf{B}$ is nonsingular. Now denote by $\mathbf{A}$ the $I \times J$ matrix whose i,j element is K_{ij}, where $K_{ij} = 0$ or 1 is the number of times the ith treatment occurs in the jth block, and form $\mathbf{B} = \mathbf{AA}'$, so that by (5.2.3) and (5.2.4), $\mathbf{B}$ has the form (5.2.6a), and is

[13] A slight modification of one given by R. C. Bose (1949b) for the inequality (5.2.6) of R. A. Fisher (1940).

hence nonsingular. By Theorem 7 of App. II, rank $\mathbf{B}$ = rank $\mathbf{A}$, and so rank $\mathbf{A} = I$. But since $\mathbf{A}$ is $I \times J$, rank $\mathbf{A} \leqq J$, and this proves (5.2.6).

The three conditions (5.2.1), (5.2.5), and (5.2.6) are necessary but not sufficient for the existence of a balanced incomplete-blocks design with given I, J, r, k, λ. A detailed listing of such designs sufficient for most practical purposes is given in Cochran and Cox (1957), Ch. 11 (in this listing the blocks are represented as rows instead of columns). If for the desired number of treatments and block size no balanced incomplete-blocks design with a suitable[14] number of blocks is listed, one should consider using a partially balanced design with two associate classes: The *number of associate classes* is the number of different $\{\lambda_{ii'}\}$ with $i \neq i'$ in (5.2.3); partial balance, which we shall not define here, insures that the analysis remains relatively simple. These designs are listed, and their analysis explained, in Bose, Clatworthy, and Shrikhande (1954). After a design (balanced or partially balanced) is selected, each of the following should be assigned in random order: numbering of the treatments, numbering of the blocks, and positions within blocks.

The analysis of the incomplete-blocks design under the normal-theory fixed-effects model is a little more complicated than that of the other designs we have treated so far except for the two-way layout with unequal numbers of observations in the cells, of which this is a special case. The estimate of a treatment effect is no longer of the form of the observed treatment mean minus the grand mean, since the block effects do not enter in the same way into all the observed treatment means: for example, a treatment may be favored by occurring only in blocks with high block effects.

Define the IJ numbers $\{K_{ij}\}$ as above, so that $K_{ij} = 1$ if the ith treatment occurs in the jth block, and $K_{ij} = 0$ otherwise, and let $\{y_{ij}\}$ be the observations, where i, j now runs through the set D for which $K_{ij} = 1$. The mathematical model is then

$$(5.2.7) \quad \Omega: \begin{cases} y_{ij} = \mu + \alpha_i + \beta_j + e_{ij} \text{ for } (i, j) \in D, \\ \text{the } rI \text{ random variables } \{e_{ij}\} \text{ are independently } N(0, \sigma^2), \\ \alpha_. = \beta_. = 0. \end{cases}$$

Here $\{\alpha_i\}$ are the treatment effects and $\{\beta_j\}$ the block effects, and it is seen that the model assumes zero interactions in addition to the usual normal-theory assumptions.

The analysis will be given first for the incomplete-blocks design, and then specialized to the balanced incomplete-blocks design. It falls out

[14] For given I and k there obviously always exists a balanced incomplete-block design with J equal to the binomial coefficient C_k^I.

as a special case of the analysis we made in sec. 4.4 of the two-way layout with zero interactions and unequal numbers in the cells, where K_{ij} denoted the number of observations in the i,j cell; the present Ω-assumptions coincide with the ω-assumptions of sec. 4.4, and the present $\{y_{ij}\}$ correspond to the former $\{y_{ij1}\}$. The quantities g_i and h_j defined by (4.4.5) are the totals of the observations for the ith treatment and for the jth block, respectively, and are more briefly called

$$g_i = i\text{th } \textit{treatment total},$$
$$h_j = j\text{th } \textit{block total},$$

while the G_i and H_j defined by (4.4.6) now reduce to $G_i = r$ and $H_j = k$. After we recall that

$$\sum_j K_{ij}K_{i'j} = \lambda_{ii'} \tag{5.2.8}$$

by (5.2.3), we see that we may write the equations (4.4.8) and (4.4.9) for the estimates $\{\hat{\alpha}_i\}$ of the treatment effects in the form

$$\sum_{i'}(r\delta_{ii'} - k^{-1}\lambda_{ii'})\hat{\alpha}_{i'} = \mathscr{G}_i, \tag{5.2.9}$$

where

$$\mathscr{G}_i = g_i - k^{-1}\sum_j K_{ij}h_j.$$

This quantity is called

$$\mathscr{G}_i = i\text{th } \textit{adjusted treatment total},$$

the adjustment consisting in subtracting from the treatment total g_i the sum of the block averages h_j/k for the blocks in which the ith treatment appears. Let us call

(5.2.10) T_i = sum of block totals in which the ith treatment appears,

so the "adjustment" is $-k^{-1}T_i$,

$$\mathscr{G}_i = g_i - k^{-1}T_i. \tag{5.2.11}$$

The error SS given by (4.4.13) becomes

$$SS_e = \sum_{(i,j)\in D} y_{ij}^2 - \sum_i \mathscr{G}_i\hat{\alpha}_i - k^{-1}\sum_j h_j^2,$$

and may be written

$$SS_e = SS_{\text{"tot"}} - SS_{\text{tr.el.bl}} - SS_{\text{bl.ig.tr}}, \tag{5.2.12}$$

where

$$SS_{\text{"tot"}} = \sum_{(i,j)\in D} y_{ij}^2 - \mathscr{C},$$

$\mathscr{C}$ is the "correction term for the grand mean"

(5.2.13) $$\mathscr{C} = \left(\sum_i g_i\right)^2/n = \left(\sum_j h_j\right)^2/n,$$

n is the total number of observations,

$$n = rI = kJ,$$

(5.2.14) $$\mathrm{SS}_{\mathrm{tr.el.bl}} = \sum_i \mathscr{G}_i \hat{\alpha}_i$$

is called the SS for "treatments, eliminating blocks," and

$$\mathrm{SS}_{\mathrm{bl.ig.tr}} = k^{-1}\sum_j h_j^2 - \mathscr{C}$$

is called the SS for "blocks, ignoring treatments."

The hypothesis of main interest is H_A: all $\alpha_i = 0$. The present $H_A \cap \Omega$ is identical with the ω_1 of sec. 4.4, where we calculated the numerator SS of the test statistic $\mathscr{F}$ to be $\mathscr{S}_{\omega_1} - \mathrm{SS}_e$ with $\mathscr{S}_{\omega_1}$ given by (4.4.15*a*), that is

$$\mathscr{S}_{\omega_1} = \sum_{(i,j)\in D} y_{ij}^2 - k^{-1}\sum_j h_j^2,$$

and hence

(5.2.15) $$\mathscr{S}_{\omega_1} = \mathrm{SS}_{\text{"tot"}} - \mathrm{SS}_{\mathrm{bl.ig.tr}}.$$

From (5.2.12) and (5.2.15) we see that the numerator SS is therefore $\mathrm{SS}_{\mathrm{tr.el.bl}}$. In sec. 4.4 we noted the numbers of d.f. of the test statistic were $I-1$ and $n-I-J+1$.

We shall specialize to the case of the balanced incomplete-blocks design in the remainder of this section. In this case the equation (5.2.9) for the estimated treatment effects $\{\hat{\alpha}_i\}$ have very simple solutions: Let us write the equations in the form

$$(r - k^{-1}\lambda_{ii})\hat{\alpha}_i - k^{-1}\sum_{\substack{i' \\ i' \neq i}} \lambda_{ii'}\hat{\alpha}_{i'} = \mathscr{G}_i,$$

and substitute (5.2.4) to get

$$(r - k^{-1}r)\hat{\alpha}_i - k^{-1}\lambda \sum_{\substack{i' \\ i' \neq i}} \hat{\alpha}_{i'} = \mathscr{G}_i.$$

Because of the side condition $\hat{\alpha}_. = 0$,

(5.2.16) $$\sum_{\substack{i' \\ i' \neq i}} \hat{\alpha}_{i'} = -\hat{\alpha}_i,$$

and so the equations reduce to

$$r\mathscr{E}\hat{\alpha}_i = \mathscr{G}_i,$$

where

$$\mathscr{E} = \frac{rk - r + \lambda}{rk} = \frac{(k-1)I}{k(I-1)}, \tag{5.2.17}$$

the last equality following from (5.2.5). The number $\mathscr{E}$ is called the *efficiency factor* of the design for reasons to appear later, and is <1 because the block size k is less than the number I of treatments. Thus

$$\hat{\alpha}_i = \mathscr{G}_i/(r\mathscr{E}), \tag{5.2.18}$$

and

$$\mathrm{SS}_{\mathrm{tr.el.bl}} = \sum_i \mathscr{G}_i^2/(r\mathscr{E}),$$

where $\mathscr{G}_i$ is given by (5.2.11).

We have now justified the analysis of variance shown in Table 5.2.1. The alternative expression for $\mathrm{SS}_{\mathrm{tr.el.bl}}$ will be explained below. The expected mean square for treatments, eliminating blocks, shown in the table is easily obtained by applying our usual rule to

$$\mathrm{MS}_{\mathrm{tr.el.bl}} = r\mathscr{E}\sum_i \hat{\alpha}_i^2/(I-1),$$

and defining, as usual,

$$\sigma_A^2 = \sum_i \alpha_i^2/(I-1).$$

TABLE 5.2.1

ANALYSIS OF VARIANCE OF BALANCED INCOMPLETE BLOCKS

Source	SS	d.f.	E(MS)
Blocks, ignoring treatments	$k^{-1}\sum_j h_j^2 - \mathscr{C}$	$J-1$	——
Treatments, eliminating blocks	$(r\mathscr{E})^{-1}\sum_i \mathscr{G}_i^2 = (r\mathscr{E})^{-1}(\sum_i \tilde{\mathscr{G}}_i^2 - IC^2)$	$I-1$	$\sigma^2 + r\mathscr{E}\sigma_A^2$
Error	by subtraction	$n-I-J+1$	σ^2
"Total"	$\sum_{(i,j)\in D} y_{ij}^2 - \mathscr{C}$	$n-1$	——

To test the hypothesis H_B: all $\beta_j = 0$, usually of less interest than H_A, we let $\omega_2 = \Omega \cap H_B$. Then by analogy with (4.4.15a)

$$\mathscr{S}_{\omega_2} = \sum_{(i,j)\in D} y_{ij}^2 - r^{-1}\sum_i g_{i\cdot}^2.$$

The numerator SS of the F-statistic for testing H_B, namely $\mathscr{S}_{\omega_2} - \mathrm{SS}_e$, called the SS for "blocks eliminating treatments," is thus

$$\mathrm{SS}_{\mathrm{bl.el.tr}} = k^{-1}\sum_j h_j^2 + (r\mathscr{E})^{-1}\sum_i \mathscr{G}_i^2 - r^{-1}\sum_i g_{i\cdot}^2. \tag{5.2.19}$$

The S-method of multiple comparison is easily applied. To do so we need the formula for the variance of an estimated contrast $\hat{\psi} = \Sigma_i c_i \hat{\alpha}_i$, where $\Sigma_i c_i = 0$. Lemma 1 at the end of sec. 9.2 is applicable since the $\{\hat{\alpha}_i\}$ have equal variances and equal correlation coefficients because of the symmetrical way in which the treatments appear in the design. The lemma says that Var $(\hat{\psi}) = V\Sigma_i c_i^2$, where

$$V = E\left\{\sum_i (\hat{\alpha}_i - \hat{\alpha}_.)^2/(I-1)\right\} - \sum_i (\alpha_i - \alpha_.)^2/(I-1)$$

$$= E\left\{\sum_i \hat{\alpha}_i^2/(I-1)\right\} - \sum_i \alpha_i^2/(I-1)$$

$$= E\{(r\mathscr{E})^{-1}\text{MS}_{\text{tr.el.bl}}\} - \sigma_A^2 = (r\mathscr{E})^{-1}\sigma^2.$$

On comparing the result

$$\text{(5.2.20)} \qquad \text{Var}\,(\hat{\psi}) = \sigma^2 \sum_i c_i^2/(r\mathscr{E})$$

with the corresponding result that we would have in a randomized-blocks design with the same treatments, the same number r of replicates, and the same σ^2 (which would mean that nothing was gained by using the smaller block size k), namely Var $(\hat{\psi}) = \sigma^2\Sigma_i c_i^2/r$, we now see why $\mathscr{E}$ is called the efficiency factor.

To apply the T-method of multiple comparison in the generalized form of Theorem 3 of sec. 3.6, we need the common variance of the $\{\hat{\alpha}_i\}$, say $\sigma_{\hat{\alpha}}^2$, and the common covariance, say $\rho\sigma_{\hat{\alpha}}^2$. If we apply the formula for the variance of a linear combination to the quantity $\Sigma_i \hat{\alpha}_i = 0$ we get

$$0 = \sum_i \sum_{i'} \text{Cov}\,(\hat{\alpha}_i, \hat{\alpha}_{i'}) = I\sigma_{\hat{\alpha}}^2 + (I^2 - I)\rho\sigma_{\hat{\alpha}}^2,$$

and hence $\rho = -1/(I-1)$. If we apply formula (5.2.20) to $\hat{\psi} = \hat{\alpha}_1 - \hat{\alpha}_2$ we get

$$2\sigma^2/(r\mathscr{E}) = \text{Var}\,(\hat{\psi}) = 2\sigma_{\hat{\alpha}}^2 - 2\rho\sigma_{\hat{\alpha}}^2 = 2I(I-1)^{-1}\sigma_{\hat{\alpha}}^2,$$

and consequently

$$\text{(5.2.21)} \qquad \begin{aligned} \text{Var}\,(\hat{\alpha}_i) &= \sigma^2(I-1)/(Ir\mathscr{E}), \\ \text{Cov}\,(\hat{\alpha}_i, \hat{\alpha}_{i'}) &= -\sigma^2/(Ir\mathscr{E}) \qquad (i \neq i'). \end{aligned}$$

It follows that the constant $(a^2-b)^{1/2}$ needed in Theorem 3 of sec. 3.6 is $(r\mathscr{E})^{-1/2}$.

The adjusted treatment totals $\{\mathscr{G}_i\}$ were calculated from the unadjusted totals $\{g_i\}$ by subtracting $k^{-1}T_i$ from g_i, where T_i is the sum of the block totals containing the ith treatment. If there are more blocks that contain the treatment than do not, i.e., if $r > J-r$, or, $2r > J$, it is easier to compute

(5.2.22) $\tilde{T}_i$ = sum of the block totals not containing the ith treatment.

In this case the treatment totals $\{g_i\}$ may be adjusted in a different way by using adjusted totals $\{\tilde{\mathscr{G}}_i\}$, where

$$\tilde{\mathscr{G}}_i = g_i + k^{-1}\tilde{T}_i. \tag{5.2.23}$$

Since $T_i + \tilde{T}_i$ equals the grand total $\Sigma_j h_j = \Sigma_i g_i$,

$$\tilde{\mathscr{G}}_i = \mathscr{G}_i + C, \tag{5.2.24}$$

where

$$C = \sum_j h_j/k = \sum_i g_i/k. \tag{5.2.25}$$

Let

$$\hat{\tilde{\alpha}}_i = \tilde{\mathscr{G}}_i/(r\mathscr{E}). \tag{5.2.26}$$

These may be regarded as estimates of treatment effects $\{\tilde{\alpha}_i\}$ subject to a different side condition; they are related to the $\{\alpha_i\}$ by the equations $\tilde{\alpha}_i = \alpha_i + (r\mathscr{E})^{-1} E(C)$, where $E(C) = J\mu$. The LS estimates $\hat{\psi} = \Sigma_i c_i \hat{\alpha}_i$ of contrasts $\psi = \Sigma_i c_i \alpha_i = \Sigma_i c_i \tilde{\alpha}_i$ $(\Sigma_i c_i = 0)$ may also be written $\hat{\psi} = \Sigma_i c_i \hat{\tilde{\alpha}}_i$. The expression $\Sigma_i \mathscr{G}_i^2$ entering $SS_{\text{tr.el.bl}}$ and $SS_{\text{bl.el.tr}}$ may be replaced by $\Sigma_i \tilde{\mathscr{G}}_i^2 - IC^2$, as may be seen by squaring (5.2.24), summing on i, and using the relation $\Sigma_i \mathscr{G}_i = 0$. The resulting formulas are

$$SS_{\text{tr.el.bl}} = (r\mathscr{E})^{-1}\left(\sum_i \tilde{\mathscr{G}}_i^2 - IC^2\right), \tag{5.2.27}$$

$$SS_{\text{bl.el.tr}} = k^{-1}\sum_j h_j^2 + (r\mathscr{E})^{-1}\left(\sum_i \tilde{\mathscr{G}}_i^2 - IC^2\right) - r^{-1}\sum_i g_i^2, \tag{5.2.28}$$

where C is given by (5.2.25).

If there are block × treatment interactions, the main effects for treatments are confounded with the interactions, given a particular randomization, but when we consider the unconditional expectations over the randomizations the confounding vanishes; this may be shown by an argument resembling that used with the Latin square (sec. 5.1).

Balancing the Position of the Treatments in the Blocks

In the three examples given at the beginning of this section one would want to allow for position within the block: the third of three chocolate puddings tasted is in a disadvantageous position, or, the rear wheels of a car are subject to different wear from the front, and the right from the left, etc. It would be preferable to "balance out" the position and "eliminate" the position effect in such cases rather than "randomizing it out," where it "swells the error." If we treat position in block as a third factor we then have an incomplete three-way layout with the three factors respectively at I, J, and k levels. If we superimpose on the requirements of the balanced incomplete-blocks design the one that each treatment appear equally often in each of the k positions, say m times in each position, this means the number of blocks must be m times the number of treatments,

$$J = mI, \tag{5.2.29}$$

which is equivalent to $r = km$. It has been proved[15] that for any balanced incomplete-blocks design that satisfies (5.2.29) there exists a rearrangement into another balanced incomplete-blocks design with the same I, J, r, k, λ which has the desired position balance. After the design is chosen the randomization is then restricted to the numbering of the levels of each of the three factors.

The example (5.2.2) may be converted into the desired kind of design where each treatment occurs the same number of times in each position by ordering the treatments within the blocks as follows:

(5.2.30)

3	4	5	6	7	1	2
5	6	7	1	2	3	4
6	7	1	2	3	4	5
7	1	2	3	4	5	6

This design might be suitable for the comparison of seven brands of tires if the seven blocks consist of the four tire positions on each of seven cars. An example of another doubly balanced design, which would be useful for the comparison of ten chocolate puddings with 30 tasters each of whom tastes three puddings, is

(5.2.31)

1	1	1	2	2	2	3	3	3	4	4	4	5	5	5	6	6	6	7	7	7	8	8	8	9	9	9	10	10	10
2	4	10	3	5	6	1	4	7	2	5	8	1	3	9	2	7	8	3	8	9	4	9	10	5	6	10	1	6	7
9	7	4	10	8	5	9	6	1	10	7	2	8	6	3	1	8	1	2	9	2	3	10	3	4	4	6	5	7	5

If 60 tasters were to be used, a good doubly balanced design with 60 blocks of three could be obtained by replicating the above design with the positions reversed. Tables of such designs are given by Shrikhande[16] (1951). If $m = 1$, as in the example (5.2.30), the design is called a

[15] Hartley, Shrikhande, and Taylor (1953).

[16] He speaks of "row balance" instead of position balance: The rows of course are not those we used earlier, which corresponded to the I treatments, but they correspond to the k positions in the blocks.

Youden square, or an *incomplete Latin square*, because the $k \times J$ rectangle such as (5.2.30) can be extended to a $J \times J$ Latin square by adding suitable rows.

The analysis is similar to that for incomplete blocks, except that from the error SS previously used one subtracts a SS for positions with $k-1$ d.f. to get a new error SS with $n-I-J-k+2$ d.f., the SS for positions being

$$J^{-1} \sum_{p=1}^{k} P_p^2 - \mathscr{C},$$

where P_p is the total of observations in the pth position, and $\mathscr{C}$ is given by (5.2.13). The shortcomings of the design are similar to those discussed for the Latin square.

Recovery of Interblock Information

We shall treat this for the case of balanced incomplete blocks; it is easily extended to the position-balanced designs just discussed. The technique[17] is applicable only when the number J of blocks exceeds the number I of treatments. The reader may feel that trying to find more efficient estimates of the treatment contrasts than the preceding ones must be a waste of time, and he would be correct if we retained the fixed-effects model we have used thus far, since under this model the estimates are optimum by the Gauss–Markoff theorem in sec. 1.4. The new model consists in modifying the block effects $\{\beta_j\}$ in (5.2.7) so that they are random variables $\{b_j\}$. It will be assumed that they are a random sample of J from an infinite population, i.e., that the $\{b_j\}$ are independently and identically distributed (we shall also assume them to be normal, but this does not affect the weighting problem we will encounter below). This assumption is appropriate if the blocks can be regarded as a random sample from a large finite population. This might be the case in example (i) at the beginning of this section if the subjects were all mailmen (perhaps, right-handed mailmen?), in example (ii) if the tasters were all male American college students, and in example (iii) if the cars were all of the same make and model.[18] The new mathematical model is not usually realistic in the agricultural case, where the $\{b_j\}$ are random variables only because they are a random permutation of the fixed block effects of the J blocks actually used in the experiment; in this case a randomization model like the one for the randomized-blocks design in sec. 9.1 would fit better, but this has not yet been developed. We shall also assume the block

[17] Invented by Yates (1940).

[18] The analysis can be extended to the case where the blocks divide into sets of similar blocks; see Cochran and Cox (1957), sec. 11.5.

effects $\{b_j\}$ to be statistically independent of the errors $\{e_{ij}\}$. The model is then

$$(5.2.32)\quad \Omega': \begin{cases} y_{ij} = \mu' + \alpha_i + b_j + e_{ij} \text{ for } (i, j) \in D, \\ \text{the } \{e_{ij}\} \text{ are independently } N(0, \sigma_e^2) \text{ and independent} \\ \quad \text{of the } \{b_j\}, \\ \text{the } \{b_j\} \text{ are independently } N(0, \sigma_B^2), \\ \alpha_. = 0. \end{cases}$$

We cannot assume that $b_. = 0$ because the $\{b_j\}$ are independent. We write σ_e^2 instead of σ^2 for the error variance to conform to the notation we will employ in Part II of this book; in comparing Ω' with the earlier Ω of (5.2.7) the reader may henceforth imagine σ^2 in Ω replaced by σ_e^2.

At this point the reader may welcome a brief summary of the somewhat lengthy remainder of this section (which he may then decide to skip on a first reading of the book). First we shall argue that the "intrablock" estimates $\{\hat{\alpha}_i\}$ obtained above under Ω have the same distribution under Ω' as under Ω. Next we shall obtain another set of "interblock" estimates $\{\hat{\alpha}_i'\}$; these will be derived by the device of treating the block totals as though they were observations in a fixed-effects model (the distribution of the block totals turns out to satisfy the assumptions on the distribution of the observations made in the general theory of Chs. 1 and 2 if the error variance σ^2 in Chs. 1 and 2 is replaced by $k^2\sigma_B^2 + k\sigma_e^2$). Then we shall show that the two sets $\{\hat{\alpha}_i\}$ and $\{\hat{\alpha}_i'\}$ are statistically independent, which suggests using weighted averages of these two sets to try to get more efficient estimates and tests; complications arise here because the "weights" used are calculated from the data.

We now proceed to prove under the assumptions Ω', that the above estimates $\{\hat{\alpha}_i\}$, the sums of squares $SS_{\text{tr.el.bl}}$ and SS_e defined above, and hence the test statistic $MS_{\text{tr.el.bl}}/MS_e$, all have the same distributions as under Ω. This might be done by substituting into the formulas defining these statistics

$$y_{ij} = \mu + \alpha_i + B_j + e_{ij},$$

where B_j stands for either β_j or b_j, and verifying that the $\{B_j\}$ drop out of these expressions, so that they cannot be affected by the assumptions made on the $\{B_j\}$. An easier and more sophisticated way of arguing this is to note that the joint distribution of these statistics obtained under Ω may be regarded as the conditional distribution under Ω', given the $\{b_j\}$, the constants μ and $\{\beta_j\}$ in Ω corresponding to the fixed values of the $\{b_j\}$ according to $\beta_j = b_j - b_.$ and $\mu = \mu' + b_.$. Since the conditional distribution does not depend on the values of μ and the $\{\beta_j\}$, and hence not on the $\{b_j\}$, it must be the same as the unconditional distribution of the $\{\hat{\alpha}_i\}$, $SS_{\text{tr.el.bl}}$, and SS_e.

Another set of unbiased estimates $\{\hat{\alpha}'_i\}$ will now be obtained from the block totals $\{h_j\}$. Substituting from (5.2.32) for the $\{y_{ij}\}$ in the block totals $\{h_j = \Sigma_i K_{ij} y_{ij}\}$, we find

$$h_j = \sum_i K_{ij}(\mu' + \alpha_i + b_j + e_{ij})$$

or

$$h_j = k\mu' + \sum_i K_{ij}\alpha_i + f_j, \tag{5.2.32a}$$

where

$$f_j = kb_j + \sum_i K_{ij}e_{ij}.$$

Since the last term is the sum of the k errors e_{ij} appearing in the block total, we see that the $\{f_j\}$ are independently $N(0, \sigma_f^2)$, where

$$\sigma_f^2 = k^2\sigma_B^2 + k\sigma_e^2. \tag{5.2.32b}$$

We now see from (5.2.32a) that the J block totals $\{h_j\}$ satisfy the assumptions made on the n observations $\{y_i\}$ in the general theory of Chs. 1 and 2 for fixed-effects models. We shall denote by $\{\hat{\alpha}'_i\}$ the estimates of the $\{\alpha_i\}$ obtained by applying the general theory as though the $\{h_j\}$ were the observations. We have to minimize

$$\mathscr{S}' = \sum_j\left(h_j - k\mu' - \sum_i K_{ij}\alpha_i\right)^2.$$

The normal equations, obtained by equating $\partial\mathscr{S}'/\partial\mu'$ and $\partial\mathscr{S}'/\partial\alpha_i$ to zero, are

$$\begin{aligned} k^2J\hat{\mu}' &= k\sum_j h_j \\ kr\hat{\mu}' + \sum_{i'}\lambda_{ii'}\hat{\alpha}'_{i'} &= T_i \qquad (i = 1, \cdots, I), \end{aligned} \tag{5.2.33}$$

where T_i is defined by (5.2.10) and $\lambda_{ii'}$ by (5.2.3). Substituting (5.2.4) and the analog of (5.2.16), we find that the latter equation becomes

$$kr\hat{\mu}' + (r-\lambda)\hat{\alpha}'_i = T_i.$$

The estimates are thus

$$\hat{\alpha}'_i = (r-\lambda)^{-1}(T_i - rJ^{-1}\sum_j h_j). \tag{5.2.34}$$

If $\hat{\psi}'$ is the estimate of the contrast $\psi = \Sigma_i c_i\alpha_i$ $(\Sigma_i c_i = 0)$ obtained from these estimates, $\hat{\psi}' = \Sigma_i c_i\hat{\alpha}'_i$, we may also write

$$\hat{\psi}' = \sum c_i\hat{\alpha}''_i,$$

where

$$\hat{\alpha}''_i = T_i/(r - \lambda). \tag{5.2.35}$$

If $2r > J$ it is easier to calculate

$$\hat{\alpha}_i''' = -\tilde{T}_i/(r-\lambda), \tag{5.2.36}$$

where $\tilde{T}_i$ is defined by (5.2.22), and use $\hat{\psi}' = \Sigma_i c_i \hat{\alpha}_i'''$. The "error SS" when the block totals are treated as the observations will be denoted by SS_f, and may be calculated from (1.3.10) and the right sides of the normal equations (5.2.33) as

$$SS_f = \sum_j h_j^2 - \hat{\mu}' k \sum_j h_j - \sum_i \hat{\alpha}_i' T_i.$$

Substituting (5.2.34), and utilizing $\Sigma_i T_i = \Sigma_i \Sigma_j K_{ij} h_j = k \Sigma_j h_j$ and (5.2.5), we find this becomes

$$SS_f = \sum_j h_j^2 - \frac{\sum_i T_i^2}{r-\lambda} + \frac{I(k-1)\left(\sum_j h_j\right)^2}{J(I-k)}. \tag{5.2.37}$$

If $J > I$, then by the general theory of Ch. 2 the following estimate of σ_f^2

$$\hat{\sigma}_f^2 = SS_f/(J-I) \tag{5.2.38}$$

is distributed as $\sigma_f^2 \chi_{J-I}^2/(J-I)$, and is independent of the $\{\hat{\alpha}_i'\}$, and hence of $\hat{\psi}'$.

We shall now prove that under Ω' the new estimate $\hat{\psi}' = \Sigma_i c_i \hat{\alpha}_i''$ is statistically independent of the former estimate[19] $\hat{\psi} = \Sigma_i c_i \hat{\alpha}_i$, no matter what the values of the $\{c_i\}$ subject to $\Sigma_i c_i = 0$. We consider first the covariance of the treatment total g_i with the block total h_j:

$$g_i = \sum_{j'} K_{ij'} y_{ij'}, \qquad h_j = \sum_{i'} K_{i'j} y_{i'j}.$$

In substituting $y_{ij} = \mu' + \alpha_i + b_j + e_{ij}$ to calculate the covariance we may assume that $\mu' = 0$ and all $\alpha_i = 0$, since this does not affect the covariance. Then $\{y_{ij}\}$, $\{g_i\}$, and $\{h_j\}$ all have zero means and

$$\text{Cov}(g_i, h_j) = E(g_i h_j) = \sum_{j'} \sum_{i'} K_{ij'} K_{i'j} E(y_{ij'} y_{i'j}). \tag{5.2.39}$$

Now

$$E(y_{ij'} y_{i'j}) = E[(b_{j'} + e_{ij'})(b_j + e_{i'j})]$$

and, since the $\{b_j, e_{ij}\}$ are completely independent with zero means and variances σ_B^2 and σ_e^2,

$$E(b_j b_{j'}) = \delta_{jj'} \sigma_B^2,$$
$$E(e_{ij'} e_{i'j}) = \delta_{ii'} \delta_{jj'} \sigma_e^2,$$
$$E(b_{j'} e_{i'j}) = E(e_{ij'} b_j) = 0,$$

hence

$$E(y_{ij'} y_{i'j}) = \delta_{jj'}(\sigma_B^2 + \delta_{ii'} \sigma_e^2).$$

[19] The estimate $\hat{\psi}'$ is called the *interblock estimate*, and $\hat{\psi}$, the *intrablock* estimate.

Substituting this in (5.2.39) gives us

$$\operatorname{Cov}(g_i, h_j) = K_{ij}\sum_{i'} K_{i'j}(\sigma_B^2+\delta_{ii'}\sigma_e^2) = K_{ij}(k\sigma_B^2+\sigma_e^2) = K_{ij}\sigma_f^2/k.$$

Still assuming that $E(y_{ij}) = 0$, we calculate next the covariance of the adjusted treatment mean $\mathscr{G}_i$ with the block total h_j:

$$\mathscr{G}_i = g_i - k^{-1}T_i = g_i - k^{-1}\sum_{j'} K_{ij'}h_{j'},$$

$$\operatorname{Cov}(\mathscr{G}_i, h_j) = E(\mathscr{G}_i h_j) = E[(g_i - k^{-1}\sum_{j'} K_{ij'}h_{j'})h_j]$$

$$= E(g_i h_j) - k^{-1}\sum_{j'} K_{ij'}E(h_{j'}h_j).$$

We have already noted that the $\{h_j\}$ are independently $N(0, \sigma_f^2)$, so $E(h_j h_{j'}) = \delta_{jj'}\sigma_f^2$, and

$$\operatorname{Cov}(\mathscr{G}_i, h_j) = K_{ij}k^{-1}\sigma_f^2 - k^{-1}\sum_{j'} K_{ij'}\delta_{jj'}\sigma_f^2 = K_{ij}k^{-1}\sigma_f^2 - k^{-1}K_{ij}\sigma_f^2 = 0.$$

Since the $\{\mathscr{G}_i, h_j\}$ have a joint normal distribution, this means the $\{\mathscr{G}_i\}$, and hence the $\{\hat{\alpha}_i\}$, are independent of the $\{h_j\}$. But $\hat{\psi}$ is a function only of the $\{\hat{\alpha}_i\}$, and $\hat{\psi}'$ only of the $\{h_j\}$, and hence $\hat{\psi}$ and $\hat{\psi}'$ are independent.

We have earlier calculated

(5.2.40) $$\operatorname{Var}(\hat{\psi}) = \sigma_e^2 \sum_i c_i^2/(r\mathscr{E})$$

under Ω, and have noted that the $\{\hat{\alpha}_i\}$ have the same distribution under Ω' as under Ω; so (5.2.40) must be true under Ω'. To calculate $\operatorname{Var}(\hat{\psi}')$ we need $\operatorname{Cov}(\hat{\alpha}_i'', \hat{\alpha}_{i'}'')$: For this calculation we again assume that μ' and $\{\alpha_i\}$ are zero:

$$(r-\lambda)^2 \operatorname{Cov}(\hat{\alpha}_i'', \hat{\alpha}_{i'}'') = E(T_i T_{i'}) = E\left(\sum_j K_{ij}h_j \sum_{j'} K_{i'j'}h_{j'}\right)$$

$$= \sum_j \sum_{j'} K_{ij}K_{i'j'}\delta_{jj'}\sigma_f^2 = \sum_j K_{ij}K_{i'j}\sigma_f^2 = \lambda_{ii'}\sigma_f^2,$$

where we may write $\lambda_{ii'} = \lambda + \delta_{ii'}(r-\lambda)$. Then

$$\operatorname{Var}(\hat{\psi}') = \operatorname{Var}\left(\sum_i c_i\hat{\alpha}_i''\right) = \sum_i \sum_{i'} c_i c_{i'} \operatorname{Cov}(\hat{\alpha}_i'', \hat{\alpha}_{i'}'')$$

$$= (r-\lambda)^{-2}\sum_i \sum_{i'} c_i c_{i'}[\lambda + \delta_{ii'}(r-\lambda)]\sigma_f^2 = \sum_i c_i^2\sigma_f^2/(r-\lambda).$$

If (the symbols w, w' are defined differently by different writers) we define

(5.2.41) $$w = r\mathscr{E}/\sigma_e^2, \qquad w' = (r-\lambda)/\sigma_f^2,$$

then, since $\hat{\psi}$ and $\hat{\psi}'$ are statistically independent, the linear combination of $\hat{\psi}$ and $\hat{\psi}'$, with expected value ψ, which has minimum variance is

$$(5.2.42) \qquad (w\hat{\psi} + w'\hat{\psi}')/(w + w'),$$

and its variance is $\Sigma_i c_i^2/(w+w')$. We shall define the estimate $\hat{\psi}^*$ as

$$(5.2.43) \qquad \hat{\psi}^* = (\hat{w}\hat{\psi} + \hat{w}'\hat{\psi}')/(\hat{w} + \hat{w}'),$$

where $\hat{w}$ and $\hat{w}'$ are estimates of w and w' obtained by replacing σ_e^2 and σ_f^2 in (5.2.41) by certain estimates: We take $\hat{w} = r\mathscr{E}/\text{MS}_e$. One might[20] take $\hat{w}' = (r-\lambda)/\hat{\sigma}_f^2$ if $J > I$, where $\hat{\sigma}_f^2$ is defined by (5.2.38). The usual choice of $\hat{w}'$ is

$$(5.2.44) \qquad \hat{w}' = (r - \lambda)/\hat{\hat{\sigma}}_f^2,$$

where $\hat{\hat{\sigma}}_f^2$ is an unbiased estimate of σ_f^2, in general more precise than $\hat{\sigma}_f^2$, obtained from $\text{MS}_{\text{bl.el.tr}} = (J-1)^{-1}\text{SS}_{\text{bl.el.tr}}$ defined by (5.2.19): We will show below that under Ω'

$$(5.2.45) \qquad E(\text{MS}_{\text{bl.el.tr}}) = \sigma_e^2 + (n-I)(J-1)^{-1}\sigma_B^2.$$

Hence an unbiased estimate of σ_B^2 is

$$\sigma_B^2 = (J-1)(n-I)^{-1}(\text{MS}_{\text{bl.el.tr}} - \text{MS}_e),$$

and it follows from (5.2.32*b*) that an unbiased estimate of σ_f^2 is

$$(5.2.46) \qquad \hat{\hat{\sigma}}_f^2 = k(n-I)^{-1}[k(J-1)\text{MS}_{\text{bl.el.tr}} - (I-k)\text{MS}_e].$$

This is used in (5.2.44).

To derive (5.2.45) from (5.2.19) it is convenient to assume that μ' and all $\alpha_i = 0$. This may be justified by arguing as follows that $E(\text{MS}_{\text{bl.el.tr}})$ does not depend on μ' or $\{\alpha_i\}$: Write $\text{SS}_{\text{bl.el.tr}} = \text{SS}_B$. Then under Ω, SS_B is distributed as σ_e^2 times a noncentral chi-square variable with noncentrality parameter which we shall call δ_B, and the expected value of SS_B under Ω is

$$(5.2.47) \qquad E(\sigma_e^2\chi_{I-1,\delta_B}'^2) = \sigma_e^2(I-1 + \delta_B^2).$$

Now δ_B is the noncentrality parameter of the statistic $\mathscr{F}$ for testing H_B: all $\beta_j = 0$ under Ω, and hence, by the remark below (2.7.3) about the power of the test, δ_B cannot depend on μ or $\{\alpha_i\}$. But (5.2.47) is the conditional

[20] This would have the advantage of making $\hat{\psi}^*$ unbiased, because $\hat{\psi}$, $\hat{\psi}'$, $\hat{w}$, $\hat{w}'$ would then be statistically independent. Although the joint distribution of $\hat{\psi}$, $\hat{\psi}'$, $\hat{w}$, $\hat{w}'$ is then quite simple, it does not seem to lead to exact tests and exact interval estimation based on the family (5.2.43). Of course the exact methods based on taking $\hat{w}' = 0$, i.e., ignoring the interblock information, are still valid under Ω' as well as under Ω, but no longer possess the optimum properties they had under Ω.

expected value, given the $\{b_j\}$, of SS_B under Ω', and since this depends only on σ_e^2 and $\{\beta_j = b_j - b_.\}$, therefore the unconditional expected value, obtained by taking the expected value of the right member of (5.2.47) with respect to the distribution of the $\{b_j\}$, which are independently $N(0, \sigma_B^2)$, can depend only on σ_e^2 and σ_B^2. If we now assume that $\mu' = 0$ and $\alpha_i = 0$ for the calculation, then $\{h_j\}$, $\{\mathscr{G}_i\}$, and $\{g_i\}$ all have zero means, and so the expected values of their squares are equal to their variances. Utilizing this in (5.2.19) we get

$$(5.2.48) \qquad E(SS_B) = k^{-1}J \operatorname{Var}(h_j) + (r\mathscr{E})^{-1}I \operatorname{Var}(\mathscr{G}_i) - r^{-1}I \operatorname{Var}(g_i).$$

Now

$$\operatorname{Var}(h_j) = \sigma_f^2 = k(k\sigma_B^2 + \sigma_e^2),$$
$$\operatorname{Var}(g_i) = r(\sigma_B^2 + \sigma_e^2),$$
$$\operatorname{Var}(\mathscr{G}_i) = \operatorname{Var}(r\mathscr{E}\hat{\alpha}_i) = r\mathscr{E}\sigma_e^2(I-1)/I,$$

the last equality following from (5.2.21). Substituting these expressions into (5.2.48), we find

$$E(SS_B) = (n-I)\sigma_B^2 + (J-1)\sigma_e^2,$$

and division by $J-1$ then gives (5.2.45).

We mention that a somewhat less ad hoc derivation of the estimate $\hat{\psi}^*$ is possible: It consists in showing that if w/w' were known then (5.2.42) would be the LS estimate[21] of ψ. The sum of squares (1.5.2) that would have to be minimized for this purpose may be shown to be equal to

$$(5.2.49) \qquad \sum_i \sum_{i'} \sum_j K_{ij} K_{i'j} (A + \delta_{ii'})(y_{ij} - \mu' - \alpha_i)(y_{i'j} - \mu' - \alpha_{i'}),$$

where

$$A = -\sigma_B^2/(k\sigma_B^2 + \sigma_e^2) = -k^{-1} + r\mathscr{E}(r - \lambda)^{-1}w'/w.$$

For making statistical inferences about the $\{\alpha_i\}$ it is convenient to define quantities $\{\hat{\alpha}_i^*\}$ as follows: If $2r \leq J$, we use the definition

$$\hat{\alpha}_i^* = (\hat{w}\hat{\alpha}_i + \hat{w}'\hat{\alpha}_i'')/(\hat{w} + \hat{w}'),$$

where $\hat{\alpha}_i$ is given by (5.2.18) and $\hat{\alpha}_i''$ by (5.2.35); if $2r > J$, we use the definition

$$\hat{\alpha}_i^* = (\hat{w}\hat{\hat{\alpha}}_i + \hat{w}'\hat{\alpha}_i''')/(\hat{w} + \hat{w}'),$$

where $\hat{\hat{\alpha}}_i$ is given by (5.2.26) and $\hat{\alpha}_i'''$ by (5.2.36). In either case

$$\hat{w} = r\mathscr{E}/MS_e, \qquad \hat{w}' = (r - \lambda)/\hat{\hat{\sigma}}_f^2,$$

[21] In general if the incomplete blocks are not balanced the two derivations lead to different results, and then those based on minimizing (5.2.49) are to be preferred; see Sprott (1956).

where $\hat{\sigma}_f^2$ is defined by (5.2.46), but $\text{MS}_{\text{bl.el.tr}}$ involved there is calculated from (5.2.19) if $2r \leqq J$, and from (5.2.28) if $2r > J$. The estimates $\{\hat{\psi}^*\}$ of the contrasts $\{\psi\}$, where $\psi = \Sigma_i c_i \alpha_i$ and $\Sigma_i c_i = 0$, may be calculated as

$$\hat{\psi}^* = \sum_i c_i \hat{\alpha}_i^*,$$

and have the same values for the two definitions of the $\{\hat{\alpha}_i^*\}$.

Statistical inferences about the $\{\alpha_i\}$ may be made by the approximation obtained by pretending that the estimates $\hat{w}$ and $\hat{w}'$ of the weights w and w' are equal to the unknown true values defined by (5.2.41), in other words, as though for all $\{c_i\}$ with $\Sigma_i c_i = 0$, $\hat{\psi}^*$ were $N(\psi, \Sigma_i c_i^2/(w+w'))$ with w and w' set equal to $\hat{w}$ and $\hat{w}'$. This leads to the approximate test of H_A: all $\alpha_i = 0$ which rejects H_A if

$$(\hat{w} + \hat{w}')\sum_i(\hat{\alpha}_i^* - \hat{\alpha}_.^*)^2 > (I-1)F_{\alpha;I-1,\infty}. \tag{5.2.50}$$

The right member of (5.2.50) may be written $\chi^2_{\alpha;I-1}$. This test will reject H_A if and only if some estimated contrast $\hat{\psi}^*$ exists which is significantly different from zero according to the approximate S-method of multiple comparison which states that for all contrasts $\{\psi\}$ the probability is approximately $1-\alpha$ that they simultaneously satisfy

$$\hat{\psi}^* - S\hat{\sigma}_{\hat{\psi}^*} \leqq \psi \leqq \hat{\psi}^* + S\hat{\sigma}_{\hat{\psi}^*},$$

where S^2 equals the right member of (5.2.50) and $\hat{\sigma}^2_{\hat{\psi}^*} = \Sigma_i c_i^2/(\hat{w}+\hat{w}')$. An approximate T-method may be based on the statement that the probability is approximately $1-\alpha$ that all $\{\psi\}$ simultaneously satisfy

$$|\psi - \hat{\psi}^*| \leqq (\hat{w} + \hat{w}')^{1/2} q_{\alpha;I,\infty}\left(\frac{1}{2}\sum_i |c_i|\right).$$

Since these approximate methods do not allow for the sampling error of w and w', one may expect that in the multiple-comparison methods the probability of all statements being correct is somewhat less than $1-\alpha$, and that the probability of a type-I error in the test is somewhat greater than α.

In order to decide whether recovery of interblock information is worth while in practice we might define the fraction of additional information available from recovery as one less than the ratio of Var $(\hat{\psi})$ to the variance of the quantity (5.2.42), or as

$$\frac{w'}{w} = \frac{r-\lambda}{rk\mathscr{E}} \cdot \frac{\sigma_e^2}{\sigma_e^2 + k\sigma_B^2} = \frac{I-k}{(k-1)I} \cdot \frac{\sigma_e^2}{\sigma_e^2 + k\sigma_B^2},$$

and estimate it as

$$\frac{\hat{w}'}{\hat{w}} = \frac{(r-1)(I-k)\text{MS}_e}{(k-1)[k(J-1)\text{MS}_{\text{bl.el.tr}} - (I-k)\text{MS}_e]}. \tag{5.2.51}$$

We may decide to forego the recovery calculation if[22] the estimated fraction (5.2.51) of additional information is below a certain level, say 10% (in particular if it is negative).

5.3. NESTED DESIGNS

We now consider for the first time a possible relationship of factors in an incomplete layout, called nesting, which requires us to re-examine our concepts of main effects and interactions. The levels of a factor C are said to be *nested* within the levels of a factor A, or, more briefly, C is nested within A, if every level of C appears with only a single level of A in the observations. This means that if A is present at I levels, then the levels of C fall into I sets of $J_1, J_2, \cdots, J_I$ levels, respectively, such that the ith set appears only with the ith level of A. For example, in an experiment to study the effect of certain factors on the corrosion resistance of tin plate, the tin plate comes in long coils, and is to be subjected to different anneals. It is not feasible to divide a coil to give it several anneals, but the whole coil must be given the same anneal. Suppose that there are I anneals, which will be called the I levels of factor A, and suppose that J_1 coils are given the first anneal, J_2 the second, $\cdots$, J_I the Ith. We shall call the $\Sigma_i J_i$ coils the $\Sigma_i J_i$ levels of factor C. Here C is nested within A. Although it would be more realistic to treat the effects of C in this example as random effects, and indeed random-effects models or mixed models are generally more realistic for nested designs, we shall postpone such treatment to Chs. 7 and 8, and consider here only fixed-effects models. The sums of squares, and estimates of fixed effects if any, used with the other models are those derived under the fixed-effects models. We shall be able to introduce the new subscript notation customary with nested designs, which will be useful later in a quick method of calculating expected mean squares in the other models (sec. 8.2).

To characterize further the relation between two factors involved in each of a set of observations it is convenient to call the factors *crossed* if neither is nested within the other, and in that case to say that they are *completely crossed* if every level of one appears with every level of the other, i.e., if they appear in a complete two-way layout when other factors are ignored, and otherwise *partly crossed.*

The new subscript notation employs a double (or possibly multiple, with more factors) subscript to number the levels of a nested factor. Thus in the above example the J_i coils with the ith anneal are numbered $(i, 1)$, $(i, 2), \cdots, (i, J_i)$. With this numbering the nature of the incompleteness

[22] The operating characteristic of this intuitively reasonable procedure is unknown.

of the layout in a nested design is shown in Table 5.3.1, where the crosses indicate the cells in which an observation is taken. The true cell mean where A is at the ith level and C at the (i, j) level will be denoted simply by η_{ij} instead of $\eta_{i,(i,j)}$. It must be remembered however that for different i

TABLE 5.3.1

THE NESTED DESIGN AS AN INCOMPLETE LAYOUT

Levels of A: $i =$	Levels of C: $i, j =$ 1, 1	1, 2	⋮	$1, J_1$	2, 1	2, 2	⋮	$2, J_2$	⋮	I, 1	I, 2	⋮	I, J_I
1	X	X	X	X									
2					X	X	X	X					
⋮													
I										X	X	X	X

and the same j the η_{ij} "have nothing in common" as they do in the complete layout: There is nothing common to the first coil with the first anneal and the first coil with the second anneal in the above example. If there were something common to all the first coils, something to all the second, etc., for instance, if the first coils came from a first source, etc., then C would not be nested within A but crossed with A.

We now consider the possibility of expressing the true cell mean η_{ij} as the sum of a general mean μ, an effect α_i of the ith level of A, and an effect γ_{ij} of the (i, j) level of C,

$$\eta_{ij} = \mu + \alpha_i + \gamma_{ij}. \tag{5.3.1}$$

The quantities on the right will be uniquely defined in terms of the $\{\eta_{ij}\}$ if we impose side conditions (summations on j are understood to be from 1 to J_i)

$$\sum_j w_{ij}\gamma_{ij} = 0, \qquad \sum_i v_i\alpha_i = 0,$$

where $\{v_i\}$, $\{w_{ij}\}$ are arbitrary weights, that is, they are nonnegative numbers such that some $v_i > 0$, and for each i some $w_{ij} > 0$; so that without loss of generality we can assume

$$\sum_i v_i = 1, \qquad \sum_j w_{ij} = 1 \qquad \text{for each } i.$$

For, suppose the $\{\eta_{ij}\}$ can be expressed this way: Then multiplying (5.3.1) by w_{ij} and summing on j, we get

$$\sum_j w_{ij}\eta_{ij} = \mu + \alpha_i,$$

and, multiplying this by v_i and summing on i, we get

$$\sum_i \sum v_i w_{ij}\eta_{ij} = \mu,$$

and hence μ, $\{\alpha_i\}$, $\{\gamma_{ij}\}$ are determined uniquely as

$$(5.3.2)\quad \mu = \sum_i \sum_j v_i w_{ij}\eta_{ij}, \qquad \alpha_i = \sum_j w_{ij}\eta_{ij} - \mu, \qquad \gamma_{ij} = \eta_{ij} - \mu - \alpha_i.$$

On the other hand, it is clear that an arbitrary set of cell means $\{\eta_{ij}\}$ can be expressed this way by defining μ, $\{\alpha_i\}$, $\{\gamma_{ij}\}$ from (5.3.2). This means that, if we think of the $\{\alpha_i\}$ as "main effects" of A and the $\{\gamma_{ij}\}$ as "main effects" of C, there is *no need*, as there is when C is crossed with A, *to introduce interaction terms* to express the cell mean in the general case. Of course the "main effects" defined in this way are not in general the same as would be defined in terms of the cell means of a conceptual complete layout obtained by associating cell means $\{\eta_{i,(i',j)}\}$ with all $I\Sigma_i J_i$ cells of Table 5.3.1.

In a three-factor experiment various combinations of nesting and crossing are possible. An example of an experiment with three factors C, B, and T, in which T is nested within B, and B within C, is one designed to study the variability of a certain brand of facial tissue, in which boxes of tissue are bought in I cities (factor C), J_i boxes (factor B) in the ith city. Then K_{ij} tissues (factor T) are taken from the (i, j) box, i.e., the jth box from the ith city, and tested for some quality characteristic, there being M_{ijk} measurements on the (i, j, k) tissue, i.e., on the kth tissue from the (i, j) box. Suitable notation for this example is indicated by the model equation

$$(5.3.3)\qquad y_{ijkm} = \mu + \gamma_i + \beta_{ij} + \tau_{ijk} + e_{ijkm},$$

for the measurements $\{y_{ijkm}\}$. Side conditions for this (unrealistic) fixed-effects model will be considered below (in equations (5.3.13)). The analysis for this example will be derived below.

An example with three factors A, C, and L, in which C is nested within A, and L is crossed with A and C, can be developed from the earlier one where factor A corresponded to anneals and C to coils, by imagining that on each coil after it is annealed specimens for test are taken from three locations (factor L) on each coil, say the head, tail, and middle of the coil. If there are I anneals, J coils with each anneal, and K ($K = 3$) locations, if Q specimens are taken at each location on each coil, and if y_{ijkq} is the measurement[23] on the qth specimen from the kth location on the jth coil with the ith anneal, then in the fixed-effects model we would express the $\{y_{ijkq}\}$ as

$$y_{ijkq} = \eta_{ijk} + e_{ijkq}. \tag{5.3.4}$$

Then without any loss of generality the $\{\eta_{ijk}\}$ can be written

$$\eta_{ijk} = \mu + \alpha_i^A + \alpha_{ij}^C + \alpha_k^L + \alpha_{ik}^{AL} + \alpha_{ijk}^{CL}, \tag{5.3.5}$$

where

$$\alpha_{.}^A = \alpha_{i.}^C = \alpha_{.}^L = \alpha_{i.}^{AL} = \alpha_{.k}^{AL} = \alpha_{ij.}^{CL} = \alpha_{i.k}^{CL} = 0 \tag{5.3.6}$$

for all i, j, k: Here α_{ij}^C is called the main effect of the (i, j) coil, and α_{ijk}^{CL} its interaction with the kth location, etc. It can be argued as above that there is no need to consider $A \times C$ and $A \times C \times L$ interactions. This is indicated automatically by the notation, since an $A \times C$ interaction term in (5.3.5) would have subscripts i, (i, j) and could be absorbed into α_{ij}^C. The decomposition satisfying (5.3.5) and (5.3.6) follows from the definitions

$$\begin{aligned}
\mu &= \eta_{...}, \\
\alpha_i^A &= \eta_{i..} - \eta_{...}, \\
\alpha_{ij}^C &= \eta_{ij.} - \eta_{i..}, \\
\alpha_k^L &= \eta_{..k} - \eta_{...}, \\
\alpha_{ik}^{AL} &= \eta_{i.k} - \eta_{i..} - \eta_{..k} + \eta_{...}, \\
\alpha_{ijk}^{CL} &= \eta_{ijk} - \eta_{ij.} - \eta_{i.k} + \eta_{i..}.
\end{aligned} \tag{5.3.7}$$

To motivate the last formula we may think of defining within the ith anneal the interaction of the (i, j) coil with the kth location, and then consider that it is not appropriate to average a property of the (i, j) coil over i, that is, over the anneals in this example, for fixed j, this having no physical meaning.

An example with three factors A, P, and C, in which A and P are crossed and C is nested within $A \times P$, can also be developed from the example of anneals A and coils C by imagining a third factor P to correspond to different methods of pickling the anneals. Suppose there are N pickles, and J coils are treated with each of the IN anneal–pickle combinations. The INJ levels of the factor C would now be numbered with a triple-subscript notation, the jth coil nested within the (i, n)

[23] Measured pack life (time to bulge) under controlled temperature of a can of prunes whose bottom is the specimen of tin plate. Larger measurements indicate higher corrosion resistance.

anneal–pickle combination being numbered *inj*. Then, if Q specimens are taken from the same location on each coil, we would put

$$(5.3.8) \qquad y_{ijnq} = \mu + \alpha_i^A + \alpha_{inj}^C + \alpha_n^P + \alpha_{in}^{AP} + e_{ijnq},$$

where

$$(5.3.9) \qquad \alpha_{.}^A = \alpha_{in.}^C = \alpha_{.}^P = \alpha_{i.}^{AP} = \alpha_{.n}^{AP} = 0$$

for all i, n, and the meanings of the terms in (5.3.8) are indicated by the notation.

If we now consider an experiment including all four factors A, P, C, and L, we see that A and P are crossed, C is nested within $A \times P$, and L is crossed with A, P, and C. We would be led to

$$(5.3.10) \qquad y_{ijknq} = \mu + \alpha_i^A + \alpha_n^P + \alpha_{inj}^C + \alpha_k^L + \alpha_{in}^{AP} + \alpha_{ik}^{AL} + \alpha_{nk}^{PL} + \alpha_{injk}^{CL} + \alpha_{ink}^{APL} + e_{ijknq},$$

where

$$(5.3.11) \qquad \alpha_{.}^A = \alpha_{.}^P = \alpha_{in.}^C = \alpha_{.}^L = \alpha_{i.}^{AP} = \alpha_{.n}^{AP} = \alpha_{i.}^{AL} = \alpha_{.k}^{AL} = \alpha_{n.}^{PL} = \alpha_{.k}^{PL} = \alpha_{inj.}^{CL} = \alpha_{in.k}^{CL} = \alpha_{in.}^{APL} = \alpha_{i.k}^{APL} = \alpha_{.nk}^{APL} = 0$$

for all i, j, k, n.

As a further illustration of the notion of nesting, we remark that for any design if we think of a "factor" D corresponding to "error," then D is always nested within the cells corresponding to the treatment combinations used: This is a more realistic formulation of the "factor" D than the one of sec. 4.6 as a fictitious factor crossed with all other factors.

We shall derive now for some of these examples the various SS's usually considered. Rules of thumb for defining the SS's, and calculating them, their numbers of d.f., and their expected values, will be given in sec. 8.2.

To apply the general theory of estimation of Ch. 1 it is convenient to use a single symbol η with appropriate subscripts to denote the sum of the effects (other than errors) entering into an observation, as in (5.3.5). The LS estimates of the η's will be obvious: namely each $\hat{\eta}$ is the mean of all the observations whose expected value is η. The effects are unique linear functions of the η's, and their LS estimates are then the same linear functions of the $\hat{\eta}$'s. This will be illustrated in the two examples that we shall now treat in detail.

A Completely Nested Experiment

Let us consider again the three-factor completely nested example concerning cities, boxes, and tissues, given above (5.3.3). Let

$$y_{ijkm} = \eta_{ijk} + e_{ijkm},$$

where

$$\eta_{ijk} = \mu + \gamma_i + \beta_{ij} + \tau_{ijk}. \tag{5.3.12}$$

Suppose that we impose the side conditions

$$\sum_i u_i\gamma_i = 0, \quad \sum_j v_{ij}\beta_{ij} = 0 \text{ for all } i, \quad \sum_k w_{ijk}\tau_{ijk} = 0 \text{ for all } i, j, \tag{5.3.13}$$

where $\{u_i\}$, $\{v_{ij}\}$, $\{w_{ijk}\}$ are nonnegative weights such that

$$\sum_i u_i = 1, \quad \sum_j v_{ij} = 1 \text{ for all } i, \quad \sum_k w_{ijk} = 1 \text{ for all } i, j.$$

Then

$$\begin{aligned}
\eta_{ijk} &= \mu + \gamma_i + \beta_{ij} + \tau_{ijk}, \\
\sum_k w_{ijk}\eta_{ijk} &= \mu + \gamma_i + \beta_{ij}, \\
\sum_j\sum_k v_{ij}w_{ijk}\eta_{ijk} &= \mu + \gamma_i, \\
\sum_i\sum_j\sum_k u_i v_{ij} w_{ijk}\eta_{ijk} &= \mu,
\end{aligned}$$

from which we find by subtracting successive equations that

$$\begin{aligned}
\tau_{ijk} &= \eta_{ijk} - \sum_{k'} w_{ijk'}\eta_{ijk'}, \\
\beta_{ij} &= \sum_k w_{ijk}\eta_{ijk} - \sum_{j'}\sum_k v_{ij'}w_{ij'k}\eta_{ij'k}, \\
\gamma_i &= \sum_j\sum_k v_{ij}w_{ijk}\eta_{ijk} - \sum_{i'}\sum_j\sum_k u_{i'}v_{i'j}w_{i'jk}\eta_{i'jk}, \\
\mu &= \sum_i\sum_j\sum_k u_i v_{ij} w_{ijk}\eta_{ijk}.
\end{aligned} \tag{5.3.14}$$

Under the usual Ω-assumptions, namely (5.3.12), (5.3.13), and the assumption that the $\{e_{ijkm}\}$ are independently $N(0, \sigma^2)$, we have to minimize

$$\mathscr{S} = \sum_{i=1}^{I}\sum_{j=1}^{J_i}\sum_{k=1}^{K_{ij}}\sum_{m=1}^{M_{ijk}} (y_{ijkm} - \eta_{ijk})^2. \tag{5.3.15}$$

The LS estimates are evidently

$$\hat{\eta}_{ijk} = y_{ijk.},$$

and the error SS is

$$\text{SS}_e = \sum_i\sum_j\sum_k\sum_m (y_{ijkm} - y_{ijk.})^2,$$

and is easily seen to have $\Sigma_i\Sigma_j\Sigma_k(M_{ijk}-1)$ d.f. If we wanted the LS estimates of μ, $\{\gamma_i\}$, $\{\beta_{ij}\}$, $\{\tau_{ijk}\}$, they would be obtained by replacing the $\{\eta_{ijk}\}$ in (5.3.14) by the $\{\hat{\eta}_{ijk}\}$. Of more interest are the sums of squares for the factors C, B, and T, which we shall now derive.

Consider the hypotheses

$$H_C\text{: all } \gamma_i = 0, \qquad H_B\text{: all } \beta_{ij} = 0, \qquad H_T\text{: all } \tau_{ijk} = 0,$$

and let

$$\omega_C = H_C \cap \Omega, \qquad \omega_B = H_B \cap \Omega, \qquad \omega_T = H_T \cap \Omega.$$

The truth or falsity of these hypotheses depends on the weights to an extent which we now note: If H_T is true for any system of weights, it is true for any other, for from the first of the equations (5.3.14) we see that if the true means for the tissues in the i,j box are equal, i.e., if $\eta_{ij1} = \eta_{ij2} = \cdots = \eta_{ijn_{ij}}$, then $\tau_{ij1} = \tau_{ij2} = \cdots = \tau_{ijn_{ij}} = 0$, and conversely, regardless of the system of weights. The hypothesis H_B may be interpreted in terms of the weighted box means

$$B_{ij} = \sum_k w_{ijk}\eta_{ijk}$$

as saying that, for each i, $B_{i1} = B_{i2} = \cdots = B_{in_i}$. Whether these equalities hold depends on the weights $\{w_{ijk}\}$ used to define the weighted box means $\{B_{ij}\}$, but not on the weights $\{u_i\}$ or $\{v_{ij}\}$. Similarly the hypothesis H_C says that certain weighted city means

$$C_i = \sum_j v_{ij}B_{ij} = \sum_j \sum_k v_{ij}w_{ijk}\eta_{ijk}$$

are all equal, and the truth of this hypothesis depends on the weights $\{v_{ij}\}$ and $\{w_{ijk}\}$ but not on the $\{u_i\}$. We shall not discuss which weighting system is appropriate to a real situation since the present model is usually not realistic.

For minimizing $\mathscr{S}$ under these different ω's it is convenient to specialize the weights as follows: Let

$$(5.3.16) \qquad n_{ij} = \sum_k M_{ijk}, \qquad n_i = \sum_j \sum_k M_{ijk}, \qquad n = \sum_i \sum_j \sum_k M_{ijk},$$

so that n_{ij} is the number of observations on the (i,j) box, n_i the number for the ith city, and n the total number. We take the weights proportional to the numbers of observations,

$$(5.3.17) \qquad w_{ijk} = M_{ijk}/n_{ij}, \qquad v_{ij} = n_{ij}/n_i, \qquad u_i = n_i/n.$$

Write (5.3.15) in the form

$$(5.3.18) \quad \mathscr{S} = \sum_i \sum_j \sum_k \sum_m [(y_{ijkm}-\hat{\eta}_{ijk}) + (\hat{\mu}-\mu) + (\hat{\gamma}_i-\gamma_i) + (\hat{\beta}_{ij}-\beta_{ij}) + (\hat{\tau}_{ijk}-\tau_{ijk})]^2.$$

Then

$$\mathscr{S} = \text{SS}_e + n(\hat{\mu}-\mu)^2 + \sum_i n_i(\hat{\gamma}_i-\gamma)^2 + \sum_i\sum_j n_{ij}(\hat{\beta}_{ij}-\beta_{ij})^2 + \sum_i\sum_j\sum_k M_{ijk}(\hat{\tau}_{ijk}-\tau_{ijk})^2, \tag{5.3.19}$$

the sums of cross products vanishing because of the choice of weights (5.3.17) in the side conditions (5.3.13), for instance,

$$\sum_i\sum_j\sum_k\sum_m(\hat{\beta}_{ij}-\beta_{ij})(\hat{\tau}_{ijk}-\tau_{ijk}) = \sum_i\sum_j(\hat{\beta}_{ij}-\beta_{ij})\sum_k M_{ijk}(\hat{\tau}_{ijk}-\tau_{ijk}) = 0.$$

The minimum of $\mathscr{S}$ under ω_C, ω_B, or ω_T is now obvious, for example

$$\mathscr{S}_{\omega_C} = \text{SS}_e + \sum_i n_i\hat{\gamma}_i^2.$$

It follows that the numerator SS of the F-statistic for testing H_C under Ω, namely $\text{SS}_C = \mathscr{S}_{\omega_C} - \text{SS}_e$, is

$$\text{SS}_C = \sum_i n_i\hat{\gamma}_i^2. \tag{5.3.20}$$

Similarly the numerator SS's for testing H_B and H_T are

$$\text{SS}_B = \sum_i\sum_j n_{ij}\hat{\beta}_{ij}^2, \tag{5.3.21}$$

$$\text{SS}_T = \sum_i\sum_j\sum_k M_{ijk}\hat{\tau}_{ijk}^2. \tag{5.3.22}$$

The usual identity relating the SS's may be obtained by noting from (5.3.15) that if we set all the symbols μ, $\{\gamma_i\}$, $\{\beta_{ij}\}$, $\{\tau_{ijk}\}$ equal to zero, then $\mathscr{S}$ is the total sum of squares. If we use this in (5.3.19) we get

$$\text{SS}_{\text{"tot"}} = \text{SS}_e + \text{SS}_C + \text{SS}_B + \text{SS}_T,$$

where

$$\text{SS}_{\text{"tot"}} = \sum_i\sum_j\sum_k\sum_m y_{ijkm}^2 - n\hat{\mu}^2. \tag{5.3.23}$$

For numerical computation it is convenient to introduce the following unweighted averages[24] of *observations*:

$$\begin{aligned} \bar{y}_{ij..} &= \sum_k\sum_m y_{ijkm}/n_{ij} = \sum_k M_{ijk}y_{ijk.}/n_{ij}, \\ \bar{y}_{i...} &= \sum_j\sum_k\sum_m y_{ijkm}/n_i = \sum_j n_{ij}\bar{y}_{ij..}/n_i, \\ \bar{y}_{....} &= \sum_i\sum_j\sum_k\sum_m y_{ijkm}/n = \sum_i n_i\bar{y}_{i...}/n. \end{aligned} \tag{5.3.24}$$

[24] A bar is put over $y_{ij..}$ because without the bar the dot notation might suggest that it means the unweighted average of the $y_{ijk.}$ over k; similarly for the bars over $y_{i...}$ and $y_{....}$.

The estimates $\{\hat{\mu}, \hat{\gamma}_i, \hat{\beta}_{ij}, \hat{\tau}_{ijk}\}$ may be expressed in terms of these averages, by replacing the $\{\eta_{ijk}\}$ in (5.3.14) by the $\{y_{ijk\cdot}\}$. Substituting the resulting expressions into (5.3.20), (5.3.21), (5.3.22), and (5.3.23), we then get the formulas indicated in Table 5.3.2. The numbers of d.f. in the first three lines of the table may be obtained by considering the numbers of independent restrictions imposed respectively by the hypotheses H_C, H_B, and H_T. The expected mean squares have not been included in the table because if calculated under the unrealistic fixed-effects model they have no practical interest. Whether the three F-statistics for testing H_C, H_B, and

TABLE 5.3.2

ANALYSIS OF VARIANCE OF A THREE-FACTOR COMPLETELY NESTED EXPERIMENT

Source	SS	d.f.
C	$SS_C = \sum_i n_i \bar{y}^2_{i\cdots} - n\bar{y}^2_{\cdots\cdot}$	$I-1$
B (within C)	$SS_B = \sum_i \sum_j n_{ij}\bar{y}^2_{ij\cdot\cdot} - \sum_i n_i \bar{y}^2_{i\cdots}$	$\sum_i (J_i - 1)$
T (within B)	$SS_T = \sum_i \sum_j \sum_k M_{ijk} y^2_{ijk\cdot} - \sum_i \sum_j n_{ij}\bar{y}^2_{ij\cdot\cdot}$	$\sum_i \sum_j (K_{ij} - 1)$
Error	$SS_e = \sum_i \sum_j \sum_k \sum_m y^2_{ijkm} - \sum_i \sum_j \sum_k M_{ijk} y^2_{ijk\cdot}$	$\sum_i \sum_j \sum_k (M_{ijk} - 1)$
"Total"	$SS_{\text{"tot"}} = \sum_i \sum_j \sum_k \sum_m y^2_{ijkm} - ny^2_{\cdots\cdot}$	$n-1$

H_T under this model are appropriate under a more realistic model will be considered later (sec. 7.6).

An Experiment with Both Nesting and Crossing of Factors

We consider further the three-factor experiment introduced above (5.3.4), with anneals as the I levels of factor A, coils as the IJ levels of factor C, locations as the K levels of factor L, and with Q specimens from each location on each coil. The Ω-assumptions are (5.3.4), (5.3.5), (5.3.6), and the assumption that the $\{e_{ijkq}\}$ are independently $N(0, \sigma^2)$. We again have the LS estimates

$$\hat{\eta}_{ijk} = y_{ijk\cdot},$$

and the error SS is

$$SS_e = \sum_i \sum_j \sum_k \sum_q y^2_{ijkq} - Q \sum_i \sum_j \sum_k y^2_{ijk\cdot}.$$

The LS estimates of the various effects are obtained by replacing the $\{\eta_{ijk}\}$ by the $\{\hat{\eta}_{ijk}\}$ in (5.3.7).

The equation (5.3.5) suggests our considering the hypotheses H_A, H_C, H_L, H_{AL}, H_{CL}, where H_A is the hypothesis that all $\alpha_i^A = 0$, H_C that all $\alpha_{ij}^C = 0$, etc., and we define ω_A, ω_C, ω_L, ω_{AL}, ω_{CL} as $\omega_A = H_A \cap \Omega$, $\omega_C = H_C \cap \Omega$, etc. By expanding

$$\mathscr{S} = \sum_i \sum_j \sum_k \sum_q (y_{ijkq} - \mu - \alpha_i^A - \alpha_{ij}^C - \text{etc.})^2$$

in the form

$$\mathscr{S} = \sum_i \sum_j \sum_k \sum_q [(y_{ijkq} - \hat{\eta}_{ijk}) + (\hat{\mu} - \mu) + (\hat{\alpha}_i^A - \alpha_i^A) + \text{etc.}]^2,$$

we get

(5.3.25)

$$\mathscr{S} = \text{SS}_e + IJKQ(\hat{\mu}-\mu)^2 + JKQ\sum_i(\hat{\alpha}_i^A - \alpha_i^A)^2 + KQ\sum_i\sum_j(\hat{\alpha}_{ij}^C - \alpha_{ij}^C)^2$$
$$+ IJQ\sum_k(\hat{\alpha}_k^L - \alpha_k^L)^2 + JQ\sum_i\sum_k(\hat{\alpha}_{ik}^{AL} - \alpha_{ik}^{AL})^2 + Q\sum_i\sum_j\sum_k(\alpha_{ijk}^{CL} - \alpha_{ijk}^{CL})^2.$$

From this the minimum of $\mathscr{S}$ under each of ω_A, ω_C, etc., is obvious, and we may deduce that the numerator SS's for testing H_A, H_C, etc., under Ω are

$$\text{SS}_A = JKQ\sum_i(\hat{\alpha}_i^A)^2,$$

$$\text{SS}_C = KQ\sum_i\sum_j(\hat{\alpha}_{ij}^C)^2,$$

$$\text{SS}_L = IJQ\sum_k(\hat{\alpha}_k^L)^2,$$

$$\text{SS}_{AL} = JQ\sum_i\sum_k(\hat{\alpha}_{ik}^{AL})^2,$$

$$\text{SS}_{CL} = Q\sum_i\sum_j\sum_k(\hat{\alpha}_{ijk}^{CL})^2.$$

Putting μ, $\{\alpha_i^A\}$, $\{\alpha_{ij}^C\}$, etc. all equal to zero in the identity (5.3.25) we get the identity

$$\text{SS}_{\text{"tot."}} = \text{SS}_e + \text{SS}_A + \text{SS}_C + \text{SS}_L + \text{SS}_{AL} + \text{SS}_{CL}.$$

Formulas for these sums of squares suitable for numerical computation are given in Table 5.3.3. The number of d.f. shown for SS_e may be calculated as $n-r$, where $n = IJKQ$ is the total number of observations, and $r = IJK$ is the number of $\{\eta_{ijk}\}$. The numbers of d.f. for the other SS's in the table are equal to the numbers of independent restrictions imposed by the corresponding hypotheses.

TABLE 5.3.3

ANALYSIS OF VARIANCE OF A THREE-FACTOR EXPERIMENT WITH NESTING AND CROSSING

Source		SS	d.f.
Main effects	A	$JKQ\sum_i y_{i..}^2 - IJKQy_{...}^2$	$I-1$
	C	$KQ\sum_i\sum_j y_{ij..}^2 - JKQ\sum_i y_{i...}^2$	$I(J-1)$
	L	$IJQ\sum_k y_{..k.}^2 - IJKQy_{....}^2$	$K-1$
Interactions	$A\times L$	$JQ\sum_i\sum_k y_{i.k.}^2 - JKQ\sum_i y_{i...}^2 - IJQ\sum_k y_{..k.}^2 + IJKQy_{....}^2$	$(I-1)(K-1)$
	$C\times L$	$Q\sum_i\sum_j\sum_k y_{ijk.}^2 - KQ\sum_i\sum_j y_{ij..}^2 - JQ\sum_i\sum_k y_{i.k.}^2 + JKQ\sum_i y_{i...}^2$	$I(J-1)(K-1)$
Error		$\sum_i\sum_j\sum_k\sum_q y_{ijkq}^2 - Q\sum_i\sum_j\sum_k y_{ijk.}^2$	$IJK(Q-1)$
"Total"		$\sum_i\sum_j\sum_k\sum_q y_{ijkq}^2 - IJKQy_{....}^2$	$IJKQ-1$

PROBLEMS

5.1. An experiment to compare six different legume intercycle crops was conducted in Hawaii according to a Latin-square design. Table A gives the yields of legumes A, B, $\cdot\cdot\cdot$, F in 10-gram units (net weight) per 1/3000 acre,

TABLE* A

B 220	*F* 98	*D* 149	*A* 92	*E* 282	*C* 169
A 74	*E* 238	*B* 158	*C* 228	*F* 48	*D* 188
D 118	*C* 279	*F* 118	*E* 278	*B* 176	*A* 65
E 295	*B* 222	*A* 54	*D* 104	*C* 213	*F* 163
C 187	*D* 90	*E* 242	*F* 96	*A* 66	*B* 122
F 90	*A* 124	*C* 195	*B* 109	*D* 79	*E* 211

* From p. 247 of *Statistical Theory in Research* by R. L. Anderson and T. A. Bancroft, McGraw-Hill, New York, 1952. Reproduced with the kind permission of the authors and the publisher.

three months after planting. (*a*) Carry out the analysis of variance. (*b*) Use the *T*-method with $\alpha = 0.10$ to compare pairwise the three legumes with the highest yields.

5.2. The responses of five pairs of monkeys to a certain kind of stimulus under five different conditions during five periods consisting of successive weeks were observed according to the Latin-square design in Table B. The numbers are the total numbers of responses, the letters denote the conditions. Analyze the data.

TABLE* B

Animals	Period 1	2	3	4	5
1	194 *B*	369 *D*	344 *C*	380 *A*	693 *E*
2	202 *D*	142 *B*	200 *A*	356 *E*	473 *C*
3	335 *C*	301 *A*	439 *E*	338 *B*	528 *D*
4	515 *E*	590 *C*	552 *B*	677 *D*	546 *A*
5	184 *A*	421 *E*	355 *D*	284 *C*	366 *B*

* Data from Query no. 113, edited by G. W. Snedecor, *Biometrics*, Vol. 11, 1955, p. 112. Reproduced with the kind permission of the editors.

5.3. In a test to compare detergents with respect to a certain characteristic a large stack of dinner plates soiled in a specified standard way is prepared, and the detergents are tested in blocks of three, there being in each block three basins with different detergents and three dishwashers who rotate after washing each plate. The measurements in Table C are the numbers of plates washed before the foam disappears from the basin. Use the *T*-method with 0.90 confidence coefficient on the intrablock estimates to decide which pairs of detergents differ significantly.

TABLE C

Detergent	Block No. 1	2	3	4	5	6	7	8	9	10
A	27	28	30	31	29	30				
B	26	26	29				30	21	26	
C	30			34	32		34	31		33
D		29		33		34	31		33	31
E			26		24	25		23	24	26

5.4. In an experiment designed to compare seven thermometers $A, B, \cdots, G$ it was possible to immerse only three of them at a time into a bath, the temperature of which was slowly and constantly rising. Seven blocks of three were used. Readings were taken in the order indicated in Table D. Each thermometer was read once first, second, and third in its block. Time intervals between readings in each block were kept fixed. Readings were made in thousandths of a degree, just above 30° C. Only the last two figures are given, e.g., the entry *A* 56 corresponds to a reading of 30.056° on thermometer *A*, etc. (*a*) Analyze the data as an incomplete Latin square. (*b*) With the *S*-method with $\alpha = 0.10$,

TABLE* D

Order of Reading	Block No. 1	2	3	4	5	6	7
I	*A* 56	*E* 16	*B* 41	*F* 46	*C* 54	*G* 34	*D* 50
II	*B* 31	*F* 41	*C* 53	*G* 32	*D* 43	*A* 68	*E* 32
III	*D* 35	*A* 58	*E* 24	*B* 46	*F* 50	*C* 60	*G* 38

* From Table 54, p. 102 of *Statistical Methods for Chemists* by W. J. Youden, John Wiley, New York, 1951. Reproduced with the kind permission of the author and the publisher.

which pairs of thermometers show no significant difference? Would the T-method pick up any of these differences? (*c*) With the S-method with $\alpha = 0.10$ answer the following question: Suppose thermometers A, F, G are from one manufacturer and the others from another. Does the average reading of the three thermometers from one manufacturer differ significantly from the average of those from the other? (This question is not to be confused with whether the average of the production of one manufacturer differs from that of the other, which involves mixed-model concepts and would require much larger samples from each manufacturer for its investigation.)

5.5. Use formula (5.2.51) to decide whether the calculation for recovery of interblock information would be worth while for the data of Problem 5.3.

5.6. Table E gives coded values of strain measurements on each of four seals made on each of the four heads of each of five sealing machines. Calculate the MS's for machines, for heads within machines, and for error. Save your results for analysis in Ch. 7.

TABLE* E

Machine *A*				*B*				*C*				*D*				*E*			
Head				Head				Head				Head				Head			
1	2	3	4	5	6	7	8	9	10	11	12	13	14	15	16	17	18	19	20
6	13	1	7	10	2	4	0	0	10	8	7	11	5	1	0	1	6	3	3
2	3	10	4	9	1	1	3	0	11	5	2	0	10	8	8	4	7	0	7
0	9	0	7	7	1	7	4	5	6	0	5	6	8	9	6	7	0	2	4
8	8	6	9	12	10	9	1	5	7	7	4	4	3	4	5	9	3	2	0

* From Table II on p. 14 of "Fundamentals of Analysis of Variance, Part III" by C. R. Hicks in *Industrial Quality Control*, Vol. 13, no. 4, 1956. Reproduced with the kind permission of the author and the editor.

5.7. Table F gives coded breaking strengths measured on six tissues (all in the same column) from different numbers (2, 3, 4) of boxes all of the same brand, bought in each of three cities in an investigation of brand variability. Calculate the mean squares for cities, boxes within cities, and tissues within boxes (error). Save your results for analysis in Ch. 7.

TABLE* F

City								
1		2			3			
1.59	1.72	2.44	2.27	2.46	1.36	1.59	1.73	1.53
1.80	1.40	2.11	2.70	2.21	1.43	1.50	1.74	1.41
1.72	2.02	2.41	2.36	2.50	1.48	1.50	1.65	1.64
1.69	1.75	2.48	2.36	2.37	1.55	1.49	1.58	1.51
1.71	1.95	2.36	2.16	2.24	1.53	1.47	1.49	1.52
1.83	1.61	2.36	2.04	2.25	1.39	1.63	1.70	1.36

* These "data" were manufactured by using random normal deviates with parameter values estimated from real data not suitable for the purpose of this problem.

5.8. In an investigation of the variability of the strength of tire cord, preliminary to the establishment of control of cord-testing laboratories, the data in Table G were gathered from two plants using different manufacturing processes to make nominally the same kind of cord. Eight bobbins of cord were selected at random from each plant. Adjacent pairs of breaks (to give as nearly as possible "duplicate" measurements) were made at 500-yard intervals over the length of each bobbin. The coded raw data are the measured strengths recorded

TABLE* G

	Distance	0 yd		500 yd		1000 yd		1500 yd		2000 yd		2500 yd	
	Adjacent Breaks	1	2	1	2	1	2	1	2	1	2	1	2
Plant I	Bobbin 1	−1	−5	−2	−8	−2	3	−3	−4	0	−1	−12	4
	2	1	10	1	2	2	2	10	−4	−4	3	4	8
	3	2	−3	5	−5	1	−1	−6	1	2	5	7	5
	4	6	10	1	5	0	5	−2	−2	1	1	5	9
	5	−1	−8	5	−10	1	−5	1	−4	5	−5	3	6
	6	−1	−10	−8	−8	−2	2	0	−3	−8	−1	−2	−4
	7	−9	−2	5	−2	7	−2	−2	−2	−1	2	10	5
	8	0	2	−5	−2	5	3	10	−1	4	1	7	−1
Plant II	Bobbin 9	10	8	−5	6	2	13	7	15	17	14	18	11
	10	9	12	6	15	15	12	18	16	13	10	9	11
	11	0	8	12	6	2	0	5	4	18	8	6	8
	12	5	9	2	16	15	5	21	18	15	11	18	15
	13	−1	−1	11	19	12	10	1	20	13	9	4	6
	14	7	16	15	11	12	12	8	12	22	11	12	21
	15	−5	1	−2	10	12	15	2	13	10	10	7	5
	16	10	9	10	15	9	16	12	11	18	20	11	15

* From Table I on p. 4 of "Establishing control of tire cord testing laboratories" by F. Akutowicz and H. M. Truax, *Industrial Quality Control*, Vol. 13, no. 2, 1956. Reproduced with the kind permission of the authors and the editor.

in 0.1-lb deviations from 21.5 lb. Calculate the entries for the analysis-of-variance table except those in the $E(\text{MS})$ column, to be added in Ch. 8.

5.9. Derive a test for nonadditivity in the Latin-square design by applying the result of Problem 4.19 with the function $\mathbf{z} = f(\hat{\boldsymbol{\eta}}_\omega)$ defined by $z_{ijk} = c_1(\hat{\eta}_{ijk} - c_2)^2$, where c_1 and c_2 are conveniently chosen constants, and $\hat{\eta}_{ijk} = \hat{\mu} + \hat{\alpha}_i^A + \hat{\alpha}_j^B + \hat{\alpha}_k^C$.

5.10. Apply the test derived in Problem 5.9 to the data of Problem 5.2.

CHAPTER 6

The Analysis of Covariance

6.1. INTRODUCTION

The applications of the general theory of Chs. 1 and 2 may be classified into three different kinds, namely those associated with the names analysis of variance, analysis of covariance, and regression analysis. The boundaries between the three kinds are not very sharp or universally agreed to, and we shall not attempt here to make the distinctions precise.

Of the factors (we use the word in the technical sense of Chs. 4 and 5) that are varied in an experiment or series of observations, some may be qualitative, like variety of grain, and others quantitative, like temperature. A quantitative factor may be treated either qualitatively or quantitatively in the mathematical model: Suppose for example that temperature is one factor varied in an experiment and that all the observations are made at only five different temperatures. If the actual value of the temperature enters the formula used for the expected value of the ith observation, for instance if it contains the terms $\gamma_1 t_i + \gamma_2 t_i^2$, where t_i is the temperature at which the ith observation is taken, then we may say the factor is treated quantitatively. On the other hand, the actual values of the temperature might not enter the formula used but only five main effects for temperature, and possibly interactions, just as with qualitative factors. In this case[1] we may say that the quantitative factor is treated qualitatively.

The distinction can now be made that in the *analysis of variance* all factors are treated qualitatively, in *regression analysis* all factors are quantitative and treated quantitatively, whereas in the *analysis of covariance* some factors are present that are treated qualitatively and some that are

[1] If α_j is the main effect of the jth level of temperature, then the sum $\Sigma_{j=1}^5 x_{ji}\alpha_j$ would enter the formula for the expected value of the ith observation, where $x_{ji} = 1$ if the ith observation is at the jth level of temperature and $x_{ji} = 0$ otherwise. The coefficients $\{x_{ji}\}$ can of course be regarded as functions of the temperature, $x_{ji} = f_i(t_i)$: Their distinction from the coefficients in the quantitative case, as in the above example where they are t_i and t_i^2, is that the functions $\{x_{ji}\}$ are used only as counter variables or indicator variables (sec. 1.2).

treated quantitatively. Of course we continue to assume as before in this book that in every case the unknown parameters in the formula for the expected value of an observation enter linearly. Examples of the three kinds of problems are the following: for analysis of variance, all the situations considered in Chs. 3, 4, and 5; for regression analysis, Examples 1 and 2 of sec. 1.2; for analysis of covariance, the examples given below.

We remark that in practical applications where a quantitative factor is treated qualitatively and the levels correspond to equal steps of the quantitative factor (for example, if the factor is temperature, and the levels are 600°, 650°, 700°, 750°, 800° F), if the main effects for the factor are found to be significant, then a very useful technique for further analysis of these effects consists in employing the contrasts for linear effect, quadratic effect, etc. The values of these contrasts, which except for known constant factors are the coefficients in orthogonal polynomials fitted by least squares, can be calculated quickly from Table XXIII of Fisher and Yates (1943), and inferences about them (together with all other contrasts) can be made by the S-method. However, if the resulting polynomial regression equation is to be used for prediction, it is important to consider the relative importance of the following two kinds of error in deciding on the degree of the polynomial to be kept: The error of working with one of higher degree than necessary versus the error of an inadequate fit from one of too low degree. If the former error is the less harmful then it is not wise to test whether a coefficient is significantly different from zero by the S-criterion because of the relatively high probability involved for the latter error.[2] Different polynomial fittings for a quantitative factor can also be made for different levels or treatment combinations of the other factors, and the differences of the linear, quadratic, etc., contrasts associated with these polynomials can be analyzed as appropriate "orthogonal d.f." partitioned out of an interaction SS (sec. 4.8 and Problems 4.17, 4.18).

The term *independent variable*, sometimes used, refers to a factor treated quantitatively in analysis of covariance or regression analysis. In this terminology the observations $\{y_i\}$ of the general theory are then said to be on the dependent variable. The "independent variables" are also called *concomitant variables*, and we shall use this terminology.

A simple example of analysis of covariance with one concomitant variable is the following:[3] In order to compare the quality of several kinds of starches (wheat starch, potato starch, etc.), an experiment is

[2] The multiple-decision problem of choosing the degree of the polynomial is an example of a curve-fitting problem for which no theoretically satisfactory solution has yet been developed; other examples involve the choice between different kinds of functions, for example, polynomial versus exponential.

[3] Freeman (1942).

made in which the breaking strength of starch films is measured. An ordinary analysis of variance as a one-way layout shows highly significant differences in breaking strength among the different kinds of starches. If, however, the thickness of the starch film is taken into account, it appears that most of the variability between breaking strengths has its origin in differences in film thickness. In this example, if y_{ij} is the jth measurement on the breaking strength (dependent variable) of the ith kind of starch, if this measurement has been performed on a starch film of thickness z_{ij} (concomitant variable), and if we assume that the "true" breaking strength varies linearly with the thickness, the model equation might be taken as

$$(6.1.1) \qquad y_{ij} = \mu + \alpha_i + \gamma z_{ij} + e_{ij} \qquad (i = 1, \cdots, I; j = 1, \cdots, J_i),$$

where μ is an additive constant, α_i is the main effect due to the ith kind of starch, γ is the regression coefficient for the dependence of breaking strength on thickness, and e_{ij} is an error term; the use of γ instead of γ_i in (6.1.1) of course implies that we assume that the linear relations have the same slope for each starch.

It will be helpful to the reader to have in mind a few more examples when he reads the general theory of the next section. If in the above example we wished to consider quadratic instead of linear regression on the thickness, the model equation would be modified to read

$$(6.1.2) \qquad y_{ij} = \mu + \alpha_i + \gamma z_{ij} + \delta z_{ij}^2 + e_{ij}.$$

The analysis is the same as that for linear regression on two concomitant variables, in this case thickness and (thickness)^2. If in a two-way layout with one observation per cell on the dependent variable y and one on a concomitant variable z, we assume zero interactions between the row and column factors, and linear regression on z, the model equation would be

$$(6.1.3) \qquad y_{ij} = \mu + \alpha_i + \beta_j + \gamma z_{ij} + e_{ij};$$

if there are two concomitant variables z and w, zero interactions, and linear regression,

$$(6.1.4) \qquad y_{ij} = \mu + \alpha_i + \beta_j + \gamma z_{ij} + \delta w_{ij} + e_{ij};$$

here (y_{ij}, z_{ij}) or (y_{ij}, z_{ij}, w_{ij}) are the observations in the i,j cell, μ is an additive constant, $\{\alpha_i\}$ are the main effects for rows, $\{\beta_j\}$ for columns, and γ and δ are the regression coefficients.

In all the above examples the additive constant μ is not the general mean: If this is defined as the (unweighted) average of the expected values of the averages of the observations in each cell, and if we assume as usual that the main effects satisfy $\Sigma_i \alpha_i = 0$ and $\Sigma_j \beta_j = 0$, then the general mean is $\mu + \gamma \bar{z}$ in (6.1.1), and $\mu + \gamma \bar{z} + \delta \overline{z^2}$ in (6.1.2), where

$$\bar{z} = I^{-1} \sum_i J_i^{-1} \sum_j z_{ij}, \qquad \overline{z^2} = I^{-1} \sum_i J_i^{-1} \sum_j z_{ij}^2,$$

$\mu + \gamma z_{..}$ in (6.1.3), and $\mu + \gamma z_{..} + \delta w_{..}$ in (6.1.4). If it is desired to preserve the symbol μ for the general mean, then instead of (6.1.1) we would use

$$y_{ij} = \mu + \alpha_i + \gamma(z_{ij} - \bar{z}) + e_{ij},$$

and similarly for the other examples.

The analysis of covariance was introduced[4] in a somewhat narrower sense than that defined above, as a device for simulating control of factors not possible or feasible to control in an experiment: Thus the estimates of the yields of varieties of grain in a comparative agricultural trial might be "adjusted" to allow for the differing numbers of shoots on the plots or the plots' yields on a previous year's uniformity trial (where all plots are given the same treatment), and the resulting estimates would within sampling errors be the same as those that would be obtained if all plots had equal numbers of shoots or equal yields on the uniformity trial—granting the validity of the regression model assumed or implied. The methods of statistical inference used here of course take into account that the covariance matrix of the adjusted yields, and the error estimate are different from those in the corresponding analysis of variance with no "adjustment."

To conform to the general theory of Chs. 1 and 2 the symbols $\{z_{ij}\}$ and $\{w_{ij}\}$ in the above model equations should denote constants, the random variables being the $\{e_{ij}\}$ and $\{y_{ij}\}$. In some applications it may be more realistic to think of the observations on the concomitant variables as values taken on by random variables, rather than as constant values selected by the experimenter.[5] If we then modify the underlying assumptions so that the distribution assumed under them—this distribution depends on the values of the concomitant variables—is conditional, given the observed values of the concomitant variables, then the distribution theory derived is also conditional, as are the significance levels, powers, and confidence coefficients. If these conditional underlying assumptions are assumed to hold for all possible values of the observations on the concomitant variables, then, regardless of the joint distribution of

[4] By R. A. Fisher (1932), sec. 49.1.

[5] The case where the constant values selected and observed by the experimenter differ from "true values" entering the model equations, because of random measurement errors, is discussed in sec. 6.5.

observations on the concomitant variables, the conditional significance levels and conditional confidence coefficients are constant, and hence the same unconditionally. No such simple statement can be made for the power, which would be the expected value of the conditional power calculated relative to the joint distribution of the observations on the concomitant variables.

Although the validity of the statistical inferences can thus be extended under the above assumptions to the case where the concomitant variables are random variables, such applications of the analysis of covariance require discretion, for the reason that the inferences may be correct answers to the wrong questions. An example would be that above about the starch films, if we imagine that the breaking strength of films is the property of crucial importance in choosing a starch, and that the manufacturing process is such that the thickness cannot be controlled and is different for the different starches. If we assumed only that in the specimens tested we had a random sample of breaking strengths from each of the starches, then, regardless of the nature of the dependence of the strength on the thickness, an ordinary analysis of variance of the strengths would be relevant, while the possible equality of the $\{\alpha_i\}$ in (6.1.1) would be largely of academic interest.

Developing this point further, we note that data of the type that can be formally subjected to the analysis of covariance with h concomitant variables can also be formally subjected to an m-variate analysis of variance, where $m = h+1$, the basic elements needed for the two computations being the same (namely the elements denoted by $\{m_{tv,\Omega}\}$ and $\{m_{tv,\omega}\}$ in sec. 6.2). Now *m-variate analysis of variance* is the extension of the univariate analysis of variance treated in this book to the case where the observations instead of being independent one-dimensional random variables are independent m-dimensional vector random variables, our assumption of constant variance of the observations being replaced by that of a constant covariance matrix, and the vectors $\mathbf{y}^{n\times 1}$ and $\boldsymbol{\beta}^{p\times 1}$ in the Ω-assumption $E(\mathbf{y}) = \mathbf{X}'\boldsymbol{\beta}$ being replaced respectively by $n\times m$ and $p\times m$ matrices. Although we shall not develop the theory of m-variate analysis of variance in this book, we shall want to discuss briefly the relation of its underlying assumptions to those of the analysis of covariance. The reader is referred to T. W. Anderson (1958) and C. R. Rao (1952) for the theory of m-variate analysis of variance, and to Tukey (1949*b*) for the bivariate case. To date few practical applications have been made.

The simplest case would be that in which the observations are obtained in pairs $\{(y_{ij}, z_{ij})\}$ in a one-way layout. To justify a bivariate analysis of variance, for each i ($i = 1, \cdots, I$) the pairs $\{(y_{ij}, z_{ij})\}$ must be a random sample from a bivariate population (the populations for different i being in general different); thus in experiments where the values of a concomitant variable z are controlled by the experimenter this model is not relevant. Furthermore, the covariance matrices of the I bivariate populations must be assumed equal. To simplify the discussion, we shall also assume the bivariate populations to be normal. Denote the mean of the ith bivariate population by (μ_{yi}, μ_{zi}). Familiar univariate analyses of the $\{y_{ij}\}$ alone or the $\{z_{ij}\}$ alone would yield tests of the hypotheses H_y: $\mu_{y1} = \mu_{y2} = \cdots = \mu_{yI}$ or H_z: $\mu_{z1} = \mu_{z2} = \cdots = \mu_{zI}$. To test the hypothesis $H_y \cap H_z$, i.e., that H_y and H_z are both true, would require a bivariate analysis of variance.

As remarked above in a more general context, one may by a conditional-probability approach obtain mathematically correct statistical inferences from an analysis of covariance of the $\{y_{ij}\}$, treating the $\{z_{ij}\}$ as values taken on by a concomitant variable. The underlying assumptions for such an analysis would be

$$y_{ij} = \beta_i + \gamma z_{ij} + e_{ij}, \tag{6.1.5}$$

where, conditionally, given the $\{z_{ij}\}$, the $\{e_{ij}\}$ are independently $N(0, \sigma^2)$. The usual hypothesis of main interest tested in the analysis of covariance would be

$$H: \quad \beta_1 = \beta_2 = \cdots = \beta_I.$$

To what does this correspond in the assumed bivariate situation?

If y and z have a bivariate normal distribution with means μ_y, μ_z, variances σ_y^2, σ_z^2, and covariance $\rho\sigma_y\sigma_z$, then the conditional distribution of y, given z, is normal with mean $\mu_y+\gamma(z-\mu_z)$ and variance $(1-\rho^2)\sigma_y^2$, where $\gamma = \rho\sigma_y/\sigma_z$. The line $y = \mu_y+\gamma(z-\mu_z)$ is the regression line of y on z, and $(1-\rho^2)\sigma_y^2$ may be called the variance of y about the regression line. Thus in the above bivariate situation the regression line of y on z in the ith population is $y = \mu_{yi}+\gamma(z-\mu_{zi})$, and the variance of y about this line is $(1-\rho^2)\sigma_y^2$. It follows that the meaning of the $\{\beta_i\}$ in (6.1.5) is given by the relation $\beta_i = \mu_{yi}-\gamma\mu_{zi}$, the meaning of σ^2 by $\sigma^2 = (1-\rho^2)\sigma_y^2$, and the meaning of the hypothesis H is that, for the I populations, the regression lines for y on z, constrained to be parallel by the underlying assumptions, are in fact identical. The analysis of covariance yields a correct test of H, but H may or may not be of any interest in various applications.

If the $\{y_{ij}\}$ were scores in mathematics of graduating seniors in I different colleges on similar examinations, and the $\{z_{ij}\}$ were mathematics scores of the same students on another set of similar examinations taken as entering freshmen, then the hypothesis H might be of considerable interest: Because of the linearity of regression implied by the underlying assumptions, in each school the expected senior score of each student can be interpreted as the sum of two terms: A term proportional to his entering score, and a term characterizing the school. The assumed parallelism of the regression lines means that the constant of proportionality is the same for all the schools (an assumption that would deserve careful weighing if this were a real instead of a hypothetical example). The hypothesis H states that the terms characterizing the schools are the same, i.e., that the differences of the performances of the students at the different schools can be attributed entirely to their different competences as measured by their scores as entering freshmen, and that after allowance for this, the effect of the schools was the same. On the other hand, if the $\{y_{ij}\}$ were senior scores in mathematics and the $\{z_{ij}\}$ were senior scores in English it would be difficult to see any sense in testing H.

It is sometimes said that the analysis of covariance is valid only if the treatments do not affect the values of the concomitant variables. In the general bivariate situation described above, the "treatments" are at I levels, corresponding to the I populations. That the treatments do not affect the values of the concomitant variables might be interpreted to mean that for each i the distribution of the $\{z_{ij}\}$ is the same, or, because of the other underlying assumptions, that $\mu_{z1} = \mu_{z2} = \cdots = \mu_{zI}$. Given this further assumption, the hypothesis H tested by the analysis of covariance would then be that of the identity of the I bivariate distributions. The dictum that the analysis of covariance can be used only in this case would thus confine it to a very restricted situation. However, the word "affect" does not always have a clear meaning in probabilistic contexts, and if we think of the two examples mentioned above and assume that the distributions of the scores $\{z_{ij}\}$ are not the same in all schools, so that by the above interpretation the treatments do affect the values of z, then in the first example we would have the schools "affecting" the entering freshman grades—which violates the everyday meaning of the word—whereas in the second example the schools would "affect" the graduating senior English grades—which does not violate it. We shall cut this semantic knot by discarding the dictum that led to it and reiterating our above viewpoint: The general bivariate situation we have considered is a special case of the still more general situation where the concomitant variables are random variables and the underlying assumptions are satisfied conditionally, given the values of these variables. Then the analysis of covariance can be applied to get tests of hypotheses that have the correct significance level, or interval estimates with the correct confidence coefficient, but the sense of using these tests or estimates must be considered separately in each application.

Whenever factors are treated quantitatively, there is the danger of extrapolating results beyond the range of the data: When the experimenter controls the value of concomitant variables he should choose with an eye to the ranges that the concomitant variables may assume in later applications of his results. We remark that a model allowing for random errors in controlled variables is treated in sec. 6.5.

We see that the need for care in the application and interpretation of the analysis of covariance is even greater than in the analysis of variance. This need in both fields has been nicely formulated by M. G. Kendall[6]—in his terminology "analysis of variance" includes "analysis of covariance"—as follows:

> . . . we would emphasize that the analysis of variance, like other statistical techniques, is not a mill which will grind out results automatically without care or forethought on the part of the operator. It is a rather delicate instrument which can be called into play when precision is needed, but requires skill as well as enthusiasm to apply to the best advantage. The reader who roves among the literature of the subject will sometimes find elaborate analyses applied to data in order to prove something which was almost obvious from careful inspection right from the start; or he will find results stated without qualification as "significant" without any attempt at critical appreciation. This is not the occasion to deliver a homily on the necessity for self-discipline in the use of advanced theoretical techniques, but the analysis of variance would provide quite a good text for a discourse on that interesting subject.

6.2. DERIVING THE FORMULAS FOR AN ANALYSIS OF COVARIANCE FROM THOSE FOR A CORRESPONDING ANALYSIS OF VARIANCE

Suppose that data are collected according to some standard design such as the one-, two-, or higher-way layout, the Latin square, incomplete blocks, or perhaps a design we have not specifically treated in this book, but that with every observation on a dependent variable y we also get an observation on each of a set of h concomitant variables $z_1, \cdots, z_h$. Suppose further that for analyzing the data we are willing to assume that $E(y_i)$, the expected value of an observation, is equal to the sum of two expressions, the first, one that would be used in an ordinary analysis of variance, and the second, a linear combination of the values of the concomitant variables with regression coefficients $\{\gamma_j\}$: The first would be $\Sigma_{j=1}^{p} x_{ji}\beta_j$, where (x_{ji}) is the matrix $\mathbf{X}'$ and $\{\beta_j\}$ are the effects that would be used in the analysis of variance, while the second would be $\Sigma_{j=1}^{h} z_{ji}\gamma_j$, where z_{ji} is the value of the concomitant variable z_j observed with y_i. Examples were mentioned in sec. 6.1. With suitably defined matrix

[6] From p. 245 of *The Advanced Theory of Statistics*, Vol. II, by M. G. Kendall, Charles Griffin, London (1946). Reproduced with the kind permission of the author and the publisher.

notation we would then have $E(\mathbf{y}) = \mathbf{X}'\boldsymbol{\beta} + \mathbf{Z}'\boldsymbol{\gamma}$. We could of course apply the general theory of Chs. 1 and 2 to analyze the data, the number $p+h$ of parameters playing the part of the number p of the general theory, and the parameters $\{\beta_1, \cdots, \beta_p; \gamma_1, \cdots, \gamma_h\}$ playing the part of $\{\beta_1, \cdots, \beta_p\}$ of the general theory. The purpose of this section[7] is to do instead the following: Let us call the "corresponding analysis-of-variance problem" the case of the ordinary analysis of variance for the design considered, i.e., the case where the terms $\Sigma_{j=1}^{h} z_{ji}\gamma_j$ are not included in $E(y_i)$, or more formally the case where we assume all $\gamma_j = 0$. In this section we learn how to modify the calculations for the corresponding analysis-of-variance problem to yield the desired analysis of covariance. This requires that the solution of the corresponding analysis-of-variance problem be available: If it is, the new method is usually simpler than applying directly the general theory of Chs. 1 and 2.

It will be convenient in this chapter to denote the underlying assumptions in the analysis-of-covariance problem by $\tilde{\Omega}$ and to reserve the symbol Ω for the underlying assumptions in the corresponding analysis-of-variance problem. Suppose then that $\tilde{\Omega}$ and Ω are

$$\tilde{\Omega}: \quad \mathbf{y}^{n\times 1} \text{ is } N(\mathbf{X}'\boldsymbol{\beta}^{p\times 1} + \mathbf{Z}'\boldsymbol{\gamma}^{h\times 1}, \sigma^2\mathbf{I}),$$
$$\Omega: \quad \mathbf{y}^{n\times 1} \text{ is } N(\mathbf{X}'\boldsymbol{\beta}^{p\times 1}, \sigma^2\mathbf{I}).$$

The components $\{\beta_1, \cdots, \beta_p\}$ of $\boldsymbol{\beta}$ are the effects in the corresponding analysis-of-variance problem; they may be main effects, interactions, block effects, or other effects. The components $\{\gamma_1, \cdots, \gamma_h\}$ are regression coefficients on h concomitant variables $\{z_1, \cdots, z_h\}$ respectively[8] in the analysis-of-covariance problem. The jth column of $\mathbf{Z}'$, which we shall denote by

$$\boldsymbol{\zeta}_j = (z_{j1}, z_{j2}, \cdots, z_{jn})',$$

consists of the values taken on by the jth concomitant variable z_j, z_{ji} being the value of z_j occurring with the ith observation y_i on the dependent variable y.

The reader who wishes to skip the derivation of the new method may

[7] Most of the results of this section were obtained in a different way by Rao (1946).

[8] Sometimes we may wish to permit a concomitant variable to enter the model equation with different regression coefficients; for example, in the problem of comparing starches in sec. 6.1, the regression coefficient of strength on thickness of film might conceivably be different for the different starches (in that case it might also be reasonable to take the "constant term" $\mu+\alpha_i$ for each starch equal to zero). The theory of this section could be applied by defining a number of artificial concomitant variables equal to the number of different regression coefficients, the values of the rth variable being zero except in the observations with the rth regression coefficient, but it is probably usually simpler to fall back on the general theory of Ch. 2.

proceed from here to the subsection headed "Computations"; he will then need to know that $\mathbf{Q}_\Omega$ and $\mathbf{Q}_\omega$ respectively denote the matrices of $\mathscr{S}_\Omega$ and $\mathscr{S}_\omega$ expressed as quadratic forms in the observations $\{y_1, \cdots, y_n\}$, $\mathscr{S}_\Omega$ and $\mathscr{S}_\omega$ having their usual connotations in connection with testing a hypothesis H_β about the $\{\beta_j\}$ in the corresponding analysis-of-variance problem.

If we denote the jth column of $\mathbf{X}'$ by $\boldsymbol{\xi}_j$, so that

$$\boldsymbol{\xi}_j = (x_{j1}, x_{j2}, \cdots, x_{jn})',$$

and $E(\mathbf{y})$ by $\boldsymbol{\eta}$, then $\boldsymbol{\eta} = \mathbf{X}'\boldsymbol{\beta} + \mathbf{Z}'\boldsymbol{\gamma}$ may be written

$$\boldsymbol{\eta} = \beta_1\boldsymbol{\xi}_1 + \cdots + \beta_p\boldsymbol{\xi}_p + \gamma_1\boldsymbol{\zeta}_1 + \cdots + \gamma_h\boldsymbol{\zeta}_h. \tag{6.2.1}$$

It will be convenient to denote the space spanned by $\{\boldsymbol{\xi}_1, \cdots, \boldsymbol{\xi}_p\}$ by V_Ω, instead of V_r as heretofore, and the space spanned by $\{\boldsymbol{\xi}_1, \cdots, \boldsymbol{\xi}_p, \boldsymbol{\zeta}_1, \cdots, \boldsymbol{\zeta}_h\}$ by $V_{\tilde{\Omega}}$. If r is the dimension of V_Ω, *we shall assume that the dimension of $V_{\tilde{\Omega}}$ is $r + h$*; this is equivalent to assuming that, if $r = \text{rank } \mathbf{X}'$, then $r+h$ is the rank of the composite matrix $(\mathbf{X}', \mathbf{Z}')$, whose columns are $\{\boldsymbol{\xi}_1, \cdots, \boldsymbol{\xi}_p, \boldsymbol{\zeta}_1, \cdots, \boldsymbol{\zeta}_h\}$. The reason for this assumption is that we want the regression coefficients $\{\gamma_j\}$ to be uniquely determined for any given $\boldsymbol{\eta}$ in (6.2.1), so that they are estimable parametric functions. To prove that the $\{\gamma_j\}$ are uniquely determined under our assumption, imagine selecting a basis $\{\boldsymbol{\alpha}_1, \cdots, \boldsymbol{\alpha}_r\}$ for V_Ω, replacing the $\boldsymbol{\xi}$'s in (6.2.1) by the appropriate linear combinations of the $\boldsymbol{\alpha}$'s, and collecting the coefficients of the $\boldsymbol{\alpha}$'s to get an equation of the form

$$\boldsymbol{\eta} = c_1\boldsymbol{\alpha}_1 + \cdots + c_r\boldsymbol{\alpha}_r + \gamma_1\boldsymbol{\zeta}_1 + \cdots + \gamma_h\boldsymbol{\zeta}_h. \tag{6.2.2}$$

Since $\{\boldsymbol{\alpha}_1, \cdots, \boldsymbol{\alpha}_r, \boldsymbol{\zeta}_1, \cdots, \boldsymbol{\zeta}_h\}$ is a basis for $V_{\tilde{\Omega}}$, the coordinates $\{\gamma_j\}$ must be unique.

We shall assume that t side conditions $\mathbf{H}'\boldsymbol{\beta} = \mathbf{0}$ are imposed, where $\mathbf{H}'$ is $t \times p$, which make the $\{\beta_j\}$ unique under Ω. That the side conditions make the $\{\beta_j\}$ unique under Ω will usually be fairly obvious in the applications; necessary and sufficient conditions for this are given in Theorem 3 of sec. 1.4. There will then exist a unique solution $\hat{\boldsymbol{\beta}}_\Omega$ which satisfies the normal equations and the side conditions

$$\mathbf{X}\mathbf{X}'\hat{\boldsymbol{\beta}}_\Omega = \mathbf{X}\mathbf{y}, \qquad \mathbf{H}'\hat{\boldsymbol{\beta}}_\Omega = \mathbf{0}. \tag{6.2.3}$$

Suppose that this solution is $\hat{\boldsymbol{\beta}}_\Omega = \mathbf{A}\mathbf{y}$; then, substituting this in (6.2.3) and noting that the results are identically true for all $\mathbf{y}$, we find the relations

$$\mathbf{X}\mathbf{X}'\mathbf{A} = \mathbf{X}, \qquad \mathbf{H}'\mathbf{A} = \mathbf{0} \tag{6.2.4}$$

which will be useful later.

We shall denote LS estimates of $\boldsymbol{\beta}$ and $\boldsymbol{\gamma}$ under $\tilde{\Omega}$ by $\hat{\boldsymbol{\beta}}_{\tilde{\Omega}}$ and $\hat{\boldsymbol{\gamma}}_{\tilde{\Omega}}$; $\hat{\boldsymbol{\gamma}}_{\tilde{\Omega}}$ is unique, since the $\{\gamma_j\}$ are estimable, and we shall show later that $\hat{\boldsymbol{\beta}}_{\tilde{\Omega}}$ may be chosen uniquely to satisfy the side conditions $\mathbf{H}'\hat{\boldsymbol{\beta}}_{\tilde{\Omega}} = \mathbf{0}$.

Now $\hat{\boldsymbol{\eta}}_{\Omega} = \mathbf{X}'\hat{\boldsymbol{\beta}}_{\Omega}$ is the projection of $\mathbf{y}$ on V_{Ω}; we may write $\hat{\boldsymbol{\eta}}_{\Omega} = \mathbf{P}_{\Omega}\mathbf{y}$, where $\mathbf{P}_{\Omega} = \mathbf{X}'\mathbf{A}$. We need the relations

$$\mathbf{P}_{\Omega}^2 = \mathbf{P}_{\Omega} \quad \text{and} \quad \mathbf{P}_{\Omega}' = \mathbf{P}_{\Omega}.$$

These are true of any matrices of transformations that are projections: The first is obvious since if we project $\hat{\boldsymbol{\eta}}_{\Omega}$ on V_{Ω} we again get $\hat{\boldsymbol{\eta}}_{\Omega}$; hence $\mathbf{P}_{\Omega}(\mathbf{P}_{\Omega}\mathbf{y}) = \mathbf{P}_{\Omega}\mathbf{y}$ for all $\mathbf{y}$. The second may be shown from (6.2.4) by multiplying by $\mathbf{A}'$ on the left to get $\mathbf{A}'\mathbf{X}\mathbf{X}'\mathbf{A} = \mathbf{P}_{\Omega}'$ and then taking transposes.[9] If we write $\hat{\boldsymbol{\eta}}_{\tilde{\Omega}} = \mathbf{X}'\hat{\boldsymbol{\beta}}_{\tilde{\Omega}} + \mathbf{Z}'\hat{\boldsymbol{\gamma}}_{\tilde{\Omega}} = \mathbf{P}_{\tilde{\Omega}}\mathbf{y}$ for the projection of $\mathbf{y}$ on $V_{\tilde{\Omega}}$, then we have $\mathbf{P}_{\Omega}(\mathbf{P}_{\tilde{\Omega}}\mathbf{y}) = \mathbf{P}_{\Omega}\mathbf{y}$, or

$$\mathbf{P}_{\Omega}\mathbf{P}_{\tilde{\Omega}} = \mathbf{P}_{\Omega},$$

since $\hat{\boldsymbol{\eta}}_{\Omega}$ may be obtained by first projecting $\mathbf{y}$ on $V_{\tilde{\Omega}}$ and then projecting the result on V_{Ω}. If we define

$$\mathbf{Q}_{\Omega} = \mathbf{I} - \mathbf{P}_{\Omega}, \qquad \mathbf{Q}_{\tilde{\Omega}} = \mathbf{I} - \mathbf{P}_{\tilde{\Omega}}, \tag{6.2.5}$$

then, for any vector $\mathbf{y}$, $\mathbf{Q}_{\Omega}\mathbf{y}$ and $\mathbf{Q}_{\tilde{\Omega}}\mathbf{y}$ are the projections of $\mathbf{y}$ on the ortho-complements (see end of App. I) of V_{Ω} and $V_{\tilde{\Omega}}$, respectively, and we find that

$$\mathbf{Q}_{\Omega}^2 = \mathbf{Q}_{\Omega}, \quad \mathbf{Q}_{\Omega}' = \mathbf{Q}_{\Omega}, \quad \text{and} \quad \mathbf{Q}_{\Omega}\mathbf{Q}_{\tilde{\Omega}} = \mathbf{Q}_{\tilde{\Omega}},$$

by substituting (6.2.5).

We shall first calculate $\hat{\boldsymbol{\gamma}}_{\tilde{\Omega}}$: Multiply

$$\mathbf{Q}_{\tilde{\Omega}}\mathbf{y} = \mathbf{y} - \mathbf{X}'\hat{\boldsymbol{\beta}}_{\tilde{\Omega}} - \mathbf{Z}'\hat{\boldsymbol{\gamma}}_{\tilde{\Omega}} \tag{6.2.6}$$

on the left by $\mathbf{Q}_{\Omega}$ to get

$$\mathbf{Q}_{\tilde{\Omega}}\mathbf{y} = \mathbf{Q}_{\Omega}\mathbf{y} - \mathbf{Q}_{\Omega}\mathbf{Z}'\hat{\boldsymbol{\gamma}}_{\tilde{\Omega}}, \tag{6.2.6a}$$

since $\mathbf{Q}_{\Omega}\mathbf{X}'\hat{\boldsymbol{\beta}}_{\tilde{\Omega}} = \mathbf{0}$, $\mathbf{X}'\hat{\boldsymbol{\beta}}_{\tilde{\Omega}}$ being a linear combination of the $\{\boldsymbol{\xi}_j\}$ and thus in V_{Ω}. Since $\mathbf{Q}_{\tilde{\Omega}}\mathbf{y}$ is orthogonal to $V_{\tilde{\Omega}}$ and hence to the $\{\boldsymbol{\zeta}_j\}$, and since the $\{\boldsymbol{\zeta}_j'\}$ are the rows of $\mathbf{Z}$, therefore $\mathbf{Z}\mathbf{Q}_{\tilde{\Omega}}\mathbf{y} = \mathbf{0}$,

$$\mathbf{Z}(\mathbf{Q}_{\Omega}\mathbf{y} - \mathbf{Q}_{\Omega}\mathbf{Z}'\hat{\boldsymbol{\gamma}}_{\tilde{\Omega}}) = \mathbf{0}$$

[9] Professor Robert Wijsman pointed out to me how the symmetry of $\mathbf{P}_{\Omega}$, the matrix of a projection transformation, may be easily proved directly without appeal to (6.2.4): By choosing as coordinate system the orthonormal basis obtained by starting with an orthonormal basis for V_{Ω} and extending this to an orthonormal basis for V_n, $\mathbf{P}_{\Omega}$ is made to take the diagonal form $\mathbf{D}$ with the first r elements on the diagonal equal to 1, the rest to 0. In any other coordinate system $\mathbf{P}_{\Omega}$ is then of the form $\mathbf{T}'\mathbf{D}\mathbf{T}$, where $\mathbf{T}$ is orthogonal, and this is symmetric since $\mathbf{D}$ is symmetric.

from (6.2.6*a*), or

$$\text{(6.2.7)} \qquad \mathbf{ZQ}_\Omega\mathbf{Z}'\hat{\boldsymbol{\gamma}}_{\tilde{\Omega}} = \mathbf{ZQ}_\Omega\mathbf{y}.$$

That the $h \times h$ matrix $\mathbf{ZQ}_\Omega\mathbf{Z}'$ is nonsingular may be argued as follows: Since $\mathbf{ZQ}_\Omega\mathbf{Z}' = (\mathbf{Q}_\Omega\mathbf{Z}')'(\mathbf{Q}_\Omega\mathbf{Z}')$, therefore rank $\mathbf{ZQ}_\Omega\mathbf{Z}'$ = rank $\mathbf{Q}_\Omega\mathbf{Z}'$ by Theorem 7 of App. II. The h columns $\{\mathbf{Q}_\Omega\boldsymbol{\zeta}_j\}$ of $\mathbf{Q}_\Omega\mathbf{Z}'$ must be linearly independent, since, if we substitute $\boldsymbol{\zeta}_j = \mathbf{P}_\Omega\boldsymbol{\zeta}_j + \mathbf{Q}_\Omega\boldsymbol{\zeta}_j$ for each $\boldsymbol{\zeta}_j$ in (6.2.2), we see that any vector in $V_{\tilde{\Omega}}$ is a linear combination of vectors $\{\boldsymbol{\alpha}_j\}$ and $\{\mathbf{P}_\Omega\boldsymbol{\zeta}_j\}$ in V_Ω and $\{\mathbf{Q}_\Omega\boldsymbol{\zeta}_j\}$ orthogonal to V_Ω, so that the $\{\mathbf{Q}_\Omega\boldsymbol{\zeta}_j\}$ must span the orthocomplement of V_Ω in $V_{\tilde{\Omega}}$, which by Lemma 10 of App. I has dimension $(r+h) - r = h$.

We may now solve (6.2.7) for $\hat{\boldsymbol{\gamma}}_{\tilde{\Omega}}$,

$$\text{(6.2.8)} \qquad \hat{\boldsymbol{\gamma}}_{\tilde{\Omega}} = (\mathbf{ZQ}_\Omega\mathbf{Z}')^{-1}\mathbf{ZQ}_\Omega\mathbf{y}.$$

The matrix $\mathbf{Q}_\Omega$ is given by the solution of the corresponding analysis-of-variance problem as the symmetric matrix of the quadratic form $\mathscr{S}_\Omega$, the error SS under Ω:

$$\mathscr{S}_\Omega = \|\mathbf{y}-\hat{\boldsymbol{\eta}}\|^2 = \|\mathbf{y}-\mathbf{P}_\Omega\mathbf{y}\|^2 = \|\mathbf{Q}_\Omega\mathbf{y}\|^2$$
$$= (\mathbf{Q}_\Omega\mathbf{y})'(\mathbf{Q}_\Omega\mathbf{y}) = \mathbf{y}'\mathbf{Q}_\Omega\mathbf{y}.$$

To calculate $\hat{\boldsymbol{\beta}}_{\tilde{\Omega}}$ we note that $\mathbf{Q}_{\tilde{\Omega}}\mathbf{y}$ is orthogonal to the $\{\boldsymbol{\xi}_j\}$ because it is orthogonal to $V_{\tilde{\Omega}}$, and hence $\mathbf{XQ}_{\tilde{\Omega}}\mathbf{y} = \mathbf{0}$. Substituting (6.2.6) in this, we get

$$\text{(6.2.9)} \qquad \mathbf{XX}'\hat{\boldsymbol{\beta}}_{\tilde{\Omega}} = \mathbf{X}(\mathbf{y}-\mathbf{Z}'\hat{\boldsymbol{\gamma}}_{\tilde{\Omega}}).$$

Recalling that the system (6.2.3) has the unique solution $\hat{\boldsymbol{\beta}}_\Omega = \mathbf{Ay}$, we see that (6.2.9) has a unique solution satisfying $\mathbf{H}'\hat{\boldsymbol{\beta}}_{\tilde{\Omega}} = \mathbf{0}$, and it is

$$\hat{\boldsymbol{\beta}}_{\tilde{\Omega}} = \mathbf{A}(\mathbf{y}-\mathbf{Z}'\hat{\boldsymbol{\gamma}}_{\tilde{\Omega}}).$$

For later use we write this

$$\text{(6.2.10)} \qquad \hat{\boldsymbol{\beta}}_{\tilde{\Omega}} = \mathbf{Ay} - \sum_{j=1}^{h} \hat{\gamma}_{j,\tilde{\Omega}}\mathbf{A}\boldsymbol{\zeta}_j.$$

We still need to calculate $\mathscr{S}_{\tilde{\Omega}}$, the error SS in the analysis-of-covariance problem. If we denote by H_γ the hypothesis that $\boldsymbol{\gamma} = \mathbf{0}$, then $\Omega = H_\gamma \cap \tilde{\Omega}$, and hence $\mathscr{S}_\Omega - \mathscr{S}_{\tilde{\Omega}}$ is the numerator SS of the statistic $\mathscr{F}$ for testing H_γ under $\tilde{\Omega}$. But by sec. 2.7 we know that this numerator SS may also be calculated as

$$\text{(6.2.11)} \qquad \hat{\boldsymbol{\gamma}}'_{\tilde{\Omega}}(\sigma^{-2}\boldsymbol{\Sigma}_{\hat{\gamma}})^{-1}\hat{\boldsymbol{\gamma}}_{\tilde{\Omega}},$$

where $\boldsymbol{\Sigma}_{\hat{\gamma}}$ denotes the covariance matrix of $\hat{\boldsymbol{\gamma}}_{\tilde{\Omega}}$, calculated under $\tilde{\Omega}$. Now from (6.2.8),

$$\boldsymbol{\Sigma}_{\hat{\gamma}} = (\mathbf{ZQ}_\Omega\mathbf{Z}')^{-1}\mathbf{ZQ}_\Omega\boldsymbol{\Sigma}_y\mathbf{Q}_\Omega\mathbf{Z}'(\mathbf{ZQ}_\Omega\mathbf{Z}')^{-1},$$

where $\mathbf{\Sigma}_y = \sigma^2\mathbf{I}$. Since $\mathbf{Q}_\Omega^2 = \mathbf{Q}_\Omega$, this gives

(6.2.12) $$\mathbf{\Sigma}_{\hat{\gamma}} = \sigma^2(\mathbf{Z}\mathbf{Q}_\Omega\mathbf{Z}')^{-1}.$$

Substituting in (6.2.11) and equating to $\mathscr{S}_\Omega - \mathscr{S}_{\tilde{\Omega}}$ gives us the desired result

$$\mathscr{S}_{\tilde{\Omega}} = \mathscr{S}_\Omega - \hat{\boldsymbol{\gamma}}'_{\tilde{\Omega}}(\mathbf{Z}\mathbf{Q}_\Omega\mathbf{Z}')\hat{\boldsymbol{\gamma}}_{\tilde{\Omega}},$$

which we may also write

(6.2.13) $$\mathscr{S}_{\tilde{\Omega}} = \mathscr{S}_\Omega - \hat{\boldsymbol{\gamma}}'_{\tilde{\Omega}}(\mathbf{Z}\mathbf{Q}_\Omega\mathbf{y})$$

because of (6.2.7).

Finally, suppose that H_β is a hypothesis about the $\{\beta_j\}$ for which we wish to calculate "the" F-test under $\tilde{\Omega}$, knowing "the" F-test of H_β under Ω. If H_β is defined by q linearly independent parametric functions of the $\{\beta_j\}$ which are estimable under $\tilde{\Omega}$, then they are estimable under Ω, since an estimate that is unbiased under $\tilde{\Omega}$ is of course unbiased under $\Omega \subset \tilde{\Omega}$. Denote by ω and $\tilde{\omega}$ the assumptions

$$\omega = H_\beta \cap \Omega, \qquad \tilde{\omega} = H_\beta \cap \tilde{\Omega}.$$

Since $\omega = H_\gamma \cap \tilde{\omega}$, where H_γ is the hypothesis $\boldsymbol{\gamma} = \mathbf{0}$, we may calculate $\mathscr{S}_{\tilde{\omega}}$ from $\mathscr{S}_\omega$ by replacing Ω and $\tilde{\Omega}$ by ω and $\tilde{\omega}$ in the above theory for calculating $\mathscr{S}_{\tilde{\Omega}}$ from $\mathscr{S}_\Omega$. If we write

$$\mathscr{S}_\omega = \mathbf{y}'\mathbf{Q}_\omega\mathbf{y},$$

where $\mathbf{Q}_\omega$ is symmetric, it follows that $\mathbf{Z}\mathbf{Q}_\omega\mathbf{Z}'$ is nonsingular by an argument similar to that above for $\mathbf{Z}\mathbf{Q}_\Omega\mathbf{Z}'$, utilizing a basis for the $(r-q)$-dimensional space V_ω in the analog of (6.2.2) for vectors in the $(r+h-q)$-dimensional space $V_{\tilde{\omega}}$. Then

$$\mathscr{S}_{\tilde{\omega}} = \mathscr{S}_\omega - \hat{\boldsymbol{\gamma}}'_{\tilde{\omega}}(\mathbf{Z}\mathbf{Q}_\omega\mathbf{y}),$$

where $\hat{\boldsymbol{\gamma}}_{\tilde{\omega}}$ is the solution of

$$\mathbf{Z}\mathbf{Q}_\omega\mathbf{Z}'\hat{\boldsymbol{\gamma}}_{\tilde{\omega}} = \mathbf{Z}\mathbf{Q}_\omega\mathbf{y},$$

analogously to (6.2.13) and (6.2.7).

Computations

We need the values of $\frac{1}{2}(h+1)(h+2)$ quantities that we shall denote by $\{m_{tv,\Omega}\}$, where t and v each stand for any of the $h+1$ variables $y, z_1, \cdots, z_h$, and $m_{tv,\Omega} = m_{vt,\Omega}$. They are defined as

$$m_{tv,\Omega} = \mathbf{t}'\mathbf{Q}_\Omega\mathbf{v},$$

where $\mathbf{t}$ and $\mathbf{v}$ each stand for any of the $h+1$ vectors $\mathbf{y}, \boldsymbol{\zeta}_1, \cdots, \boldsymbol{\zeta}_h$. They are calculated from the identities used in the corresponding analysis-of-variance problem to calculate $\mathscr{S}_\Omega = \mathbf{y}'\mathbf{Q}_\Omega\mathbf{y}$, i.e., the identities that give

$\mathscr{S}_\Omega$ as the result of adding and subtracting certain sums of squares in the values of the dependent variable y. Thus if $\mathscr{S}_\Omega$ involves a SS

$$\sum_k C_k[L_k(y)]^2,$$

where $L_k(y)$ is a linear form in the values of y, then m_{tv} involves in the same way the expression

$$\sum_k C_k\, L_k(t)\, L_k(v).$$

If the reader does not understand this prescription for calculating the $\{m_{tv,\Omega}\}$ he should immediately proceed to sec. 6.4 to see how it is applied.

The $h \times h$ matrix $\mathbf{ZQ}_\Omega\mathbf{Z}'$ will be denoted by $\mathbf{M}_\Omega$, and the $h \times 1$ vector $\mathbf{ZQ}_\Omega\mathbf{y}$ by $\mathbf{m}_\Omega$. The i,j element of $\mathbf{M}_\Omega$ is $m_{z_iz_j,\Omega}$, and the $i,1$ element of $\mathbf{m}_\Omega$ is $m_{z_iy,\Omega}$.

We now form and solve the following h equations for the h regression coefficients $\{\hat{\gamma}_{j,\tilde{\Omega}}\}$,

$$\mathbf{M}_\Omega \hat{\boldsymbol{\Upsilon}}_{\tilde{\Omega}} = \mathbf{m}_\Omega,$$

which is the new notation for (6.2.7). There is some question, if $h > 1$, whether we should solve the equations by calculating the inverse matrix $\mathbf{M}_\Omega^{-1}$ (see footnote[10] concerning methods of calculation). If we know that later we will want to estimate all variances and covariances of the $\{\hat{\gamma}_{j,\tilde{\Omega}}\}$, then we should calculate $\mathbf{M}_\Omega^{-1}$ now, since by (6.2.12) the estimated covariance matrix of $\hat{\boldsymbol{\Upsilon}}_{\tilde{\Omega}}$ under $\tilde{\Omega}$ is $s^2\mathbf{M}_\Omega^{-1}$, where s^2 is the error MS under $\tilde{\Omega}$.

Having calculated the vectors $\mathbf{m}_\Omega$ and $\boldsymbol{\Upsilon}_{\tilde{\Omega}}$, we now form their inner product

$$\mathbf{m}'_\Omega \hat{\boldsymbol{\Upsilon}}_{\tilde{\Omega}} = m_{z_1y,\Omega}\hat{\gamma}_{1,\tilde{\Omega}} + \cdot\cdot\cdot + m_{z_hy,\Omega}\hat{\gamma}_{h,\tilde{\Omega}},$$

and subtract it from the analysis-of-variance error SS to get the analysis-of-covariance error SS,

$$\mathscr{S}_{\tilde{\Omega}} = \mathscr{S}_\Omega - \mathbf{m}'_\Omega \hat{\boldsymbol{\Upsilon}}_{\tilde{\Omega}}, \tag{6.2.14}$$

by (6.2.13); the number of d.f. for $\mathscr{S}_{\tilde{\Omega}}$ is $n-r-h$.

The LS estimates of $\{\beta_1, \cdot\cdot\cdot, \beta_p\}$ will usually be wanted. If so, we may calculate them by noting that in the analysis-of-variance problem each $\hat{\beta}_{i,\Omega}$ $(i = 1, \cdot\cdot\cdot, p)$ is a known linear form in the values of the dependent variable y, say

$$\hat{\beta}_{i,\Omega} = l_i(y), \tag{6.2.15}$$

[10] If $h \leq 3$, the calculation of the inverse matrix from its expression in terms of cofactors (Theorem 4 of App. II) is quite feasible. If $h > 3$, the Gauss–Doolittle method may be used; see Rao (1952), pp. 30–31, or Dwyer (1951), p. 191.

and constructing the same linear form in each of the concomitant variables, namely $l_i(z_j)$ for $j = 1, \cdots, h$. Then the estimate of β_i under $\tilde{\Omega}$ is

$$(6.2.16) \qquad \hat{\beta}_{i,\tilde{\Omega}} = l_i(y) - \sum_{j=1}^{h} \hat{\gamma}_{j,\tilde{\Omega}}\, l_i(z_j),$$

as a consequence of (6.2.10). Because of (6.2.16) the term $-\hat{\gamma}_{j,\tilde{\Omega}}\, l_i(z_j)$ is called the correction to $\hat{\beta}_{i,\Omega}$ for regression on z_j. For each β_i that it is desired to estimate under $\tilde{\Omega}$, we therefore compute the value (6.2.15) of its estimate under Ω, and the values of the same linear form in the concomitant variables, and then apply (6.2.16).

If H_β is a hypothesis which puts q linearly independent linear restrictions on the $\{\beta_j\}$ and we wish to calculate the test of H_β under $\tilde{\Omega}$ from the known test of H_β under Ω, we first go through the following steps analogous to the above calculation of $\mathscr{S}_{\tilde{\Omega}}$ from $\mathscr{S}_\Omega$: We start with the identities for calculating $\mathscr{S}_\omega$ in the analysis-of-variance problem, where $\mathscr{S}_\omega$ is the sum of the numerator and denominator SS's of the statistic $\mathscr{F}$ for testing H_β under Ω. Using the same identities, we calculate the $\frac{1}{2}(h+1)(h+2)$ quantities $\{m_{tv,\omega}\}$, where $m_{tv,\omega}$ is defined as $\mathbf{t}'\mathbf{Q}_\omega\mathbf{v}$ if $\mathscr{S}_\omega = \mathbf{y}'\mathbf{Q}_\omega\mathbf{y}$. Let $\mathbf{M}_\omega$ denote the $h \times h$ matrix whose i,j element is $m_{z_i z_j,\omega}$, and $\mathbf{m}_\omega$ the $h \times 1$ vector whose $i,1$ element is $m_{z_i y,\omega}$. This time there is no further use for the inverse matrix $\mathbf{M}_\omega^{-1}$, and so we merely solve the system of h equations in h unknowns $\{\hat{\gamma}_{1,\tilde{\omega}}, \cdots, \hat{\gamma}_{h,\tilde{\omega}}\}$

$$\mathbf{M}_\omega \hat{\boldsymbol{\gamma}}_{\tilde{\omega}} = \mathbf{m}_\omega$$

to get the h elements of the vector $\hat{\boldsymbol{\gamma}}_{\tilde{\omega}}$. These are then used as in (6.2.14) to calculate

$$\mathscr{S}_{\tilde{\omega}} = \mathscr{S}_\omega - \mathbf{m}_\omega' \hat{\boldsymbol{\gamma}}_{\tilde{\omega}}.$$

The numerator SS of the statistic $\mathscr{F}$ for testing H_β under $\tilde{\Omega}$ is now evaluated as $\mathscr{S}_{\tilde{\omega}} - \mathscr{S}_{\tilde{\Omega}}$, the denominator SS is $\mathscr{S}_{\tilde{\Omega}}$, and the numbers of d.f. are q and $n-r-h$.

Confidence sets, or tests of hypotheses, involving only the regression coefficients $\{\gamma_j\}$ can be obtained by using the method of confidence ellipsoids of secs. 2.3 and 2.4. For this purpose some or all the elements of the covariance matrix of the estimates $\{\hat{\gamma}_{j,\tilde{\Omega}}\}$ will be needed; they are most easily obtained from (6.2.12) which says that the covariance matrix is $\sigma^2\mathbf{M}_\Omega^{-1}$. If the estimates $\{\hat{\beta}_{j,\tilde{\Omega}}\}$ are calculated, their variances and covariances may also be wanted. If the variances and covariances are calculated from the formula (6.2.16) for the $\{\hat{\beta}_{j,\tilde{\Omega}}\}$ it is helpful to note that the two sets of random variables $\{l_i(y)\}$ and $\{\hat{\gamma}_{j,\tilde{\Omega}}\}$ entering the formula are statistically independent under $\tilde{\Omega}$. This may be seen as follows: The LS estimate $\hat{\gamma}_{j,\tilde{\Omega}}$ is unbiased under $\tilde{\Omega}$, and hence also under Ω, that is, its expected value under Ω is zero. Thus this linear form belongs to the error space (end of sec. 1.6) under Ω, while any $l_i(y) = \hat{\beta}_{i,\Omega}$ belongs to the estimation space under Ω, and so the two sets of linear forms are orthogonal. Hence they are independent under $\tilde{\Omega}$ as well as Ω (all that matters is that the observations be independently normal with equal variances). There seems to be no easy way of handling inferences involving both the $\{\beta_j\}$ and the $\{\gamma_j\}$, but this problem does not arise as frequently as those we have solved; if it does one can of course fall back on the general theory of Chs. 1 and 2.

6.3. AN EXAMPLE WITH ONE CONCOMITANT VARIABLE

We consider a one-way layout with one concomitant variable. A physical situation where the analysis could be appropriate is described in connection with (6.1.1). The general assumptions will be taken in the form[11]

$$\tilde{\Omega}: \begin{cases} y_{ij} = \beta_i + \gamma z_{ij} + e_{ij} & (i = 1, \cdots, I;\ j = 1, \cdots, J_i), \\ \text{the } \{e_{ij}\} \text{ are independently } N(0, \sigma^2). \end{cases}$$

The corresponding analysis of variance under the assumptions Ω where $\gamma = 0$ was derived in sec. 3.1. The error SS there was found to be

$$\mathscr{S}_\Omega = \sum_i \sum_j (y_{ij} - y_{i.})^2, \tag{6.3.1}$$

and may be calculated from the identity

$$\mathscr{S}_\Omega = \sum_i \sum_j y_{ij}^2 - \sum_i J_i y_{i.}^2. \tag{6.3.2}$$

To apply the theory of sec. 6.2 we have to calculate the three quantities $m_{yy,\Omega}$, $m_{yz,\Omega}$, and $m_{zz,\Omega}$. To define $m_{tv,\Omega}$ we regard (6.3.1) as a sum of squares of linear forms $L(y) = y_{ij} - y_{i.}$ and replace $[L(y)]^2$ by $L(t)\,L(v)$, so that

$$m_{tv,\Omega} = \sum_i \sum_j (t_{ij} - t_{i.})(v_{ij} - v_{i.}).$$

To calculate the $\{m_{tv,\Omega}\}$ we use the same device on the identity (6.3.2),

$$m_{tv,\Omega} = \sum_i \sum_j t_{ij} v_{ij} - \sum_i J_i t_{i.} v_{i.};$$

[11] In this example and that of the next section we should add for mathematical completeness the rank condition on the matrix $(\mathbf{X}', \mathbf{Z}')$ stated after (6.2.1). In this example it means that for at least one i we do not have $z_{i1} = z_{i2} = \cdot \cdot \cdot = z_{iJ_i}$. In the case of the starch application if the condition is violated the effect of thickness could be absorbed into (would be confounded with) the effect of starch. In practice, it would be most unusual for the rank condition to be violated, and if it were, our attention would be called to it when we tried to solve the equations for the regression coefficients $\{\hat{\gamma}_{j,\tilde{\Omega}}\}$, for we would then not find a unique solution.

that is, we calculate them from the formulas

$$m_{yy,\Omega} = \sum_i \sum_j y_{ij}^2 - \sum_i J_i y_{i.}^2,$$

$$m_{zz,\Omega} = \sum_i \sum_j z_{ij}^2 - \sum_i J_i z_{i.}^2,$$

$$m_{zy,\Omega} = \sum_i \sum_j z_{ij} y_{ij} - \sum_i J_i z_{i.} y_{i.}.$$

The next step is to solve the equation $\mathbf{M}_\Omega \hat{\boldsymbol{\gamma}}_{\tilde{\Omega}} = \mathbf{m}_\Omega$, in the present case,

$$m_{zz,\Omega} \hat{\gamma}_{\tilde{\Omega}} = m_{zy,\Omega},$$

for

$$\hat{\gamma}_{\tilde{\Omega}} = m_{zy,\Omega} / m_{zz,\Omega}.$$

The error SS under $\tilde{\Omega}$ is now available as $\mathscr{S}_\Omega - \mathbf{m}'_\Omega \hat{\boldsymbol{\gamma}}_{\tilde{\Omega}}$, or

$$\mathscr{S}_{\tilde{\Omega}} = \mathscr{S}_\Omega - m_{zy,\Omega} \hat{\gamma}_{\tilde{\Omega}} = m_{yy,\Omega} - m_{zz,\Omega}^{-1} m_{zy,\Omega}^2.$$

The number of d.f. is $n-I-1$, where $n = \Sigma_i J_i$.

The LS estimate of β_i under Ω is

$$\hat{\beta}_{i,\Omega} = y_{i.}.$$

The correction for regression on z is hence $-\hat{\gamma}_{\tilde{\Omega}} z_{i.}$, and the LS estimate under $\tilde{\Omega}$ is thus

$$\hat{\beta}_{i,\tilde{\Omega}} = y_{i.} - \hat{\gamma}_{\tilde{\Omega}} z_{i.}. \tag{6.3.3}$$

To test the hypothesis H_γ: $\gamma = 0$, or to construct a confidence interval for γ, we need Var $(\hat{\gamma}_{\tilde{\Omega}})$, and by the above general theory in which $\sigma^2 \mathbf{M}_\Omega^{-1}$ is the covariance matrix of $\hat{\boldsymbol{\gamma}}_{\tilde{\Omega}}$, this is

$$\text{Var}\,(\hat{\gamma}_{\tilde{\Omega}}) = \sigma^2 / m_{zz,\Omega} = \sigma^2 / \sum_i \sum_j (z_{ij} - z_{i.})^2. \tag{6.3.4}$$

The test and confidence interval can then be based on the fact that the ratio $(\hat{\gamma}_{\tilde{\Omega}} - \gamma)/(m_{zz,\Omega}^{-1/2} s)$, where $s^2 = \mathscr{S}_{\tilde{\Omega}}/(n-I-1)$, has the t-distribution with $n-I-1$ d.f. under $\tilde{\Omega}$.

Suppose H_β is the hypothesis

$$H_\beta: \quad \beta_1 = \beta_2 = \cdots = \beta_I.$$

Denote $H_\beta \cap \Omega$ by ω. We may recall from the corresponding analysis-of-variance problem that

$$\mathscr{S}_\omega = \sum_i \sum_j (y_{ij} - \bar{y})^2,$$

where $\bar{y} = \Sigma_i \Sigma_j y_{ij}/n$, or may calculate it by adding $\mathscr{S}_\Omega$ to $\Sigma_i n_i (y_{i.} - \bar{y})^2$, the numerator SS for testing H_β under Ω. In any case, $\mathscr{S}_\omega$ is calculated from the identity

$$\mathscr{S}_\omega = \sum_i \sum_j y_{ij}^2 - n\bar{y}^2.$$

Proceeding for ω as we did above for Ω, we calculate

$$m_{yy,\omega} = \sum_i \sum_j y_{ij}^2 - n\bar{y}^2,$$

$$m_{zz,\omega} = \sum_i \sum_j z_{ij}^2 - n\bar{z}^2,$$

$$m_{zy,\omega} = \sum_i \sum_j z_{ij} y_{ij} - n\bar{z}\bar{y},$$

$$\mathscr{S}_{\tilde{\omega}} = m_{yy,\omega} - m_{zz,\omega}^{-1} m_{yz,\omega}^2.$$

The numerator SS for testing H_β under $\tilde{\Omega}$ is now available as $\mathscr{S}_{\tilde{\omega}} - \mathscr{S}_{\tilde{\Omega}}$, the denominator SS is $\mathscr{S}_{\tilde{\Omega}}$, and the numbers of d.f. are $I-1$ and $n-I-1$.

If the hypothesis H_β is rejected, one can determine by the S-method which of the contrasts in the $\{\beta_i\}$ are responsible for this. To apply the S-method, we need the variance of $\hat{\psi}_{\tilde{\Omega}} = \Sigma_i c_i \hat{\beta}_{i,\tilde{\Omega}}$, the estimate under $\tilde{\Omega}$ of the contrast $\psi = \Sigma_i c_i \beta_i$, where $\Sigma_i c_i = 0$. From (6.3.3),

$$\hat{\psi}_{\tilde{\Omega}} = \sum_i c_i y_{i.} - \hat{\gamma}_{\tilde{\Omega}} \sum_i c_i z_{i.},$$

and, from the remark at the end of sec. 6.2, $\hat{\gamma}_{\tilde{\Omega}}$ is statistically independent of the $\{y_{i.}\}$ under $\tilde{\Omega}$; thus

$$\begin{aligned} \text{Var}(\hat{\psi}_{\tilde{\Omega}}) &= \sum_i c_i^2 \,\text{Var}(y_{i.}) + \left(\sum_i c_i z_{i.}\right)^2 \text{Var}(\hat{\gamma}_{\tilde{\Omega}}) \\ &= \sigma^2 \left\{ \sum_i \frac{c_i^2}{J_i} + \frac{\left(\sum_i c_i z_{i.}\right)^2}{\sum_i \sum_j (z_{ij} - z_{i.})^2} \right\}, \end{aligned}$$

with the help of (6.3.4). The T-method is not applicable because the estimates $\{\hat{\beta}_{i,\tilde{\Omega}}\}$ defined by (6.3.3) will not in general have equal variances (or equal covariances), even if the $\{J_i\}$ are equal.

6.4. AN EXAMPLE WITH TWO CONCOMITANT VARIABLES

This section is written so that it may be read immediately after, or along with, sec. 6.2, and may seem somewhat repetitious if read after sec. 6.3.

Suppose that in a two-way layout with one observation per cell on the dependent variable y the values of two concomitant variables z and w are also given. If we assume linear regression on z and w, and no interaction between the factors corresponding to rows and columns, the underlying assumptions will be[12]

$$\tilde{\Omega}: \begin{cases} y_{ij} = \mu + \alpha_i + \beta_j + \gamma z_{ij} + \delta w_{ij} + e_{ij}, \\ \sum_i \alpha_i = 0, \qquad \sum_j \beta_j = 0, \\ \text{the } \{e_{ij}\} \text{ are independently } N(0, \sigma^2), \end{cases}$$

where y_{ij}, z_{ij}, w_{ij} are respectively the values of y, z, w observed in the i,j cell, and γ and δ are the regression coefficients on z and w. The present analysis would also cover the case of quadratic regression on a single concomitant variable z if we set $w_{ij} = z_{ij}^2$. The corresponding analysis of variance under the assumptions Ω where $\gamma = \delta = 0$ was derived in sec. 4.2. The error SS there was found to be

$$\mathscr{S}_\Omega = \sum_i \sum_j (y_{ij} - y_{i.} - y_{.j} + y_{..})^2, \tag{6.4.1}$$

and may be calculated from the identity

$$\mathscr{S}_\Omega = \sum_i \sum_j y_{ij}^2 - J \sum_i y_{i.}^2 - I \sum_j y_{.j}^2 + IJy_{..}^2. \tag{6.4.2}$$

We have to calculate the six quantities $\{m_{tv,\Omega}\}$ where t and v each stand for y, z, or w. To define $m_{tv,\Omega}$ we regard (6.4.1) as a sum of squares of linear forms $L(y) = y_{ij} - y_{i.} - y_{.j} + y_{..}$ and replace $[L(y)]^2$ by $L(t)\,L(v)$, so that

$$m_{tv,\Omega} = \sum_i \sum_j (t_{ij} - t_{i.} - t_{.j} + t_{..})(v_{ij} - v_{i.} - v_{.j} + v_{..}).$$

To calculate the $\{m_{tv,\Omega}\}$ we use the same device on the identity (6.4.2),

$$m_{tv,\Omega} = \sum_i \sum_j t_{ij} v_{ij} - J \sum_i t_{i.} v_{i.} - I \sum_j t_{.j} v_{.j} + IJt_{..} v_{..},$$

that is, we calculate them from the formulas

$$m_{yy,\Omega} = \sum_i \sum_j y_{ij}^2 - J \sum_i y_{i.}^2 - I \sum_j y_{.j}^2 + IJy_{..}^2,$$

$$m_{zz,\Omega} = \sum_i \sum_j z_{ij}^2 - J \sum_i z_{i.}^2 - I \sum_j z_{.j}^2 + IJz_{..}^2,$$

$$m_{zy,\Omega} = \sum_i \sum_j z_{ij} y_{ij} - J \sum_i z_{i.} y_{i.} - I \sum_j z_{.j} y_{.j} + IJz_{..} y_{..},$$

etc.

[12] See the footnote at the beginning of sec. 6.3 concerning a rank condition needed to complete the $\tilde{\Omega}$-assumptions.

The next step is to solve the equations $\mathbf{M}_\Omega \hat{\mathbf{\Upsilon}}_{\tilde{\Omega}} = \mathbf{m}_\Omega$ of the general theory. In the present example

$$\mathbf{M}_\Omega = \begin{pmatrix} m_{zz,\Omega} & m_{zw,\Omega} \\ m_{wz,\Omega} & m_{ww,\Omega} \end{pmatrix}, \qquad \hat{\mathbf{\Upsilon}}_{\tilde{\Omega}} = \begin{pmatrix} \hat{\gamma}_{\tilde{\Omega}} \\ \hat{\delta}_{\tilde{\Omega}} \end{pmatrix}, \qquad \mathbf{m}_\Omega = \begin{pmatrix} m_{zy,\Omega} \\ m_{wy,\Omega} \end{pmatrix},$$

and so we have to solve the system

$$(6.4.3) \qquad \begin{aligned} m_{zz,\Omega}\hat{\gamma}_{\tilde{\Omega}} + m_{zw,\Omega}\hat{\delta}_{\tilde{\Omega}} &= m_{zy,\Omega}, \\ m_{wz,\Omega}\hat{\gamma}_{\tilde{\Omega}} + m_{ww,\Omega}\hat{\delta}_{\tilde{\Omega}} &= m_{wy,\Omega} \end{aligned}$$

for $\hat{\gamma}_{\tilde{\Omega}}$ and $\hat{\delta}_{\tilde{\Omega}}$. The error SS under $\tilde{\Omega}$ is now available as the $\mathscr{S}_\Omega - \mathbf{m}'_\Omega \hat{\mathbf{\Upsilon}}_{\tilde{\Omega}}$ of the general theory, or, in the present example,

$$(6.4.4) \qquad \mathscr{S}_{\tilde{\Omega}} = m_{yy,\Omega} - m_{zy,\Omega}\hat{\gamma}_{\tilde{\Omega}} - m_{wy,\Omega}\hat{\delta}_{\tilde{\Omega}}.$$

The number of d.f. is $(I-1)(J-1) - 2$.

The LS estimate of the ith-row main effect under Ω is

$$(6.4.5) \qquad \hat{\alpha}_{i,\Omega} = y_{i.} - y_{..}.$$

To get the corresponding estimate under $\tilde{\Omega}$ we "correct for regression" by forming the same linear forms in z and w, namely $z_{i.} - z_{..}$ and $w_{i.} - w_{..}$, multiply by $\hat{\gamma}_{\tilde{\Omega}}$ and $\hat{\delta}_{\tilde{\Omega}}$, respectively, and subtract from (6.4.5) to get

$$\hat{\alpha}_{i,\tilde{\Omega}} = y_{i.} - y_{..} - \hat{\gamma}_{\tilde{\Omega}}(z_{i.} - z_{..}) - \hat{\delta}_{\tilde{\Omega}}(w_{i.} - w_{..}).$$

In the same way we get

$$(6.4.6) \qquad \begin{aligned} \hat{\beta}_{j,\tilde{\Omega}} &= y_{.j} - y_{..} - \hat{\gamma}_{\tilde{\Omega}}(z_{.j} - z_{..}) - \hat{\delta}_{\tilde{\Omega}}(w_{.j} - w_{..}), \\ \hat{\mu}_{\tilde{\Omega}} &= y_{..} - \hat{\gamma}_{\tilde{\Omega}} z_{..} - \hat{\delta}_{\tilde{\Omega}} w_{..}. \end{aligned}$$

To test any of the hypotheses

$$H_\gamma\colon\ \gamma = 0, \qquad H_\delta\colon\ \delta = 0, \qquad H_{\gamma,\delta}\colon\ \gamma = \delta = 0,$$

or to get a confidence interval for γ or δ, or a confidence ellipsoid for (γ, δ), we may proceed according to the method of secs. 2.3 and 2.4 and will then need some or all of the elements of $\mathbf{M}_\Omega^{-1}$, since $\sigma^2\mathbf{M}_\Omega^{-1}$ is the covariance matrix of $(\hat{\gamma}_{\tilde{\Omega}}, \hat{\delta}_{\tilde{\Omega}})'$. Since $\mathbf{M}_\Omega$ is 2×2, its inverse is easily found by the method of cofactors (Theorem 4 of App. II) to be

$$(6.4.7) \qquad \mathbf{M}_\Omega^{-1} = \begin{pmatrix} M^{-1}m_{ww,\Omega} & -M^{-1}m_{zw,\Omega} \\ -M^{-1}m_{zw,\Omega} & M^{-1}m_{zz,\Omega} \end{pmatrix},$$

where

$$M = m_{zz,\Omega}m_{ww,\Omega} - m_{zw,\Omega}^2.$$

Thus

$$
\text{(6.4.8)} \qquad \begin{aligned}
\operatorname{Var}(\hat{\gamma}_{\tilde{\Omega}}) &= M^{-1}m_{ww,\Omega}\sigma^2,\\
\operatorname{Var}(\hat{\delta}_{\tilde{\Omega}}) &= M^{-1}m_{zz,\Omega}\sigma^2,\\
\operatorname{Cov}(\hat{\gamma}_{\Omega}, \hat{\delta}_{\Omega}) &= -M^{-1}m_{zw,\Omega}\sigma^2.
\end{aligned}
$$

For example, to find a confidence interval for γ we use the fact that under $\tilde{\Omega}$ the ratio $(\hat{\gamma}_{\tilde{\Omega}}-\gamma)/Cs$ has the t-distribution with $(I-1)(J-1)-2$ d.f., if $\operatorname{Var}(\hat{\gamma}_{\tilde{\Omega}}) = C^2\sigma^2$ and $s^2 = \mathscr{S}_{\tilde{\Omega}}/(IJ-I-J-1)$. The value of C^2 is seen from the first of the formulas (6.4.8) to be $M^{-1}m_{ww,\Omega}$.

We shall apply the general theory of sec. 6.2 to derive the test of the hypothesis H_β about column effects,

$$H_\beta: \quad \beta_1 = \beta_2 = \cdots = \beta_J = 0,$$

under $\tilde{\Omega}$; the test of the corresponding hypothesis about row effects would of course be similar. Denote $H_\beta \cap \Omega$ by ω. We may recall from the corresponding analysis-of-variance problem that

$$\mathscr{S}_\omega = \sum_i \sum_j (y_{ij}-y_{i.})^2,$$

or may calculate it by adding $\mathscr{S}_\Omega$ to $I\Sigma_j(y_{.j}-y_{..})^2$, the numerator SS for testing H_β under Ω. In any case, $\mathscr{S}_\omega$ is calculated from the identity

$$\mathscr{S}_\omega = \sum_i \sum_j y_{ij}^2 - J\sum_i y_{i.}^2.$$

Proceeding for ω as we did above for Ω, we calculate the six values of

$$m_{tv,\omega} = \sum_i \sum_j t_{ij}v_{ij} - J\sum_i t_{i.}v_{i.},$$

where t and v each stand for y, z, or w. We solve next for $\hat{\gamma}_{\tilde{\omega}}$ and $\hat{\delta}_{\tilde{\omega}}$ the equations $\mathbf{M}_\omega \hat{\mathbf{\Upsilon}}_{\tilde{\omega}} = \mathbf{m}_\omega$, which look like (6.4.3) with Ω replaced by ω. The quantity $\mathscr{S}_{\tilde{\omega}}$ is then given by the analog of (6.4.4) as

$$\mathscr{S}_{\tilde{\omega}} = m_{yy,\omega} - m_{zy,\omega}\hat{\gamma}_{\tilde{\omega}} - m_{wy,\omega}\hat{\delta}_{\tilde{\omega}}.$$

The numerator SS for testing H_β under $\tilde{\Omega}$ is now available as $\mathscr{S}_{\tilde{\omega}}-\mathscr{S}_{\tilde{\Omega}}$, the denominator SS is $\mathscr{S}_{\tilde{\Omega}}$, and the numbers of d.f. are $J-1$ and $IJ-I-J-1$.

If the hypothesis H_β is rejected, one can determine by the S-method which of the contrasts in the $\{\beta_j\}$ are responsible for this. To apply the S-method we need the variance of $\hat{\psi} = \Sigma_j c_j \hat{\beta}_{j,\tilde{\Omega}}$, the estimate under $\tilde{\Omega}$ of the contrast $\psi = \Sigma_j c_j \beta_j$, where $\Sigma_j c = 0$. From (6.4.6),

$$\hat{\psi}_{\tilde{\Omega}} = \sum_j c_j y_{.j} - \hat{\gamma}_{\tilde{\Omega}} \sum_j c_j z_{.j} - \hat{\delta}_{\tilde{\Omega}} \sum_j c_j w_{.j},$$

which might be written as

$$\hat{\psi}_{\tilde{\Omega}} = l(y) - \hat{\gamma}_{\tilde{\Omega}}\, l(z) - \hat{\delta}_{\tilde{\Omega}}\, l(w),$$

where $l(y)$ is the linear form

$$l(y) = \sum_j c_j y_{.j},$$

to indicate how the estimate of ψ under Ω is "corrected for regression." From the remark at the end of sec. 6.2, $\hat{\gamma}_{\tilde{\Omega}}$ and $\hat{\delta}_{\tilde{\Omega}}$ are statistically independent of the $\{y_{.j}\}$ under $\tilde{\Omega}$, and hence

$$\begin{aligned}\text{Var}\,(\hat{\psi}_{\tilde{\Omega}}) = \text{Var}\,[l(y)] + [l(z)]^2\,\text{Var}\,(\hat{\gamma}_{\tilde{\Omega}}) \\ + [l(w)]^2\,\text{Var}\,(\hat{\delta}_{\tilde{\Omega}}) + 2l(z)\,l(w)\,\text{Cov}\,(\hat{\gamma}_{\tilde{\Omega}}, \hat{\delta}_{\tilde{\Omega}}),\end{aligned}$$

where $\text{Var}\,[l(y)] = \sigma^2\Sigma_j c_j^2/I$, and the remaining terms are evaluated from (6.4.8). The T-method is not applicable because the estimates $\{\hat{\beta}_{j,\tilde{\Omega}}\}$ defined by (6.4.6) will not in general have equal variances (or equal covariances).

6.5. LINEAR REGRESSION ON CONTROLLED VARIABLES SUBJECT TO ERROR

The discussion of this section applies to regression analysis as well as the analysis of covariance. We continue to denote the dependent variable by y and to suppose that there are h concomitant variables $z_1, \cdots, z_h$. We have heretofore assumed that in the ith observation $(y_i, z_{1i}, \cdots, z_{hi})$ on the variables the observed value y_i differs from a "true" value η_i by a random error e_i,

$$y_i = \eta_i + e_i, \tag{6.5.1}$$

with $E(e_i) = 0$; we shall now similarly permit the observed value z_{ji} to differ from a "true" value $\tilde{z}_{ji}$ by a random error f_{ji},

$$z_{ji} = \tilde{z}_{ji} + f_{ji}, \tag{6.5.2}$$

with $E(f_{ji}) = 0$ $(j = 1, \cdots, h;\ i = 1, \cdots, n)$.

The underlying assumptions of sec. 6.2, aside from normality, could be written

$$\begin{cases} y_i = \eta_i + e_i, \\ \text{the } \{e_i\} \text{ are statistically independent,} \\ E(e_i) = 0, \qquad \text{Var}\,(e_i) = \sigma^2, \\ \eta_i = \sum_{j=1}^{p} x_{ji}\beta_j + \sum_{j=1}^{h} z_{ji}\gamma_j. \end{cases} \tag{6.5.3}$$

In the case of regression analysis all $x_{ji} = 0$. In the new model of this section the linear relation between the dependent and the concomitant variables will be between the *true* values, so that the last line of (6.5.3) is replaced by

(6.5.4) $$\eta_i = \sum_{j=1}^{p} x_{ji}\beta_j + \sum_{j=1}^{h} \tilde{z}_{ji}\gamma_j.$$

We shall assume that the $(h+1)\times(h+1)$ covariance matrix of the errors

(6.5.5) $$(e_i, f_{1i}, f_{2i}, \cdots, f_{hi})'$$

in the ith observation is the same for all i, and denote it by

(6.5.6) $$\begin{pmatrix} \sigma_{00} & \sigma_{01} & \cdots & \sigma_{0h} \\ \sigma_{10} & \sigma_{11} & \cdots & \sigma_{1h} \\ \cdot & \cdot & \cdots & \cdot \\ \sigma_{h0} & \sigma_{h1} & \cdots & \sigma_{hh} \end{pmatrix}.$$

We have already assumed that $E(e_i) = 0$, $E(f_{ji}) = 0$. Finally, we shall assume that the n vectors of errors (6.5.5) are statistically independent.

The model is still not sufficiently complete until we indicate how the observed values $\{z_{ji}\}$ or the true values $\{\tilde{z}_{ji}\}$ of the concomitant variables are selected. In some cases it may be appropriate to consider the $\{\tilde{z}_{ji}\}$ as sampled from a certain distribution independent of the distribution of the errors. In such a model the problems of statistical inference are in general very difficult. The problems become relative simple, and the model is widely applicable to actual experiments, if we assume that the observed values $\{z_{ji}\}$ are constant values selected beforehand by the experimenter.

We refer to the variables $z_1, \cdots, z_h$ in this case as *controlled variables subject to error*.[13] (If any z_j are controlled without error this is included as a special case where the corresponding σ_{jj} are zero.) The $\{z_{ji}\}$ in (6.5.2) are then preassigned known constants, the random variables in (6.5.2) being the errors $\{f_{ji}\}$ and the true values

(6.5.7) $$\tilde{z}_{ji} = z_{ji} - f_{ji},$$

the joint distribution of the $\{\tilde{z}_{ji}\}$ being determined by the assumptions we have made on the joint distribution of the $\{f_{ji}\}$, together with the values of the constants $\{z_{ji}\}$; this distribution would then be completely determined if we added the normality assumption on the distribution of the $\{f_{ji}\}$.

[13] The basic idea of this section is due to Berkson (1950), who developed it in connection with fitting straight lines. An extension was made by Scheffé (1958) to the case where the two parameters of the true line are treated as random effects. Although the model of this section is notationally similar to certain others, used for example in economics, involving errors in all variables and linear relations between the true values, the distribution assumptions are very different: instead of the correlation coefficient of the error f_{ji} and the true value $\tilde{z}_{ji}$ being 0, it is -1 in the present model.

The relative simplicity attained is a consequence of the model's reducing to the previous underlying assumptions (6.5.3) with a different $\sigma^2 = \text{Var}(y_i)$: If we substitute (6.5.7) in (6.5.4), we have

$$\eta_i = \sum_{j=1}^{p} x_{ji}\beta_j + \sum_{j=1}^{h} (z_{ji} - f_{ji})\gamma_j,$$

and hence from (6.5.1),

$$y_i = \sum_{j=1}^{p} x_{ji}\beta_j + \sum_{j=1}^{h} z_{ji}\gamma_j + e'_i,$$

where

(6.5.8) $$e'_i = e_i - \sum_{j=1}^{h} \gamma_j f_{ji}.$$

Since e'_i depends only on the ith vector of errors (6.5.5), it follows from the assumed independence of these vectors that the $\{e'_i\}$ are independent. Furthermore, from (6.5.8), $E(e'_i) = 0$ and

(6.5.9) $$\text{Var}(e'_i) = \sigma_{00} - 2\sum_{j=1}^{h} \gamma_j\sigma_{j0} + \sum_{j=1}^{h}\sum_{j'=1}^{h} \gamma_j\gamma_{j'}\sigma_{jj'}$$

does not depend on i. The assumptions (6.5.3) are thus satisfied with $\{e_i\}$ replaced by $\{e'_i\}$, and σ^2 by (6.5.9). It follows[14] that all the methods of inference are valid as though the observed values of the concomitant variables were true values observed without error, except that if the (unknown) error variance of the dependent variable y, denoted by the parameter σ^2 in the former theory and by σ_{00} in the present, appears in an inference in the former theory, it is to be replaced by the (unknown) constant (6.5.9); in particular (6.5.9) is estimated exactly like σ^2 in the former theory.

We consider now in more detail the applicability of this model. The chief requirement is that the experimenter run the experiment by choosing beforehand (in practice, perhaps after some preliminary observations) what values of the independent variables will be used in the experiment; this is perhaps the usual situation in experimentation in physical science, frequent in biological science, but infrequent in social science (where data usually do not come from experiments, anyway). More precisely, the concomitant variables are controlled so as to bring the observations on them (meter readings, nominal dosages, etc.) to preassigned values. The assumptions $E(e_i) = 0$ and $E(f_{ji}) = 0$ imply an unbiasedness of the errors,

[14] It also follows that the covariance matrix (6.5.6) is not identifiable if the errors are jointly normal, since for given $\{\gamma_j\}$ any covariance matrix giving the same value to (6.5.9) would give the same distribution to the observations: A parameter of the distribution of a sample is said to be *not identifiable* if the same distribution can be obtained for different values of the parameter.

which would be violated, for example, if an error is caused by bias in a measuring instrument such as could be removed by using a correct calibration curve for the instrument. The observations on the concomitant as well as the dependent variables must be independent for different i: Thus if replicated observations on the dependent variable were made for the same values of the concomitant variables, then to achieve the necessary independence, when a new observation is made on the dependent variable we would not leave the setting of the concomitant variables undisturbed, but would change them from their previous values and then bring them back so as to make the observations on them the same as before. Finally, the dependence on the concomitant variables must be strictly linear, the device mentioned below (6.1.2) for treating, for example, quadratic regression on a concomitant variable z_j being now ruled out unless z_j is controlled without error: For, on squaring (6.5.7) we should get a term f_{ji}^2 which would show up in e_i' and violate the condition $E(e_i') = 0$.

PROBLEMS

6.1. Table A gives the breaking strength y in grams and the thickness x in 10^{-4} inch from tests on seven types of starch film. (*a*) Ignoring the x-data, test for

TABLE* A

Wheat		Rice		Canna		Corn		Potato		Dasheen		Sweet Potato	
y	x	y	x	y	x	y	x	y	x	y	x	y	x
263.7	5.0	556.7	7.1	791.7	7.7	731.0	8.0	983.3	13.0	485.4	7.0	837.1	9.4
130.8	3.5	552.5	6.7	610.0	6.3	710.0	7.3	958.8	13.3	395.4	6.0	901.2	10.6
382.9	4.7	397.5	5.6	710.0	8.6	604.7	7.2	747.8	10.7	465.4	7.1	595.7	9.0
302.5	4.3	532.3	8.1	940.7	11.8	508.8	6.1	866.0	12.2	371.4	5.3	510.0	7.6
213.3	3.8	587.8	8.7	990.0	12.4	393.0	6.4	810.8	11.6	402.0	6.2		
132.1	3.0	520.9	8.3	916.2	12.0	416.0	6.4	950.0	9.7	371.9	5.8		
292.0	4.2	574.3	8.4	835.0	11.4	400.0	6.9	1282.0	10.8	430.0	6.6		
315.5	4.5	505.0	7.3	724.3	10.4	335.6	5.8	1233.8	10.1	380.0	6.6		
262.4	4.3	604.6	8.5	611.1	9.2	306.4	5.3	1660.0	12.7				
314.4	4.1	522.5	7.8	621.7	9.0	426.0	6.7	746.0	9.8				
310.8	5.5	555.0	8.0	735.4	9.5	382.5	5.8	650.0	10.0				
280.8	4.8	561.1	8.4	990.0	12.5	340.8	5.7	992.5	13.8				
331.7	4.8			862.7	11.7	436.7	6.1	896.7	13.3				
672.5	8.0					333.3	6.2	873.9	12.4				
496.0	7.4					382.3	6.3	924.4	12.2				
311.9	5.2					397.7	6.0	1050.0	14.1				
276.7	4.7					619.1	6.8	973.3	13.7				
325.7	5.4					857.3	7.9						
310.8	5.4					592.5	7.2						
288.0	5.4												
269.3	4.9												

* From pp. 120–121 of *Industrial Statistics* by H. A. Freeman, John Wiley, New York (1942). Reproduced with the kind permission of the author and the publisher.

differences among the starches in expected breaking strengths. (*b*) Assuming that the regression coefficient of y on x is the same for all starches, test for differences of strength among the starches after allowing for differences of thicknesses. (*c*) Letting y_{ij} and x_{ij} denote the jth measurements of y and x on the ith starch, list in adjacent columns the values of $\{y_{i.}\}$ for the seven starches and of $\{y_{i.}-\hat{\beta}(x_{i.}-\bar{x})\}$, where $\bar{x}$ is the unweighted average of the $\{x_{i.}\}$, and $\hat{\beta}$ is the estimated regression coefficient of y on x.

6.2. The data in Table B are from an experimental piggery arranged for individual feeding of six pigs in each of five pens. From each of five litters six young pigs, three males and three females, were selected and allotted to one of the pens. Three feeding treatments denoted by A, B, C, containing increasing proportions ($p_A < p_B < p_C$) of protein, were used and each given to one male and one female in each pen. The pigs were individually weighed each week for 16 weeks. For each pig the growth rate in pounds per week was calculated as the slope of a line fitted by LS, and is denoted by y in the table; the weight at the beginning of the experiment is denoted by x. (*a*) Make the analysis of covariance allowing for linear regression of y on x with the same regression coefficient for all the pigs and a possible food × sex interaction, but no other interactions. (*b*) Which pairs of the three feeding treatments differ significantly by the S-method with $\alpha = 0.10$?

TABLE* B

Pen	Variable	Food *A* Male	*A* Female	*B* Male	*B* Female	*C* Male	*C* Female
1	y	9.52	9.94	8.51	10.00	9.11	9.75
	x	38	48	39	48	48	48
2	y	8.21	9.48	9.95	9.24	8.50	8.66
	x	35	32	38	32	37	28
3	y	9.32	9.32	8.43	9.34	8.90	7.63
	x	41	35	46	41	42	33
4	y	10.56	10.90	8.86	9.68	9.51	10.37
	x	48	46	40	46	42	50
5	y	10.42	8.82	9.20	9.67	8.76	8.57
	x	43	32	40	37	40	30

* From Table 11, p. 17 of Commonwealth Bureau of Plant Breeding and Genetics Technical Communication 15, *Field Trials II: The Analysis of Covariance* by John Wishart, Cambridge Univ. Press, Cambridge, 1950. Reproduced with the kind permission of the publisher.

6.3. Table C exhibits the yearly yield y of wheat in hundredweights per acre at six British agricultural stations for three years, together with shoot height z in inches at ear emergence and number w of plants per foot at tillering. (*a*) Does

TABLE* C

Year	Variable	Place					
		Seale Hayne	Rotham-sted	Newport	Boghall	Sprow-ston	Plump-ton
1933	y	19.0	22.2	35.3	32.8	25.3	35.8
	z	25.6	25.4	30.8	33.0	28.5	28.0
	w	14.9	13.3	4.6	14.7	12.8	7.5
1934	y	32.4	32.2	43.7	35.7	28.3	35.2
	z	25.4	28.3	35.3	32.4	25.9	24.2
	w	7.2	9.5	6.8	9.7	9.2	7.5
1935	y	26.2	34.7	40.0	29.6	20.6	47.2
	z	27.9	34.4	32.5	27.5	23.7	32.9
	w	18.6	22.2	10.0	17.6	14.4	7.9

* The data are from Table 14.5.1, p. 427 of *Statistical Methods* by G. W. Snedecor, Iowa State College Press, 5th ed., 1956. Reproduced with the kind permission of the author and the publisher.

there exist at the 0.05 level a significant variation in yield from year to year over the three-year period, which is not explained by the regression of y on z and w? (b) Test at the 0.05 level the hypothesis that there is no regression on the number of plants per foot. (c) In 1934 a crop of wheat at Sprowston was observed during growth to have $z = 27$, $w = 10$. Give a point estimate of the average yield you would expect from this crop.

PART II

The Analysis of Variance in the Case of Other Models

CHAPTER 7

Random-Effects Models

7.1. INTRODUCTION

The random-effects models for the analysis of variance are also called *variance-components models*, for reasons to be seen below. The general nature of the three kinds of models—fixed-effects, random-effects, and mixed—was indicated in sec. 1.2. There is no general theory comparable to that of Chs. 1 and 2 for models other than the fixed-effects model; our knowledge of optimum properties of the statistical methods used with these models is at present very limited.[1]

The origin of the random-effects models, like that of the fixed-effects models, lies in astronomical problems; statisticians re-invented random-effects models long after they were introduced by astronomers, and then developed more complicated ones.[2]

7.2. THE ONE-WAY LAYOUT

Suppose that an experiment is tried in a factory with I workers and a machine run by a single worker, which produces small parts of some kind, that a large number of parts are produced daily on the machine, and that for any worker there is considerable day-to-day variation (for some purposes we will treat the output as though it were a continuous random variable). We shall assume the time trend of the machine is negligible during the experiment. Suppose each worker is assigned to the machine for J days during the experiment. Denote by y_{ij} the output of the ith worker the jth day he uses the machine.

[1] Some results have been published by Graybill (1954), Thompson (1955), and Herbach (1957).

[2] For a little historical background see Scheffé (1956*b*).

We shall now try to motivate the assumptions we will make concerning a *model equation* of the form

$$y_{ij} = m_i + e_{ij}, \tag{7.2.1}$$

where m_i is the "true" mean for the ith worker and e_{ij} is his "error" on the jth day. Here m_i might be regarded as an idealized daily average for the ith worker after he has reached a relatively stable period following a learning stage. The variability of the ith worker's output about his "true" mean m_i could be measured by a variance σ_i^2. We shall assume that the I workers in the experiment are a random sample from a large labor pool, which we will idealize as an infinite population of workers. Most of our results rest on this basic assumption, and it is essential in applying them to ask what is this conceptual population, if any, from which the I workers in the experiment can reasonably be regarded as a random sample.

Suppose the workers in the population are labeled by an index[3] v. It will be convenient to denote by $\mathscr{P}$ the population distribution of v, even though it does not enter the calculations directly. Let $m(v)$ and $\sigma^2(v)$ denote the "true" mean and variance of the daily output of the worker labeled v in the population. Let the labels attached to the I workers selected at random for the experiment be $\{v_1, \cdots, v_I\}$; this is then a random sample from $\mathscr{P}$, and the above m_i and σ_i^2 for the ith worker in the experiment will be $m_i = m(v_i)$ and $\sigma_i^2 = \sigma^2(v_i)$. If μ and σ_A^2 denote the mean and variance of the "true" daily outputs of the workers in the population, i.e., μ and σ_A^2 are the expected value and variance of the random variable $m(v)$ calculated with respect to the population distribution $\mathscr{P}$, then the $\{m_i\}$ are independently and identically distributed with mean μ and variance σ_A^2.

We shall now make the simplifying assumption that $\sigma^2(v)$ is the same for all workers in the population, and denote the common value by σ_e^2. We remark that this assumption is less reasonable than the preceding ones: It is likely that the variances of the workers vary as well as their "true" means; it is conceivable, for example, that the better workers, namely, those with higher values of $m(v)$, might be "steadier" and have smaller variances, or, again, that the standard deviation $\sigma(v)$ might be

[3] There is no harm in thinking of v as a real number and of $\mathscr{P}$ as the distribution of a random variable which takes on values $\{v\}$. The mathematically more advanced reader may think of $\mathscr{P}$ as a probability distribution on an abstract probability space of points $\{v\}$, and of $m(v)$ and $\sigma^2(v)$ as random variables. In sec. 7.4 is involved the product space of two probability spaces, one with the distribution $\mathscr{P}_v$ on the space of points $\{v\}$, the other with distribution $\mathscr{P}_u$ on the space of points $\{u\}$, and the random variable $m(u, v)$. Similar comments apply in later sections.

proportional to the mean $m(v)$. Nevertheless we shall adopt this assumption because no results are at present available under the more complicated model.[4] Under this assumption the $\{e_{ij}\}$ in (7.2.1) have zero means and equal variances σ_e^2; we shall assume further they are independently and identically distributed, and independently of the $\{m_i\}$.

We define the *effect of the worker* labeled v in the population to be

$$a(v) = m(v) - \mu,$$

so that the effect of the ith worker in the experiment is $a(v_i)$, which we shall write

$$a_i = m_i - \mu.$$

Then (7.2.1) becomes

$$y_{ij} = \mu + a_i + e_{ij}, \tag{7.2.2}$$

where the $\{a_i\}$, $\{e_{ij}\}$ are completely independent, with zero means, the $\{a_i\}$ are identically distributed with variance σ_A^2, and the $\{e_{ij}\}$ are identically distributed with variance σ_e^2. The variance of an observation y_{ij} is

$$\sigma_y^2 = \sigma_A^2 + \sigma_e^2,$$

and so it is appropriate to call σ_A^2 and σ_e^2 the *variance components* (i.e., the components of the variance of an observation).

About the random-effects model we have now formulated, and random-effects models in general, we remark that they differ basically from fixed-effects models in that under the underlying assumptions (i) all the observations have the same expectation, and (ii) the observations are *not* statistically independent. The statistical dependence in the above random-effects model (or its generalization where we permit the numbers $\{J_i\}$ to be unequal, where J_i is the number of measurements in the ith class) is formulated in a concept, useful in genetics, called the intraclass correlation coefficient,[5] defined as follows:

Definition: *The intraclass correlation coefficient* $\tilde{\rho}$ is the ordinary correlation coefficient between any two of the observations y_{ij} and $y_{ij'}$ $(j' \neq j)$ in the same class (i.e., with the same i),

$$\tilde{\rho} = E[(y_{ij}-\mu)(y_{ij'}-\mu)]/\sigma_y^2 = E[(a_i+e_{ij})(a_i+e_{ij'})]/\sigma_y^2 = E(a_i^2)/\sigma_y^2;$$

hence

$$\tilde{\rho} = \sigma_A^2/(\sigma_A^2 + \sigma_e^2).$$

[4] The above generalization of the model we adopt was suggested to me by Professor E. L. Lehmann.

[5] This historically important concept was introduced by R. A. Fisher (1925), Ch. 7.

We shall see below how a confidence interval for $\tilde{\rho}$ can be obtained if a normality assumption is added to the preceding assumptions.[6]

If the numbers $\{J_i\}$ are equal, as we are assuming, the one-way layout is said to be *balanced*. More generally, the complete p-way layout is called *balanced* if the numbers of observations in the cells are equal. If some factors are nested the layout may be called *balanced* if the number of levels of a nested factor is the same within each combination of those other factors within which it is nested, and the factors (if any) which are crossed are completely crossed (sec. 5.3); furthermore the number of replications (levels of an "error factor") must be the same for every combination of the (other) factors appearing in the experiment. The general procedure used to obtain tests and estimates with random-effects models and mixed models in balanced cases is to consider all the mean squares in the usual analysis-of-variance table for the same design with the fixed-effects model. However, they are in general not used in the same way. The column of expected mean squares in the table suggests which of the mean squares have to be used for testing the different hypotheses. We shall see that the denominator of the F-statistic is often different from that in the corresponding fixed-effects model. Point estimates are derived from the column of expected mean squares as follows: If each unknown variance component σ_x^2 in the expected mean squares is replaced by a symbol $\hat{\sigma}_x^2$, and the resulting expressions equated to the observed mean squares and solved for the quantities $\{\hat{\sigma}_x^2\}$, the solutions will be a set of unbiased point estimates of the $\{\sigma_x^2\}$. Although this procedure is commonly used also in the unbalanced cases, it loses there the intuitive justification it has for this writer. At the present writing, the "best" tests and estimates in the unbalanced cases of random-effects models and mixed models are not known, even in a rough intuitive sense. The basic trouble is that the distribution theory gets so much more complicated.[7] We have nothing to offer the reader on the unbalanced cases outside the fixed-effects models except for some results for the completely nested cases in sec. 7.6. However, if a layout is balanced except for unequal numbers of replications, approximate methods similar to those described in sec. 10.5 for complete layouts are at hand. We now illustrate these general remarks in the case of the one-way layout.

[6] Wald (1940) solved this problem in the case of unequal numbers $\{J_i\}$ of observations in the classes, but his solution is not easy to compute numerically.

[7] In the one-way layout, for example, there are three unknown parameters, μ, σ_A^2, and σ_e^2. In the case of balance the (minimal) number of (real) sufficient statistics is three; in the case of unbalance it is greater. The sum of squares between groups, $\Sigma_i w_i(y_{i.}-\bar{y}_.)^2$, where $\bar{y}_. = \Sigma_i w_i y_{i.}/\Sigma_i w_i$, is not distributed as a constant times a noncentral chi-square, no matter what (known) weights $w_i > 0$ are used. There is no unbiased quadratic estimate of σ_A^2 of uniformly minimum variance, etc.

Test of a Hypothesis

The hypothesis usually tested in the present model for the one-way layout is

$$H_A\colon\quad \sigma_A^2 = 0.$$

The hypothesis is true if and only if[8] all workers in the population have the same "true" mean, i.e., $m(v) = \mu$ for all v. The mean squares occurring in the analysis-of-variance table (Table 3.1.1) for the one-way layout were defined from the SS's (SS_A was there written SS_H)

$$SS_A = J\sum_i (y_{i.} - y_{..})^2,$$

$$SS_e = \sum_i \sum_j (y_{ij} - y_{i.})^2.$$

Under the present model

$$y_{i.} = \mu + a_i + e_{i.}$$

from (7.2.2), and

$$y_{..} = \mu + a_. + e_{..},$$

where of course $a_.$, unlike its counterpart $\alpha_.$ in the fixed-effects model, does not in general vanish. We thus have

$$SS_A = J\sum_i (a_i + e_{i.} - a_. - e_{..})^2, \tag{7.2.3}$$

$$SS_e = \sum_i \sum_j (e_{ij} - e_{i.})^2. \tag{7.2.4}$$

In order to obtain distribution theory on which we can base tests and confidence intervals, we now add the normality assumption, namely that the $\{a_i\}$ and $\{e_{ij}\}$ are normal. It will be seen in Ch. 10 that the effect of violation of the normality assumption on the tests and confidence intervals is much more serious in this model than in the fixed-effects model. However, as far as the expected mean squares are concerned, the results that we will obtain must be the same as though we had not added the normality assumption, since the expected value of any quadratic form in the observations can depend only on the means, variances, and covariances of the variables $\{a_i\}$, $\{e_{ij}\}$. We summarize the assumptions made thus far:

$$\Omega\colon \begin{cases} y_{ij} = \mu + a_i + e_{ij}, \\ \text{the } I + IJ \text{ random variables } \{a_i\}, \{e_{ij}\} \text{ are completely independent,} \\ \text{the } \{a_i\} \text{ are } N(0, \sigma_A^2), \\ \text{the } \{e_{ij}\} \text{ are } N(0, \sigma_e^2). \end{cases}$$

[8] Here and in other places where it will be pretty obvious, the qualification "with probability one" is to be understood.

Writing $g_i = a_i + e_{i.}$, we have

$$SS_A = J\sum_i (g_i - g_.)^2,$$

and the random variables $\{g_i\}$ are independently $N(0, \sigma_g^2)$, where $\sigma_g^2 = \sigma_A^2 + J^{-1}\sigma_e^2$. Therefore $\Sigma_i(g_i - g_.)^2/\sigma_g^2$ is a chi-square variable with $I-1$ d.f., and hence

$$SS_A = J\sigma_g^2\chi_{I-1}^2 = (J\sigma_A^2 + \sigma_e^2)\chi_{I-1}^2. \tag{7.2.5}$$

On the other hand we see from (7.2.4) that SS_e is distributed exactly as in the fixed-effects model: If we think of a fictitious fixed-effects model in which the $\{e_{ij}\}$ play the role of the observations and all the parameters in the model are zero except the error variance, which is σ_e^2, then (7.2.4) is there the error SS, which is distributed as $\sigma_e^2\chi_{\nu_e}^2$ with

$$\nu_e = I(J-1).$$

It follows that

$$SS_e = \sigma_e^2\chi_{\nu_e}^2, \tag{7.2.6}$$

$$E(MS_e) = \sigma_e^2. \tag{7.2.7}$$

We now want to argue that SS_A and SS_e are statistically independent. Consider $\{a_i - a_. + e_{i.} - e_{..}\}$ and $\{e_{i'j} - e_{i'.}\}$. It follows immediately from our assumptions that $(a_i - a_.)$ and $(e_{i'j} - e_{i'.})$ are independent. Furthermore, if we again utilize the fictitious fixed-effects model where the $\{e_{ij}\}$ are the observations, then, as we know, in the IJ-dimensional space of linear forms in the $\{e_{ij}\}$, the set $\{e_{i'j} - e_{i'.}\}$ are in the "error space" while the set $\{e_{i.} - e_{..}\}$ are in the "estimation space" of the fictitious model, and so we see that the two sets are orthogonal or statistically independent. It is now clear from (7.2.3) and (7.2.4) that SS_A and SS_e are statistically independent.

From (7.2.5), we obtain

$$E(MS_A) = E(SS_A)/(I-1) = (J\sigma_A^2 + \sigma_e^2)E(\chi_{I-1}^2)/(I-1),$$

$$E(MS_A) = J\sigma_A^2 + \sigma_e^2. \tag{7.2.8}$$

We can now write the analysis-of-variance table: It looks like Table 3.1.1 except that the E(MS) column contains the expressions (7.2.8) and (7.2.7). These suggest that the hypothesis H_A be tested by using the ratio

$$\mathfrak{F} = MS_A/MS_e, \tag{7.2.9}$$

since under H_A numerator and denominator have the same expected value. We may write

$$\mathfrak{F} = \frac{J\sigma_A^2 + \sigma_e^2}{\sigma_e^2}\,\frac{\chi_{I-1}^2}{\chi_{\nu_e}^2}\,\frac{\nu_e}{I-1} = \left(1 + J\frac{\sigma_A^2}{\sigma_e^2}\right)F_{I-1,\nu_e}, \tag{7.2.10}$$

where F_{I-1,ν_e} is a central F-variable with $I-1$ and ν_e d.f. The test consists in rejecting H_A at the α level of significance if $\mathfrak{F} \geqq F_{\alpha;I-1,\nu_e}$. The power of the test is a function of

$$\theta = \sigma_A^2/\sigma_e^2;$$

we denote the power by $\beta(\theta)$,

$$(7.2.11) \qquad \beta(\theta) = \Pr\{\mathfrak{F} \geqq F_{\alpha;I-1,\nu_e}\} = \Pr\{F_{I-1,\nu_e} \geqq F_{\alpha;I-1,\nu_e}/(1+J\theta)\}.$$

The power is thus seen to involve only the *central* F-distribution.

Test of a More General Hypothesis

The hypothesis $\sigma_A^2 = 0$ is a rather restrictive one, and gives a somewhat pathological theory, as we shall see in sec. 7.3. We now consider the more general hypothesis

$$H_A': \quad \sigma_A^2 \leqq \theta_0\sigma_e^2,$$

where $\theta_0 \geqq 0$ is a preassigned constant. The previous hypothesis H_A is included as the special case $\theta_0 = 0$. We again use the statistic $\mathfrak{F}$ defined in (7.2.9), and we reject H_A' if $\mathfrak{F} \geqq c$, where the constant c is determined by the condition that under H_A' the probability that $\mathfrak{F} \geqq c$ should be $\leqq\alpha$, and $=\alpha$ if $\sigma_A^2 = \theta_0\sigma_e^2$. With $\sigma_A^2 = \theta_0\sigma_e^2$ in (7.2.10) this condition becomes

$$\Pr\{(1+J\theta_0)F_{I-1,\nu_e} \geqq c\} = \alpha,$$

hence

$$c = (1+J\theta_0)F_{\alpha;I-1,\nu_e}.$$

The power of this test may be calculated as in (7.2.11) to be

$$(7.2.12) \qquad \beta(\theta) = \Pr\{F_{I-1,\nu_e} \geqq F_{\alpha;I-1,\nu_e}(1+J\theta_0)/(1+J\theta)\},$$

and again involves only central F. If we think of θ varying from 0 to ∞ in (7.2.12) we see that the power behaves as in Fig. 7.2.1.

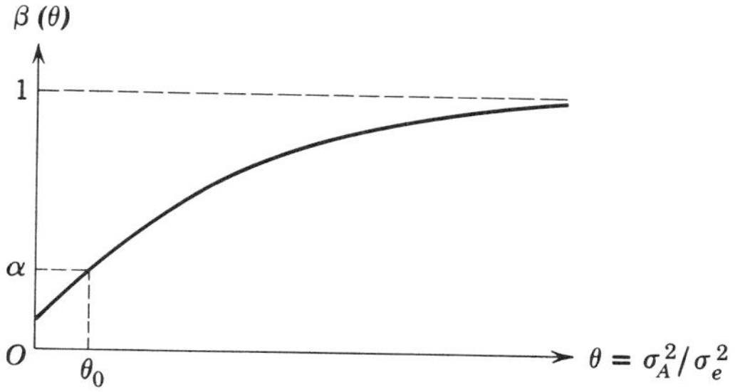

FIG. 7.2.1

Point Estimation of Variance Components

We obtain unbiased estimates of σ_A^2 and σ_e^2 by applying the general procedure mentioned above: We replace σ_A^2 and σ_e^2 in (7.2.8) and (7.2.7) by $\hat{\sigma}_A^2$ and $\hat{\sigma}_e^2$, equate the resulting expressions to MS_A and MS_e, and solve for $\hat{\sigma}_A^2$ and $\hat{\sigma}_e^2$ to get

(7.2.13) $$\hat{\sigma}_A^2 = J^{-1}(MS_A - MS_e)$$

(7.2.14) $$\hat{\sigma}_e^2 = MS_e.$$

In order to obtain the variances of these estimates, we note the following rule: If some MS is distributed as a constant times a chi-square variable divided by its number of d.f.,

$$MS = c\chi_\nu^2/\nu,$$

so

$$c = E(MS),$$

then

$$\text{Var}(MS) = 2[E(MS)]^2/\nu,$$

since Var (MS) $= c^2 \text{Var}(\chi_\nu^2/\nu) = 2c^2/\nu$. Thus, from (7.2.5–8),

(7.2.15) $$\text{Var}(\hat{\sigma}_e^2) = \text{Var}(MS_e) = 2\sigma_e^4/\nu_e,$$

and

$$\text{Var}(MS_A) = 2(J\sigma_A^2 + \sigma_e^2)^2/(I-1).$$

Because MS_A and MS_e are statistically independent we find from (7.2.13) that

$$\text{Var}(\hat{\sigma}_A^2) = J^{-2}[\text{Var}(MS_A) + \text{Var}(MS_e)],$$

and hence

(7.2.16) $$\text{Var}(\hat{\sigma}_A^2) = 2(I-1)^{-1}(\sigma_A^2 + J^{-1}\sigma_e^2)^2 + 2\nu_e^{-1}(J^{-1}\sigma_e^2)^2.$$

We further calculate that

$$\text{Cov}(\hat{\sigma}_A^2, \hat{\sigma}_e^2) = \text{Cov}[J^{-1}(MS_A - MS_e), MS_e] = -J^{-1}\text{Var}(MS_e),$$

the last equality also following from the statistical independence MS_A and MS_e, and hence

(7.2.16*a*) $$\text{Cov}(\hat{\sigma}_A^2, \hat{\sigma}_e^2) = -2J^{-1}\nu_e^{-1}\sigma_e^4.$$

This method works with the random-effects models for all the balanced designs under the normality assumption, where it will always be found that the estimated variance component is a linear combination of *independent* mean squares.

The following consideration also applies in more general situations: It may happen with positive probability that the estimate of a variance component is negative, for example (7.2.13). Since the estimated parameter is nonnegative, the estimate is sometimes modified by redefining it to be zero when it is negative, for example, in place of (7.2.13), the maximum of 0 and $\hat{\sigma}_A^2$ is used. We prefer *not* to use such modified estimates: Their distribution theory is more complicated—in particular, the simple formulas we get for the variances of the estimates under the normality assumption are no longer valid—and the modified estimates are biased.

We remind the reader that the expected mean squares that we compute for random-effects models are valid without the normality assumptions. However, the variances and zero covariances computed are not valid, the correct formulas then generally involving certain population fourth moments.[9]

Interval Estimation

A confidence interval for σ_e^2 can be obtained, as in the fixed-effects case, from (7.2.6) by using two tails of the chi-square distribution (or one tail for a one-sided interval).

To get a confidence interval for the ratio $\theta = \sigma_A^2/\sigma_e^2$ of variance components let $1-\alpha$ be the desired confidence coefficient and choose $\alpha_1 \geqq 0$ and $\alpha_2 \geqq 0$ so that $\alpha_1+\alpha_2 = \alpha$. (Ordinarily we would take $\alpha_1 = \alpha_2 = \frac{1}{2}\alpha$ or else $\alpha_1 = \alpha$, $\alpha_2 = 0$.) Denote by F'' the upper α_2 point and by F' the lower α_1 point of F_{I-1,ν_e}, that is

$$\Pr\{F_{I-1,\nu_e} < F'\} = \alpha_1, \qquad \Pr\{F_{I-1,\nu_e} > F''\} = \alpha_2.$$

Then we have

$$\Pr\{F' \leqq F_{I-1,\nu_e} \leqq F''\} = 1-\alpha. \tag{7.2.17}$$

Using (7.2.10) in the form $F_{I-1,\nu_e} = \mathfrak{F}/(1+J\theta)$, we find that (7.2.17) gives us the following confidence interval with confidence coefficient $1-\alpha$:

$$\frac{1}{J}\left(\frac{\mathfrak{F}}{F''} - 1\right) \leqq \theta \leqq \frac{1}{J}\left(\frac{\mathfrak{F}}{F'} - 1\right). \tag{7.2.18}$$

Some discussion is required because one or both end points of the interval (7.2.18) may be negative while the true value of θ is of course nonnegative. (If $\alpha_1 = \alpha$, $\alpha_2 = 0$, we use only the right inequality.)

It would be mathematically correct to modify the interval (7.2.18) so that if the left end point is negative it is replaced by zero and if the right end point is negative it is also replaced by zero.[10] It is easily verified that

[9] The variances and covariances for several important cases have been derived by Tukey (1956), (1957*a*), including the unbalanced one-way layout; see also sec. 10.3.

[10] The left end point is negative if and only if H_A is accepted by the above test at the α_2 level of significance. The right end point is negative if and only if $\mathfrak{F} < F'$, or $F_{I-1,\nu_e} < F'/(1+J\theta)$; the probability of this happening is evidently a decreasing function of θ and attains its maximum value α_1 when $\theta = 0$.

the modified interval, like (7.2.18), covers the true value of θ with probability $1-\alpha$ if $\theta > 0$, but with probability $>1-\alpha$ if $\theta = 0$. However, even though the modified interval has length $\leqq$ that of (7.2.18), and probability of covering $\geqq$ that of (7.2.18), we recommend the use of (7.2.18) for the following reasons:[11]

Although there is nothing in the formal theory of confidence intervals to justify it, most users of confidence intervals have a more or less conscious feeling that the length of a two-sided confidence interval is a measure of the error of some point estimate of the parameter. Actually many of the commonly used confidence intervals for a parameter θ are of the form

$$\hat{\theta} - A\hat{\sigma}_\theta \leqq \theta \leqq \hat{\theta} + B\hat{\sigma}_\theta, \tag{7.2.19}$$

where $\hat{\theta}$ is an intuitively plausible point estimate of θ, $\hat{\sigma}_\theta$ is a similar estimate of the standard deviation of $\hat{\theta}$, and A and B are constants which are obtained from tables and depend on the confidence coefficient and the sample size. The interval (7.2.18) is of this form: Let us take $\hat{\theta} = \hat{\sigma}_A^2/\hat{\sigma}_e^2$. Then the approximate formula

$$\text{Var}\,[f(x_1, x_2)] \sim f_1^2 \,\text{Var}\,(x_1) + 2f_1f_2 \,\text{Cov}\,(x_1, x_2) + f_2^2 \,\text{Var}\,(x_2),$$

where f_i denotes $\partial f/\partial x_i$ evaluated at $(x_1, x_2) = (\xi_1, \xi_2)$ and $\xi_i = E(x_i)$, applied to $\hat{\theta} = \hat{\sigma}_A^2/\hat{\sigma}_e^2 = J^{-1}(x_1-x_2)/x_2$ with $x_1 = \text{MS}_A$ and $x_2 = \text{MS}_e$ gives

$$\text{Var}\,(\hat{\theta}) \sim \left(\theta + \frac{1}{J}\right)^2\left(\frac{2}{I-1} + \frac{2}{\nu_e}\right),$$

and suggests the estimate

$$\hat{\sigma}_\theta = \left(\hat{\theta} + \frac{1}{J}\right)\left(\frac{2}{I-1} + \frac{2}{\nu_e}\right)^{1/2}. \tag{7.2.20}$$

We note that $\hat{\sigma}_\theta \geqq 0$. The interval (7.2.18) may now be written in the form (7.2.19) with $\hat{\sigma}_\theta$ defined by (7.2.20), and A and B by[12]

$$A = \left(1 - \frac{1}{F''}\right)\left(\frac{2}{I-1} + \frac{2}{\nu_e}\right)^{-1/2},$$

$$B = \left(\frac{1}{F'} - 1\right)\left(\frac{2}{I-1} + \frac{2}{\nu_e}\right)^{-1/2}.$$

In the light of the above discussion we see that if the interval is considerably shortened by deleting the part, if any, to the left of the origin, a

[11] In reaching this point of view I was helped by conversations with Professors Charles Kraft and J. L. Hodges Jr.

[12] For large I and ν_e, A and B respectively tend to the upper α_2 and α_1 points of $N(0, 1)$.

misleading impression of the accuracy of the estimation may result. If the interval is completely to the left of the origin one might consider translating it until it just includes the origin, to meet the above objection to shortening it. However, one might again feel on nonmathematical and intuitive grounds that an interval estimate like that from -5 to -3 is stronger evidence that the true value of a nonnegative parameter is zero than that from -2 to 0. In practice it would be well to report the values of $\hat{\sigma}_A^2$ and $\hat{\sigma}_e^2$ in addition to the interval.[13]

A confidence interval for the intraclass correlation coefficient $\tilde{\rho}$ is now easily obtained from (7.2.18) by noting that $\tilde{\rho} = (1+\theta^{-1})^{-1}$. The probability is thus $1-\alpha$ that

$$(7.2.21) \qquad (1+L^{-1})^{-1} \leqq \tilde{\rho} \leqq (1+R^{-1})^{-1},$$

where L and R are respectively the left and right members of (7.2.18). Although the true $\tilde{\rho}$ is nonnegative we prefer not to modify (7.2.21) if it covers negative values for reasons similar to those discussed above.

Approximate Confidence Intervals for Variance Components

Since the kind of confidence interval we shall now obtain for σ_A^2 is useful also in other cases of estimation of a variance component, we shall adopt a more general notation. We consider the problem of calculating a confidence interval for a parameter ϕ from two mean squares MS_1 and MS_2 independently distributed with ν_1 and ν_2 d.f., respectively, in such a way that $\nu_1 MS_1$ is $(\phi+\sigma^2)\chi_{\nu_1}^2$ and $\nu_2 MS_2$ is $\sigma^2\chi_{\nu_2}^2$. In the present application $\phi = J\sigma_A^2$, $MS_1 = MS_A$, $MS_2 = MS_e$, $\nu_1 = I-1$, $\nu_2 = I(J-1)$, $\sigma^2 = \sigma_e^2$.

We shall first solve[14] the problem of finding a lower confidence limit $f(MS_1, MS_2)$ with confidence coefficient equal to a given $1-\alpha$, at least approximately. We shall do this by formulating certain intuitively desirable properties of the function $f(MS_1, MS_2)$ and then selecting a simple solution possessing these properties.

(i) The first is an invariance property, whose consequence is to restrict $f(MS_1, MS_2)$ to the form MS_2 times a function of $\mathfrak{F}$,

$$(7.2.22) \qquad f(MS_1, MS_2) = MS_2\, g(\mathfrak{F}),$$

[13] They together with $y_{..}$ constitute a set of sufficient statistics under the normality assumption.

[14] My approach is similar to that of Bulmer (1957), my $1-\alpha$ corresponding to Bulmer's α. The conditions (ii), (iv), (v) below were imposed by Bross (1950) in a fiducial approach that I am unable to follow. His $g(\mathfrak{F})$ misbehaved rather badly with an infinite discontinuity and a change of sign, as pointed out by Tukey (1951), who proposed the linear solution (7.2.27). The solution (7.2.31) was proposed by Moriguti (1954), who imposed the condition (i), and who also showed that the solution gave a probability whose error was $O(\nu_2^{-2})$. The only adequate investigation of the error of any of the solutions was made by Bulmer (1957).

where $\mathfrak{F} = \text{MS}_1/\text{MS}_2$: Suppose that all the observations were multiplied by a positive number c, for example, by being measured on a different scale. Then the unbiased point estimates $\hat{\sigma}^2 = \text{MS}_2$ and $\hat{\phi} = \text{MS}_1 - \text{MS}_2$ would be multiplied by c^2. We impose the condition that the confidence limit $f(\text{MS}_1, \text{MS}_2)$ should then also be multiplied by c^2, i.e., we impose the condition

$$f(c^2\text{MS}_1, c^2\text{MS}_2) \equiv c^2 f(\text{MS}_1, \text{MS}_2)$$

identically in c. In particular, if we take $c^2 = 1/\text{MS}_2$, we get $f(\text{MS}_1, \text{MS}_2) = \text{MS}_2 f(\mathfrak{F}, 1)$, which is of the form (7.2.22).

(ii) The next property integrates the behavior of the $1-\alpha$ confidence interval with that of the α-level F-test of the hypothesis H: $\phi = 0$ which consists in rejecting H if $\mathfrak{F} > F_\alpha$, where $F_\alpha = F_{\alpha;\nu_1,\nu_2}$ is the upper α point of F with ν_1 and ν_2 d.f. An α-level test of H is implied by the confidence interval, consisting in rejecting H if and only if the confidence interval does not cover $\phi = 0$. We require the two tests to be equivalent; this imposes the condition that $g(\mathfrak{F}) > 0$ if and only if $\mathfrak{F} > F_\alpha$. In the light of the discussion following (7.2.18) we might consider allowing negative values of $g(\mathfrak{F})$ for some $\mathfrak{F} < F_\alpha$, but to simplify the present conditions we shall assume that $g(\mathfrak{F}) = 0$ for $\mathfrak{F} \leqq F_\alpha$.

(iii) We require that for $\mathfrak{F} \geqq F_\alpha$, $g(\mathfrak{F})$ should be an increasing[15] function of $\mathfrak{F}$. The confidence limit $\text{MS}_2\, g(\mathfrak{F})$ then shares this property with the point estimate $\hat{\phi} = \text{MS}_1 - \text{MS}_2$, which may be written $\text{MS}_2(\mathfrak{F}-1)$.

The intuitive appeal of the next two properties is less compelling: They will require that in certain limiting cases the confidence intervals coincide with the "natural" ones resulting in those cases from the distribution of $\nu_1\text{MS}_1$ being $(\phi+\sigma^2)\chi^2_{\nu_1}$, or MS_1 being $(\phi+\sigma^2)F_{\nu_1,\infty}$ so that, with $F'_\alpha = F_{\alpha;\nu_1,\infty}$ the probability is $1-\alpha$ that $\text{MS}_1 \leqq (\phi+\sigma^2)F'_\alpha$, or

$$\phi \geqq \text{MS}_2 \frac{\mathfrak{F}}{F'_\alpha} - \sigma^2. \tag{7.2.23}$$

(iv) In the limiting case $\nu_2 = \infty$ we may regard σ^2 as known and equal to MS_2. Then the interval (7.2.23) becomes

$$\phi \geqq \text{MS}_2 \left(\frac{\mathfrak{F}}{F'_\alpha} - 1\right).$$

We impose the condition that the limiting form of $g(\mathfrak{F})$ for $\nu_2 = \infty$ be

$$\frac{\mathfrak{F}}{F'_\alpha} - 1. \tag{7.2.24}$$

[15] I mean strictly increasing.

(v) Finally, suppose that $\mathfrak{F}$ is large, thus indicating[16] that ϕ is large compared with σ^2. This suggests that we then consider the limiting case $\phi \to \infty$ for fixed σ^2. For large $\mathfrak{F}$ the second term on the right of (7.2.23) is negligible compared with the first, and so we are led to require that $g(\mathfrak{F})$ behave like $\mathfrak{F}/F'_\alpha$ for large $\mathfrak{F}$, in the sense that

$$(7.2.25) \qquad g(\mathfrak{F}) = (\mathfrak{F}/F'_\alpha)[1 + h(\mathfrak{F})], \quad \text{where} \quad h(\mathfrak{F}) \to 0 \quad \text{as} \quad \mathfrak{F} \to \infty.$$

For any[17] choice of $g(\mathfrak{F})$ the probability that $\phi \geqq \text{MS}_2\, g(\mathfrak{F})$ depends on the value of

$$\theta' = \phi/\sigma^2$$

and on ν_1 and ν_2, and is easily calculated to be given by the formula

$$(7.2.26) \qquad \Pr\{\phi \geq \text{MS}_2\, g(\mathfrak{F})\} = \int_0^\infty \left[\int_0^{h(x_2)} f_1(x_1)\, dx_1\right] f_2(x_2)\, dx_2,$$

where

$$h(x_2) = (\theta' + 1)^{-1} x_2\, g^{-1}(\theta'/x_2),$$

the inverse of the function $u = g(\mathfrak{F})$ for $\mathfrak{F} \geqq F_\alpha$ is denoted by $\mathfrak{F} = g^{-1}(u)$, and $f_i(x_i)$ denotes the probability density function of $x_i = \chi^2_{\nu_i}/\nu_i$ $(i = 1, 2)$. The numerical value of the integral in brackets in (7.2.26) can be found from tables of the cumulative distribution function of $\chi^2_{\nu_1}$ (sec. 2.2), but the further integration must be carried out by numerical methods.

Any $g(\mathfrak{F})$ satisfying our conditions will make the probability (7.2.26) exactly equal to $1-\alpha$ in the three limiting cases $\theta' = 0$, $\theta' = \infty$, $\nu_2 = \infty$. The simplest function satisfying those conditions is the linear function

$$(7.2.27) \qquad g_1(\mathfrak{F}) = (\mathfrak{F} - F_\alpha)/F'_\alpha \qquad (\mathfrak{F} \geqq F_\alpha).$$

Numerical integration of (7.2.26) shows[18] that $g(\mathfrak{F}) = g_1(\mathfrak{F})$ gives the probability (7.2.26) values $> 1-\alpha$ if $\nu_2 < \infty$ and $0 < \theta' < \infty$, and hence the resulting confidence interval is too wide. As a second approximation we may try

$$(7.2.28) \qquad g(\mathfrak{F}) = a_1\mathfrak{F} + a_0 + a_{-1}\mathfrak{F}^{-1} \qquad (\mathfrak{F} \geqq F_\alpha),$$

where $a_i = a_i(\nu_1, \nu_2)$.

Our condition (v), requiring (7.2.25), implies that

$$(7.2.29) \qquad a_1 = 1/F'_\alpha$$

[16] For example, according to a confidence interval like (7.2.18).

[17] Such that $g(\mathfrak{F})$ is strictly increasing for $\mathfrak{F} >$ some $\mathfrak{F}_0$ and $g(\mathfrak{F}) = 0$ for $\mathfrak{F} \leq \mathfrak{F}_0$.

[18] This will appear reasonable from Table 7.2.1 and the inequality resulting from (7.2.32).

for all ν_1, ν_2. Condition (ii), requiring $g(F_\alpha) = 0$ implies

$$a_{-1} = -F_\alpha(a_1 F_\alpha + a_0) \tag{7.2.30}$$

for all ν_1, ν_2. Condition (iv), requiring that the limiting form of $g(\mathfrak{F})$ for $\nu_2 = \infty$ be (7.2.24), implies

$$a_1(\nu_1, \infty)\mathfrak{F} + a_0(\nu_1, \infty) + a_{-1}(\nu_1, \infty)\mathfrak{F}^{-1} = (F'_\alpha)^{-1}\mathfrak{F} - 1.$$

Substituting in this (7.2.29), (7.2.30), and also $F'_\alpha = F_\alpha$ for $\nu_2 = \infty$, we find that $a_0(\nu_1, \infty) = -1$ for all ν_1. A $g(\mathfrak{F})$ of the type (7.2.28) that has the desired properties must therefore be of the form

$$g(\mathfrak{F}) = \frac{\mathfrak{F}}{F'_\alpha} + a_0(\nu_1, \nu_2) - \frac{F_\alpha}{\mathfrak{F}}\left[\frac{F_\alpha}{F'_\alpha} + a_0(\nu_1, \nu_2)\right]$$

with $a_0(\nu_1, \infty) = -1$ for all ν_1. The simplest solution is to take $a_0 = -1$ for all ν_1, ν_2. We then have

$$g(\mathfrak{F}) = \frac{\mathfrak{F}}{F'_\alpha} - 1 - \frac{F_\alpha}{\mathfrak{F}}\left(\frac{F_\alpha}{F'_\alpha} - 1\right) \qquad (\mathfrak{F} \geqq F_\alpha). \tag{7.2.31}$$

We may verify that this choice of $g(\mathfrak{F})$ satisfies also the condition (iii) requiring it to be an increasing function.

We note that this $g(\mathfrak{F})$ gives a tighter lower confidence limit than the linear function $g_1(\mathfrak{F})$, since

$$g(\mathfrak{F}) - g_1(\mathfrak{F}) = \left(\frac{F_\alpha}{F'_\alpha} - 1\right)\left(1 - \frac{F_\alpha}{\mathfrak{F}}\right) \qquad (\mathfrak{F} > F_\alpha). \tag{7.2.32}$$

Inspection of the F-tables shows that for $\alpha \leqq 0.10$ and $\nu_2 > 2$, $F_\alpha > F'_\alpha$, and so for $\mathfrak{F} > F_\alpha$, $g(\mathfrak{F}) > g_1(\mathfrak{F})$.

Before considering the numerical evidence supporting this approximation we note that if we can formally calculate a lower confidence limit for a given small confidence coefficient, we have at hand an upper confidence limit for the complementary large confidence coefficient, since

$$\Pr\{\phi < f(\text{MS}_1, \text{MS}_2)\} = 1 - \Pr\{\phi \geqq f(\text{MS}_1, \text{MS}_2)\}. \tag{7.2.33}$$

By using an upper confidence limit with confidence coefficient $1-\alpha_1$ and a lower confidence limit with confidence coefficient $1-\alpha_2$ we then get a confidence interval between these confidence limits with confidence coefficient $1-\alpha$, where $\alpha = \alpha_1+\alpha_2$.

The excellence of the approximation (7.2.31) is indicated by Table 7.2.1, calculated by numerical integration of (7.2.26). The "exact range" in the last column is for θ' varying over the interval from 0 to ∞. The entries for nominal $1-\alpha = 0.05$ and 0.95 pertain, respectively, to approximate upper and lower 95 per cent confidence limits. The approximation appears to be quite satisfactory.

TABLE* 7.2.1

ACCURACY OF THE APPROXIMATE CONFIDENCE LIMIT FOR A VARIANCE COMPONENT

d.f.		$1-\alpha$	
ν_1	ν_2	Nominal	Exact Range
8	6	0.05	0.050–0.051
8	12		0.050–0.050
24	24		0.050–0.052
24	48		0.050–0.051
2	6	0.95	0.950–0.955
2	12		0.950–0.952
8	6		0.950–0.959
8	12		0.950–0.953
24	24		0.949–0.951
24	48		0.950–0.950

* From p. 163 of "Approximate confidence limits for components of variance" by M. G. Bulmer, *Biometrika*, Vol. 44 (1957). Here reproduced with the kind permission of the author and the editor. I rounded off the last column to three decimals.

The formulas resulting from (7.2.31), (7.2.33), and $F_{1-\alpha;\nu_1,\nu_2} = 1/F_{\alpha;\nu_2,\nu_1}$ may be written in the following detailed forms involving only upper per cent points of the F-distribution: The upper confidence limit $\text{MS}_2\, g_U(\mathfrak{F})$ for ϕ with approximate confidence coefficient $1-\alpha_1$ is given by

$$g_U(\mathfrak{F}) = F_{\alpha_1;\infty,\nu_1}\mathfrak{F} - 1 + \frac{1}{F_{\alpha_1;\nu_2,\nu_1}\mathfrak{F}}\left(1 - \frac{F_{\alpha_1;\infty,\nu_1}}{F_{\alpha_1;\nu_2,\nu_1}}\right)$$

for $\mathfrak{F} \geqq 1/F_{\alpha_1;\nu_2,\nu_1}$ and $g_U(\mathfrak{F}) = 0$ for $\mathfrak{F} \leqq 1/F_{\alpha_1;\nu_2,\nu_1}$; the lower confidence limit $\text{MS}_2\, g_L(\mathfrak{F})$ with approximate confidence coefficient $1-\alpha_2$ is given by

$$g_L(\mathfrak{F}) = \frac{\mathfrak{F}}{F_{\alpha_2;\nu_1,\infty}} - 1 - \frac{F_{\alpha_2;\nu_1,\nu_2}}{\mathfrak{F}}\left(\frac{F_{\alpha_2;\nu_1,\nu_2}}{F_{\alpha_2;\nu_1,\infty}} - 1\right)$$

for $\mathfrak{F} \geqq F_{\alpha_2;\nu_1,\nu_2}$, and $g_L(\mathfrak{F}) = 0$ for $\mathfrak{F} \leqq F_{\alpha_2;\nu_1,\nu_2}$.

These confidence limits can be seriously invalidated by nonnormality, especially of the random effects for which MS_1 is the mean square; the reason is indicated in sec. 10.2.

7.3. ALLOCATION OF MEASUREMENTS

Since in a model where a factor is treated as a random-effects factor the interest is not in the values of the individual effects but in the variance of the population from which they are assumed to come, the question arises as to how large a sample to take from each such population. In many cases the total number of measurements will be roughly fixed by various cost considerations, and the question then is how to allocate the measurements among the various populations of effects; we here include the "population of errors." In this section we consider the problem for the one-way layout: What is the relative importance of increasing I or J in improving the accuracy of the point estimates of the variance components, and the power of the test? A solution of the problem of optimum allocation is obtained for point estimation (but not for interval estimation or tests, where the problem is still unsolved).

From (7.2.15) we see that Var $(\hat{\sigma}_e^2)$ tends to zero for large I or J, since $\nu_e = I(J-1)$, implying that the error of the estimate of σ_e^2 can be made as small as desired by sufficiently increasing either I or J. The situation is different for σ_A^2: From (7.2.16), Var $(\hat{\sigma}_A^2)$ tends to $2\sigma_A^4/(I-1)$ for large J. Hence, increasing J is of negligible benefit after a certain point, in estimating σ_A^2. This suggests that the power of the test concerning σ_A^2 might tend for large J and fixed I to a limit smaller than unity. We shall see in a moment that this does occur when the hypothesis tested is H_{θ_0}: $\sigma_A^2 \leqq \theta_0\sigma_e^2$ with $\theta_0 > 0$, but that for the hypothesis H_0: $\sigma_A^2 = 0$ the power against any alternative does tend to unity when J becomes infinite with I fixed.

The power of the test of H_{θ_0}, given by (7.2.12), may be written

$$\beta(\theta) = \Pr\left\{\frac{\chi^2_{I-1}}{\nu_e^{-1}\chi^2_{\nu_e}} \geqq \frac{\theta_0 + J^{-1}}{\theta + J^{-1}}(I-1)F_{\alpha;I-1,\nu_e}\right\}.$$

For large J, $\nu_e^{-1}\chi^2_{\nu_e}$ tends to unity in probability (Problem IV.3*b*); hence[19] the distribution of the left member of the inequality and of $(I-1)F_{I-1,\nu_e}$ converges to that of χ^2_{I-1}. Thus the right member has the limit $\theta_0\theta^{-1}\chi^2_{\alpha;I-1}$, and the power $\beta(\theta)$ tends to the limit

$$\Pr\{\chi^2_{I-1} \geqq \theta_0\theta^{-1}\chi^2_{\alpha;I-1}\},$$

which is not equal to unity unless $\theta_0 = 0$. This limit is the same as the power of the standard chi-square test of the hypothesis $\sigma_A^2 \leqq C$ based on a sample of I from a normal population of unknown variance σ_A^2 if

[19] Cramér (1946), sec. 20.6.

$C = \theta_0 \sigma_e^2$ is a known constant (in the limiting case, where J is infinite, σ_e^2 may be regarded as known).

Optimum Allocation for Point Estimation

Suppose that the total number n of measurements is fixed. The problem is then the best choice of I and J with $IJ = n$. A possible criterion is to choose them so as to minimize Var $(\hat{\sigma}_A^2)$ or Var $(\hat{\sigma}_e^2)$. To simplify matters we shall assume n even. Then I may vary over the divisors of n from 2 to $n/2$ (with $I = 1$ we cannot estimate σ_A^2, and with $I = n$ neither σ_A^2 nor σ_e^2); thus for $n = 100$ the possible values of I are 2, 4, 5, 10, 20, 25, 50. Now (7.2.15), written as

$$\text{Var}\,(\hat{\sigma}_e^2) = 2\sigma_e^4/(n-I),$$

shows that Var $(\hat{\sigma}_e^2)$ increases by a factor less than 2 as I increases from 2 to $n/2$. On the other hand, (7.2.16) shows that Var $(\hat{\sigma}_A^2)$ may vary much more with I (as may be seen, for example, by calculating the ratio of its values for $I = n/2$ and $I = 2$ to be $n/4$ in the limiting case $\sigma_A^2 = 0$), and so we shall minimize[20] Var $(\hat{\sigma}_A^2)$.

With $J = n/I$, (7.2.16) may be written

(7.3.1) $$\text{Var}\,(\hat{\sigma}_A^2) = 2\sigma_e^4 \left[\frac{\theta^2}{I-1} + \frac{2\theta}{n}\frac{I}{I-1} + \frac{n-1}{n^2}\frac{I^2}{(I-1)(n-I)}\right],$$

where

$$\theta = \sigma_A^2/\sigma_e^2.$$

To minimize the function of I in brackets in (7.3.1), which we shall call V, we shall first let I vary continuously over the interval $1 < I < n$. All three terms in the expression for V are positive and continuous, and V becomes infinite as I approaches 1 or n, hence V has at least one minimum in the interval. If we set $dV/dI = 0$ we get

(7.3.2)

$$I^2(n^2\theta^2+2n\theta-n^2+1) - 2I(n^3\theta^2+2n^2\theta-n^2+n) + (n^4\theta^2+2n^3\theta) = 0.$$

The solutions of this quadratic equation in I may be simplified[21] to

$$I_1 = n^2\theta/(n\theta - n + 1), \qquad I_2 = (n^2\theta + 2n)/(n\theta + n + 1).$$

Now for $0 \leq \theta < 1-n^{-1}$, $I_1 \leq 0$, and for $\theta > 1-n^{-1}$, $I_1 > n$, while for $\theta = 1-n^{-1}$, I_1 is infinite. Hence in every case the root I_1 must be rejected because it does not lie in the interval $1 < I < n$. The function V therefore has a single minimum in the interval, which must be attained at[22] I_2.

[20] This problem was solved by Hammersley (1949).

[21] It is easier to verify by substitution of I_1 and I_2 that $(I-I_1)(I-I_2) = 0$ is equivalent to (7.3.2).

[22] We notice that for large n we have asymptotically $I_2 \sim n\theta/(\theta+1) = \tilde{\rho}n$.

If I_2 is a possible value of I, that is, a divisor of n between 2 and $n/2$ (inclusive), then our problem is solved. If not, suppose first that I_2 is in the interval $2 < I < n/2$. Then it is necessary to compute $\text{Var}(\hat{\sigma}_A^2)$, using for I both the largest divisor of n which is $< I_2$ and the smallest which is $> I_2$. The one of these two values of I that gives the smaller value to $\text{Var}(\hat{\sigma}_A^2)$ is the solution.

If $I_2 \leqq 2$, the value that has to be taken is $I = 2$. This happens when

$$\theta \leqq 2n^{-1}(n-2)^{-1}.$$

It means it is best to take only two groups ($I = 2$) when θ is close to zero. On the other hand, if $I_2 \geqq n/2$ we must take $I = n/2$. This happens for

$$\theta \geqq 1-3n^{-1}.$$

It means that, if θ is $\geqq 1$, the best choice is to design our experiment with only two observations per group ($J = 2$).

Unfortunately, as in many problems of optimum design, the solution depends on the value of an unknown parameter, in this case θ: We have to substitute for θ in the above formulas for the minimizing I (but not elsewhere!) an estimate based on preliminary information or a guess.

7.4. THE COMPLETE TWO-WAY LAYOUT

An example of a two-way layout may be obtained by introducing different machines into the experiment with workers in a factory described in sec. 7.2. It is convenient[23] now to designate the workers as factor B and the machines as factor A. The factor A would appropriately be treated as having fixed effects in many cases, for example if the machines were of different makes. The present treatment of factor A as having random effects would be appropriate if the machines are all of the same make and model, and if the following essential condition is satisfied: that the machines in the experiment can reasonably be regarded as a random sample from some population about which we wish to make our statistical inferences, rather than making them about the particular machines in the experiment. We shall idealize the population of machines as infinite. This would be acceptable for example if the machines in the experiment were a randomly selected sample from a relatively large number of such machines in the factory.

Suppose that there are I machines and J workers in the experiment and that each worker is assigned to each machine for K days. We shall

[23] Because when we consider the corresponding mixed model in sec. 8.1 we will encounter a certain "vector of true means" which we prefer to handle, like other vectors in this book, as a column vector rather than as a row vector.

include the case $K = 1$ where there is one observation per cell; the subscript k may be dropped in this case. Denote by y_{ijk} the output of the jth worker the kth day he uses the ith machine. We shall arrive at the structure

$$y_{ijk} = m_{ij} + e_{ijk}, \tag{7.4.1}$$

where m_{ij} is the "true" mean for the jth worker on the ith machine, e_{ijk} is an "error," and the joint distribution of the $\{m_{ij}\}$ and $\{e_{ijk}\}$ will be determined partly as the consequence of a certain underlying mathematical model, which seems natural and acceptable to the author, and partly by certain simplifying assumptions.

Suppose that we label the workers in the population of workers with an index v as in sec. 7.2, and that $\mathscr{P}_v$ now denotes the population distribution. Let the machines in the population of machines be labeled by u, and let $\mathscr{P}_u$ denote the corresponding distribution. We shall assume u and v statistically independent, corresponding to the combination of a randomly chosen machine with an independently and randomly chosen worker. Denote by $m(u, v)$ the "true" output of the worker labeled v on the machine labeled u. As in the similar situation discussed in sec. 7.2, we will make the simplifying assumption that the variance $\sigma^2(u, v)$ of the daily output of the (u, v) combination about the "true" mean $m(u, v)$ is a constant σ_e^2 not depending on (u, v).

The general mean of $m(u, v)$ *in the* (bivariate) *population* of men and machines is

$$\mu = m(.,.),$$

where replacing u by a dot in $m(u, v)$ signifies that the mean has been taken over machines in the population of machines, that is, the expected value with respect to the distribution $\mathscr{P}_u$ over u, and, similarly, replacing v by a dot signifies that the mean has been taken over the population of workers, i.e., with respect to $\mathscr{P}_v$ over v. The mean of the "true" output on the machine labeled u over the population of workers is $m(u, .)$, and the amount by which this exceeds the general mean,

$$a(u) = m(u, .) - m(.,.), \tag{7.4.2}$$

we define as the main effect *in the population* of the machine labeled u. Similarly, the main effect of the worker labeled v is defined as

$$b(v) = m(., v) - m(.,.). \tag{7.4.3}$$

The interaction *in the population* of the machine labeled u and the worker labeled v is defined as

$$c(u, v) = m(u, v) - m(u, .) - m(., v) + m(.,.), \tag{7.4.4}$$

and its meaning as a differential effect in the population is entirely analogous to that elucidated in the discussion of sec. 4.1, which applies to a finite collection of machines and workers.

We now have the structure

$$m(u, v) = \mu + a(u) + b(v) + c(u, v), \tag{7.4.5}$$

by definition of the terms on the right, and we need to inquire into their joint distribution. From (7.4.2), (7.4.3), and (7.4.4) we find the mean values

$$a(.) = 0, \qquad b(.) = 0,$$
$$c(u, .) = 0 \text{ for all } u,$$
$$c(., v) = 0 \text{ for all } v,$$

where replacing u or v by a dot has the meaning indicated above. These are the analogs of the relations $\alpha_. = 0$, etc., for a finite collection (sec. 4.1). The random effects $a(u)$, $b(v)$, $c(u, v)$ thus have zero means. We shall now prove that they are uncorrelated (i.e., any pair of the three have zero correlation coefficients).

Actually, $a(u)$ and $b(v)$ are statistically independent, because u and v are. The zero covariance of $c(u, v)$ with $a(u)$ and with $b(v)$ may appear a bit more surprising since the random variables $c(u, v)$ and $a(u)$ are both functions of the random quantity u, and $c(u, v)$ and $b(v)$, of v. To see that the covariance of $c(u, v)$ with $a(u)$ is zero, we note that it equals the expected value of $a(u)\, c(u, v)$, which may be calculated as the expected value of $f(u)$, where $f(u)$ is the conditional expected value of $a(u)\, c(u, v)$, given u. But in this conditional calculation u may be treated as a constant:

$$f(u) = E(a(u)\, c(u, v)|u) = a(u)\, E(c(u, v)|u) = a(u)\, c(u, .).$$

Thus $f(u) = 0$ for all u and hence $E(f(u)) = 0$. Similarly we may prove $b(v)$ and $c(u, v)$ uncorrelated.

For later use we define the variance components σ_A^2, σ_B^2, and σ_{AB}^2 as

$$\sigma_A^2 = \text{Var}(a(u)), \qquad \sigma_B^2 = \text{Var}(b(v)), \qquad \sigma_{AB}^2 = \text{Var}(c(u, v)).$$

Let the I machines in the experiment be labeled by $\{u_1, \cdots, u_I\}$ and the J workers by $\{v_1, \cdots, v_J\}$. Then we assume that the $\{u_i\}$ are a random sample of I from $\mathscr{P}_u$, and the $\{v_j\}$ an independent random sample of J from $\mathscr{P}_v$. The "true" mean m_{ij} in (7.4.1) is then $m_{ij} = m(u_i, v_j)$, and since, as discussed above, the "error variance" about $m(u_i, v_j)$ is assumed to have the same value σ_e^2 for all u_i and v_j, the $\{e_{ijk}\}$ in (7.4.1) have zero means and a common variance σ_e^2. We add the further simplifying assumptions that the $\{e_{ijk}\}$ are independently and identically distributed, and independently of the $\{m_{ij}\}$. According to (7.4.5) the $\{m_{ij}\}$ have the structure[24]

$$m_{ij} = \mu + a_i + b_j + c_{ij},$$

where

$$a_i = a(u_i), \qquad b_j = b(v_j), \qquad c_{ij} = c(u_i, v_j). \tag{7.4.6}$$

Since the bivariate distribution of (u_i, v_j) is the same as that of (u, v) above, the $\{a_i\}$ are identically distributed, likewise the $\{b_j\}$, and the $\{c_{ij}\}$, with

$$E(a_i) = E(b_j) = E(c_{ij}) = 0,$$

$$\text{Var}(a_i) = \sigma_A^2, \qquad \text{Var}(b_j) = \sigma_B^2, \qquad \text{Var}(c_{ij}) = \sigma_{AB}^2,$$

and c_{ij} is uncorrelated with a_i and with b_j. We see also from (7.4.6), that, since the $I+J$ quantities $\{u_i\}$ and $\{v_j\}$ are completely independent, so are the $I+J$ main effects $\{a_i\}$, $\{b_j\}$, and that c_{ij} is statistically independent of $a_{i'}$ for $i' \neq i$, of $b_{j'}$ for $j' \neq j$, and of $c_{i'j'}$ for $i' \neq i$ and $j' \neq j$.

We shall now show that the set of $I+J+IJ$ effects $\{a_i\}$, $\{b_j\}$, $\{c_{ij}\}$ are completely uncorrelated. For this it remains only to prove that c_{ij} and $c_{i'j'}$ are uncorrelated if $i' = i$ and $j' \neq j$, or if $i' \neq i$ and $j' = j$. Consider the first case: The covariance of c_{ij} and $c_{ij'}$ is the expected value of $c(u_i, v_j)\, c(u_i, v_{j'})$. If we take this expected value first conditionally, given u_i, and call the result of this $g(u_i)$, then, in the conditional calculation, u_i may be treated as a constant, and $c(u_i, v_j)$ and $c(u_i, v_{j'})$ as independent since v_j and $v_{j'}$ are conditionally independent because they are unconditionally independent. This gives

$$g(u_i) = E(c(u_i, v_j)|u_i)\, E(c(u_i, v_{j'})|u_i).$$

Now since the bivariate distribution of (u_i, v_j) is the same as that of (u, v) above, and since $E(c(u, v)|u) = c(u, .) = 0$ for all u, therefore $E(c(u_i, v_j)|u_i) = 0$ for all u_i, hence $g(u_i) = 0$ for all u_i, and therefore

$$\text{Cov}(c_{ij}, c_{ij'}) = E(g(u_i)) = 0.$$

Similarly it can be shown that $\text{Cov}(c_{ij}, c_{i'j}) = 0$ for $i' \neq i$.

If we add the normality assumption, that the effects $\{a_i\}$, $\{b_j\}$, $\{c_{ij}\}$, $\{e_{ijk}\}$ are jointly normal, this together with the relation that these effects are completely uncorrelated then forces them to be *completely independent*! That the interaction $c(u, v)$ between the machine labeled u and the worker labeled v is independent of the machine or the worker may violate our intuition of interaction. It suggests that the normality assumption is in

[24] The model resulting from the distribution obtained below for the effects $\{a_i\}$, $\{b_j\}$, $\{c_{ij}\}$ is a special case of one introduced by Tukey (1949*c*).

this case not innocuous. We shall adopt it nevertheless for the purpose of deriving interval estimates and tests. As usual, the expected mean squares calculated and the unbiasedness of the point estimates derived from them are valid without the normality assumption. The assumptions are now in the form

$$(7.4.7) \qquad \Omega: \begin{cases} y_{ijk} = \mu + a_i + b_j + c_{ij} + e_{ijk}, \\ \text{the } \{a_i\},\ \{b_j\},\ \{c_{ij}\}, \text{ and } \{e_{ijk}\} \text{ are independently normal,} \\ \text{with zero means and respective variances } \quad \sigma_A^2, \sigma_B^2, \sigma_{AB}^2, \sigma_e^2. \end{cases}$$

The four SS's with subscripts A, B, AB, and e—called those for A main effects, B main effects, interactions, and error—are defined in terms of the observations $\{y_{ijk}\}$ as in Table 4.3.1 and are of course computed by the same identities as in sec. 4.3. If there is only one observation per cell ($K = 1$), we do not employ SS_e (which would be 0 with 0 d.f.). If we substitute (7.4.7) into the definitions of the SS's we get

$$\begin{aligned} SS_A &= JK\sum_i (a_i - a_. + c_{i.} - c_{..} + e_{i..} - e_{...})^2, \\ SS_B &= IK\sum_j (b_j - b_. + c_{.j} - c_{..} + e_{.j.} - e_{...})^2, \\ SS_{AB} &= K\sum_i \sum_j (c_{ij} - c_{i.} - c_{.j} + c_{..} + e_{ij.} - e_{i..} - e_{.j.} + e_{...})^2, \\ SS_e &= \sum_i \sum_j \sum_k (e_{ijk} - e_{ij.})^2. \end{aligned}$$

We note that the B main effects are not involved in SS_A but the interactions are; similarly for the A main effects and SS_B, while SS_e involves only the errors.

We shall show that under Ω the four SS's are statistically independent, and we shall do this by arguing that the following nine sets of variables are statistically independent: (i) $\{a_i - a_.\}$, (ii) $\{b_j - b_.\}$, (iii) $\{c_{i.} - c_{..}\}$, (iv) $\{c_{.j} - c_{..}\}$, (v) $\{c_{ij} - c_{i.} - c_{.j} + c_{..}\}$, (vi) $\{e_{i..} - e_{...}\}$, (vii) $\{e_{.j.} - e_{...}\}$, (viii) $\{e_{ij.} - e_{i..} - e_{.j.} + e_{...}\}$, (ix) $\{e_{ijk} - e_{ij.}\}$. Since the totality of variables in all the sets are jointly normal, it suffices to show that the sets are pairwise independent. The set (i) is independent of (ii) because the $\{a_i - a_.\}$ are functions of the $\{a_i\}$ only, the $\{b_j - b_.\}$ are functions of the $\{b_j\}$ only, and the $\{a_i\}$ are independent of the $\{b_j\}$ by the Ω-assumptions. By similar reasoning, (i) is independent of all subsequent sets, (ii) is independent of all subsequent sets, and (iii), (iv), (v) are each independent of (vi), (vii), (viii), (ix). To see that the sets (iii), (iv), (v) are pairwise independent consider a fictitious two-way layout with one observation per cell under the fixed-effects model, where the $\{c_{ij}\}$ play the role of the observations[25] with error variance σ^2_{AB} and all parameters except σ^2_{AB} are zero; then (iii), (iv), (v) are, respectively, the linear forms spanning three of the four mutually orthogonal spaces noted in sec. 4.2, and so the three sets are statistically independent. Similarly, comparing the $\{e_{ijk}\}$ with observations in a fictitious two-way layout under the fixed-effects model with K observations per cell, we see that the four sets (vi), (vii), (viii), (ix) are statistically independent. A similar argument would show that the four SS's are also statistically independent of the grand mean

$$y_{\cdot\cdot\cdot} = \mu + a_{\cdot} + b_{\cdot} + c_{\cdot\cdot} + e_{\cdot\cdot\cdot}.$$

If we let

$$g_i = a_i + c_{i\cdot} + e_{i\cdot\cdot},$$

the $\{g_i\}$ are independently $N(0, \sigma_g^2)$, where $\sigma_g^2 = \sigma_A^2 + J^{-1}\sigma^2_{AB} + J^{-1}K^{-1}\sigma_e^2$, and hence $SS_A = JK\Sigma_i(g_i - g_{\cdot})^2$ is distributed as $JK\sigma_g^2$ times a chi-square variable with $I-1$ d.f. From this it follows that $SS_A = E(MS_A)\chi^2_{I-1}$, where

$$E(MS_A) = JK\sigma_g^2 = \sigma_e^2 + K\sigma^2_{AB} + JK\sigma^2_A.$$

The distribution of SS_B may be found similarly. To treat SS_{AB} let

$$h_{ij} = c_{ij} + e_{ij\cdot},$$

so the $\{h_{ij}\}$ are independently $N(0, \sigma_h^2)$ with $\sigma_h^2 = \sigma^2_{AB} + K^{-1}\sigma_e^2$, and SS_{AB} equals K times

(7.4.8) $$\sum_i \sum_j (h_{ij} - h_{i\cdot} - h_{\cdot j} + h_{\cdot\cdot})^2.$$

But (7.4.8) is distributed like the residual SS in a fictitious two-way layout where the $\{h_{ij}\}$ are the observations under the fixed-effects model with all parameters zero except σ_h^2. Finally, the distribution of SS_e is like that of the error SS in a fictitious two-way layout where the $\{e_{ijk}\}$ are the observations, etc. In this way we find that each of the four SS's is distributed as the corresponding $E(MS)$ times a chi-square variable with the number of d.f. and the $E(MS)$ shown in Table 7.4.1. This distribution theory tells us how we may test the usual hypotheses.

In the one-way layout the test of the hypothesis H_A: $\sigma_A^2 = 0$ was the same as in the fixed-effects model, but here it is different. Under H_A, SS_A is $(\sigma_e^2 + K\sigma^2_{AB})\chi^2_{I-1}$ and SS_{AB} is $(\sigma_e^2 + K\sigma^2_{AB})\chi^2_{\nu_{AB}}$, where $\nu_{AB} = (I-1)(J-1)$, and so MS_A/MS_{AB} is $F_{I-1,\nu_{AB}}$. Note that, whereas in the fixed-effects model MS_A/MS_e has the F-distribution under H_A and

[25] Actually the $\{c_{ij}\}$ are not observable, but their distribution is the same as that of the observations in the fictitious case.

TABLE 7.4.1

EXPECTED MEAN SQUARES IN RANDOM-EFFECTS MODEL FOR TWO-WAY LAYOUT

SS	d.f.	E(MS)
SS_A	$I-1$	$\sigma_e^2 + K\sigma_{AB}^2 + JK\sigma_A^2$
SS_B	$J-1$	$\sigma_e^2 + K\sigma_{AB}^2 + IK\sigma_B^2$
SS_{AB}	$(I-1)(J-1)$	$\sigma_e^2 + K\sigma_{AB}^2$
SS_e	$IJ(K-1)$	σ_e^2

MS_A/MS_{AB} does not (unless $\sigma_{AB}^2 = 0$), here MS_A/MS_{AB} has the F-distribution under H_A and MS_A/MS_e does not (unless $\sigma_{AB}^2 = 0$). We therefore reject H_A at the α level of significance if $MS_A/MS_{AB} > F_{\alpha;I-1,\nu_{AB}}$. The power of the test may be found as in sec. 7.2, from the relation

$$\frac{MS_A}{MS_{AB}} = \frac{\sigma_e^2 + K\sigma_{AB}^2 + JK\sigma_A^2}{\sigma_e^2 + K\sigma_{AB}^2} F_{I-1,\nu_{AB}} \tag{7.4.9}$$

valid under Ω. The power of all the tests in this section depends only on the *central* F-distribution. As in sec. 4.2, we may use (7.4.9) to obtain a test of the more general hypothesis $\theta \leq \theta_0$, where $\theta = \sigma_A^2/(\sigma_e^2+K\sigma_{AB}^2)$, or to obtain a confidence interval for θ. An approximate confidence interval for σ_A^2 is yielded by the method at the end of sec. 7.2. Inferences about σ_B^2 may be made similarly.

If $K > 1$ we can make inferences about σ_{AB}^2 by using the ratio MS_{AB}/MS_e, whose distribution under Ω is that of $(\sigma_e^2+K\sigma_{AB}^2)/\sigma_e^2$ times a central F-variable with ν_{AB} and $IJ(K-1)$ d.f.

Point estimates for the variance components σ_A^2, σ_B^2, σ_{AB}^2, and σ_e^2 are easily derived from Table 7.4.1 by the method of sec. 7.2. For example, we have

$$\hat{\sigma}_A^2 = J^{-1}K^{-1}(MS_A - MS_{AB}), \qquad \hat{\sigma}_{AB}^2 = K^{-1}(MS_{AB} - MS_e). \tag{7.4.10}$$

(If $K = 1$ we cannot estimate σ_e^2 and σ_{AB}^2 separately, but only $\sigma_e^2+\sigma_{AB}^2$.) It is then also easy to obtain the formulas for the variances of these estimates by the method used in sec. 7.2. From these formulas we could show that Var $(\hat{\sigma}_A^2)$ tends to zero if and only if I becomes infinite, also that Var $(\hat{\sigma}_{AB}^2)$ tends to zero if I or J become infinite, but not if K becomes infinite with fixed I and J. Such behavior of the point estimates suggests that the power of certain tests may tend to a limit less than unity if the number of measurements becomes large in certain ways, and this may be examined as in sec. 7.2; for example, if $K = 1$, this is true, as J becomes infinite for fixed I, of the power of the test of the hypothesis $\theta \leqq \theta_0$, where $\theta = \sigma_A^2/(\sigma_e^2+\sigma_{AB}^2)$, and $\theta_0 > 0$.

When data are analyzed by the theory of this section an analysis-of-variance table like Table 4.3.1 should be made with the E(MS) column replaced from Table 7.4.1.

7.5. THE COMPLETE THREE- AND HIGHER-WAY LAYOUTS

We shall use the notation of sec. 4.5 for factors, numbers of levels, SS's, and MS's. An example of a three-way layout might be obtained from the example of machines and workers of sec. 7.4 by adding a factor C referring to batches of material. If y_{ijkm} is the mth observation in the i,j,k cell[26] we shall assume that

$$y_{ijkm} = m_{ijk} + e_{ijkm},$$

where the errors $\{e_{ijkm}\}$ are independently and identically distributed with zero means and variance σ_e^2, and independently of the "true" cell means $\{m_{ijk}\}$. If for each of the three factors the levels are a random sample from a population, and the three samplings are done independently, then it may be shown by proceeding as in sec. 7.4 that without loss of generality we may write

$$(7.5.1) \quad m_{ijk} = \mu + a_i^A + a_j^B + a_k^C + a_{ij}^{AB} + a_{jk}^{BC} + a_{ik}^{AC} + a_{ijk}^{ABC},$$

where the random effects symbolized by the a's are completely uncorrelated and have zero means; and the $\{a_i^A\}$ are identically distributed with variance σ_A^2, similarly for $\{a_j^B\}$ and $\sigma_B^2, \cdots, \{a_{ijk}^{ABC}\}$ and σ_{ABC}^2. If the normality assumption is added (the remarks about this in sec. 7.4 are again pertinent) the assumptions are then

$$\Omega: \begin{cases} y_{ijkm} = m_{ijk} + e_{ijkm}, \text{ where } m_{ijk} \text{ satisfies (7.5.1), and the } \{a_i^A\}, \\ \{a_j^B\}, \{a_k^C\}, \{a_{ij}^{AB}\}, \{a_{jk}^{BC}\}, \{a_{ik}^{AC}\}, \{a_{ijk}^{ABC}\}, \{e_{ijkm}\} \text{ are independently} \\ \text{normal with zero means and variances } \sigma_A^2, \sigma_B^2, \cdots, \sigma_{ABC}^2, \sigma_e^2, \\ \text{respectively.} \end{cases}$$

Proceeding as in sec. 7.4 we may show that the eight SS's (seven if $M = 1$) defined in sec. 4.5 are independently distributed, each as the corresponding E(MS) times a chi-square variable, with the E(MS) and number of d.f. shown in Table 7.5.1. The rules which permit writing the expected mean squares without any calculations may be generalized for the higher-way layouts in the balanced case under the random-effects model by inspecting Table 7.5.1. They are included in sec. 8.2. The

[26] We may drop the subscript m if the number M of observations per cell is one.

TABLE 7.5.1

EXPECTED MEAN SQUARES IN RANDOM-EFFECTS MODEL FOR THREE-WAY LAYOUT

SS	d.f.	E(MS)
SS_A	$I-1$	$\sigma_e^2+M\sigma_{ABC}^2+KM\sigma_{AB}^2+JM\sigma_{AC}^2+JKM\sigma_A^2$
SS_B	$J-1$	$\sigma_e^2+M\sigma_{ABC}^2+KM\sigma_{AB}^2+IM\sigma_{BC}^2+IKM\sigma_B^2$
SS_C	$K-1$	$\sigma_e^2+M\sigma_{ABC}^2+IM\sigma_{BC}^2+JM\sigma_{AC}^2+IJM\sigma_C^2$
SS_{AB}	$(I-1)(J-1)$	$\sigma_e^2+M\sigma_{ABC}^2+KM\sigma_{AB}^2$
SS_{BC}	$(J-1)(K-1)$	$\sigma_e^2+M\sigma_{ABC}^2+IM\sigma_{BC}^2$
SS_{AC}	$(I-1)(K-1)$	$\sigma_e^2+M\sigma_{ABC}^2+JM\sigma_{AC}^2$
SS_{ABC}	$(I-1)(J-1)(K-1)$	$\sigma_e^2+M\sigma_{ABC}^2$
SS_e	$IJK(M-1)$	σ_e^2

SS's continue to be distributed independently as the E(MS) times a chi-square variable.

With the three-way layout we encounter for the first[27] time the difficulty in the random-effects models for the higher-way layouts that even under the normality assumption there may not be available exact F-tests of some of the hypotheses usually tested. There is no difficulty about testing the three-factor interactions (i.e., testing the hypothesis $\sigma^2_{ABC} = 0$) or the two-factor interactions; thus the $A \times B \times C$ interactions are tested with the ratio $\text{MS}_{ABC}/\text{MS}_e$, the $A \times B$ interactions with $\text{MS}_{AB}/\text{MS}_{ABC}$, etc. However, the situation is different when testing the main effects, as we shall now see.

Approximate F-Test

Suppose that we wish to test H_A: $\sigma^2_A = 0$ (H_B and H_C would of course be treated similarly). If we are willing to assume that $\sigma^2_{AB} = 0$, then an exact F-test of H_A can be based on the statistic $\text{MS}_A/\text{MS}_{AC}$. In this case SS_{AB} could be pooled with SS_{ABC} since they have the same E(MS). Similarly, if we are willing to assume $\sigma^2_{AC} = 0$ we may test H_A with $\text{MS}_A/\text{MS}_{AB}$, and pool SS_{AC} with SS_{ABC}. If we were willing to assume other variance components to be zero, the reader would have no difficulty deducing exact tests, if any, of the standard hypotheses, and pooling procedures from the table obtained from Table 7.5.1 by deleting in it the variance components assumed to be zero. However, if we are unwilling to assume $\sigma^2_{AB} = 0$ or $\sigma^2_{AC} = 0$, then no exact test of H_A can be found from the table.

An approximate F-test[28] is usually made by the following method: Write

$$\tau = \sigma^2_e + M\sigma^2_{ABC} + KM\sigma^2_{AB} + JM\sigma^2_{AC},$$

so $E(\text{MS}_A) = \tau + JKM\sigma^2_A$. Then τ may be expressed as the following linear combination of E(MS)'s in Table 7.5.1:

$$\tau = E(\text{MS}_{AB}) + E(\text{MS}_{AC}) - E(\text{MS}_{ABC}),$$

and hence has the unbiased estimate

$$\hat{\tau} = \text{MS}_{AB} + \text{MS}_{AC} - \text{MS}_{ABC}.$$

[27] The difficulty would have occurred with the two-way layout if we had considered testing the hypothesis $\mu = 0$.

[28] This is an obvious consequence of the approximation to the distribution of a linear combination of chi-square variables used by Satterthwaite (1946) and credited by him to H. F. Smith. Numerical evidence which indicates that this is an excellent approximation when the coefficients in the linear combination are all positive is given by Box (1954*a*), p. 294. A further approximation is involved in the present application by the estimation of these coefficients.

Write the mean squares on the right as $\hat{\tau}_1$, $\hat{\tau}_2$, $\hat{\tau}_3$, respectively, so $\hat{\tau} = \hat{\tau}_1 + \hat{\tau}_2 - \hat{\tau}_3$. Then the $\hat{\tau}_i$ $(i = 1, 2, 3)$ are independently distributed as $\tau_i \chi^2_{\nu_i}/\nu_i$, where $\tau_i = E(\hat{\tau}_i)$, and ν_i is the number of d.f. of $\hat{\tau}_i$. Next, we try to approximate $\hat{\tau}$ by a random variable of the form $\tau\chi^2_\nu/\nu$, where ν is determined so that $\hat{\tau}$ and $\tau\chi^2_\nu/\nu$ have the same variance (they already have the same mean). This gives the condition

$$2\tau^2/\nu = 2\sum_1^3(\tau_i^2/\nu_i),$$

or

$$\nu = \tau^2/\sum_1^3(\tau_i^2/\nu_i).$$

This determines ν, but unfortunately in terms of unknown parameters. We estimate ν with

$$\hat{\nu} = \hat{\tau}^2/\sum_{i=1}^{3}(\hat{\tau}_i^2/\nu_i).$$

The approximate F-test for testing H_A is then obtained by using the ratio $\text{MS}_A/\hat{\tau}$, proceeding as though it were distributed as

$$\frac{JKM\sigma_A^2 + \tau}{\tau}\,\frac{\nu}{I-1}\,\frac{\chi^2_{I-1}}{\chi^2_\nu} = \frac{JKM\sigma_A^2 + \tau}{\tau} F_{I-1,\nu},$$

and as though the constant ν were equal to its estimate $\hat{\nu}$. This also leads to an approximation for the power, based on the central F-distribution.

The method is obviously general. It may be used whenever we want to test a hypothesis H_x: $\sigma_x^2 = 0$, such that MS_x is of the form $c\sigma_x^2 + \tau$, and there is no expected mean square in the table equal to τ. We then consider a linear combination of mean squares whose expected value equals τ, approximate it as $\tau\chi^2_\nu/\nu$, etc.

7.6. A NESTED DESIGN

As an example of the random-effects model for a nested design we will consider an experiment with three factors, T nested within B, and B within C. The reader may think of the illustration in sec. 5.3 concerning variability of a given brand of tissues, in which C refers to cities, B to boxes, and T to tissues. We shall use the same notation for the various sample sizes, namely that C is at I levels; that within the ith level of C, B is at J_i levels; that within the i,j level of B (the jth level of B within the ith level of C), T is at K_{ij} levels; and that M_{ijk} $(M_{ijk} \geqq 1)$ measurements are made at the i,j,k level of T (the kth level of T within the jth level of B

within the ith level of C). (Later we will specialize $K_{ij} = K$, $M_{ijk} = M$.) If y_{ijkm} denotes the mth of these M_{ijk} measurements, we shall assume

$$y_{ijkm} = m_{ijk} + e_{ijkm}, \tag{7.6.1}$$

where the "errors" $\{e_{ijkm}\}$ are independently and identically distributed with zero means and variance σ_e^2, and independently of the "true" means $\{m_{ijk}\}$. Appropriate distribution assumptions on the $\{m_{ijk}\}$ may be motivated as follows:

Suppose that there is a population of levels of C, labeled by the index u, from which the I levels of C in the experiment are sampled. All populations will be idealized as infinite populations. Call $\mathscr{P}_u$ the population distribution of the index u. We next imagine that for every u there is a population of levels of B, labeled by a pair of indices (u, v), and that $\mathscr{P}_{v|u}$ is the population distribution of v for given u. Similarly, for every (u, v) we suppose that there is a population of levels of T, labeled by a triple of indices (u, v, w), and that $\mathscr{P}_{w|u,v}$ is the population distribution of w for given (u, v). Let $m(u, v, w)$ be the "true" mean for the element labeled (u, v, w), corresponding to successive random choices of u according to $\mathscr{P}_u$, then v according to $\mathscr{P}_{v|u}$, then w according to $\mathscr{P}_{w|u,v}$. Denote by $m(u, v, .)$ the (conditional) mean of $m(u, v, w)$ for fixed (u, v), calculated according to $\mathscr{P}_{w|u,v}$; by $m(u, ., .)$ the (conditional) mean of $m(u, v, .)$ for given u, calculated according to $\mathscr{P}_{v|u}$; and by $m(., ., .)$ the mean of $m(u, ., .)$ calculated according to $\mathscr{P}_u$. Define

$$\begin{aligned} t(u, v, w) &= m(u, v, w) - m(u, v, .), \\ b(u, v) &= m(u, v, .) - m(u, ., .), \\ c(u) &= m(u, ., .) - m(., ., .), \\ \mu &= m(., ., .), \end{aligned}$$

so

$$m(u, v, w) = \mu + c(u) + b(u, v) + t(u, v, w). \tag{7.6.2}$$

The three random variables $c(u)$, $b(u, v)$, and $t(u, v, w)$ have zero means, since by their definition

$$\begin{aligned} c(.) = 0, \quad b(u, .) &= 0 \quad \text{for all } u, \\ t(u, v, .) &= 0 \quad \text{for all } u, v, \end{aligned} \tag{7.6.3}$$

where replacing w by a dot in $t(u, v, w)$ indicates that the (conditional) mean has been taken with respect to $\mathscr{P}_{w|u,v}$, etc., and so the unconditional expected values $b(., .)$ and $t(., ., .)$ must also be zero. The variances of $c(u)$, $b(u, v)$, and $t(u, v, w)$ will be denoted by σ_C^2, σ_B^2, and σ_T^2, respectively. The three random variables are uncorrelated. This may be proved as follows for $b(u, v)$ and $t(u, v, w)$, and similarly for the other two pairs: The covariance of $b(u, v)$ and $t(u, v, w)$ is the expected value of their product; if we take this first conditionally, given (u, v), we get

$$E(b(u, v)\, t(u, v, w)|u, v) = b(u, v)\, E(t(u, v, w)|u, v) = 0,$$

because of (7.6.3), and so the unconditional expectation is also zero.

Formula (7.6.2) resolves the difference of the true mean $m(u, v, w)$ from μ into parts which may be regarded as the respective effects of the factors C, B, T; thus, in the example of cities, boxes, and tissues, $c(u)$ is the effect of the city labeled u, $b(u, v)$ is the effect of the box labeled (u, v) within the city labeled u, and $t(u, v, w)$ is the effect of the tissue labeled (u, v, w) within the box labeled (u, v). Since the mean box effect in the city labeled u, namely $b(u, .)$, is zero, a measure of the magnitude of the box effects in the city labeled u is the conditional variance of $b(u, v)$, given u, which we shall write as $\text{Var}\,(b|u)$. However, in the present model we are not primarily interested in any particular city, but rather in the population of cities. This suggests using the mean over cities (i.e., the expected value according to $\mathscr{P}_u$) of $\text{Var}\,(b|u)$. Ordinarily this would not be the same as $\text{Var}\,(b(u, v)) = \sigma_B^2$; it happens to coincide with σ_B^2 in the present case because the last term in the general relation (see Problem 7.5)

$$\text{Var}\,(b) = E(\text{Var}\,(b \mid u)) + \text{Var}\,(E(b \mid u))$$

vanishes, since $E(b|u) = b(u, .) = 0$ for all u. There is a similar justification of σ_T^2 as a measure of the magnitude of the effects of tissues within boxes.

Let $\{u_1, \cdots, u_I\}$ be the labels of the I levels of C selected in the experiment; these are supposed to be a random sample from $\mathscr{P}_u$. Let $\{(u_i, v_1), \cdots, (u_i, v_{J_i})\}$ be the labels of the J_i levels of B selected within the ith level of C in the experiment; these are supposed to be a random sample from $\mathscr{P}_{v|u_i}$ $(i = 1, \cdots, I)$. Finally let $\{(u_i, v_j, w_1), \cdots, (u_i, v_j, w_{K_{ij}})\}$ be the labels of the K_{ij} levels of T selected within the jth level of B within the ith level of C in the experiment; they are supposed to be a random sample from $\mathscr{P}_{w|u_i, v_j}$. The $\{m_{ijk}\}$ in (7.6.1) then have the structure

$$m_{ijk} = \mu + c_i + b_{ij} + t_{ijk}, \tag{7.6.4}$$

where

$$c_i = c(u_i), \qquad b_{ij} = b(u_i, v_j), \qquad t_{ijk} = t(u_i, v_j, w_k). \tag{7.6.5}$$

Since t_{ijk} is distributed like $t(u, v, w)$ above, its mean is zero and its variance is σ_T^2; similarly, b_{ij} and c_i have zero means and variances σ_B^2 and σ_C^2, respectively. The $\{c_i\}$, $\{b_{ij}\}$, $\{t_{ijk}\}$ are completely uncorrelated; this may be verified by arguments of a kind by now familiar to the reader.

The normality assumption seems less obnoxious here than with the complete layouts, since now in general there are no interactions needed, whose treatment might violate our intuitions. If we make it we have

$$\Omega: \begin{cases} y_{ijkm} = \mu + c_i + b_{ij} + t_{ijk} + e_{ijkm} \quad (i = 1, \cdots, I; \\ j = 1, \cdots, J_i; \quad k = 1, \cdots, K_{ij}; \quad m = 1, \cdots, M_{ijk}), \\ \text{where the } \{c_i\}, \{b_{ij}\}, \{t_{ijk}\}, \{e_{ijkm}\} \text{ are independently normal} \\ \text{with zero means and variances } \sigma_C^2, \sigma_B^2, \sigma_T^2, \sigma_e^2, \text{ respectively.} \end{cases}$$

We shall now specialize to the case where

$$\text{(7.6.6)} \qquad \text{all } K_{ij} = K, \qquad \text{all } M_{ijk} = M.$$

The general case[29] is treated further in the appendix to this section, where the expected mean squares are derived. The restriction (7.6.6) would be a reasonable one in the example of cities, boxes, and tissues, since there is no reason for taking different numbers of measurements per tissue or different numbers of tissues per box, but it might be desired to have larger samples of boxes from the larger cities in the sample of cities.

The SS's found in sec. 5.3 may be written

$$\text{(7.6.7)} \qquad \begin{aligned} SS_C &= KM\sum_i J_i(y_{i\cdot\cdot\cdot} - \bar{y}_{\cdot\cdot\cdot\cdot})^2, \\ SS_B &= KM\sum_i\sum_j (y_{ij\cdot\cdot} - y_{i\cdot\cdot\cdot})^2, \\ SS_T &= M\sum_i\sum_j\sum_k (y_{ijk\cdot} - y_{ij\cdot\cdot})^2, \\ SS_e &= \sum_i\sum_j\sum_k\sum_m (y_{ijkm} - y_{ijk\cdot})^2, \end{aligned}$$

where the symbols $y_{ijk\cdot}$, $y_{ij\cdot\cdot}$, $y_{i\cdot\cdot\cdot}$ have their usual meaning, and $\bar{y}_{\cdot\cdot\cdot\cdot}$ is the unweighted average of all the observations or the weighted average of observed means for the levels of C,

$$\bar{y}_{\cdot\cdot\cdot\cdot} = \sum_i\sum_j\sum_k\sum_m y_{ijkm}/n = \sum_i J_i y_{i\cdot\cdot\cdot} \Big/ \sum_i J_i,$$

where $n = KM\Sigma_i J_i$. If in (7.6.7) we substitute

$$y_{ijkm} = \mu + c_i + b_{ij} + t_{ijk} + e_{ijkm},$$

[29] In the general case the SS's, other than SS_e, are not distributed as a constant times chi-square, and they are not statistically independent, except that SS_e is independent of the other SS's. In the restricted case (7.6.6) the SS's, other than SS_e, are not independent of $y_{\cdot\cdot\cdot\cdot}$.

we get

$$
\begin{aligned}
\mathrm{SS}_C &= KM\sum_i J_i(c_i-\bar{c}_. + b_{i.}-\bar{b}_{..} + t_{i..}-\bar{t}_{...} + e_{i...}-\bar{e}_{....})^2,\\
\mathrm{SS}_B &= KM\sum_i\sum_j(b_{ij}-b_{i.} + t_{ij.}-t_{i..} + e_{ij..}-e_{i...})^2,\\
\mathrm{SS}_T &= M\sum_i\sum_j\sum_k(t_{ijk}-t_{ij.} + e_{ijk.}-e_{ij..})^2,\\
\mathrm{SS}_e &= \sum_i\sum_j\sum_k\sum_m(e_{ijkm}-e_{ijk.})^2,
\end{aligned}
\tag{7.6.8}
$$

where a bar over a quantity means that the dot replacing the subscript i is a weighted average over i with weights $\{J_i\}$, for example,

$$\bar{t}_{...} = \sum_i J_i t_{i..}\Big/\sum_i J_i,$$

and otherwise the dot notation has its usual meaning.

To prove that the four SS's are statistically independent it is sufficient, because of (7.6.8), to argue that the following ten sets of normal variables are independent: (i) $\{c_i-\bar{c}_.\}$, (ii) $\{b_{i.}-\bar{b}_{..}\}$, (iii) $\{b_{ij}-b_{i.}\}$, (iv) $\{t_{i..}-\bar{t}_{...}\}$, (v) $\{t_{ij.}-t_{i..}\}$, (vi) $\{t_{ijk}-t_{ij.}\}$, (vii) $\{e_{i...}-\bar{e}_{....}\}$, (viii) $\{e_{ij..}-e_{i...}\}$, (ix) $\{e_{ijk.}-e_{ij..}\}$, (x) $\{e_{ijkm}-e_{ijk.}\}$. The four sets of sets (I) $\{(i)\}$, (II) $\{(ii), (iii)\}$, (III) $\{(iv), (v), (vi)\}$, (IV) $\{(vii), (viii), (ix), (x)\}$ are independent by our Ω-assumptions.

To see that the four sets in (IV) are independent consider the distribution theory in a fictitious fixed-effects model where the observations, which we shall denote by $\{y'_{ijkm}\}$ to distinguish them from the actual observations $\{y_{ijkm}\}$, have the structure defined in sec. 5.3. By using the method of nested ω's with the chain of hypotheses H'_T, $H'_T \cap H'_B$, $H'_T \cap H'_B \cap H'_C$, where H'_T, H'_B, H'_C denote, respectively, the hypotheses H_T, H_B, H_C of sec. 5.3 for the fictitious fixed-effects model, we would find that the following four sets of linear forms in the observations are statistically independent:

$$\{y'_{i...}-\bar{y}'_{....}\},\quad \{y'_{ij..}-y'_{i...}\},\quad \{y'_{ijk.}-y'_{ij..}\},\quad \{y'_{ijkm}-y'_{ijk.}\}. \tag{7.6.9}$$

Now the $\{e_{ijkm}\}$ are distributed like the $\{y'_{ijkm}\}$ of the fictitious fixed-effects model with all the parameters

$$\mu,\quad \{\gamma_i\},\quad \{\beta_{ij}\},\quad \{\tau_{ijk}\} \tag{7.6.10}$$

equal to zero. Thus the four sets (IV) of the random-effects model are distributed like the four sets (7.6.9) of the fictitious fixed-effects model and hence are statistically independent. Furthermore, SS_e is seen from (7.6.8) to be distributed like the error SS in the fictitious model, namely as $\sigma_e^2\chi^2_{\nu_e}$, where $\nu_e = K(M-1)\Sigma_i J_i$.

To prove the three sets (III) statistically independent, consider the special case of the fictitious fixed-effects model where $M = 1$. Then $y'_{ijk.} = y'_{ijk1}$, and the independence of the first three sets in (7.6.9) gives us the independence of the three sets (III) if we note the $\{t_{ijk}\}$ are distributed like the $\{y'_{ijk1}\}$ when all the parameters (7.6.10) are zero.

The set (II) may be treated similarly by considering a fictitious fixed-effects model with $K = M = 1$.

If we let

$$f_i = c_i + b_{i.} + t_{i..} + e_{i...},$$

$$g_{ij} = b_{ij} + t_{ij.} + e_{ij..},$$

$$h_{ijk} = t_{ijk} + e_{ijk.},$$

then the $\{h_{ijk}\}$ are independently $N(0, \sigma_h^2)$, with

$$\sigma_h^2 = \sigma_T^2 + M^{-1}\sigma_e^2,$$

the $\{g_{ij}\}$ are independently $N(0, \sigma_g^2)$, with

$$\sigma_g^2 = \sigma_B^2 + K^{-1}\sigma_T^2 + (KM)^{-1}\sigma_e^2, \tag{7.6.11}$$

and the $\{f_i\}$ are independently normal with zero means and

$$\text{Var}(f_i) = \sigma_C^2 + J_i^{-1}\sigma_g^2. \tag{7.6.12}$$

It follows that $H_{ij} = \Sigma_k(h_{ijk} - h_{ij.})^2$ is $\sigma_h^2\chi^2_{K-1}$, and hence $SS_T = M\Sigma_i\Sigma_j H_{ij}$ is $M\sigma_h^2\chi^2_{\nu_T}$ or $(\sigma_e^2 + M\sigma_T^2)\chi^2_{\nu_T}$, where $\nu_T = (K-1)\Sigma_i J_i$. Similarly, SS_B is $(\sigma_e^2 + M\sigma_T^2 + KM\sigma_B^2)\chi^2_{\nu_B}$, where $\nu_B = \Sigma_i(J_i - 1)$.

However, SS_C will in general not be distributed as a constant times a chi-square variable, since

$$SS_C = KM\sum_i J_i(f_i - \bar{f}_.)^2, \tag{7.6.13}$$

where the $\{f_i\}$ are distributed as stated above (7.6.12), and $\bar{f}_. = \Sigma_i J_i f_i / \Sigma_i J_i$. However, if $\sigma_C^2 = 0$, then $SS_C = (\sigma_e^2 + M\sigma_T^2 + KM\sigma_B^2)\chi^2_{I-1}$, or if all $J_i = J$, then $SS_C = (\sigma_e^2 + M\sigma_T^2 + KM\sigma_B^2 + JKM\sigma_C^2)\chi^2_{I-1}$. The latter chi-square distribution may be derived like the preceding ones, the former as follows: Consider a fictitious one-way layout with "observations" $\{y'_{ij}\}$, I classes, and J_i observations in the ith class. If $\sigma_C^2 = 0$, the $\{f_i\}$ are distributed like the observed means $\{y'_{i.}\}$ in the fixed-effects model if the error variance is the σ_g^2 of (7.6.11), and the true means are zero. The numerator SS for testing the hypothesis that the true means are equal in the fictitious layout is $\Sigma_i J_i(y'_{i.} - \bar{y}'_{..})^2$, and is distributed as $\sigma_g^2\chi^2_{I-1}$. Hence $SS_C = KM\Sigma J_i(f_i - \bar{f}_.)^2$

is distributed as $KM\sigma_g^2\chi_{I-1}^2 = (\sigma_e^2+M\sigma_T^2+KM\sigma_B^2)\chi_{I-1}^2$ when $\sigma_C^2 = 0$. Under Ω the mean and variance of SS_C are[30]

$$(7.6.14) \qquad E(SS_C) = (I-1)(\sigma_e^2+M\sigma_T^2+KM\sigma_B^2) + KM(A_1-A_1^{-1}A_2)\sigma_C^2,$$

$$(7.6.15) \quad \text{Var}\,(SS_C) = 2(KM)^2[(A_2-2A_1^{-1}A_3+A_1^{-2}A_2^2)\sigma_C^4 + 2(A_1-A_1^{-1}A_2)\sigma_C^2\sigma_g^2 + (I-1)\sigma_g^4],$$

where

$$(7.6.16) \qquad A_m = \sum_i J_i^m,$$

and σ_g^2 is defined by (7.6.11). Formulas (7.6.14) and (7.6.15) follow from the lemma in the appendix to this section.

The analysis-of-variance table is formed by specializing Table 5.3.2 to the case

$$M_{ijk} = M, \quad K_{ij} = K, \quad n_{ij} = KM, \quad n_i = J_i KM, \quad n = KM\sum_i J_i,$$

and completing it with an E(MS) column consisting of

$$(7.6.17) \qquad \begin{aligned} E(MS_C) &= \sigma_e^2 + M\sigma_T^2 + KM\sigma_B^2 + A\sigma_C^2, \\ E(MS_B) &= \sigma_e^2 + M\sigma_T^2 + KM\sigma_B^2, \\ E(MS_T) &= \sigma_e^2 + M\sigma_T^2, \\ E(MS_e) &= \sigma_e^2, \end{aligned}$$

where

$$A = KM(I-1)^{-1}(A_1-A_1^{-1}A_2),$$

and A_1, A_2 are defined by (7.6.16), so that $A = JKM$ if all $J_i = J$. The first of the formulas (7.6.17) follows from (7.6.14), the others from the chi-square distributions deduced above.

F-tests of the hypotheses H_C: $\sigma_C^2 = 0$, H_B: $\sigma_B^2 = 0$, H_T: $\sigma_T^2 = 0$ are made by using the appropriate ratio of mean squares suggested by (7.6.17), for example, under H_B, $E(MS_B) = E(MS_T)$, and this suggests using MS_B/MS_T to test H_B. These will be "exact" F-tests, and, except for the test of H_C in the case of unequal $\{J_i\}$, the power is easily expressed in terms of the central F-distribution, because under Ω the four SS's are independently distributed, each as the corresponding E(MS) times a chi-square variable, except for SS_C in the case of unequal $\{J_i\}$, where the chi-square distribution holds only under $H_C \cap \Omega$. The power of the test of H_C in the case of unequal $\{J_i\}$ could be approximated in terms of the central F-distribution by approximating SS_C as a constant times a chi-square variable, by fitting the first two moments, namely as $E(SS_C)\chi_\nu^2/\nu$ with

[30] The distribution of SS_C under Ω is that of some linear combination of independent chi-square variables. This is true for any quadratic form in jointly normal variables with zero means (see Problem V.2).

$\nu = 2[E(SS_C)]^2/\text{Var}(SS_C)$, where $E(SS_C)$ and $\text{Var}(SS_C)$ are given by (7.6.14) and (7.6.15).

Unbiased estimates for the variance components σ_C^2, σ_B^2, σ_T^2, and σ_e^2 may be obtained as usual by solving equations similar to (7.6.17) for the estimates. In the example of cities, boxes, and tissues, one would want also an estimate of

$$\sigma_y^2 = \sigma_C^2 + \sigma_B^2 + \sigma_T^2 + \sigma_e^2, \tag{7.6.18}$$

which is a measure of the variability of quality of the given brand of tissue.[31] All these estimates will be linear combinations of the four independent MS's, and so formulas for their variances are easily obtained from the above results for the distributions of the corresponding SS's under Ω. These formulas involve unknown parameters; replacing these by their estimates would then give estimated variances. For the estimate of (7.6.18), inspection of the four contributions from the four MS's to the estimated variance of the estimate in a past experiment would usually suggest improvements in the allocation of the samples in future experiments of the same kind. A similar remark applies in most situations where variance components are estimated.

An exact confidence interval for σ_e^2 may be derived from the chi-square distribution of σ_e^2 SS_e. Approximate confidence intervals for the other variance components may be obtained by the method at the end of sec. 7.2, except for σ_C^2 when the $\{J_i\}$ are unequal. The above method for approximating the power of the test of H_C in this case is of no help here because it involves the value of parameters which are specified by the alternative in the power calculation, but are now unknown. A procedure to obtain an interval estimate for σ_C^2 still rougher than the approximate power calculation, would be to pretend that SS_C is $E(MS_C)\chi^2_{I-1}$, and then continue as at the end of sec. 7.2; this would be the more dubious the more unequal the $\{J_i\}$. A satisfactory interval for σ_y^2 seems difficult to construct; until such is available it is probably best only to report the estimate and its estimated standard deviation.

Appendix on Means and Variances of Certain Quadratic Forms

The following lemma will be used:

Lemma: If $S = \Sigma_\nu a_\nu(x_\nu - \bar{x}_.)^2$, where $\bar{x}_. = \Sigma_\nu a_\nu x_\nu / a$, $a = \Sigma_\nu a_\nu$, and the $\{x_\nu\}$ are independent random variables with $E(x_\nu) = 0$ and $\text{Var}(x_\nu) = \sigma_\nu^2$, then

$$E(S) = V_1 - a^{-1}V_2, \tag{7.6.19}$$

[31] If σ_e^2 were due mainly to variation of the measuring instrument rather than variation over a single tissue, a better measure might be obtained by deleting σ_e^2 in (7.6.18); however, the result should then not be denoted by σ_y^2.

and

$$\text{Var}(S) = W_2 - 2a^{-1}W_3 + a^{-2}W_4 + 2a^{-2}V_2^2, \tag{7.6.20}$$

where

$$\begin{aligned} V_m &= \sum_\nu a_\nu^m \sigma_\nu^2 && \text{for} \quad m = 1, 2, \\ W_m &= \sum_\nu a_\nu^m(\gamma_{2,\nu} + 2)\sigma_\nu^4 && \text{for} \quad m = 2, 3, \\ W_4 &= \sum_\nu a_\nu^4 \gamma_{2,\nu}\sigma_\nu^4, \end{aligned} \tag{7.6.21}$$

and $\gamma_{2,\nu} = \sigma_\nu^{-4}E(x_\nu^4) - 3$. If the $\{x_\nu\}$ are normal, then $\gamma_{2,\nu} = 0$, and so

$$W_m = 2\sum_\nu a_\nu^m \sigma_\nu^4 \text{ for } m = 2, 3; \quad W_4 = 0. \tag{7.6.22}$$

Formula (7.6.19) may be derived by taking expected values in the last member of the equality

$$S = \sum_\nu a_\nu x_\nu^2 - a\bar{x}_.^2 = \sum_\nu a_\nu x_\nu^2 - a^{-1}\sum_\nu\sum_{\nu'} a_\nu a_{\nu'} x_\nu x_{\nu'}.$$

To derive formula (7.6.20) one may start from $\text{Var}(S) = E(S^2) - [E(S)]^2$. On squaring the above expression for S one gets a sum of fourth-degree terms in the $\{x_\nu\}$, and so one needs the value of $E(x_\nu x_{\nu'} x_{\nu''} x_{\nu'''})$. This is zero except in the four cases

$$\nu = \nu' = \nu'' = \nu''', \quad \nu = \nu' \neq \nu'' = \nu''', \quad \nu = \nu'' \neq \nu' = \nu''',$$
$$\nu = \nu''' \neq \nu' = \nu'',$$

where it is $(\gamma_{2,\nu} + 3)\sigma_\nu^4$, $\sigma_\nu^2\sigma_{\nu''}^2$, $\sigma_\nu^2\sigma_{\nu'}^2$, $\sigma_\nu^2\sigma_{\nu'}^2$, respectively. Sums of the type $\Sigma\Sigma_{\nu\neq\nu'} a_\nu^m a_{\nu'}^{m'} \sigma_\nu^2 \sigma_{\nu'}^2$ are encountered, and are evaluated by adding and subtracting $\Sigma_\nu a_\nu^{m+m'}\sigma_\nu^4$ to obtain $V_m V_{m'} - \Sigma_\nu a_\nu^{m+m'}\sigma_\nu^4$. Further details are rather tedious.

From the lemma we may calculate the expected values of the SS's of Table 5.3.2 under the Ω-assumptions of this section (normality is of course irrelevant) to be

$$\begin{aligned} E(\text{SS}_C) &= \sigma_e^2(I-1) + \sigma_T^2\left(\sum_i \frac{T_i}{n_i} - \frac{T}{n}\right) + \sigma_B^2\left(\sum_i \frac{B_i}{n_i} - \frac{B}{n}\right) + \sigma_C^2\left(n - \frac{C}{n}\right), \\ E(\text{SS}_B) &= \sigma_e^2\sum_i (J_i - 1) + \sigma_T^2\left(\sum_i\sum_j \frac{T_{ij}}{n_{ij}} - \sum_i \frac{T_i}{n_i}\right) + \sigma_B^2\left(n - \sum_i \frac{B_i}{n_i}\right), \\ E(\text{SS}_T) &= \sigma_e^2\sum_i\sum_j (K_{ij} - 1) + \sigma_T^2\left(n - \sum\sum_j \frac{T_{ij}}{n_{ij}}\right), \\ E(\text{SS}_e) &= \sigma_e^2\sum_i\sum_j\sum_k (M_{ijk} - 1), \end{aligned} \tag{7.6.23}$$

where

$$T_{ij} = \sum_k M_{ijk}^2, \qquad T_i = \sum_j T_{ij}, \qquad T = \sum_i T_i,$$

$$B_i = \sum_j n_{ij}^2, \qquad B = \sum_i B_i,$$

$$C = \sum_i n_{i\cdot}^2.$$

We shall illustrate these calculations by obtaining the above formula for $E(\text{SS}_B)$: Write $\text{SS}_B = \Sigma_i \text{SS}_{B,i}$, where

(7.6.24) $$\text{SS}_{B,i} = \sum_j n_{ij}(\bar{y}_{ij\cdot\cdot} - \bar{y}_{i\cdot\cdot\cdot})^2.$$

Now

(7.6.25) $$\bar{y}_{ij\cdot\cdot} = \sum_k \sum_m y_{ijkm}/n_{ij} = \mu + c_i + b_{ij} + \bar{t}_{ij\cdot} + \bar{e}_{ij\cdot\cdot},$$

where

(7.6.26) $$\bar{t}_{ij\cdot} = \sum_k \sum_m t_{ijk}/n_{ij} = \sum_k M_{ijk} t_{ijk}/n_{ij}, \qquad \bar{e}_{ij\cdot\cdot} = \sum_k \sum_m e_{ijkm}/n_{ij};$$

and

(7.6.27) $$\bar{y}_{i\cdot\cdot\cdot} = \sum_j \sum_k \sum_m y_{ijkm}/n_i = \mu + c_i + \bar{b}_{i\cdot} + \bar{t}_{i\cdot\cdot} + \bar{e}_{i\cdot\cdot\cdot},$$

where

$$\bar{b}_{i\cdot} = \sum_j \sum_k \sum_m b_{ij}/n_i = \sum_j n_{ij} b_{ij}/n_i,$$

$$\bar{t}_{i\cdot\cdot} = \sum_j \sum_k \sum_m t_{ijk}/n_i = \sum_j n_{ij} \bar{t}_{ij\cdot}/n_i,$$

$$\bar{e}_{i\cdot\cdot\cdot} = \sum_j \sum_k \sum_m e_{ijkm}/n_i = \sum_j n_{ij} \bar{e}_{ij\cdot\cdot}/n_i.$$

Substituting (7.6.25) and (7.6.27) into (7.6.24) gives

(7.6.28) $$\text{SS}_{B,i} = \sum n_{ij}(g_{ij} - \bar{g}_{i\cdot})^2,$$

where

$$g_{ij} = b_{ij} + \bar{t}_{ij\cdot} + \bar{e}_{ij\cdot\cdot}, \qquad \bar{g}_{i\cdot} = \sum_j n_{ij} g_{ij}/n_i.$$

Under Ω the $\{g_{ij}\}$ are statistically independent with zero means and

$$\text{Var}(g_{ij}) = \sigma_B^2 + \text{Var}(\bar{t}_{ij\cdot}) + \text{Var}(\bar{e}_{ij\cdot\cdot}).$$

From (7.6.26) we see that this may be evaluated as

(7.6.29) $$\text{Var}(g_{ij}) = \sigma_B^2 + n_{ij}^{-2} T_{ij} \sigma_T^2 + n_{ij}^{-1} \sigma_e^2.$$

We apply the lemma to (7.6.28) with ν replaced by j, a_ν by n_{ij}, x_ν by g_{ij}, so $\bar{x}_.$ is to be replaced by $\bar{g}_{i.}$, a by $\Sigma_j n_{ij} = n_i$, and σ_ν^2 by (7.6.29). The result is

$$E(\text{SS}_{B,i}) = \sum_{j=1}^{J_i} n_{ij}(\sigma_B^2 + n_{ij}^{-2}T_{ij}\sigma_T^2 + n_{ij}^{-1}\sigma_e^2) - n_i^{-1}\sum_j n_{ij}^2(\sigma_B^2 + n_{ij}^{-2}T_{ij}\sigma_T^2 + n_{ij}^{-1}\sigma_e^2)$$

$$= \sigma_B^2\left(n_i - \frac{B_i}{n_i}\right) + \sigma_T^2\left(\sum_j \frac{T_{ij}}{n_{ij}} - \frac{T_i}{n_i}\right) + \sigma_e^2(J_i - 1),$$

and summing this over i we get (7.6.23).

We shall illustrate the use of the other part of the lemma by calculating Var (SS$_C$) in the special case where $K_{ij} = K$, $M_{ijk} = M$, and the observations are normal, so we may use (7.6.22). (Nonnormality does affect the variances of the SS's.) We write (7.6.13) in the form

$$S = \text{SS}_C/(KM) = \sum_i J_i(f_i - \bar{f}_.)^2,$$

where the $\{f_i\}$ are independently normal with zero means, variances given by (7.6.12), and $\bar{f}_. = \Sigma_i J_i f_i / \Sigma_i J_i$. We now apply the lemma with ν replaced by i, a_ν by J_i, x_ν by f_i, and σ_ν^2 by (7.6.12). From (7.6.21) and (7.6.22) we calculate

$$V_2 = \sum_i J_i^2(\sigma_C^2 + J_i^{-1}\sigma_g^2) = A_2\sigma_C^2 + A_1\sigma_g^2,$$

$$W_2 = 2\sum_i J_i^2(\sigma_C^4 + 2J_i^{-1}\sigma_C^2\sigma_g^2 + J_i^{-2}\sigma_g^4) = 2(A_2\sigma_C^4 + 2A_1\sigma_C^2\sigma_g^2 + I\sigma_g^4),$$

$$W_3 = 2\sum_i J_i^3(\sigma_C^4 + 2J_i^{-1}\sigma_C^2\sigma_g^2 + J_i^{-2}\sigma_g^4) = 2(A_3\sigma_C^4 + 2A_2\sigma_C^2\sigma_g^2 + A_1\sigma_g^4),$$

$$W_4 = 0.$$

Substituting these values and $a = A_1$ into (7.6.20) gives Var (S), and multiplying this by $(KM)^2$ yields (7.6.15).

PROBLEMS

7.1. In a study of routine production in a cannery each operator of apricot-cutting machines was observed for five two-minute periods. Three different sizes of fruit were being used on three different production lines (the size decreases as the size number increases). Table A summarizes the data separately for each size, the notation corresponding to that of the model equation $y_{ij} = \mu + a_i + e_{ij}$, where y_{ij} is the number of apricots per minute cut by the ith operator in the jth period that she was observed, with $J = 5$ and I shown in the table. (*a*) Calculate, separately for each size, estimates of μ, σ_A, σ_e in the forms

TABLE A

Size	I	$y_{..}$	MS_A	MS_e
2	9	53.17	59.72	1.144
3	17	52.26	68.20	2.537
4	17	47.32	78.96	4.926

$\hat{\theta} \pm \hat{\sigma}_{\hat{\theta}}$, where $\hat{\theta}$ is the point estimate and $\hat{\sigma}_{\hat{\theta}}$ its estimated standard deviation under normal theory. (*b*) Since I estimates of $\hat{\sigma}_e$ could be made with each size from the five tests on the I operators, a direct estimate of the standard deviation of $\hat{\sigma}_e$ was possible from the sample variance of these I estimates. For sizes 2, 3, 4 these direct estimates were 0.18, 0.26, 0.32. Compare these with the corresponding normal-theory estimates, and deduce how the shape of the distribution of deviations of an operator from her own average seems to differ from a normal distribution.

7.2. The analysis-of-variance table, Table B, summarizes the results of four successive experiments (factor B) with the same sample of 25 races (factor A) of the common fruitfly (*Drosophila melanogaster*), 12 females being sampled from each race for each experiment. The observations were the number of eggs laid by a female on the fourth day of laying. (*a*) Fill in the E(MS) column.

TABLE* B

Source	d.f.	MS	E(MS)
A = races	24	3,243	
B = experiments	3	46,659	
$A \times B$	72	459	
Error	1100	231	

* From Table 2 on p. 9 of "The estimation of variance components in analysis of variance" by S. L. Crump, *Biometrics Bulletin*, Vol. 2, 1946. Reproduced with the kind permission of the author and the editor.

(*b*) Test each of H_A, H_B, H_{AB} at the 0.025 level. (*c*) Calculate point estimates of the variance components σ_A^2, σ_B^2, σ_{AB}^2, σ_e^2. (*d*) Estimate the variances of these estimates. (*e*) Calculate (two-sided) 95 per cent confidence intervals for each of the variance components.

7.3. Using the MS's calculated in Problem 5.6, test the significance of differences between machines and of differences between heads within machines.

7.4. (*a*) From the mean squares calculated in Problem 5.7, estimate σ_C^2, σ_B^2, σ_e^2, and $\sigma_y^2 = \sigma_C^2 + \sigma_B^2 + \sigma_e^2$, where the subscripts C, B, e, y refer respectively to cities, boxes within cities, tissues within boxes, and individual measurements. (*b*) Estimate the variance of each of the four estimates in (*a*). (*c*) Assuming that the main purpose of the experiment was to estimate σ_y^2, make suggestions for a better allocation of measurements in similar experiments in the future.

7.5. Prove that if random variables b and u have a joint distribution then $\text{Var}(b) = E(\text{Var}(b|u)) + \text{Var}(E(b|u))$. [*Hint:* Write $E(b|u) = f(u)$, $E(b) = \mu_b$, so $E(f(u)) = \mu_b$, and take expectations in the identity $(b - \mu_b)^2 = [b - f(u)]^2 + [f(u) - \mu_b]^2 + 2[f(u) - \mu_b][b - f(u)]$. In calculating the expectations of the first and last of the three terms on the right, take first the conditional expectation, given u.]

CHAPTER 8

Mixed Models

8.1. A MIXED MODEL FOR THE TWO-WAY LAYOUT

In this chapter we shall first treat in detail the mixed model in the case of two factors, and then formulate general rules in the case of more factors for the definition of the SS's, their computation, their numbers of d.f., their E(MS)'s, and (approximate) F-tests based on them.

An example of a two-way layout in which it is appropriate to treat one of the factors as having fixed effects and the other as having random effects can be obtained by modifying the example of sec. 7.4 concerning machines and workers so that, while the workers are still regarded as a random sample from a large population, the machines are not, the interest being in the individual performance of the machines. This would be the case if some of the machines in the experiment were of different makes.

We shall use the notation of sec. 7.4 for factors (A refers to machines, B to workers), numbers of levels, and subscripts for the levels. We again permit $K = 1$. We again assume that the output of the jth worker on the kth day that he is assigned to the ith machine has the structure

$$y_{ijk} = m_{ij} + e_{ijk}, \tag{8.1.1}$$

where the "errors" $\{e_{ijk}\}$ are independently and identically distributed with zero means and variance σ_e^2, and independently of the "true" means $\{m_{ij}\}$. We shall now attempt to motivate reasonable assumptions about the distribution of the $\{m_{ij}\}$, and from these to deduce the distribution of main effects and interactions.[1]

Again labeling the workers in the population by an index v with the population distribution $\mathscr{P}_v$, we shall denote the "true" output of the worker labeled v on the ith machine by $m(i, v)$. Here v is a random quantity, corresponding to random selection of the worker according to $\mathscr{P}_v$, but i

[1] The model was introduced in this way by Scheffé (1956a); see pp. 35–36 for citations to related work of others.

is not,[2] referring to the particular machine labeled i in the experiment. The I random variables $\{m(i, v)\}$ are the components of a vector random variable $\mathbf{m} = \mathbf{m}(v)$ whose multivariate distribution is really the basic concept of the present model. The vector random variable

$$\mathbf{m} = \mathbf{m}(v) = (m(1, v), m(2, v), \cdots, m(I, v))' \tag{8.1.2}$$

is generated by the population of workers, the worker labeled v in the experiment carrying the value $\mathbf{m}(v)$ of the vector. The main effects and interactions will now be defined in terms of the random vector $\mathbf{m}$.

The vector of means $E(\mathbf{m})$ for (8.1.2) will give us the "true" means for the machines; i.e., we define the "true" mean for the ith machine to be

$$\mu_i = m(i, .), \tag{8.1.3}$$

where replacing v by a dot signifies that the mean has been taken over the population of workers, i.e., the expected value of $m(i, v)$ has been taken with respect to $\mathscr{P}_v$. The general mean is defined as the arithmetic average of (8.1.3) over the I machines,

$$\mu = \mu_. = m(., .),$$

where replacing i by a dot signifies that the arithmetic average has been taken over i. The amount by which this is exceeded by the "true" mean for the ith machine,

$$\alpha_i = \mu_i - \mu_. = m(i, .) - m(., .),$$

is called the main effect of the ith machine. The "true" mean for the worker labeled v is defined as the average of his I "true" means on the I machines, namely $m(., v)$; and the excess of this over the general mean,

$$b(v) = m(., v) - m(., .), \tag{8.1.4}$$

is called the main effect of the worker labeled v in the population. The main effect of the worker labeled v, specific to the ith machine, might be defined as $m(i, v) - m(i, .)$, and the excess of this above its average (8.1.4) over the machines,

$$c_i(v) = m(i, v) - m(i, .) - m(., v) + m(., .)$$

[2] I find it confusing to derive the mixed model, as often done, from a more general model in which i and v are both random, but the v's in the experiment are a sample of J from an infinite population, while the i's in the experiment are a sample of I from a finite population of I. It can be argued that, because of the symmetry of the MS's in the I levels, the E(MS)'s will then be the same as though the i's were not sampled, but in general the distribution theory would have to be different, since in the actual situation there is nothing corresponding to equal probability for the $I!$ permutations of the labels $\{1, \cdots, I\}$ on the machines.

is called the interaction of the ith machine and the worker labeled v in the population. We now have

$$m(i, v) = \mu + \alpha_i + b(v) + c_i(v). \tag{8.1.5}$$

From their definitions the main effects and interactions in the population are seen to satisfy the conditions

$$\sum_i \alpha_i = 0, \quad E(b(v)) = 0, \quad \sum_i c_i(v) = 0 \quad \text{for all } v,$$

$$E(c_i(v)) = 0 \quad \text{for all } i;$$

these are the analogs of the side conditions (4.1.10) for a finite set of workers as well as a finite set of machines.

The random effects $\{b(v), c_1(v), \cdots, c_I(v)\}$ are not independent; their variances and covariances are functions of the covariance matrix of the vector random variable $\mathbf{m}$: If the elements of this covariance matrix are

$$\sigma_{ii'} = \text{Cov}\,(m(i, v), m(i', v)),$$

then we may calculate from the definitions of the random effects that

$$b(v) = I^{-1}\sum_i m(i, v) - \mu,$$

$$\text{Var}\,(b(v)) = I^{-2}\sum_i\sum_{i'} \text{Cov}\,(m(i, v), m(i', v)) = I^{-2}\sum_i\sum_{i'}\sigma_{ii'},$$

$$\text{Var}\,(b(v)) = \sigma_{\cdot\cdot}. \tag{8.1.6}$$

Since $c_i(v) = m(i, v) - m(., v) - \mu_i + \mu$, the value of $\text{Cov}\,(c_i(v), c_{i'}(v))$ will not depend on the $\{\mu_i\}$, and so in its calculation we may pretend that all $\mu_i = 0$. Then its value is the expectation of

$$\begin{aligned}
&[m(i, v) - m(., v)][m(i', v) - m(., v)] \\
&\quad = m(i, v)\, m(i', v) - m(i, v)\, m(., v) - m(., v)\, m(i', v) + [m(., v)]^2 \\
&\quad = m(i, v)\, m(i', v) - I^{-1}\sum_{i''} m(i, v)\, m(i'', v) - I^{-1}\sum_{i''} m(i'', v)\, m(i', v) \\
&\qquad\qquad + I^{-2}\sum_{i''}\sum_{i'''} m(i'', v)\, m(i''', v);
\end{aligned}$$

hence

$$\text{Cov}\,(c_i(v), c_{i'}(v)) = \sigma_{ii'} - I^{-1}\sum_{i''}\sigma_{ii''} - I^{-1}\sum_{i''}\sigma_{i''i'} + I^{-2}\sum_{i''}\sum_{i'''}\sigma_{i''i'''},$$

or

$$\text{Cov}\,(c_i(v), c_{i'}(v)) = \sigma_{ii'} - \sigma_{i.} - \sigma_{.i'} + \sigma_{\cdot\cdot}. \tag{8.1.7}$$

We note that because of the symmetry of the matrix $(\sigma_{ii'})$, $\sigma_{ii'} = \sigma_{i'i}$, $\sigma_{i.} = \sigma_{.i}$. In a similar way we may calculate

$$\text{Cov}\,(b(v), c_i(v)) = \sigma_{i.} - \sigma_{..}.$$

We adopt the following definitions of the symbols σ_A^2, σ_B^2, σ_{AB}^2:

$$\sigma_A^2 = (I-1)^{-1}\sum_i \alpha_i^2, \tag{8.1.8}$$

$$\sigma_B^2 = \text{Var}\,(b(v)),$$

$$\sigma_{AB}^2 = (I-1)^{-1}\sum_i \text{Var}\,(c_i(v)); \tag{8.1.9}$$

they may be motivated by starting with the definitions above Table 4.3.1 for a finite set of workers and considering the limiting case as the number of workers becomes infinite. The quantities σ_B^2 and σ_{AB}^2 may be expressed in terms of the covariance matrix of $\mathbf{m}(v)$ as

$$\sigma_B^2 = \sigma_{..}, \tag{8.1.10}$$

$$\sigma_{AB}^2 = (I-1)^{-1}\sum_i(\sigma_{ii}-\sigma_{..}). \tag{8.1.11}$$

Formula (8.1.10) follows from (8.1.6), and (8.1.11) from (8.1.7) with $i = i'$, and $\Sigma_i(\sigma_{ii}-2\sigma_{i.}+\sigma_{..}) = \Sigma_i(\sigma_{ii}-\sigma_{..})$.

We note that $\sigma_B^2 = 0$ if and only if $b(v) = 0$ for all v, that is, if and only if the basic vector $\mathbf{m}(v)$ has a degenerate distribution satisfying $\Sigma_i m_i(v) = \text{constant} = I\mu$. Also, $\sigma_{AB}^2 = 0$ if and only if $\text{Var}\,(c_i(v)) = 0$ for all i, or $m(i, v) = m(., v)+\alpha_i$, that is, except for additive constants $\{\alpha_i\}$, the random variables $m(i, v)$ are identical (not just identically distributed). Some further insight into our definitions may be obtained by considering the highly symmetric case where the covariance matrix of $\mathbf{m}(v)$ satisfies

$$\sigma_{ii'} = \rho\sigma^2 \text{ if } i' \neq i, \quad \sigma_{ii} = \sigma^2. \tag{8.1.12}$$

Then from (8.1.10), (8.1.11), and (8.1.12)

$$\sigma_B^2 = \sigma^2 I^{-1}[1 + \rho(I-1)],$$

$$\sigma_{AB}^2 = \sigma^2(1-\rho),$$

where[3] $-(I-1)^{-1} \leqq \rho \leqq 1$. These relations are graphed in Fig. 8.1.1. *We do not recommend* that the assumption (8.1.12) ordinarily be made in applications, where there usually exists no real symmetry corresponding to it; thus, in an example of machines and men, two machines might be very similar to each other (perhaps of the same make and model), but

[3] The first inequality may be derived as a consequence of (8.1.12) and the fact that the covariance matrix $(\sigma_{ii'})$ must be positive indefinite.

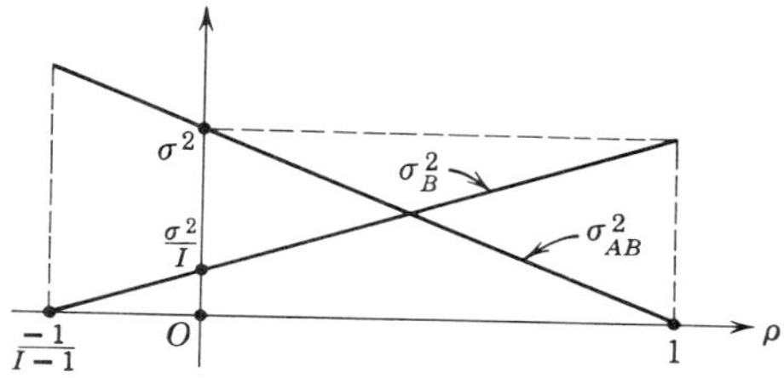

FIG. 8.1.1. Variance components σ_B^2 and σ_{AB}^2 in symmetric case (8.1.12)

very different from the other machines. A further objection is that the analog of (8.1.12) in the case of a finite set of workers, covered in sec. 4.3, would be the fulfilment of certain conditions[4] for which there seems to be nothing in most applications to justify their assumption.

If the J workers in the experiment are a random sample from $\mathscr{P}_v$ with labels $\{v_1, \cdots, v_J\}$, then the "true" mean m_{ij} in (8.1.1) is $m(i, v_j)$, and so the J vectors $(m_{1j}, m_{2j}, \cdots, m_{Ij})$ with I components, or

$$(m(1, v_j), m(2, v_j), \cdots, m(I, v_j))' \qquad (j = 1, \cdots, J),$$

are independently distributed like (8.1.2). From (8.1.5) we may also write

$$m_{ij} = \mu + \alpha_i + b_j + c_{ij},$$

where

$$b_j = b(v_j), \qquad c_{ij} = c_i(v_j),$$

and so the J vectors $(b_j, c_{1j}, \cdots, c_{Ij})'$ with $I+1$ components are independently distributed like

$$(b(v), c_1(v), \cdots, c_I(v))'.$$

Suppose we denote by $\sigma^2_{A|J\text{ workers}}$ the σ_A^2 that would be defined for the $I \times J$ layout of the I machines and those J workers actually used in the experiment, as defined by (4.3.7a) and (4.1.9), with analogous definitions of $\sigma^2_{B|J\text{ workers}}$ and $\sigma^2_{AB|J\text{ workers}}$. These three σ^2's are then random variables whose values depend on which set of J workers is sampled from the population of workers. The reader may verify that $\sigma_B^2 = 0$ implies $\sigma^2_{B|J\text{ workers}} = 0$ for all sets of J workers, and that $\sigma_{AB}^2 = 0$ has a similar implication, but that $\sigma_A^2 = 0$ does not.[5]

If we now add the normality assumption, namely that the vector random

[4] Scheffé (1956a), p. 27.

[5] A similar situation is considered below (9.1.23).

variable $\mathbf{m}(v)$ has a multivariate normal distribution and that the $\{e_{ijk}\}$ are normal, then we may express the Ω-assumptions in two equivalent ways:

$$\Omega: \begin{cases} y_{ijk} = m_{ij} + e_{ijk}, \text{ the } J \text{ vector random variables } (m_{1j}, \cdots, m_{Ij})' \\ \text{are independently } N(\boldsymbol{\mu}, \boldsymbol{\Sigma}_m), \text{ where } \boldsymbol{\mu} = (\mu_1, \cdots, \mu_I)' \text{ and} \\ \boldsymbol{\Sigma}_m = (\sigma_{ii'}), \text{ and are independent of the } \{e_{ijk}\}, \text{ which are} \\ \text{independently } N(0, \sigma_e^2), \end{cases}$$

or

$$(8.1.13)\quad \Omega: \begin{cases} y_{ijk} = \mu + \alpha_i + b_j + c_{ij} + e_{ijk}, \text{ where } \alpha_. = 0,\ c_{.j} = 0 \\ \text{for all } j, \text{ the } \{b_j\}, \{c_{ij}\}, \{e_{ijk}\} \text{ are jointly normal, the } \{e_{ijk}\} \\ \text{are independently } N(0, \sigma_e^2) \text{ and independent of the } \{b_j\} \\ \text{and } \{c_{ij}\}, \text{ which have zero means and the following} \\ \text{variances and covariances, defined in terms of an } I \times I \\ \text{covariance matrix with elements } \{\sigma_{ii'}\}: \\ \qquad \operatorname{Cov}(b_j, b_{j'}) = \delta_{jj'}\sigma_{..}, \\ \qquad \operatorname{Cov}(c_{ij}, c_{i'j'}) = \delta_{jj'}(\sigma_{ii'} - \sigma_{i.} - \sigma_{.i'} + \sigma_{..}), \\ \qquad \operatorname{Cov}(b_j, c_{ij'}) = \delta_{jj'}(\sigma_{i.} - \sigma_{..}). \end{cases}$$

The only restriction on the $\{\sigma_{ii'}\}$ is that they be elements of a symmetric positive indefinite matrix. In either statement of Ω the unknown parameters are σ_e^2, the elements $\{\sigma_{ii'}\}$ of the covariance matrix $\boldsymbol{\Sigma}_m$, and the means $\{\mu_i\}$, which are written $\{\mu+\alpha_i\}$ in the second form.

It is helpful with mixed models to use the notational convention in the model equations that fixed effects are denoted by Greek letters and random effects by Latin letters; we observe this convention in (8.1.13) and elsewhere in this chapter.

If we substitute (8.1.13) into the four SS's defined in Table 4.3.1 we get

$$(8.1.14)\qquad SS_A = JK\sum_i(\alpha_i + c_{i.} + e_{i..} - e_{...})^2,$$

$$(8.1.15)\qquad SS_B = IK\sum_j(b_j - b_. + e_{.j.} - e_{...})^2,$$

$$(8.1.16)\qquad SS_{AB} = K\sum_i\sum_j(c_{ij} - c_{i.} + e_{ij.} - e_{i..} - e_{.j.} + e_{...})^2,$$

$$(8.1.17)\qquad SS_e = \sum_i\sum_j\sum_k(e_{ijk} - e_{ij.})^2,$$

since $c_{.j} = 0$ and hence $c_{..} = 0$. These four SS's are pairwise independent except for the pair SS_B, SS_{AB}. We shall prove the independence of the pair SS_A, SS_{AB}; the independence of the other pairs may be verified similarly.

Let us write

$$SS_A = JK\sum_{i'} L_{i'}^2,$$

$$SS_{AB} = K\sum_i \sum_j L_{ij}^2,$$

where

$$L_{i'} = A_{i'} + B_{i'}, \qquad L_{ij} = A_{ij} + B_{ij},$$

$$A_{i'} = \alpha_{i'} + c_{i'.}, \qquad B_{i'} = e_{i'..} - e_{...},$$

$$A_{ij} = c_{ij} - c_{i.}, \qquad B_{ij} = e_{ij.} - e_{i..} - e_{.j.} + e_{...}.$$

Then it suffices because of the joint normality of the set $\{L_{i'}, L_{ij}\}$ to prove that $\text{Cov}\,(L_{i'}, L_{ij}) = 0$ for all i', i, j. Now, any B just defined is independent of any A because of our assumption that the set $\{e_{ijk}\}$ is independent of the set $\{m_{ij}\}$. Furthermore, $B_{i'}$ and B_{ij} are orthogonal by the now familiar argument using a fictitious fixed-effects model. Hence, it remains only to show that $\text{Cov}\,(A_{i'}, A_{ij}) = 0$:

$$\begin{aligned}\text{Cov}\,(A_{i'}, A_{ij}) &= E[c_{i'.}(c_{ij}-c_{i.})] = E[J^{-1}\sum_{j'} c_{i'j'}(c_{ij} - J^{-1}\sum_{j''} c_{ij''})] \\ &= J^{-1}\sum_{j'} E(c_{i'j'}c_{ij}) - J^{-2}\sum_{j'}\sum_{j''} E(c_{i'j'}c_{ij''}) \\ &= J^{-1}E(c_{i'}(v)\, c_i(v)) - J^{-1}E(c_{i'}(v)\, c_i(v)) = 0\end{aligned}$$

since $E(c_{ij}c_{i'j'}) = \delta_{jj'}E(c_i(v)\, c_{i'}(v))$.

By applying to (8.1.17) the argument concerning a fictitious fixed-effects case, we conclude that SS_e is $\sigma_e^2\chi^2_{IJ(K-1)}$. If we set $b_j+e_{.j.} = f_j$ in (8.1.15), we get $SS_B = IK\Sigma_j(f_j-f_.)^2$, where the $\{f_j\}$ are independently $N(0, \sigma_f^2)$ with $\sigma_f^2 = \sigma_B^2+I^{-1}K^{-1}\sigma_e^2$, and hence SS_B is $IK\sigma_f^2\chi^2_{J-1}$, or $(\sigma_e^2+IK\sigma_B^2)\chi^2_{J-1}$. It follows that

$$E(MS_e) = \sigma_e^2, \tag{8.1.18}$$

$$E(MS_B) = \sigma_e^2 + IK\sigma_B^2. \tag{8.1.19}$$

For different i the $\{c_{i.}\}$ in (8.1.14) and the $\{c_{ij}\}$ in (8.1.16) in general have unequal variances and are correlated, and this suggests that SS_A and SS_{AB} are not distributed as a constant times a noncentral (this includes central) chi-square variable, and this may be shown to be true for $I > 2$. However, if the hypothesis H_{AB}: $\sigma_{AB}^2 = 0$ is true, then all c_{ij} and $c_{i.}$ in (8.1.16) vanish, and so by the familiar argument SS_{AB} is $\sigma_e^2\chi^2_{(I-1)(J-1)}$. To calculate $E(MS_A)$ and $E(MS_{AB})$ we will use the following easily derived rules:

Lemma: If $\{x_1, \cdots, x_N\}$ are independently and identically distributed random variables with variance σ_x^2,

(8.1.20) $$\text{Var}(x_.) = N^{-1}\sigma_x^2,$$

(8.1.21) $$\text{Var}(x_n - x_.) = (1-N^{-1})\sigma_x^2,$$

(8.1.22) $$\sum_n E(x_n - x_.)^2 = (N-1)\sigma_x^2.$$

It is convenient now to define

(8.1.23) $$\hat{\alpha}_i = y_{i..} - y_{...};$$

so

$$\hat{\alpha}_i = \alpha_i + c_{i.} + e_{i..} - e_{...},$$

and hence

$$E(\hat{\alpha}_i) = \alpha_i,$$

$$\text{Var}(\hat{\alpha}_i) = \text{Var}(c_{i.}) + \text{Var}(e_{i..} - e_{...}).$$

Applying to this (8.1.20) and (8.1.21), we find

$$\text{Var}(\hat{\alpha}_i) = J^{-1}\,\text{Var}(c_i(v)) + (1-I^{-1})\,\text{Var}(e_{i..}),$$

(8.1.24) $$\text{Var}(\hat{\alpha}_i) = J^{-1}[\text{Var}(c_i(v)) + K^{-1}(1-I^{-1})\sigma_e^2].$$

Writing

$$SS_A = JK\sum_i \hat{\alpha}_i^2,$$

we may substitute (8.1.24) in

$$E(SS_A) = JK\sum_i E(\hat{\alpha}_i^2) = JK\sum_i [\text{Var}(\hat{\alpha}_i) + \alpha_i^2]$$

to get

$$E(SS_A) = K\sum_i \text{Var}(c_i(v)) + (I-1)\sigma_e^2 + JK\sum_i \alpha_i^2.$$

Using the definitions (8.1.8) and (8.1.9) we then find

(8.1.25) $$E(MS_A) = \sigma_e^2 + K\sigma_{AB}^2 + JK\sigma_A^2.$$

If we take the expectation of (8.1.16) we get

$$E(SS_{AB}) = K\sum_i\sum_j E(c_{ij} - c_{i.})^2 + K\,E\left[\sum_i\sum_j (e_{ij.} - e_{i..} - e_{.j.} + e_{...})^2\right].$$

By the familiar argument the second term on the right is $\sigma_e^2(I-1)(J-1)$. By (8.1.22), $\Sigma_j E(c_{ij} - c_{i.})^2 = (J-1)\,\text{Var}(c_i(v))$, and hence the first term is

$$K\sum_i (J-1)\,\text{Var}(c_i(v)) = K(J-1)(I-1)\sigma_{AB}^2,$$

by (8.1.9). This gives

(8.1.26) $$E(MS_{AB}) = \sigma_e^2 + K\sigma_{AB}^2.$$

The formulas we have now derived for the expected mean squares are collected in Table 8.1.1. They are as usual valid without the normality assumption. They lead to the following unbiased estimates if $K > 1$:

(8.1.27) $$\hat{\sigma}_B^2 = (IK)^{-1}(\text{MS}_B - \text{MS}_e),$$

(8.1.28) $$\hat{\sigma}_{AB}^2 = K^{-1}(\text{MS}_{AB} - \text{MS}_e),$$

(8.1.29) $$\hat{\sigma}_e^2 = \text{MS}_e.$$

TABLE 8.1.1

ANALYSIS-OF-VARIANCE TABLE

Source	d.f.	MS	E(MS)
Main effects of A (fixed)	$I-1$	MS_A	$\sigma_e^2+K\sigma_{AB}^2+JK\sigma_A^2$
Main effects of B (random)	$J-1$	MS_B	$\sigma_e^2+IK\sigma_B^2$
$A\times B$ interactions	$(I-1)(J-1)$	MS_{AB}	$\sigma_e^2+K\sigma_{AB}^2$
Error	$IJ(K-1)$	MS_e	σ_e^2

We shall now find point estimates for the remaining parameters of the model. To estimate μ_i and α_i we may use the same estimates as in the fixed-effects model, namely $y_{i..}$ and $\hat{\alpha}_i$ defined by (8.1.23); the latter was shown to be unbiased below (8.1.23), and the former can be similarly proved unbiased. We consider next the distribution of the J vectors of column means

(8.1.30) $$(y_{1j.}, y_{2j.}, \cdots, y_{Ij.})'.$$

From (8.1.1),

(8.1.31) $$y_{ij.} = m_{ij} + e_{ij.},$$

and so the J vectors (8.1.30) are independently and identically distributed like a vector random variable which we shall denote by $\mathbf{z}$, which is evidently normal, and for which we shall now calculate $E(\mathbf{z})$ and $\boldsymbol{\Sigma}_z$. From (8.1.31),

$$E(y_{ij.}) = E(m_{ij}) = \mu_i,$$

and hence $E(\mathbf{z}) = \boldsymbol{\mu}$. Furthermore, if we denote the i,i' element of $\boldsymbol{\Sigma}_z$ by $\tau_{ii'}$,

$$\tau_{ii'} = \text{Cov}\,(y_{ij.}, y_{i'j.}) = E([(m_{ij}-\mu_i) + e_{ij.}][(m_{i'j}-\mu_{i'}) + e_{i'j.}])$$
$$= \text{Cov}\,(m_{ij}, m_{i'j}) + E(e_{ij.}e_{i'j.}),$$

(8.1.32) $$\tau_{ii'} = \sigma_{ii'} + \delta_{ii'}K^{-1}\sigma_e^2.$$

An unbiased estimate $\hat{\tau}_{ii'}$ of $\tau_{ii'}$ is the sample covariance of the ith row of cell means $\{y_{i1.}, y_{i2.}, \cdots, y_{iJ.}\}$ with the i'th row,

$$(8.1.33) \qquad \hat{\tau}_{ii'} = (J-1)^{-1} \sum_j (y_{ij.}-y_{i..})(y_{i'j.}-y_{i'..}),$$

and hence if $K > 1$ an unbiased estimate of the element $\sigma_{ii'}$ of the covariance matrix of the basic vector $\mathbf{m}$ of the model is

$$(8.1.34) \qquad \hat{\sigma}_{ii'} = \hat{\tau}_{ii'} - \delta_{ii'}K^{-1}\hat{\sigma}_e^2.$$

We remark that if we estimate σ_B^2 and σ_{AB}^2 by substituting the estimates (8.1.34) in (8.1.10) and (8.1.11) we get the same estimates we got before in (8.1.27) and (8.1.28). All the point estimates we have found remain unbiased when the normality assumption is dropped.

Confidence intervals for σ_e^2 can be based on the chi-square distribution of SS_e/σ_e^2 if $K > 1$. It is possible[6] to get confidence intervals based on the t-distribution for a particular α_i, a particular μ_i, or a particular difference $\alpha_i - \alpha_{i'}$. In applications this is apt to suffer from the same objections discussed at the end of sec. 2.3 concerning multiple confidence intervals based on t, and a multiple-comparison method is usually preferable.

If $K > 1$ the quotient of $(IK\sigma_B^2+\sigma_e^2)^{-1}MS_B$ by MS_e/σ_e^2 has the F-distribution with $J-1$ and $IJ(K-1)$ d.f., and so confidence intervals for σ_B^2/σ_e^2 are available in an obvious way, as well as tests of the hypotheses $\sigma_B^2 = 0$ or $\sigma_B^2/\sigma_e^2 \leq c$, and the power of the tests can easily be expressed in terms of the central F-distribution.

The hypothesis H_{AB}: $\sigma_{AB}^2 = 0$ may be tested with the statistic MS_{AB}/MS_e, which has under H_{AB} the F-distribution with $(I-1)(J-1)$ and $IJ(K-1)$ d.f. However, the power is not expressible in terms of the central or noncentral F-distribution, since SS_{AB} is not distributed as a constant times a chi-square variable when H_{AB} is false.[7]

Even though MS_A and MS_{AB} are statistically independent and under the hypothesis

$$H_A: \text{ all } \alpha_i = 0$$

have the same expected values, their quotient does not in general have the F-distribution under H_A. An exact test of this hypothesis can be based on Hotelling's T^2-statistic.[8]

Before developing this we remark that it is not clear at present whether in practice the use of this exact test instead of the approximate[9] F-test

[6] Scheffé (1956*a*), p. 32.
[7] The power of the test has been studied by Imhof (1958).
[8] Hotelling (1931).
[9] This approximate F-test and the approximate S- and T-methods can be proved to be exact if the symmetry condition (8.1.12) holds.

suggested by Table 8.1.1, based on referring $\text{MS}_A/\text{MS}_{AB}$ to the F-tables with $I-1$ and $(I-1)(J-1)$ d.f., is worth the extra computational labor involved. If the approximate test is used and the hypothesis is rejected we could follow it with an approximate S- or T-method of multiple comparison in which the $\{y_{i..}\}$ in an estimated contrast $\Sigma_i c_i y_{i..}$ $(\Sigma_i c_i = 0)$ are treated as though they were independent with equal variances obtained by deleting σ_A^2 from $E(\text{MS}_A)$ in Table 8.1.1 and dividing by JK.

To calculate the T^2-statistic, and, in case we find it significant, to make multiple comparisons, we construct a rectangular table with $R = I-1$ rows and J columns, the entry in the rth row and jth column being

(8.1.34a) $$d_{rj} = y_{rj.} - y_{Ij.},$$

and we compute the R means $\{d_{r.}\}$ and the $\frac{1}{2}R(R+1)$ sums of products

(8.1.35) $$a_{rr'} = \sum_j (d_{rj}-d_{r.})(d_{r'j}-d_{r'.}) = \sum_j d_{rj}d_{r'j} - Jd_{r.}d_{r'.}.$$

The J vectors

(8.1.36) $$\mathbf{d}^{(j)} = (d_{1j}, d_{2j}, \cdots, d_{Rj})'$$

are independently $N(\boldsymbol{\xi}, \boldsymbol{\Sigma}_d)$ where the rth component of $\boldsymbol{\xi}$ is

$$\xi_r = \alpha_r - \alpha_I,$$

and $\boldsymbol{\Sigma}_d$ can be shown to be nonsingular.[10] An unbiased estimate of $\boldsymbol{\Sigma}_d$ is

(8.1.37) $$\hat{\boldsymbol{\Sigma}}_d = (J-1)^{-1}\mathbf{A},$$

where the matrix $\mathbf{A} = (a_{rr'})$ is defined by (8.1.35). Denoting by $\bar{\mathbf{d}}$ the vector whose rth element is $d_{r.}$, we see from (V.5) of App. V that

(8.1.38) $$J(J-1)(\bar{\mathbf{d}}-\boldsymbol{\xi})'\mathbf{A}^{-1}(\bar{\mathbf{d}}-\boldsymbol{\xi})$$

is distributed as Hotelling's T^2, or as

(8.1.39) $$(J-1)(I-1)(J-I+1)^{-1}F_{I-1,J-I+1}.$$

It is evidently necessary to assume that $J \geqq I$, and thus, in the above example, that the number of workers in the experiment is $\geqq$ the number of machines.

From (8.1.38) and (8.1.39) we see that

(8.1.40) $$C(\bar{\mathbf{d}}-\boldsymbol{\xi})'\,\mathbf{A}^{-1}(\bar{\mathbf{d}}-\boldsymbol{\xi}),$$

where

$$C = J(J-I+1)/(I-1),$$

[10] If $\boldsymbol{\Sigma}_d$ were singular the $\{d_{1j}, \cdots, d_{Rj}\}$ would have to satisfy a linear relation.

has the F-distribution with $I-1$ and $J-I+1$ d.f. under Ω. Since this depends only on the J vectors (8.1.36) being independently and identically distributed with an R-variate normal distribution, it remains valid, and so do all the resulting statistical inferences, if the J vectors $(e_{ij.}, \cdots, e_{Ij.})'$ are independently and identically distributed in an I-variate normal distribution, as would be the case for example if in place of a common error variance σ_e^2 we have different error variances $\{\sigma_{e,i}^2\}$ for the I machines.

Under the hypothesis H_A, $\boldsymbol{\xi} = \mathbf{0}$, and so, from (8.1.40), the statistic

$$\mathfrak{F} = C\bar{\mathbf{d}}'\mathbf{A}^{-1}\bar{\mathbf{d}}$$

is $F_{I-1,J-I+1}$ under H_A. The T^2-test of H_A consists in rejecting H_A at the α level of significance if $\mathfrak{F} > F_{\alpha;I-1,J-I+1}$. To calculate the test statistic $\mathfrak{F}$ it is not actually necessary to invert the $R\times R$ matrix $\mathbf{A}$, but we may use instead the relation (V.2) of App. V, namely

$$\bar{\mathbf{d}}'\mathbf{A}^{-1}\bar{\mathbf{d}} = \frac{|\mathbf{A}+\bar{\mathbf{d}}\bar{\mathbf{d}}'|}{|\mathbf{A}|} - 1,$$

which requires evaluating only two $R\times R$ determinants.

The above form of the T^2-test appears to lack symmetry, since the Ith row plays a distinguished role. It is not difficult to show that if instead of the $\{d_{r.} = y_{r..} - y_{I..}\}$ any other basis is used for the $(I-1)$-dimensional space spanned by the differences $\{y_{i..} - y_{i'..}\}$, we would obtain the same test.[11]

The power of this test can be expressed in terms of the noncentral F-distribution. By (V.6, 7), $\mathfrak{F}$ is distributed under Ω as $F'_{I-1,J-I+1;\delta}$, where

$$\delta^2 = J\boldsymbol{\xi}'\boldsymbol{\Sigma}_d^{-1}\boldsymbol{\xi},$$

and the r,r' element of $\boldsymbol{\Sigma}_d$ is

$$\text{Cov}\,(d_{rj}, d_{r'j}) = \tau_{rr'} - \tau_{rI} - \tau_{r'I} + \tau_{II}.$$

A confidence ellipsoid for $\boldsymbol{\xi}$ is afforded by the F-distribution of the random variable (8.1.40): The probability is $1-\alpha$ that

$$C(\boldsymbol{\xi}-\bar{\mathbf{d}})'\mathbf{A}^{-1}(\boldsymbol{\xi}-\bar{\mathbf{d}}) \leqq F_{\alpha;I-1,J-I+1}. \tag{8.1.41}$$

This ellipsoid is centered at $\bar{\mathbf{d}}$. It differs in an important way from the former confidence ellipsoid of sec. 2.3, in that in the former case the shape and orientation, which depend on the constant matrix $\mathbf{B}$, were fixed, whereas in the present case the shape and orientation, which depend on

[11] A symmetric form of the T^2-statistic (and of the noncentrality parameter δ^2 below) was given by Hsu (1938c), but this form would involve more numerical calculation.

the random matrix **A**, are random. However, we can still apply the derivation used in sec. 3.5 for the S-method of multiple comparison, using the present confidence ellipsoid of random shape and orientation instead of the former one: The basis of the derivation, namely that a point lies inside an ellipsoid if and only if it lies between all pairs of parallel planes of support, is of course still correct.

Let $\psi = \Sigma_i c_i \alpha_i$ be any contrast among the $\{\alpha_i\}$, so $\Sigma_i c_i = 0$. Then also $\psi = \Sigma_i c_i \mu_i$. Its estimate may be written $\hat{\psi} = \Sigma_i c_i \hat{\alpha}_i$ or $\hat{\psi} = \Sigma_i c_i y_{i..}$. We may also write

$$\psi = \sum_1^I c_i(\alpha_i - \alpha_I) = \sum_1^{I-1} c_i \xi_i = \mathbf{h}'\boldsymbol{\xi},$$

if we let $\mathbf{h}$ denote the vector $(c_1, \cdots, c_{I-1})'$, which is subject to no restrictions. By applying the method of sec. 3.5 to the ellipsoid (8.1.41) we find that the probability is $1-\alpha$ that simultaneously for all $\mathbf{h}$

$$|\mathbf{h}'\boldsymbol{\xi} - \mathbf{h}'\bar{\mathbf{d}}| \leqq (C^{-1}F_\alpha \mathbf{h}'\mathbf{A}\mathbf{h})^{1/2}, \tag{8.1.42}$$

where F_α denotes $F_{\alpha;I-1,J-I+1}$. Now $\hat{\psi}$ may also be expressed as $\mathbf{h}'\bar{\mathbf{d}}$; hence

$$\text{Var}(\hat{\psi}) = \mathbf{h}'\boldsymbol{\Sigma}_{\bar{d}}\mathbf{h} = J^{-1}\mathbf{h}'\boldsymbol{\Sigma}_d\mathbf{h},$$

and Var $(\hat{\psi})$ has the unbiased estimate

$$\hat{\sigma}^2_{\hat{\psi}} = J^{-1}(J-1)^{-1}\mathbf{h}'\mathbf{A}\mathbf{h} \tag{8.1.43}$$

by (8.1.37). Thus (8.1.42) may be written

$$|\psi - \hat{\psi}| \leqq S\hat{\sigma}_{\hat{\psi}},$$

where $S^2 = C^{-1}F_\alpha J(J-1)$, or

$$S^2 = (I-1)(J-1)(J-I+1)^{-1}F_{\alpha;I-1,J-I+1}. \tag{8.1.44}$$

We have now proved that the probability is $1-\alpha$ that all contrasts $\psi = \Sigma_1^I c_i \alpha_i = \Sigma_1^I c_i \mu_i$, where $\Sigma_1^I c_i = 0$, simultaneously satisfy

$$\hat{\psi} - S\hat{\sigma}_{\hat{\psi}} \leqq \psi \leqq \hat{\psi} + S\hat{\sigma}_{\hat{\psi}},$$

where the constant S is given by (8.1.44). The estimate $\hat{\psi}$ may be calculated as $\Sigma_1^I c_i y_{i..}$ or as $\Sigma_1^{I-1} c_r d_{r.}$. The estimate $\hat{\sigma}^2_{\hat{\psi}}$ may be calculated in terms of the quadratic form $\mathbf{h}'\mathbf{A}\mathbf{h}$ in (8.1.43) with matrix $\mathbf{A} = (a_{rr'})$ defined by (8.1.35), and $\mathbf{h} = (c_1, \cdots, c_{I-1})'$, or by the following alternative method[12] not requiring calculation of the $\{a_{rr'}\}$ or the $\{d_{r.}\}$: Let $\hat{\psi}_j$

[12] Suggested to me by Professor J. W. Tukey.

be the estimate of ψ from the jth column of the table of cell means $\{y_{ij.}\}$,

$$\hat{\psi}_j = \sum_{i=1}^{I} c_i y_{ij.},$$

so $\hat{\psi} = \hat{\psi}_.$; then $\hat{\sigma}^2_{\hat{\psi}}$ may be calculated as

$$\hat{\sigma}^2_{\hat{\psi}} = J^{-1}(J-1)^{-1} \sum_{j=1}^{J} (\hat{\psi}_j - \hat{\psi})^2, \tag{8.1.45}$$

i.e., as J^{-1} times the sample variance of the $\{\hat{\psi}_j\}$. The proof of (8.1.45) may be made by writing

$$\hat{\psi}_j = \sum_1 c_i(y_{ij.} - y_{Ij.}) = \sum_r c_r d_{rj},$$

$$\sum_j (\hat{\psi}_j - \hat{\psi})^2 = \sum_j \left[\sum_r c_r(d_{rj} - d_{r.}) \right]^2 = \sum_j \sum_r \sum_{r'} c_r c_{r'} (d_{rj} - d_{r.})(d_{r'j} - d_{r'.}),$$

and then summing first on j in the last expression. Whenever the T^2-test rejects H_A, the S-criterion will find some contrasts significantly different from zero, and conversely; thus the S-method may be used to find which contrasts are responsible for the T^2-test rejecting H_A in the same way that we used it formerly to follow up the F-test. If the calculations for the T^2-test of H_A have already been made, then (8.1.43) is probably faster to use than (8.1.45) for calculating $\hat{\sigma}^2_{\hat{\psi}}$.

8.2. MIXED MODELS FOR HIGHER-WAY LAYOUTS

In this section we shall set up and motivate the model for two examples, one a complete four-way layout, and the other a four-way layout with both crossing and nesting of factors; from these examples the way to set up the model equation in any given case will be clear. We shall also give rules,[13] valid in the general balanced case, for the definition of the SS's, their computation, their numbers of d.f., and their E(MS)'s; these rules will be based on the model equation; and we shall illustrate them. The use of an analysis-of-variance table, constructed according to these rules, for performing F-tests of various hypotheses is explained at the end of the section.

The complete four-way layout considered will have two factors, A and B, with fixed effects and two, C and D, with random effects. Let y_{ijknq} denote the qth observation in the cell where A is at the ith level, B at the jth, C at the kth, and D at the nth. The number Q of observations per cell may be one. We shall assume that

$$y_{ijknq} = m_{ijkn} + e_{ijknq},$$

[13] These are adapted from Bennett and Franklin (1954), sec. 7.6.

where the errors $\{e_{ijknq}\}$ are independently distributed with zero means and equal variance σ_e^2, and are statistically independent of the true cell means $\{m_{ijkn}\}$.

We shall be led to a resolution of the true cell means, resulting in the following *model equation*:

$$
\begin{aligned}
y_{ijknq} = \mu &+ \alpha_i^A + \alpha_j^B + a_k^C + a_n^D + \alpha_{ij}^{AB} + a_{ik}^{AC} \\
&+ a_{in}^{AD} + a_{jk}^{BC} + a_{jn}^{BD} + a_{kn}^{CD} + a_{ijk}^{ABC} \\
&+ a_{ijn}^{ABD} + a_{ikn}^{ACD} + a_{jkn}^{BCD} + a_{ijkn}^{ABCD} + e_{ijknq}.
\end{aligned}
\tag{8.2.1}
$$

In particular applications we may be willing to assume that some of the interactions in (8.2.1) are zero. In (8.2.1), μ and the α's are constants, while the a's are random variables. A main effect or interaction is written as an α if all the factors in its superscripts are fixed-effect factors, otherwise it is written as an a.

The rules to be given later for defining the SS's, etc., necessitate writing the model equation but do not require writing the side conditions for the effects. It will be convenient to call the subscript denoting the levels of a factor briefly the "subscript of the factor." The side conditions for the complete layout may then be stated as follows: If any α or a contains the subscript of a fixed-effects factor and is summed on this subscript (i or j in the example), the sum is zero, and this is true for all values of the other subscripts (if any); thus

$$
\alpha_.^A = \alpha_{i.}^{AB} = \alpha_{.j}^{AB} = a_{.k}^{AC} = a_{.jk}^{ABC} = a_{.kn}^{ACD} = a_{.jkn}^{ABCD} = 0, \quad \text{etc.}
\tag{8.2.2}
$$

for all i, j, k, n, but in general

$$
a_.^C \neq 0, \quad a_{i.}^{AC} \neq 0, \quad a_{ij.}^{ABC} \neq 0, \quad \text{etc.}
$$

The a's all have zero means, i.e., $E(a) = 0$. The variance of an a does not depend on its subscripts of random-effects factors but in general will depend on its subscripts (if any) of fixed-effects factors, and will be denoted by σ^2 with subscripts consisting of the superscripts (capitals) on a, followed by the subscripts (lower case) on a (if any) of fixed-effects factors; thus

$$
\begin{aligned}
&\operatorname{Var}(a_k^C) = \sigma_C^2 && \text{for all } k, \\
&\operatorname{Var}(a_{ik}^{AC}) = \sigma_{AC,i}^2 && \text{for all } k, \\
&\operatorname{Var}(a_{kn}^{CD}) = \sigma_{CD}^2 && \text{for all } k, n, \\
&\operatorname{Var}(a_{ijk}^{ABC}) = \sigma_{ABC,ij}^2 && \text{for all } k, \\
&\operatorname{Var}(a_{ikn}^{ACD}) = \sigma_{ACD,i}^2 && \text{for all } k, n, \\
&\operatorname{Var}(a_{ijkn}^{ABCD}) = \sigma_{ABCD,ij}^2 && \text{for all } k, n, \quad \text{etc.}
\end{aligned}
\tag{8.2.3}
$$

We shall not set forth here the covariances of the a's. They are determined[14] by the function $m(i, j, u, v)$ to be introduced below and the distributions $\mathscr{P}_u$ and $\mathscr{P}_v$ of u and v.

In the formulas for the E(MS)'s we shall encounter only σ^2's with subscripts denoting factors but not factor levels. If its subscripts involve only fixed-effects factors, σ^2 is defined in the familiar way,

$$\sigma_A^2 = (I-1)^{-1} \sum_i (\alpha_i^A)^2, \quad \text{etc.,}$$

$$\sigma_{AB}^2 = (I-1)^{-1}(J-1)^{-1} \sum_i \sum_j (\alpha_{ij}^{AB})^2.$$

If only random-effects factors are involved, the σ^2 has already been defined; thus σ_C^2, σ_{CD}^2 are given in (8.2.3). If both kinds of factors are involved, the σ^2 of the kind defined in (8.2.3) depends on the subscripts of the fixed-effects factors, and for each of these we sum on the subscript and divide by one less than its upper limit; thus

$$\sigma_{AC}^2 = (I-1)^{-1} \sum_i \sigma_{AC,i}^2,$$

$$\sigma_{ABC}^2 = (I-1)^{-1}(J-1)^{-1} \sum_i \sum_j \sigma_{ABC,ij}^2,$$

$$\sigma_{ACD}^2 = (I-1)^{-1} \sum_i \sigma_{ACD,i}^2,$$

$$\sigma_{ABCD}^2 = (I-1)^{-1}(J-1)^{-1} \sum_i \sum_j \sigma_{ABCD,ij}^2, \quad \text{etc.}$$

As our second example let us consider the four-factor experiment mentioned in connection with (5.3.10): Here factor A corresponds to I anneals, and factor P corresponds to N pickling solutions, these two factors being crossed. Within each anneal–pickle combination J coils are nested, the J coils corresponding to factor C. Factor L corresponds to K locations completely crossed with A, P, and C. In this experiment the coils are sampled from a large population, and so C is treated as a random-effects factor. In order to have another random-effects factor in our illustration let us suppose that the pickling solutions are not of different types but that a random variation is suspected from batch to batch of pickling solutions because of some uncontrolled causes, so that P is treated as a random-effects factor. However, we are interested individually in the different methods of annealing and the different locations, and so A and L are treated as fixed-effects factors.

[14] They are linear functions of the elements of three covariance matrices, namely, the covariance matrices of the IJ random variables $\{m(i, j, u, v)\}$, of the IJ variables $\{m(i, j, u, .)\}$, and the IJ variables $\{m(i, j, ., v)\}$.

We may formally obtain the model equation (to be motivated below) by replacing by a's those α's in (5.3.10) which have random-effects factors (C and P in the present case) among their superscripts, giving

$$(8.2.4) \qquad y_{ijknq} = \mu + \alpha_i^A + a_n^P + a_{inj}^C + \alpha_k^L + a_{in}^{AP} + \alpha_{ik}^{AL} + a_{nk}^{PL} + a_{injk}^{CL} + a_{ink}^{APL} + e_{ijknq}.$$

Since writing the correct model equation, like (8.2.4), is the most important step in applying the rules below, let us here review how the terms are formed. The level of a nested factor is for this purpose indicated by two or more subscripts. In the present case J coils are nested within each anneal–pickle combination, and so a coil is identified by three subscripts, the jth coil within the i,n anneal–pickle combination by inj. In (8.2.4) the main effects of the factors are written with all the subscripts necessary to identify the level of the factor. We then consider all two-factor interactions, the subscripts indicating the levels of both factors. For the $A \times C$ interaction the subscripts would thus be i and inj. But an effect with subscripts inj has already been included, namely a_{inj}^C, and so we do not include an $A \times C$ interaction term. It would be reasonable to write the term included either as a_{inj}^C or a_{inj}^{AC}, or still in other ways; all the effects with subscripts inj are confounded in this type of experiment, and we collect them into a single term. This will become clearer when the effects are defined below in motivating the model. Similarly we decide to omit from (8.2.4) terms with superscripts PC, APC, ACL, PCL, $APCL$.

We now consider the side conditions. We recall that even when all factors were treated as fixed-effects factors in (5.3.10), not all sums of terms obtained by summing over a subscript were zero: No sum over i or n was zero for a term with subscript j. If we call j the "subscript of the nested factor" (*its level* being the injth), we may state the following rule: Sums of terms containing the subscript of a nested factor are not zero when taken over subscripts of factors within which it is nested. (In the terminology to be introduced with the rules below, we may say that sums over "dead" subscripts are not zero.) Besides this we will now obtain no side conditions[15] in which the sum is over the subscript of a random-effects factor (j or n in this case). This leaves as the only side conditions

$$(8.2.5) \quad \alpha_{\cdot}^A = \alpha_{\cdot}^L = a_{\cdot n}^{AP} = \alpha_{i\cdot}^{AL} = \alpha_{\cdot k}^{AL} = a_{n\cdot}^{PL} = a_{inj\cdot}^{CL} = a_{in\cdot}^{APL} = a_{\cdot nk}^{APL} = 0$$

for all i, j, k, n.

The a's in (8.2.4) all have zero means. The variance of an a does not

[15] The analogs of these conditions hold in the populations which are sampled but not for the terms obtained in the experiment.

depend on its subscripts of random-effects factors, but will in general depend on its subscripts of fixed-effects factors; thus

(8.2.6)
$$\begin{aligned} \operatorname{Var}(a_n^P) &= \sigma_P^2 && \text{for all } n, \\ \operatorname{Var}(a_{inj}^C) &= \sigma_{C,i}^2 && \text{for all } n, j, \\ \operatorname{Var}(a_{injk}^{CL}) &= \sigma_{CL,ik}^2 && \text{for all } n, j, \quad \text{etc.} \end{aligned}$$

The E(MS)'s involve only σ^2's without subscripts for factor levels. If the σ^2 defined in (8.2.6) has such subscripts, they are subscripts of fixed-effects factors, and we get rid of them as in the definitions of σ_C^2 and σ_{CL}^2 among the following:

$$\begin{aligned} \sigma_A^2 &= (I-1)^{-1}\sum(\alpha_i^A)^2, \\ \sigma_C^2 &= I^{-1}\sum_i \sigma_{C,i}^2, \\ \sigma_{AL}^2 &= (I-1)^{-1}(K-1)^{-1}\sum_i\sum_k(\alpha_{ik}^{AL})^2, \\ \sigma_{CL}^2 &= I^{-1}(K-1)^{-1}\sum_i\sum_k \sigma_{CL,ik}^2, \text{ etc.} \end{aligned}$$

The coefficients of the above sums are products of numerical factors corresponding to the indices of summation, the numerical factors being given by the following rule: Suppose r is the index of summation, has range 1 to R, and is the subscript of factor F. Then if F is present in the subscript of σ^2, the numerical factor is $(R-1)^{-1}$, otherwise, R^{-1}. Later we shall see that the resulting coefficient is the reciprocal of the number of d.f. of the SS with the same subscripts as σ^2.

As usual we assume the errors $\{e_{ijknq}\}$ in (8.2.4) to be independent with zero means, equal variances σ_e^2, and independent of all a's. We shall not list the covariances of the a's; they are determined by the function $m(i, u, v, k)$ introduced below and the distribution $\mathscr{P}_u$ and $\mathscr{P}_v$ of u and v.

The reader interested only in rules for performing approximate F-tests for the usual hypotheses may skip from here to the subsection headed "Definition and calculation of SS's, numbers of d.f."

We return now to the first example: We are led to the above model for the complete four-way layout by imagining that the K levels of C in the experiment are sampled from a population of levels, the N levels of D are independently sampled from another population, but the levels of A and B are not sampled. If the population levels of C and D are respectively indexed by labels u and v, we denote by $m(i, j, u, v)$ the true response for the treatment combination consisting of the ith level of A, the jth of B, the level of C labeled u in the population, and the level of D labeled v in the population. The labels u and v are independent random quantities with probability distributions $\mathscr{P}_u$ and $\mathscr{P}_v$. The true responses $m(i, j, u, v)$ are IJ jointly distributed random variables.[16]

If we define

$$
\begin{aligned}
\mu &= m(.,.,.,.),\\
\alpha_i^A &= m(i,.,.,.) - m(.,.,.,.),\\
a^C(u) &= m(.,.,u,.) - m(.,.,.,.),\\
\alpha_{ij}^{AB} &= m(i,j,.,.) - m(i,.,.,.) - m(.,j,.,.) + m(.,.,.,.),\\
a_i^{AC}(u) &= m(i,.,u,.) - m(i,.,.,.) - m(.,.,u,.) + m(.,.,.,.),\\
a^{CD}(u,v) &= m(.,.,u,v) - m(.,.,u,.) - m(.,.,.,v) + m(.,.,.,.),\\
a_{ij}^{ABC}(u) &= m(i,j,u,.) - m(i,j,.,.) - m(i,.,u,.) - m(.,j,u,.)\\
&\quad + m(i,.,.,.) + m(.,j,.,.) + m(.,.,u,.)\\
&\quad - m(.,.,.,.), \quad \text{etc.,}
\end{aligned}
\tag{8.2.7}
$$

where replacing i or j by a dot in $m(i, j, u, v)$ means that we have averaged over i or j from 1 to I or J, respectively, and replacing u or v by a dot means that we have taken the expectation over u or v with respect to the distribution $\mathscr{P}_u$ or $\mathscr{P}_v$, then we have

$$
\begin{aligned}
m(i,j,u,v) &= \mu + \alpha_i^A + \alpha_j^B + a^C(u) + a^D(v) + \alpha_{ij}^{AB} + a_i^{AC}(u)\\
&\quad + a_i^{AD}(v) + a_j^{BC}(u) + a_j^{BD}(v) + a^{CD}(u,v)\\
&\quad + a_{ij}^{ABC}(u) + a_{ij}^{ABD}(v) + a_i^{ACD}(u,v) + a_j^{BCD}(u,v)\\
&\quad + a_{ij}^{ABCD}(u,v).
\end{aligned}
\tag{8.2.8}
$$

From the definition of the terms on the right of (8.2.8), it follows that the expected value of each term written as an a is zero, and that if any term containing i is summed on i, the sum is zero for all values of j, u, v, if any of these appear in the term, and similarly for summing on j; thus

$$
\begin{aligned}
\alpha_.^A &= \alpha_.^B = \alpha_{.j}^{AB} = \alpha_{i.}^{AB} = a_.^{AC}(u) = a_.^{AD}(v) = a_.^{BC}(u) = a_.^{BD}(v)\\
&= a_{.j}^{ABC}(u) = a_{i.}^{ABC}(u) = a_{.j}^{ABD}(v) = a_{i.}^{ABD}(v)\\
&= a_.^{ACD}(u,v) = a_.^{BCD}(u,v) = a_{.j}^{ABCD}(u,v)\\
&= a_{i.}^{ABCD}(u,v) = 0
\end{aligned}
\tag{8.2.9}
$$

for all i, j, u, v.

The K levels of C and the N levels of D employed in the experiment are regarded as independent random samples $\{u_1, \cdots, u_K\}$ and $\{v_1, \cdots, v_N\}$ from the corresponding populations of levels, that is, $\{u_1, \cdots, u_K\}$ and

[16] Assumed to have finite variances. Likewise in the other four-way layout considered.

$\{v_1, \cdots, v_N\}$ are independently distributed, the $\{u_k\}$ according to $\mathscr{P}_u$ and the $\{v_n\}$ according to $\mathscr{P}_v$. Then the true value m_{ijkn} for the observation y_{ijknq} is

$$m_{ijkn} = m(i, j, u_k, v_n).$$

From (8.2.8) now follows the model equation (8.2.1), where

$$\text{(8.2.10)} \qquad \begin{aligned} a_k^C = a^C(u_k), \qquad a_n^D = a^D(v_n), \qquad a_{ik}^{AC} = a_i^{AC}(u_k), \\ a_{ijkn}^{ABCD} = a_{ij}^{ABCD}(u_k, v_n), \quad \text{etc.,} \end{aligned}$$

and hence the a's have zero means. The side conditions (8.2.2) follow from (8.2.9). The equality of variances stated in (8.2.3) for various a's is a consequence of the definition of these a's in (8.2.10). The covariances of the a's could be calculated from the definitions (8.2.10) and (8.2.7); it is evident, for example, that a_{ik}^{AC} is independent of $a_{i'jk'n}^{ABCD}$ if $k \neq k'$.

To motivate the model stated in connection with (8.2.4) for the second example, with four factors A, P, C, L, we proceed by the following steps: We imagine an infinite population of pickling solutions from which the N pickles in the experiment are sampled, the pickles in the population being labeled by u, where u has the distribution $\mathscr{P}_u$, and an infinite population of coils from which the INJ coils in the experiment are sampled, the coils in the population being labeled by v with distribution $\mathscr{P}_v$. If the coil labeled v were subjected to the ith anneal and the pickle labeled u, then the true response at the kth location on the coil is denoted by $m(i, u, v, k)$.

It is enlightening to define all possible interactions in the populations, even though they are not all estimable from the kind of experiment considered, and to see how they contribute to the terms in the model equation (8.2.4). We define the 2^4 kinds of population effects (1 general mean, 4 kinds of main effects, 6 kinds of two-factor interactions, etc.) as in (8.2.7), thus

$$\text{(8.2.11)} \qquad \begin{aligned} \mu &= m(., ., ., .), \\ \alpha_i^A &= m(i, ., ., .) - m(., ., ., .), \\ a^P(u) &= m(., u, ., .) - m(., ., ., .), \\ a_i^{AP}(u) &= m(i, u, ., .) - m(i, ., ., .) - m(., u, ., .) \\ &\quad + m(., ., ., .), \\ a_i^{APC}(u, v) &= m(i, u, v, .) - m(i, u, ., .) - m(i, ., v, .) \\ &\quad - m(., u, v, .) + m(i, ., ., .) + m(., u, ., .) \\ &\quad + m(., ., v, .) - m(., ., ., .), \quad \text{etc.} \end{aligned}$$

From these definitions it follows that the expected value of any a is zero, that if any α or a is summed on i or k the sum is zero, and that

$$(8.2.12)\quad \begin{aligned} m(i, u, v, k) = \mu &+ \alpha_i^A + a^P(u) + a^C(v) + \alpha_k^L + a_i^{AP}(u) \\ &+ a_i^{AC}(v) + \alpha_{ik}^{AL} + a^{PC}(u, v) + a_k^{PL}(u) \\ &+ a_k^{CL}(v) + a_i^{APC}(u, v) + a_{ik}^{APL}(u) \\ &+ a_{ik}^{ACL}(v) + a_k^{PCL}(u, v) + a_{ik}^{APCL}(u, v). \end{aligned}$$

Now let $\{u_1, \cdots, u_N\}$ be the labels of the N pickles employed in the experiment, so that $\{u_n\}$ are independently distributed according to $\mathscr{P}_u$. Let v_{inj} be the label of the jth coil put on the ith anneal and nth pickle in the experiment, so that the INJ quantities $\{v_{inj}\}$ are independently distributed according to $\mathscr{P}_v$. If the mean value for the observation y_{ijknq} is denoted by m_{injk}, so that

$$(8.2.13)\quad y_{ijknq} = m_{injk} + e_{ijknq},$$

then m_{injk} is equal to the expression obtained by replacing u by u_n and v by v_{inj} in (8.2.12). If we substitute this expression for m_{injk} in (8.2.13) and define

$$(8.2.14)\quad \begin{aligned} a_n^P &= a^P(u_n), \\ a_{in}^{AP} &= a_i^{AP}(u_n), \\ a_{nk}^{PL} &= a_k^{PL}(u_n), \\ a_{ink}^{APL} &= a_{ik}^{APL}(u_n), \\ a_{inj}^C &= a^C(v_{inj}) + a_i^{AC}(v_{inj}) + a^{PC}(u_n, v_{inj}) + a_i^{APC}(u_n, v_{inj}), \\ a_{injk}^{CL} &= a_k^{CL}(v_{inj}) + a_{ik}^{ACL}(v_{inj}) + a_k^{PCL}(u_n, v_{inj}) + a_{ik}^{APCL}(u_n, v_{inj}), \end{aligned}$$

we get (8.2.4). It is clear from the definitions (8.2.14) that all a's have expected value zero. From the last two of the equations (8.2.14) and the definitions (8.2.11), we see that

$$(8.2.15)\quad \begin{aligned} a_{inj}^C &= m(i, u_n, v_{inj}, .) - m(i, u_n, ., .), \\ a_{injk}^{CL} &= m(i, u_n, v_{inj}, k) - m(i, u_n, v_{inj}, .) - m(i, u_n, ., k) \\ &\quad + m(i, u_n, ., .). \end{aligned}$$

From this expression for a_{inj}^C it is evident that for each i the JN effects a_{inj}^C are identically distributed like $m(i, u, v, .) - m(i, u, ., .)$. We denote their common variance by $\sigma_{C,i}^2$. Similarly the other statements (8.2.6) about the equality of certain variances may be verified. The side conditions (8.2.5) follow from (8.2.14) and (8.2.11), or from (8.2.15) and (8.2.11).

Definition and Calculation of SS's, Numbers of d.f.

We consider now mixed models with any number of factors. All our rules will be based on the model equation for the problem at hand. We have already indicated how this is found. If one is willing to assume all interactions of a certain kind to be zero—for example, all $A \times C$ interactions in (8.2.1), or all $A \times P \times L$ interactions in (8.2.4)—the corresponding term may be omitted from the model equation, the corresponding SS is then not calculated, and the SS for error is obtained by subtracting from the total SS about the grand mean the SS's calculated, one for each term in the model equation except μ. Where the σ^2 with the corresponding subscript appears in the E(MS) formulas—σ^2_{AC} or σ^2_{APL} in the above examples—it is deleted. Alternatively, we may put *all* the terms into the model equation, as indicated above, compute all the SS's, pool into the error SS the SS's for the interactions assumed to be zero, and where the corresponding σ^2's appear in the E(MS)'s put them equal to zero. In either case there will be a SS computed for each term except μ in the model equation adopted.

We remind the reader that when we speak of the "subscript of a factor" which is nested we mean the single symbol identifying the level of the factor within the nesting and not the complete set of symbols giving the level of the factor. (The subscript of C in (8.2.4) is j; its level is inj.)

The definition and calculation of a SS, its number of d.f., and its E(MS) may all be determined from the notation for the corresponding term in the model equations, which we shall call the *key term*. Thus for the CL interactions in the situation for which we formulated the model equation (8.2.4), the key term is a^{CL}_{injk}. It is convenient to call the subscripts live, dead, or absent in calculations from a particular key term: The *live* subscripts are those of the factors in the superscript of the key term, the *dead* subscripts are the remaining ones in the subscript (if any), the *absent* subscripts are those not appearing in the key term (but appearing on the observation on the left side of the model equation). The limits of the subscripts are likewise called the live limits, dead limits, and absent limits. The live subscripts and limits are seen to be those of the factors involved in the interaction or main effect, the dead ones to be those of factors within which these are nested, and the absent ones to be those of the other factors. In the example where the key term is a^{CL}_{injk} and the observation is y_{ijknq}, the live subscripts are j and k (those of the superscripts C and L), the dead subscripts are i and n, the absent one is q; the live limits are J and K, the dead limits are I and N, the absent one is Q. For the P main effects the key term is a^P_n, the live subscript is n, there are no dead ones, and the absent ones are i, j, k, and q.

The rules for SS's and their numbers of d.f. can be expressed in terms of the following formalism: For the key term corresponding to the SS write a symbolic product of the following factors: For each of the dead subscripts (if any) the factor is that subscript, and for each of the live subscripts the factor is that subscript minus 1; and expand this product. We call it a symbolic product because in this operation no subscript is to be replaced by a numerical value it assumes. Thus for the $C\times L$ interactions in the above example, with the key term a_{injk}^{CL}, the symbolic product is

$$in(j-1)(k-1) = injk - inj - ink + in. \tag{8.2.16}$$

In the definition of the SS the quantity squared contains a term corresponding to each term in the expanded symbolic product: The term is the symbol $\pm y$ followed by subscripts consisting of the corresponding term in the product and enough dots to bring the number of subscripts including dots up to the number of subscripts on an observation, unless the term in the symbolic product is 1, in which case the corresponding term is $\pm y$ followed by dots alone. The $\pm$ sign is the same as the sign before the corresponding term in the product. In the present example, the symbolic product (8.2.16) tells us that the quantity squared will be

$$y_{ijkn.} - y_{ij.n.} - y_{i.kn.} + y_{i..n.}.$$

This quantity is squared, summed on all the subscripts present on the key term, and multiplied by the absent limits to form the SS. In the present example this gives

$$\text{SS}_{CL} = Q\sum_i\sum_j\sum_k\sum_n(y_{ijkn.} - y_{ij.n.} - y_{i.kn.} + y_{i..n.})^2.$$

A form of the SS suitable for numerical computation may be obtained in a similar way from the symbolic product, except that the $\pm$ signs are attached to sums of y^2's with subscripts formed by the previous rule, and each sum is taken only over the subscripts present on the y^2 being summed, and then multiplied by the limits of the subscripts which have been replaced by dots. Thus in the above example the resulting form is

$$\text{SS}_{CL} = Q\sum_i\sum_j\sum_k\sum_n y_{ijkn.}^2 - KQ\sum_i\sum_j\sum_n y_{ij.n.}^2 - JQ\sum_i\sum_k\sum_n y_{i.kn.}^2 + JKQ\sum_i\sum_n y_{i..n.}^2.$$

The same sums will generally appear in the computing forms of several of the SS's.

The symbolic product associated with a SS through its key term also gives us its number of d.f. The number of d.f. is given by the symbolic product when the subscripts in it are replaced by their limits. Thus in the above example where the symbolic product is (8.2.16) the number of d.f. is $IN(J-1)(K-1)$.

We have already mentioned that the error SS is obtained by subtraction from the total SS about the grand mean or, if there is more than one observation per cell, as the SS of the observations about the cell means. (In the latter case we subsequently pool into the error SS the SS's for all interactions assumed to be zero.) For the model equation (8.2.4), the error SS is thus

$$\sum_i \sum_j \sum_k \sum_n \sum_q (y_{ijknq} - y_{ijkn.})^2$$

if $Q > 1$. We might regard this as being related to a symbolic product

$$ijkn(q-1) = ijknq - ijkn, \tag{8.2.17}$$

which would correspond to a key term written a_{ijknq}^E instead of e_{ijknq}, where E denotes a "factor" that is "error" or "replication," and q is its subscript. If the error SS is formed by subtraction, its number of d.f. is, likewise; if it is formed from the SS within cells its number of d.f. is the number of cells times one less than the number of observations per cell—in the above example, $IJKN(Q-1)$, corresponding to the symbolic product (8.2.17).

Expected Mean Squares

To form the E(MS) column of the analysis-of-variance table we construct an auxiliary table,[17] the column headings of which are the terms in the model equation, except for the general mean and the error term, and the row headings are the subscripts used. Thus for the model (8.2.4), there are in the auxiliary Table 8.2.1 nine columns headed α_i^A, a_n^P, · · ·, a_{ink}^{APL}, and five rows headed i, j, k, n, q.

The table may be rapidly filled in as follows: First, we recall which are the fixed-effects factors and their subscripts, perhaps listing them—in the present example, A and i, L and k. We start filling the entries in partially by columns. For each column heading we ask if any of the superscripts are for fixed-effects factors; if so, we put 0 in the row for the corresponding subscripts (these are the live subscripts for fixed-effects factors); for all other subscripts (if any) appearing in the column heading we put 1 in the

[17] This method is given by Bennett and Franklin (1954), p. 414, with rows and columns interchanged and for "Model III" which includes as limiting cases all the models considered here. Although the method of Bennett and Franklin is correct their derivation is not. An indication of the derivation is given in Wilk and Kempthorne (1955). An alternative formulation of the rules, and indication of their proof, is given by Cornfield and Tukey (1956), p. 932.

row for the subscript. Thus, in the column headed a^{CL}_{injk}, the superscript L refers to a fixed-effects factor, its subscript is k, and so in the row for k we put 0, while in the rows for the other subscripts appearing, namely i, n, j, we put 1. After all the columns have been treated this way we complete the table by rows by writing the limit of the row heading where-ever there is no entry.

TABLE 8.2.1

AUXILIARY TABLE FOR CALCULATING COEFFICIENTS IN E(MS)'S FOR MODEL (8.2.4)

	α^A_i	a^P_n	a^C_{inj}	α^L_k	a^{AP}_{in}	α^{AL}_{ik}	a^{PL}_{nk}	a^{CL}_{injk}	a^{APL}_{ink}
i	0	I	1	I	0	0	I	1	0
j	J	J	1	J	J	J	J	1	J
k	K	K	K	0	K	0	0	0	0
n	N	1	1	N	1	N	1	1	1
q	Q	Q	Q	Q	Q	Q	Q	Q	Q

To each column of the auxiliary table there corresponds a σ^2, namely the one whose subscripts are the superscripts of the column heading; thus to the column headed by a^{AP}_{in} there corresponds σ^2_{AP}. For any SS the E(MS) is a linear combination of σ^2_e, with coefficient 1, and the σ^2's corresponding to the columns whose headings contain all the subscripts of the key term for the SS, with coefficients (some of which may be zero) formed from those columns by the method to be explained below. In our example the key term for SS_P is a^P_n, hence $E(MS_P)$ would involve besides σ^2_e also σ^2_P, σ^2_C, σ^2_{AP}, σ^2_{PL}, σ^2_{CL}, and σ^2_{APL}; $E(MS_C)$ would involve σ^2_C, σ^2_{CL}, and σ^2_e; $E(MS_{AL})$ would involve σ^2_{AL}, σ^2_{CL}, σ^2_{APL}, and σ^2_e.

The coefficients of the σ^2's other than σ^2_e in an E(MS) are formed from the corresponding columns by taking the product of those entries in the column which fall in the rows for the absent subscripts for the SS whose E(MS) is being calculated. In other words, for this SS consider the column heading which is its key term, and imagine deleting[18] the rows corresponding to the subscripts present on this key term; then the coefficient of each σ^2 is the product of the remaining entries in the corresponding column. Continuing the above example, to find the coefficients of σ^2_P, σ^2_C, σ^2_{AP}, σ^2_{PL}, σ^2_{CL}, and σ^2_{APL} in $E(MS_P)$, we note that the subscript

[18] In practice, this is facilitated by covering the lines with pencils or with paper strips cut to a width that will cover one line.

TABLE 8.2.2

ANALYSIS-OF-VARIANCE TABLE FOR THE MODEL WITH EQUATION (8.2.4)

Source	Definition of SS	Computation of SS	d.f.	$\dot{E}$(MS)
A	$JKNQ\sum_i(y_{i....}-y_{.....})^2$	$JKNQ\sum_i y^2_{i....}-IJKNQy^2_{.....}$	$I-1$	$JKNQ\sigma^2_A+KQ\sigma^2_C$ $+JKQ\sigma^2_{AP}+\sigma^2_e$
P	$IJKQ\sum_n(y_{...n.}-y_{.....})^2$	$IJKQ\sum_n y^2_{...n.}-IJKNQy^2_{.....}$	$N-1$	$IJKQ\sigma^2_P+KQ\sigma^2_C$ $+\sigma^2_e$
C	$KQ\sum_i\sum_n\sum_j(y_{ij.n.}-y_{i..n.})^2$	$KQ\sum_i\sum_n\sum_j y^2_{ij.n.}-JKQ\sum_i\sum_n y^2_{i..n.}$	$IN(J-1)$	$KQ\sigma^2_C+\sigma^2_e$
L	$IJNQ\sum_k(y_{..k..}-y_{.....})^2$	$IJNQ\sum_k y^2_{..k..}-IJKNQy^2_{.....}$	$K-1$	$IJNQ\sigma^2_L+IJQ\sigma^2_{PL}$ $+Q\sigma^2_{CL}+\sigma^2_e$
$A\times P$	$JKQ\sum_i\sum_n(y_{i..n.}-y_{i....}$ $-y_{...n.}+y_{.....})^2$	$JKQ\sum_i\sum_n y^2_{i..n.}-JKNQ\sum_i y^2_{i....}$ $-IJKQ\sum_n y^2_{...n.}+IJKNQy^2_{.....}$	$(I-1)(N-1)$	$JKQ\sigma^2_{AP}+KQ\sigma^2_C$ $+\sigma^2_e$
$A\times L$	$JNQ\sum_i\sum_k(y_{i.k..}-y_{i....}$ $-y_{..k..}+y_{.....})^2$	$JNQ\sum_i\sum_k y^2_{i.k..}-JKNQ\sum_i y^2_{i....}$ $-IJNQ\sum_k y^2_{..k..}+IJKNQy^2_{.....}$	$(I-1)(K-1)$	$JNQ\sigma^2_{AL}+Q\sigma^2_{CL}$ $+JQ\sigma^2_{APL}+\sigma^2_e$

$P\times L$	$IJQ\sum_n\sum_k(y_{..kn.}-y_{..k..}$ $-y_{...n.}+y_{.....})^2$	$IJQ\sum_n\sum_k y^2_{..kn.}-IJNQ\sum_k y^2_{..k..}$ $-IJKQ\sum_n y^2_{...n.}+IJKNQy^2_{.....}$	$(N-1)(K-1)$	$IJQ\sigma^2_{PL}+Q\sigma^2_{CL}$ $+\sigma^2_e$
$C\times L$	$Q\sum_i\sum_n\sum_j\sum_k(y_{ijkn.}-y_{ij.n.}$ $-y_{i.kn.}+y_{i..n.})^2$	$Q\sum_i\sum_n\sum_j\sum_k y^2_{ijkn.}-KQ\sum_i\sum_j\sum_n y^2_{ij.n.}$ $-JQ\sum_i\sum_k\sum_n y^2_{i.kn.}+JKQ\sum_i\sum_n y^2_{i..n.}$	$IN(J-1)(K-1)$	$Q\sigma^2_{CL}+\sigma^2_e$
$A\times P\times L$	$JQ\sum_i\sum_n\sum_k(y_{i.kn.}-y_{i.k..}$ $-y_{i..n.}-y_{..kn.}+y_{i....}$ $+y_{..k..}+y_{...n.}$ $-y_{.....})^2$	$JQ\sum_i\sum_n\sum_k y^2_{i.kn.}-JNQ\sum_i\sum_k y^2_{i.k..}$ $-JKQ\sum_i\sum_n y^2_{i..n.}-IJQ\sum_k\sum_n y^2_{..kn.}$ $+JKNQ\sum_i y^2_{i....}+IJNQ\sum_k y^2_{..k..}$ $+IJKQ\sum_n y^2_{...n.}-IJKNQy^2_{.....}$	$(I-1)(N-1)(K-1)$	$JQ\sigma^2_{APL}+Q\sigma^2_{CL}$ $+\sigma^2_e$
Error	$\sum_i\sum_j\sum_k\sum_n\sum_q(y_{ijknq}-y_{ijkn.})^2$	$\sum_i\sum_j\sum_k\sum_n\sum_q y^2_{ijknq}-Q\sum_i\sum_j\sum_k\sum_n y^2_{ijkn.}$	$IJKN(Q-1)$	σ^2_e
"Total"	$\sum_i\sum_j\sum_k\sum_n\sum_q(y_{ijknq}-y_{.....})^2$	$\sum_i\sum_j\sum_k\sum_n\sum_q y^2_{ijknq}-IJKNQy^2_{.....}$	$IJKNQ-1$	—

of the key term a_n^P is n, we imagine deleting the row for the subscript n and multiplying the entries in the remaining rows, and we find in the column corresponding to σ_P^2 the product $IJKQ$, which is thus the coefficient of σ_P^2; in the column corresponding to σ_C^2 the product KQ, which is the coefficient of σ_C^2; the coefficient $0JKQ = 0$ for σ_{AP}^2, etc., giving the result in Table 8.2.2. For $E(\text{MS}_C)$ we imagine deleting the rows for i, n, j; for $E(\text{MS}_{AP})$ the rows for i and n, etc. Of course $E(\text{MS}_e) = \sigma_e^2$, always.

The reader is now urged to write, without looking back at (8.2.4), the model equation for the situation with four factors A, P, C, L that we have been considering, with subscripts i, n, j, k, respectively, and q for error, C nested within $A \times P$, L crossed with all, A and L being treated as fixed-effects factors, and P and C as random. After checking this against (8.2.4), he should form the auxiliary table for calculating the E(MS)'s and check this against Table 8.2.1. He should then construct the analysis-of-variance table with columns for source, definition of SS, computation of SS, d.f., and E(MS), and check this against Table 8.2.2.

The E(MS) column of the analysis-of-variance table thus constructed tells us how to perform F-tests—in general, only approximate, even under the normality assumption—of the hypotheses corresponding to each row of the table except the row for error, the numerator mean square being that of this row, the denominator the one that has the same E(MS) under the hypothesis. If for the denominator no such row exists, a linear combination of MS's is employed whose expectation equals that of the numerator E(MS) under the hypothesis. The number of d.f. assigned to this linear combination may be calculated by the method given at the end of sec. 7.5. In applying this method the MS's are treated as though they were independent and each MS were a constant times a chi-square variable, the constant being E(MS) divided by the number of d.f. Actually the independence and the chi-square distributions may not be valid without further assumptions on the model.

If only one of the factors has random effects, then exact tests and multiple comparisons based on Hotelling's T^2 are possible for the main effects of the fixed-effects factors: If J is the number of levels of the random-effects factor, say B, and I ($I \leq J$) is the number of levels of one of the fixed-effects factors, say A, then, to analyze the main effects of A by use of T^2 we would make an $R \times J$ table of differences

$$d_{rj} = \bar{y}_{rj} - \bar{y}_{Ij},$$

where $R = I-1$ and $\bar{y}_{ij}$ is mean of all observations where A is at the ith level and B at the jth. The analysis would go exactly as after (8.1.34a), with the $\{y_{ij.}\}$ there replaced by the $\{\bar{y}_{ij}\}$, and $\{\alpha_i\}$ denoting the main effects of A. However, if in a mixed model two or more of the factors have random effects the use of Hotelling's T^2 is numerically so complicated that it is unlikely ever to be applied in practice.[19]

How to modify the above rules for calculating the E(MS)'s in the case where the errors no longer have equal variance σ_e^2 is indicated in the rule at the beginning of sec. 10.4.

PROBLEMS

8.1. The coded data in Table A are flow rates of a fuel through three types of nozzle as measured by five different operators, each of whom made three determinations on each nozzle. (*a*) Analyze the data according to the mixed model of sec. 8.1. (*b*) Would the differences between nozzles be significant if the analysis were made on the basis of a fixed-effects model? (*c*) If the answer to (*b*) disagrees with the corresponding one found in (*a*), give an intuitive explanation of why differences that are significant under the one model are not under the other.

TABLE* A

Nozzle	Operator 1	2	3	4	5
A	6, 6, −15	26, 12, 5	11, 4, 4	21, 14, 7	25, 18, 25
B	13, 6, 13	4, 4, 11	17, 10, 17	−5, 2, −5	15, 8, 1
C	10, 10, −11	−35, 0, −14	11, −10, −17	12, −2, −16	−4, 10, 24

* From Table IV on p. 19 of "Fundamentals of Analysis of Variance, Part I" by C. R. Hicks in *Industrial Quality Control*, Vol. 13, no. 2, 1956. Reproduced with the kind permission of the author and the editor.

8.2. Table B gives measurements of the waterproof quality (logarithms of permeabilities in seconds; the higher the measurement the better the waterproof) of sheets of material manufactured on three different machines on nine different days. (*a*) At the 0.05 level, test each of H_A, H_B, H_{AB}, where A denotes machines and B denotes days, under the mixed model of sec. 8.1. (*b*) Calculate point estimates of all the parameters of the model, including the three correlations $\sigma_{ii'}(\sigma_{ii}\sigma_{i'i'})^{-1/2}$ for $i \neq i'$. (*c*) Apply the S-method to the differences between the three pairs of machines.

8.3. (*a*) Fill in the E(MS) column in Problem 5.8. (*b*) Make the (approximate) F-tests that this suggests. (*c*) Fit by LS a straight line to the six means at the different spacings. (*d*) Calculate a 95 per cent confidence interval for the slope in (*c*).

[19] Imhof (1958) found an exact test and multiple-comparison method based on T^2 for the case of a complete three-way layout with two random-effects factors and one fixed-effects factor.

TABLE* B

Machine	1	2	3	4	5	6	7	8	9
	1.40	1.45	1.91	1.89	1.77	1.66	1.92	1.84	1.54
1	1.35	1.57	1.48	1.48	1.73	1.54	1.93	1.79	1.43
	1.62	1.82	1.89	1.39	1.54	1.68	2.13	2.04	1.70
	1.31	1.24	1.51	1.67	1.23	1.40	1.23	1.58	1.64
2	1.63	1.18	1.58	1.37	1.40	1.45	1.51	1.63	1.07
	1.41	1.52	1.65	1.11	1.53	1.63	1.44	1.28	1.38
	1.93	1.43	1.38	1.72	1.32	1.63	1.33	1.69	1.70
3	1.67	1.77	1.69	1.53	1.49	1.61	1.80	2.25	1.37
	1.40	1.86	1.36	1.37	1.34	1.36	1.38	1.80	1.84

* From Table 16.32 on p. 472 of *Statistical Theory with Engineering Applications* by A. Hald, John Wiley, New York, 1952. Reproduced with the kind permission of the author and the publisher.

8.4. In an experiment to study the effects of M different treatments on baby chicks K cocks are sampled from each of J flocks sampled from each of I breeds of chicken. For each of the IJK cocks in the experiment a sample of MN is taken from the female chicks sired by the cock, and N of these chicks are subjected to each of the M treatments. Formulate a model equation and construct an analysis-of-variance table like Table 8.2.2 for this example.*

8.5. The following problem is considered in Davies (1956), secs. 8.23–8.26. To investigate the corrosion resistance of I aluminum alloys (factor A) in a chemical plant atmosphere, one plate of each of the I alloys is exposed for a year in each of J sites (factor B) in a plant. Each plate is then rated by each of K observers (factor C). We agree to treat all three factors as primarily fixed-effects factors; in particular, the nine alloys are *not* a sample from a population of alloys. However, we expect that there will be some variation among plates made of the same alloy, and, possibly also, of subsites (plate locations) in the same site. To allow for this, add to the model equation for a fixed-effects model the random effects a_{ij}^A and a_{ij}^B, and assume these effects independent with zero means and $\text{Var}(a_{ij}^A) = \tilde{\sigma}_A^2$, $\text{Var}(a_{ij}^B) = \tilde{\sigma}_B^2$. Show that the $E(\text{MS})$'s are given by making the following additions to the formulas for the fixed-effects model: To each of $E(\text{MS}_A)$, $E(\text{MS}_B)$, and $E(\text{MS}_{AB})$ add $K\tilde{\sigma}_A^2 + K\tilde{\sigma}_B^2$.

* Suggested by Professor Hans Abplanalp.

CHAPTER 9

Randomization Models

9.1. RANDOMIZED BLOCKS: ESTIMATION

In the models we have considered up until now we have had to assume that the errors in the observations were statistically independent with equal variance, or else that the covariance matrix of the errors was known except possibly for a scalar factor (sec. 1.5). In this chapter we cope with models not restricted by such assumptions about the errors. These models take into account a randomization employed in assigning the treatment combinations to the experimental units. Such *randomization models*[1] will be formulated for the randomized-blocks and the Latin-square designs; we shall discuss first estimation under the models, and later testing, which requires our introducing the concept of a permutation test. The reader has probably read some nonmathematical exposition (we hope at least that of sec. 4.2) which conveys the easy generalization that randomization disposes of all difficulties caused by the violation of the usual normal-theory assumptions. Precise formulations of randomization models necessitate considerable detail, and the detailed formulations together with our subsequent developments may seem rather lengthy in view of the results obtained for statistical inference. Nevertheless, the author judged it worth while to attempt a careful treatment for these two designs, because it seemed to him part of the price of an informed opinion about the above generalization, and also because an understanding of the nature of the error distribution generated by the physical act of randomization should be part of our knowledge of the basic theory of the analysis of variance.

[1] Randomization models were first formulated by Neyman (1923) for the completely randomized design, by Neyman (1935) for randomized blocks, by Welch (1937) and Pitman (1937) for the Latin square under a certain null hypothesis, and by Kempthorne (1952, 1955) and Wilk (1955) for many other designs. Of these, only Neyman (1935) and Wilk (1955) included technical errors, and these were then assumed to have zero correlations and equal variances.

The reader may find it helpful to reread the subsections at the end of sec. 4.2 headed "The Randomized-Blocks Design" and "Randomization."

Suppose then that I treatments are to be compared on IJ experimental units (agricultural plots, experimental animals, etc.) and that the IJ units are grouped into J blocks of I units each. In each of the blocks the I treatments are assigned to the I units by a randomization, independently for the J blocks, so that all $(I!)^J$ assignments of treatments to units have the same probability of being used in the experiment. Number the units in each block with $\nu = 1, 2, \cdots, I$. Let $\mu_{ij\nu}$ be the "true" response under the ith treatment on the j,ν unit (νth unit of the jth block); this conceptual quantity is regarded as the expected value of the response *if* the ith treatment were applied to the j,ν unit. We may write

$$\mu_{ij\nu} = \mu + \alpha_i + \beta_j + \gamma_{ij} + \varepsilon_{ij\nu}, \tag{9.1.1}$$

where the general mean μ, treatment main effects $\{\alpha_i\}$, block main effects $\{\beta_j\}$, and treatment–block interactions $\{\gamma_{ij}\}$ are defined in terms of the $\{\mu_{ij.}\}$ in the same way as they are defined in terms of the $\{\eta_{ij}\}$ in (4.1.9) and hence satisfy the usual side conditions, and $\varepsilon_{ij\nu}$ is defined as

$$\varepsilon_{ij\nu} = \mu_{ij\nu} - \mu_{ij.}. \tag{9.1.2}$$

We remark that the values of the general mean $\mu = \mu_{...}$ and the treatment main effects $\{\alpha_i = \mu_{i..} - \mu\}$ do not depend on how the IJ experimental units are grouped into blocks (of I units each), since they are averages over all IJ units, and that the $\{\varepsilon_{ij\nu}\}$ satisfy

$$\sum_\nu \varepsilon_{ij\nu} = 0 \tag{9.1.3}$$

for all i,j. (All summations on ν or ν' are understood to be from 1 to I.) The quantity $\varepsilon_{ij\nu}$ will be called a *unit error*; it is the unit effect of the j,ν unit specific to the ith treatment, and within the jth block.

In a conceptual experiment involving a sequence of repetitions under the same conditions (this may be impossible to realize), the observed response $y_{ij\nu}$ of the j,ν unit to the ith treatment would differ from the conceptual "true" response $\mu_{ij\nu}$ of the j,ν unit to the ith treatment on any particular trial by a *technical error* $e_{ij\nu}$, $y_{ij\nu} = \mu_{ij\nu} + e_{ij\nu}$; $e_{ij\nu}$ is regarded as a random variable, with

$$E(e_{ij\nu}) = 0 \tag{9.1.4}$$

by definition of $\mu_{ij\nu}$ as the expectation of $y_{ij\nu}$. The technical error $e_{ij\nu}$ is to be distinguished from the unit error $\varepsilon_{ij\nu}$: The unit error $\varepsilon_{ij\nu}$ is a constant equal to the difference of the true response $\mu_{ij\nu}$ of the j,ν unit to the ith treatment from the average true response $\mu_{ij.}$ of the units in the jth block to the ith treatment. The technical error $e_{ij\nu}$ is a random variable equal to the difference of the conceptual observed response $y_{ij\nu}$ from its conceptual true value $\mu_{ij\nu}$. The unit error arises from inequality of the true responses of different units in the same block j to the same treatment i, the technical error is a measurement error, the difference between an observed and the corresponding true value, an error caused by the measuring instrument or the observer. The randomization by which the treatments are assigned to the units is performed in such a way that it is independent of the technical errors $\{e_{ij\nu}\}$.

If y_{ij} denotes the observation on the ith treatment in the jth block in the experiment, then[2]

$$y_{ij} = \mu + \alpha_i + \beta_j + \gamma_{ij} + \tilde{e}_{ij} + e_{ij}, \tag{9.1.5}$$

where $\tilde{e}_{ij}$ and e_{ij} are, respectively, the $\varepsilon_{ij\nu}$ and $e_{ij\nu}$ for which $\nu = \nu(i,j)$ is the unit number to which the ith treatment got assigned in the jth block. We shall call the $\{\tilde{e}_{ij}\}$ as well as the $\{\varepsilon_{ij\nu}\}$ *unit errors*, and the $\{e_{ij}\}$ as well as the $\{e_{ij\nu}\}$ *technical errors*.

We may write the unit error $\tilde{e}_{ij}$ and technical error e_{ij} in the following convenient notation:[3]

$$\tilde{e}_{ij} = \sum_\nu d_{ij\nu}\varepsilon_{ij\nu}, \tag{9.1.6}$$

$$e_{ij} = \sum_\nu d_{ij\nu}e_{ij\nu}, \tag{9.1.7}$$

where the $\{\varepsilon_{ij\nu}\}$, defined by (9.1.2), are regarded as unknown constants and the $\{d_{ij\nu}\}$ are I^2J random variables taking on only the values 0 and 1. The random variable $d_{ij\nu}$ is defined so as to take on the value 1 if the ith treatment is assigned to the j,ν unit, and 0 otherwise. The joint distribution of the $\{d_{ij\nu}\}$ is completely determined[4] by the randomization described above and is independent of the distribution of the $\{e_{ij\nu}\}$.

[2] It is academic to attempt to classify the randomization models as fixed-, mixed-, or random-effects models. Thus if we use the notation (9.1.5), it would appear that we have a fixed-effects model, since all but the error terms in (9.1.5) are constants. However, if we defined the observation $z_{j\nu}$ to be the observation on the (j, ν) unit, it would appear that we have a mixed model, for $z_{j\nu} = \mu+a_{j\nu}+\beta_j+c_{j\nu}+\tilde{f}_{j\nu}+f_{j\nu}$, where $\{\tilde{f}_{j\nu}\}$ and $\{f_{j\nu}\}$ are error terms, $a_{j\nu} = \Sigma_i d_{ij\nu}\alpha_i$, etc., and the $\{d_{ij\nu}\}$ are the random variables introduced below.

[3] Due to Kempthorne (1952).

[4] The distribution may be thought of as follows: For different j the J sets of I^2 variables $\{d_{ij\nu}\}$ are independent. For fixed j, think of the set $\{d_{ij\nu}\}$ arranged in an $I\times I$ square with $d_{ij\nu}$ in the ith row and νth column; then the possible values for the set are the $I!$ ones in which there is exactly one 1 in each row and column and 0's elsewhere, and these $I!$ values are taken on with equal probability.

We shall want the first and second moments of the $\{d_{ij\nu}\}$. If X is a random value that takes on only the values 0 and 1, then $E(X) = \Pr\{X=1\}$, and so

$$E(d_{ij\nu}) = \Pr\{d_{ij\nu} = 1\} = 1/I, \tag{9.1.8}$$

since all I treatments have the same probability of being assigned to the j,ν unit, and thus the probability for the ith to be assigned is $1/I$.

Since for $j \neq j'$ the randomizations in the jth and j'th blocks are independent, $d_{ij\nu}$ and $d_{i'j'\nu'}$ are independent, and hence

$$E(d_{ij\nu}d_{i'j'\nu'}) = 1/I^2 \qquad (j' \neq j).$$

To calculate $E(d_{ij\nu}d_{i'j\nu'})$, note that $d_{ij\nu}d_{i'j\nu'} = 0$ or 1, and hence

$$E(d_{ij\nu}d_{i'j\nu'}) = \Pr\{d_{ij\nu}d_{i'j\nu'} = 1\} = \Pr\{d_{ij\nu} = 1, d_{i'j\nu'} = 1\},$$

or

$$E(d_{ij\nu}d_{i'j\nu'}) = \Pr\{d_{i'j\nu'} = 1 | d_{ij\nu} = 1\} \Pr\{d_{ij\nu} = 1\}. \tag{9.1.9}$$

The above conditional probability, which we shall denote by P for the moment, is that of assigning the i'th treatment to the j,ν' unit, given that the ith treatment has been assigned to the j,ν unit. We easily find that

$$P = \begin{cases} 1 & \text{if } \nu = \nu',\ i = i', \\ 0 & \text{if } \nu = \nu',\ i \neq i', \\ 0 & \text{if } \nu \neq \nu',\ i = i', \\ (I-1)^{-1} & \text{if } \nu \neq \nu',\ i \neq i', \end{cases} \tag{9.1.10}$$

the last value following from the fact that if the ith treatment is assigned to the νth unit in the jth block, then the i'th treatment cannot be assigned to the same unit, and it is then equally likely to be assigned to any of the remaining $I-1$ units in the block. The result of substituting (9.1.10) and (9.1.8) into (9.1.9) may be written

$$E(d_{ij\nu}d_{i'j\nu'}) = \begin{cases} I^{-1}\delta_{\nu\nu'} & \text{if } i = i', \\ I^{-1}(I-1)^{-1}(1-\delta_{\nu\nu'}) & \text{if } i \neq i', \end{cases} \tag{9.1.11}$$

where $\delta_{\nu\nu'}$ denotes 1 if $\nu = \nu'$, and 0 otherwise.

From (9.1.6), (9.1.3), and (9.1.8) we get

$$E(\tilde{e}_{ij}) = 0; \tag{9.1.12}$$

from (9.1.7), from the independence of the two sets $\{d_{ij\nu}\}$ and $\{e_{ij\nu}\}$, and, from (9.1.4),

$$E(e_{ij}) = 0. \tag{9.1.13}$$

If $\psi = \Sigma_i c_i \alpha_i$ is any contrast in the treatment main effects ($\Sigma_i c_i = 0$), then an unbiased estimate is

$$\hat{\psi} = \sum_i c_i y_{i.},$$

since, from (9.1.5), (9.1.12), and (9.1.13), we have $E(y_{i.}) = \mu + \alpha_i$. From the moments calculated for the $\{d_{ij\nu}\}$ it is not difficult to derive an expression for Var $(\hat{\psi})$; this expression looks complicated and involves the $\{\varepsilon_{ij\nu}\}$ and the variances and covariances of the $\{e_{ij\nu}\}$. It is not possible to find an unbiased estimate for Var $(\hat{\psi})$ without further simplifying assumptions.[5] However, an overestimate can be obtained by the following method if we assume that the technical errors in different blocks are independent: We estimate ψ separately from each block by

$$\hat{\psi}_j = \sum_i c_i y_{ij}.$$

(We shall see that the estimates are biased by the amount λ_j defined below.) The sample variance of these J estimates, namely

$$s^2 = (J-1)^{-1} \sum_j (\hat{\psi}_j - \hat{\psi}_.)^2, \tag{9.1.14}$$

provides an overestimate of Var $(\hat{\psi})$ in the sense that $E(s^2/J) \geqq$ Var $(\hat{\psi})$, since we shall show that

$$E(s^2/J) = J^{-1}(J-1)^{-1} \sum_j \lambda_j^2 + \text{Var}\,(\hat{\psi}), \tag{9.1.15}$$

where $\lambda_j = \Sigma_i c_i \gamma_{ij}$. Formula (9.1.15) is derived as follows: From (9.1.5), $\hat{\psi}_j = \psi + \lambda_j + f_j$, where $f_j = \Sigma_i c_i(\tilde{e}_{ij} + e_{ij})$. The $\{f_j\}$ have zero means, and are independent since they are associated with different blocks. Since $\hat{\psi} = \hat{\psi}_.$, therefore Var $(\hat{\psi}) = J^{-2}\Sigma_j$ Var (f_j). Write

$$\sum_j (\hat{\psi}_j - \hat{\psi}_.)^2 = \sum_j (\lambda_j + f_j - f_.)^2 = \sum_j \lambda_j^2 + \sum_j (f_j - f_.)^2 + 2\sum_j \lambda_j (f_j - f_.).$$

Now

$$E\left(\sum_j (f_j - f_.)^2\right) = E\left(\sum_j f_j^2\right) - J\,E(f_.^2) = \sum_j \text{Var}\,(f_j) - J\,\text{Var}\,(f_.)$$
$$= \sum_j \text{Var}\,(f_j) - J^{-1} \sum_j \text{Var}\,(f_j).$$

Hence

$$E\left(\sum_j (\hat{\psi}_j - \hat{\psi}_.)^2\right) = \sum_j \lambda_j^2 + (1 - J^{-1}) \sum_j \text{Var}\,(f_j) \tag{9.1.16}$$
$$= \sum_j \lambda_j^2 + (1 - J^{-1}) J^2\,\text{Var}\,(\hat{\psi}).$$

[5] Such as in the subsection below headed "Efficiency of Randomized Blocks."

If we take J^{-1} times the expectation in (9.1.14) and then substitute (9.1.16) we get (9.1.15).

From (9.1.15) we see that s^2/J is an unbiased estimate of Var $(\hat{\psi})$ if all the block–treatment interactions $\{\gamma_{ij}\}$ are zero.

Some remarks on interval estimation under a more restricted form of the present model will be found at the end of sec. 9.3.

Expected Mean Squares

The MS's for treatments and blocks will be denoted by MS_A and MS_B, respectively, and the residual, error, or interaction MS by MS_e. They are defined in sec. 4.2. It is straightforward, but very tedious, to calculate the E(MS)'s if we assume that the technical errors $\{e_{ij}\}$ are uncorrelated: A sufficient condition for this is to assume the $\{e_{ij\nu}\}$ uncorrelated,[6] for we have

$$\begin{aligned}\text{Cov}\,(e_{ij}, e_{i'j'}) &= E(e_{ij}e_{i'j'}) = E\left(\sum_\nu d_{ij\nu}e_{ij\nu}\sum_{\nu'} d_{i'j'\nu'}e_{i'j'\nu'}\right)\\ &= \sum_\nu\sum_{\nu'} E(d_{ij\nu}d_{i'j'\nu'})\, E(e_{ij\nu}e_{i'j'\nu'}),\end{aligned}$$

since the $\{d_{ij\nu}\}$ are independent of the $\{e_{ij\nu}\}$, whence, substituting first $E(e_{ij\nu}e_{i'j'\nu'}) = \delta_{ii'}\delta_{jj'}\delta_{\nu\nu'}\,\text{Var}\,(e_{ij\nu})$, and then $E((d_{ij\nu})^2) = I^{-1}$, we find

$$\text{(9.1.17)} \qquad \text{Cov}\,(e_{ij}, e_{i'j'}) = \delta_{ii'}\delta_{jj'}I^{-1}\sum_\nu \text{Var}\,(e_{ij\nu}).$$

The formulas for the E(MS)'s are easier to interpret if we resolve the unit effect $\varepsilon_{ij\nu} = \mu_{ij\nu} - \mu_{ij.}$ of the j,ν unit specific to the ith treatment, and within the jth block, into a unit main effect within the jth block,

$$\xi_{j\nu} = \mu_{.j\nu} - \mu_{.j.},$$

plus a treatment–unit interaction within the jth block,

$$\eta_{ij\nu} = \mu_{ij\nu} - \mu_{.j\nu} - \mu_{ij.} + \mu_{.j.},$$

and define the symbols σ_U^2 and σ_{AU}^2 (corresponding to a "units factor" and to its interaction with the "treatments factor"—both within blocks) as

$$\begin{aligned}\sigma_U^2 &= J^{-1}(I-1)^{-1}\sum_j\sum_\nu \xi_{j\nu}^2,\\ \sigma_{AU}^2 &= J^{-1}(I-1)^{-2}\sum_i\sum_j\sum_\nu \eta_{ij\nu}^2.\end{aligned}$$

[6] Independence of the $\{e_{ij\nu}\}$ or normality of the $\{e_{ij\nu}\}$ does *not* imply the same for the $\{e_{ij}\}$.

Then with σ_A^2, σ_B^2, σ_{AB}^2 defined by (4.3.7*a*), and σ_e^2 defined as

$$(9.1.18)\qquad \sigma_e^2 = I^{-1}J^{-1}\sum_i\sum_j \text{Var}(e_{ij}) = I^{-2}J^{-1}\sum_i\sum_j\sum_v \text{Var}(e_{ijv}),$$

the last equality following from (9.1.17), the desired formulas are[7]

$$(9.1.19)\qquad \begin{aligned} E(\text{MS}_A) &= J\sigma_A^2 + \sigma_U^2 + I^{-1}(I-2)\sigma_{AU}^2 + \sigma_e^2,\\ E(\text{MS}_B) &= I\sigma_B^2 + I^{-1}(I-1)\sigma_{AU}^2 + \sigma_e^2,\\ E(\text{MS}_e) &= \sigma_{AB}^2 + \sigma_U^2 + I^{-1}(I-2)\sigma_{AU}^2 + \sigma_e^2. \end{aligned}$$

It is interesting to note that if the treatments have no effect in the sense that the treatment main effects and the block–treatment interactions are zero ($\sigma_A^2 = \sigma_{AB}^2 = 0$), then $E(\text{MS}_A) = E(\text{MS}_e)$ even if there are treatment–unit interactions within blocks ($\sigma_{AU}^2 \neq 0$).

Let σ_x^2 denote any σ^2 employed in the E(MS) formulas for an analysis of variance, such as σ_A^2, σ_{AB}^2, etc. A design is sometimes called *unbiased* for testing $\sigma_x^2 = 0$ (this does not imply that an unbiased test exists) if there exist two MS's whose E(MS)'s are equal if $\sigma_x^2 = 0$. Let us adopt the following more satisfactory definition: A design is unbiased for testing $\sigma_x^2 = 0$ if there exist two MS's whose E(MS's) differ by $c\sigma_x^2$, where c is a known nonzero constant.[8] Since for the randomized-blocks design $E(\text{MS}_A) - E(\text{MS}_e) = J\sigma_A^2 - \sigma_{AB}^2$, we see that the design is biased for testing $\sigma_A^2 = 0$ unless we assume that $\sigma_{AB}^2 = 0$.

Hypothesis testing under the above model will be treated in sec. 9.3.

In many applications it may seem more appropriate to treat the blocks as a random-effects factor. It may be shown[9] that the formulas for the case where the blocks are treated as a random sample from an infinite population of blocks are identical with (9.1.19) except that the term σ_{AB}^2 is added to $E(\text{MS}_A)$, so that now

$$(9.1.20)\qquad E(\text{MS}_A) - E(\text{MS}_e) = J\sigma_A^2.$$

It is curious that the design is now unbiased in the above sense for testing $\sigma_A^2 = 0$. This involves an apparent paradox, whose statement and resolution may deepen our understanding of the relation of different models in which the same factor is treated respectively as a fixed-effects or

[7] Kempthorne (1952), p. 148, gives the E(MS)'s and indicates their derivation in the case where the technical errors are zero.

[8] This definition implies that an unbiased estimate of σ_x^2 exists. If x refers to a fixed-effects factor there is usually little interest in estimating σ_x^2. I wanted to incorporate into the definition the notion related to the power of a good test, that the more false the hypothesis the more the MS's should tend to differ.

[9] Wilk (1954).

random-effects factor. Given a sample of J blocks from an infinite population, the expectations (9.1.19) may be regarded as conditional, so

$$E(\text{MS}_A - \text{MS}_e | J \text{ blocks}) = J\sigma_A^2 - \sigma_{AB}^2. \tag{9.1.21}$$

Now suppose that $\sigma_A^2 = 0$ and $\sigma_{AB}^2 \neq 0$. Then $E(\text{MS}_A - \text{MS}_e | J \text{ blocks}) < 0$. Since

$$E(\text{MS}_A - \text{MS}_e) = E(E(\text{MS}_A - \text{MS}_e | J \text{ blocks})), \tag{9.1.22}$$

therefore $E(\text{MS}_A - \text{MS}_e) < 0$ if the blocks are treated as a random-effects factor. But in this case $E(\text{MS}_A - \text{MS}_e) = 0$ by (9.1.20).

What has been overlooked in obtaining the paradox is that the symbol σ_A^2 has different meanings depending on whether the blocks are treated as a fixed-effects or a random-effects factor. In the former case σ_A^2 refers to main effects of the treatments which are defined as averages over the J blocks in the experiment, in the latter case over the blocks in the population. A similar remark applies to the averaging in the definition of the interactions entering into σ_{AB}^2. Replacing the symbols in the former case by the more adequate notation $\sigma^2_{A|J \text{ blocks}}$, $\sigma^2_{AB|J \text{ blocks}}$ and dropping the earlier condition $\sigma_A^2 = 0$, we get by substituting (9.1.20) for the left member of (9.1.22), and (9.1.21) in the right member,

$$J\sigma_A^2 = E(J\sigma^2_{A|J \text{ blocks}} - \sigma^2_{AB|J \text{ blocks}}). \tag{9.1.23}$$

If $\sigma_A^2 = 0$, this does not imply that $\sigma^2_{A|J \text{ blocks}} = 0$; however if the latter is true for all sets of J blocks, then[10] $\sigma^2_{AB|J \text{ blocks}}$ must also be zero for all sets of J blocks, and this disposes of the apparent paradox.

Efficiency of Randomized Blocks

We return to the standpoint that the blocks correspond to a fixed-effects factor. We shall consider the efficiency of the randomized blocks design relative to the *completely randomized design*, in which the treatments are assigned at random to the IJ units subject only to the condition that each treatment appears J times, in such a way that all such assignments have equal probability. Simple results are obtainable only if we make this comparison under the assumption of complete additivity, which we shall define in a moment. Interesting by-products of this calculation will be expressions exhibiting the nature of the correlation of errors in randomization models and the increase of precision in the comparison of treatments obtainable by good blocking of the experimental units, these expressions

[10] This follows from (9.1.23) with $\sigma_A^2 = 0$; a direct proof for the case where the J blocks are sampled from a finite population of $\mathscr{J}$ blocks ($\mathscr{J} > J$) is indicated in Problem 9.4.

being valid under the complete additivity assumed. We remark here that although under the model the greatest precision will follow from blocking into the most homogeneous blocks, where differences of unit effects are least within blocks and greatest between blocks,[11] if the resulting block differences are large, we may expect that the treatment–block interactions will be increased in the real situation, and this is not reflected in the present additive model.

By *complete additivity* we mean that no interactions exist between units and treatments in the completely randomized design, which implies that, for any grouping of the units into blocks, the treatment–block interactions $\{\gamma_{ij}\}$ and the treatment–unit interactions $\{\eta_{ij\nu}\}$ within a block are all zero,

(9.1.24) $$\sigma^2_{AB} = 0, \qquad \sigma^2_{AU} = 0,$$

and furthermore that the technical errors are additive, in the sense that the technical error $e_{ij\nu}$, incurred when the ith treatment is applied to the j,ν unit, consists of independent components $t_{ij\nu}$ and $u_{ij\nu}$ associated respectively with the treatment and the unit,

(9.1.25) $$e_{ij\nu} = t_{ij\nu} + u_{ij\nu},$$

where $E(t_{ij\nu}) = E(u_{ij\nu}) = 0$, the $\{t_{ij\nu}\}$, $\{u_{ij\nu}\}$ are assumed completely independent of each other and the set $\{d_{ij\nu}\}$, so

(9.1.26) $$\text{Var}\,(e_{ij\nu}) = \sigma^2_{t,i} + \sigma^2_{u,j\nu},$$

where $\sigma^2_{t,i} = \text{Var}\,(t_{ij\nu})$ and $\sigma^2_{u,j\nu} = \text{Var}\,(u_{ij\nu})$. This permits, for example, that the variance of the observations on one or more of the treatments or units is unusually large compared with the rest.

The model equation in the case of complete additivity becomes

$$y_{ij} = \mu + \alpha_i + \beta_j + \tilde{e}_{ij} + e_{ij},$$

where

$$\tilde{e}_{ij} = \sum_\nu d_{ij\nu}\xi_{j\nu}$$

since all $\eta_{ij\nu} = 0$ and hence $\varepsilon_{ij\nu} = \xi_{j\nu}$, and

$$e_{ij} = \sum_\nu d_{ij\nu}e_{ij\nu} = \sum_\nu d_{ij\nu}(t_{ij\nu}+u_{ij\nu}).$$

From what has been assumed so far concerning the distribution of the $\{d_{ij\nu}\}$ and $\{e_{ij\nu}\}$ we may calculate that the unit errors $\{\tilde{e}_{ij}\}$ and technical errors $\{e_{ij}\}$ have the following variances and covariances:

(9.1.27) $$\text{Cov}\,(\tilde{e}_{ij}, \tilde{e}_{i'j'}) = \delta_{jj'}(\delta_{ii'}-I^{-1})\sigma^2_{U,j},$$

[11] In the sense that the blocking minimizes σ^2_U, and hence maximizes σ^2_B, since $\tilde{\sigma}^2_U$ is fixed in (9.1.37).

where

$$\sigma_{U,j}^2 = (I-1)^{-1}\sum_\nu \xi_{j\nu}^2,$$

(9.1.28) $$\text{Cov}\,(e_{ij}, e_{i'j'}) = \delta_{ii'}\delta_{jj'}(\sigma_{t,i}^2+\sigma_{u,j}^2),$$

where

$$\sigma_{u,j}^2 = I^{-1}\sum_\nu \sigma_{u,j\nu}^2,$$

and

(9.1.29) $$\text{Cov}\,(\tilde{e}_{ij}, e_{i'j'}) = 0.$$

To derive (9.1.27) we use (9.1.12) to write

$$\text{Cov}\,(\tilde{e}_{ij}, \tilde{e}_{i'j'}) = E(\tilde{e}_{ij}\tilde{e}_{i'j'}) = \sum_\nu\sum_{\nu'} E(d_{ij\nu}d_{i'j'\nu'})\xi_{j\nu}\xi_{j'\nu}$$
$$= \delta_{jj'}\sum_\nu\sum_{\nu'} E(d_{ij\nu}d_{i'j\nu'})\xi_{j\nu}\xi_{j\nu'}.$$

Now from (9.1.11) we get for $i = i'$,

$$\text{Cov}\,(\tilde{e}_{ij}, \tilde{e}_{ij'}) = \delta_{jj'}\sum_\nu\sum_{\nu'} I^{-1}\delta_{\nu\nu'}\xi_{j\nu}\xi_{j\nu'} = \delta_{jj'}I^{-1}\sum_\nu \xi_{j\nu}^2 = \delta_{jj'}(1-I^{-1})\sigma_{U,j}^2,$$

and for $i \neq i'$,

$$\text{Cov}\,(\tilde{e}_{ij}, \tilde{e}_{i'j'}) = \delta_{jj'}I^{-1}(I-1)^{-1}\sum_\nu\sum_{\nu'}(1-\delta_{\nu\nu'})\xi_{j\nu}\xi_{j\nu'} = -\,\delta_{jj'}I^{-1}\sigma_{U,j}^2,$$

since $\Sigma_\nu \xi_{j\nu} = 0$. These two formulas may be combined to yield (9.1.27). To derive (9.1.28), we substitute (9.1.26) into (9.1.17). Similarly, we derive

$$\text{Cov}\,(\tilde{e}_{ij}, e_{i'j'}) = \sum_\nu\sum_{\nu'} E(d_{ij\nu}d_{i'j'\nu'})\xi_{j\nu}\, E(e_{i'j'\nu'}),$$

and hence (9.1.29).

From (9.1.27) we find that the correlation coefficient for two different unit errors $\tilde{e}_{ij}$ and $\tilde{e}_{i'j}$ in the same block is

(9.1.30) $$-1/(I-1).$$

If the arrangement of the units in an experiment corresponds to the temporal order of a long sequence of observations, or the order along a long line of agricultural plots, and if blocking consists of grouping the units in successive order into sets of I, one might expect a positive correlation of units in the same block: Adjacent or nearly adjacent units tend to be more alike. The reason for the negative sign in (9.1.30) is that the calculation is for unit errors defined *within blocks*; it expresses the correlation of the differences from the block mean; whereas the anticipated positive correlation just mentioned would affect the block effects for adjacent blocks. In the extreme case $I = 2$, the unit errors in the jth block satisfy $\tilde{e}_{1j}+\tilde{e}_{2j} = 0$, and hence the correlation coefficient is -1, in agreement with (9.1.30).

Since under our additivity assumption

$$y_{ij} = \mu + \alpha_i + \beta_j + \tilde{e}_{ij} + e_{ij},$$

the mean for the ith treatment is

$$y_{i.} = \mu + \alpha_i + \tilde{e}_{i.} + e_{i.},$$

and therefore the variance of the estimate $\hat{\psi} = \Sigma_i c_i y_{i.}$ of the contrast

(9.1.31) $$\psi = \sum_i c_i \alpha_i \qquad \left(\sum_i c_i = 0\right)$$

is

(9.1.32) $$\begin{aligned} \text{Var}(\hat{\psi}) &= \sum_i \sum_{i'} c_i c_{i'} [\text{Cov}(\tilde{e}_{i.}, \tilde{e}_{i'.}) + \text{Cov}(e_{i.}, e_{i'.})] \\ &= J^{-2} \sum_i \sum_{i'} \sum_j \sum_{j'} c_i c_{i'} [\text{Cov}(\tilde{e}_{ij}, \tilde{e}_{i'j'}) + \text{Cov}(e_{ij}, e_{i'j'})]. \end{aligned}$$

Substitution of (9.1.27) and (9.1.28) into (9.1.32) gives

(9.1.33) $$\text{Var}(\hat{\psi}) = J^{-1}\left[(\sigma_U^2+\sigma_u^2)\sum_i c_i^2 + \sum_i c_i^2 \sigma_{t,i}^2\right]$$

where

(9.1.34) $$\sigma_u^2 = J^{-1}\sum_j \sigma_{u,j}^2 = I^{-1}J^{-1}\sum_j \sum_v \sigma_{u,jv}^2.$$

In contemplating the result (9.1.33) we should recall that σ_U^2 refers to the unit errors within blocks, σ_u^2 to the components of the technical errors associated with units, and $\sigma_{t,i}^2$ to the components associated with the ith treatment.

We now consider the completely randomized design. It is convenient for the present purpose of comparison with randomized blocks to continue to number the IJ units with two indices, as though they were grouped into J blocks of I, but of course no attention is paid to this fictitious blocking in the random assignment of treatments to units. Under the additivity assumptions of this subsection, if the ith treatment is applied to the j,v unit the observed response will be

(9.1.35) $$\mu + \alpha_i + \kappa_{jv} + t_{ijv} + u_{ijv},$$

where

(9.1.36) $$\kappa_{jv} = \beta_j + \xi_{jv}$$

is a unit effect (*not* within the jth block like $\xi_{j\nu}$). We note that $\Sigma_j\Sigma_\nu\kappa_{j\nu} = 0$. Then

$$\tilde{\sigma}_U^2 = (IJ-1)^{-1}\sum_j\sum_\nu \kappa_{j\nu}^2$$

is a measure of the magnitude of the unit effects, to be distinguished from σ_U^2, which measures the magnitude of the unit effects within blocks. We remark that $\tilde{\sigma}_U^2$ does not depend on how the IJ units are grouped into blocks, whereas σ_U^2 does. By squaring and summing (9.1.36) we get

$$(9.1.37) \qquad (IJ-1)\tilde{\sigma}_U^2 = I(J-1)\sigma_B^2 + J(I-1)\sigma_U^2.$$

From (9.1.35) we see that the observed ith treatment mean is

$$(9.1.38) \qquad \bar{y}_i = \mu + \alpha_i + J^{-1}\sum_j\sum_\nu f_{ij\nu}(\kappa_{j\nu}+t_{ij\nu}+u_{ij\nu}),$$

where the random variables $\{f_{ij\nu}\}$ are defined for the completely randomized design similarly to the $\{d_{ij\nu}\}$ introduced earlier for the randomized-blocks design: $f_{ij\nu} = 1$ if the ith treatment is applied to the j,v unit, 0 otherwise. We derive

$$(9.1.39) \qquad E(f_{ij\nu}f_{i'j'\nu'}) = \begin{cases} \dfrac{J-\delta_{ii'}}{(IJ-1)I} & \text{if } (j,\nu) \neq (j',\nu'), \\ I^{-1}\delta_{ii'} & \text{if } (j,\nu) = (j',\nu'), \end{cases}$$

like (9.1.11); and by calculations similar to those leading to (9.1.33), we find from (9.1.38) and (9.1.39) that if the unbiased estimate

$$\tilde{\psi} = \sum_i c_i\bar{y}_i$$

is used for the contrast (9.1.31) then

$$(9.1.40) \qquad \text{Var}(\tilde{\psi}) = J^{-1}\left[(\tilde{\sigma}_U^2+\sigma_u^2)\sum_i c_i^2 + \sum_i c_i^2\sigma_{t,i}^2\right].$$

Comparing this with (9.1.33) we see that the randomized-blocks design is more efficient than the completely randomized design, in the sense of yielding an unbiased estimate of a contrast (9.1.31) with smaller variance, if and only if

$$(9.1.41) \qquad \sigma_U^2 < \tilde{\sigma}_U^2,$$

and this condition does not depend on the contrast (9.1.31). The condition (9.1.41) is equivalent to $\sigma_B^2 > I^{-1}\tilde{\sigma}_U^2$, or $\sigma_B^2 > I^{-1}\sigma_U^2$, by (9.1.37). We show in the appendix below that if the units are grouped into blocks at random, in such a way that all groupings of the IJ units into J blocks of I have the same probability, then $E(\sigma_U^2) = IE(\sigma_B^2) = \tilde{\sigma}_U^2$. The condition (9.1.41) thus means that the blocking has achieved greater homogeneity of units within blocks, as measured by σ_U^2, than its expectation under random blocking. An exact expression for the gain possible by good blocking is given by (9.1.33), and depends on how small the measure σ_U^2 of variation of units within blocks can be made.

The efficiency of randomized blocks as compared with the completely randomized design may be defined as the fraction $\mathscr{E} = \text{Var}(\tilde{\psi})/\text{Var}(\hat{\psi})$ determined by (9.1.33) and (9.1.40), namely

$$\mathscr{E} = \frac{(\tilde{\sigma}_U^2+\sigma_u^2)\sum_i c_i^2 + \sum_i c_i^2\sigma_{t,i}^2}{(\sigma_U^2+\sigma_u^2)\sum_i c_i^2 + \sum_i c_i^2\sigma_{t,i}^2}.$$

This depends on the contrast considered, unless all $\sigma_{t,i}^2$ have a common value σ_t^2, in which case

$$\mathscr{E} = \frac{\tilde{\sigma}_U^2 + \sigma_e^2}{\sigma_U^2 + \sigma_e^2},$$

since we then have $\sigma_e^2 = \sigma_t^2+\sigma_u^2$ from (9.1.18), (9.1.26), and (9.1.34). This may also be written, by use of (9.1.37), as

$$\mathscr{E} = \frac{(J-1)(I\sigma_B^2+\sigma_e^2) + J(I-1)(\sigma_U^2+\sigma_e^2)}{(IJ-1)(\sigma_U^2+\sigma_e^2)}.$$

Unbiased estimates of the numerator and denominator of this fraction are available from MS_B and MS_{AB}, since (9.1.19) and (9.1.24) give

$$E(\text{MS}_B) = I\sigma_B^2 + \sigma_e^2, \qquad E(\text{MS}_e) = \sigma_U^2 + \sigma_e^2$$

in the case of complete additivity. We may thus estimate from the data of a randomized-blocks experiment the efficiency $\mathscr{E}$ obtained by the blocking to be

$$\hat{\mathscr{E}} = \frac{(J-1)\text{MS}_B + J(I-1)\text{MS}_e}{(IJ-1)\text{MS}_e}$$

if complete additivity and equality of the $\{\sigma_{t,i}^2\}$ are assumed. We remark that any advantage gained by blocking will be slightly offset by a loss with the randomized blocks design of $J-1$ of the $I(J-1)$ d.f. for "error" in the completely randomized design, and we recall the earlier remark that successful blocking may increase the measure σ_{AB}^2 of the block × treatment interactions.

Appendix on Expectations of σ_B^2 and σ_U^2 under Random Blocking

Suppose a finite population of $N = IJ$ units with unit effects $\{\tau_1, \cdots, \tau_N\}$, where $\Sigma_n\tau_n = 0$ and $\tilde{\sigma}_U^2 = (N-1)^{-1}\Sigma_n\tau_n^2$, is grouped into J blocks of I units each. Denote by $\kappa_{j\nu}$ the unit effect (*not* within blocks) of the νth unit in the jth block. If the blocking is random, $\kappa_{j\nu}$ equals one of $\{\tau_n\}$, each with probability $1/N$. Now $E(\kappa_{j\nu}^m) = \Sigma_n \tau_n^m \Pr\{\kappa_{j\nu} = \tau_n\} = N^{-1}\Sigma_n \tau_n^m$; hence $E(\kappa_{j\nu}) = 0$, and $\text{Var}(\kappa_{j\nu}) = E(\kappa_{j\nu}^2) = (1-N^{-1})\tilde{\sigma}_U^2$. For $\nu \neq \nu'$, the correlation coefficient of $\kappa_{j\nu}$ and $\kappa_{j\nu'}$ is $\rho = -(N-1)^{-1}$: This well-known result for sampling from finite populations may be argued by replacing I by N in (9.1.30). Hence for $\nu \neq \nu'$,

$$E(\kappa_{j\nu}\kappa_{j\nu'}) = \text{Cov}(\kappa_{j\nu}, \kappa_{j\nu'}) = \rho \,\text{Var}(\kappa_{j\nu}) = -N^{-1}\tilde{\sigma}_U^2.$$

Since $\beta_j = I^{-1}\Sigma_\nu \kappa_{j\nu}$,

$$E(\beta_j^2) = I^{-2}\sum_\nu \sum_{\nu'} E(\kappa_{j\nu}\kappa_{j\nu'}). \tag{9.1.42}$$

The I terms with $\nu = \nu'$ which appear after the summation signs in (9.1.42) have the value $(1-N^{-1})\tilde{\sigma}_U^2$, and the remaining $I(I-1)$ terms have the value $-N^{-1}\tilde{\sigma}_U^2$. Thus

$$E(\beta_j^2) = I^{-2}[I(1-N^{-1}) + I(I-1)(-N^{-1})]\tilde{\sigma}_U^2 = I^{-1}J^{-1}(J-1)\tilde{\sigma}_U^2,$$
$$E(\sigma_B^2) = (J-1)^{-1}\sum_j E(\beta_j^2) = I^{-1}\tilde{\sigma}_U^2,$$

and from (9.1.37),

$$E(\sigma_U^2) = J^{-1}(I-1)^{-1}[(IJ-1)\tilde{\sigma}_U^2 - I(J-1)\,E(\sigma_B^2)] = \tilde{\sigma}_U^2.$$

9.2. LATIN SQUARES: ESTIMATION

The method of picking a Latin square at random has been discussed in sec. 5.1. The randomization model that we shall adopt in this section will be valid regardless of how we choose the transformation set containing the square actually used, but will be based on the following property of the method of selection: All squares in the transformation set, i.e., all squares obtainable from the one actually used, by permutation of rows, columns, and numbers, had the same probability of being selected for the experiment. Different randomization models with this property would be appropriate in different situations, among which we mention the following three:[12]

(i) There are three factors A, B, C of primary interest, each at m levels, the m^2 treatment combinations employed in the experiment are selected by a Latin-square design, the observations are constituted of the "true" values of the response to these treatment combinations plus random technical errors which are independent of the randomization used in

[12] Some other cases, in which the factors are not treated as necessarily having fixed effects, are considered by Wilk and Kempthorne (1957).

picking the Latin square. This model might be appropriate in physical science experiments with three factors and no clearly defined experimental units, for example if three factors are varied in a pilot plant experiment and each observation is the result of a run on the pilot plant. (However, while the model is appropriate to this example, the design is not a good one if the factors interact.)

(ii) As above, but the m^2 three-factor treatment combinations are assigned at random to m^2 experimental units. This might be appropriate in biological experiments where the experimental units are animals. Technical errors could be incorporated into the model, similar to those in sec. 9.1.

(iii) There is only one factor, say C, of primary interest and the experiment is done with m^2 experimental units. Instead of assigning those at random we try to "eliminate" some of the heterogeneity of the experimental units by grouping them, not according to one classification as we might in a randomized-block design but according to two classifications A and B. For example the m^2 plots in the agricultural example of sec. 5.1 are classified according to row and column; in an experiment with automobile tires where the factor of primary interest is the brand of tire, we might take $m = 4$, and use four cars, the experimental units would be the 16 tires, one classification of the units being the car, and the other the tire position; or if there are m litters of animals we might use the m largest animals in each litter, one classification being litter, the other, weight order in litter.[13] Again, technical errors can be incorporated.

We shall not treat case (ii), which involves two different randomizations, one in the choice of the m^2 treatment combinations out of the m^3 possible ones by selection of a Latin square, and another in the selection of one of the $(m^2)!$ possible assignments of these treatment combinations to the experimental units. Our model will include cases (i) and (iii), which really have the same probabilistic structure but differ in our attitudes toward the factors A and B: In case (i) these are of primary interest as well as C, and we wish to determine their effects, whereas in case (iii) they are of secondary interest, having been introduced by, and perhaps even defined by, the groupings used in attempting to reduce the heterogeneity of the experimental units, like the rows and columns in the agricultural example.

Let y_{ijk} denote the response that would be observed if an observation were made on the treatment combination consisting of the ith level of A, the jth of B, and the kth of C. We may resolve this into a true value $\mu_{ijk} = E(y_{ijk})$ plus a technical error e_{ijk}, so $E(e_{ijk}) = 0$. To simplify the

[13] An analysis of covariance design with regression on weight should also be considered here.

discussion we shall make more restrictive assumptions[14] about the $\{e_{ijk}\}$ than we did with randomized blocks, namely that they are independent with equal variance σ_e^2. We have

$$(9.2.1) \quad y_{ijk} = \mu + \alpha_i^A + \alpha_j^B + \alpha_k^C + \alpha_{ij}^{AB} + \alpha_{ik}^{AC} + \alpha_{jk}^{BC} + \alpha_{ijk}^{ABC} + e_{ijk}$$

where the general mean, main effects, and interactions are defined as in sec. 4.5 (μ_{ijk} now replacing the former η_{ijk}) and satisfy the usual side conditions

$$(9.2.2) \quad \alpha_.^A = \cdots = \alpha_{i.}^{AB} = \alpha_{.j}^{AB} = \cdots = \alpha_{ij.}^{ABC} = \alpha_{i.k}^{ABC} = \alpha_{.jk}^{ABC} = 0$$

for all i, j, k. We note that, in case (iii) described above, where the factors A, B are characteristics according to which the m^2 experimental units are grouped, the main effects $\{\alpha_i^A\}$ and $\{\alpha_j^B\}$ are effects "eliminated" by the grouping (similar to the block effects in randomized blocks), the interactions $\{\alpha_{ij}^{AB}\}$ are remaining unit effects, the $\{\alpha_{ik}^{AC}\}$ and $\{\alpha_{jk}^{BC}\}$ are interactions of the treatment C with the two grouping characteristics, and the $\{\alpha_{ijk}^{ABC}\}$ are unit–treatment interactions.

By letting the randomization generate random variables $\{d_{ijk}\}$ as in sec. 9.1, $d_{ijk} = 1$ if the treatment combination consisting of the ith level of A, the jth of B, and the kth of C occurs in the experiment, and $d_{ijk} = 0$ otherwise, so that the $\{d_{ijk}\}$ are independent of the technical errors $\{e_{ijk}\}$, we get model equations for the observations: These may be conveniently written in any of the three forms

$$y_{ij.} = \sum_k d_{ijk} y_{ijk}, \quad \text{or} \quad y_{i.k} = \sum_j d_{ijk} y_{ijk}, \quad \text{or} \quad y_{.jk} = \sum_i d_{ijk} y_{ijk},$$

where the $\{y_{ijk}\}$ are to be replaced by (9.2.1), and $y_{ij.}$ is the observation on the treatment combination in the experiment containing the ith level of A and the jth of B, etc. All the statistics used in analyzing Latin-square experiments in sec. 5.1 are functions of the observed means for the different levels of the three factors, the observed general mean, and the total SS, and these have the following structure: The observed means for the ith level of A, the jth of B, and the kth of C are respectively

$$y_{i..} = m^{-1} \sum_j \sum_k d_{ijk} y_{ijk}, \qquad y_{.j.} = m^{-1} \sum_j \sum_k d_{ijk} y_{ijk},$$

$$(9.2.3) \qquad y_{..k} = m^{-1} \sum_i \sum_j d_{ijk} y_{ijk},$$

[14] The estimate (9.2.17) of a contrast is unbiased if $E(e_{ijk}) = 0$. If furthermore the $\{e_{ijk}\}$ are independent the formulas (9.2.12) and (9.2.13) for the E(MS)'s are valid if we define $\sigma_e^2 = m^{-3} \Sigma_i \Sigma_j \Sigma_k \operatorname{Var}(e_{ijk})$; these conditions are also sufficient for the result (9.2.19) below about the average variance of the $m(m-1)/2$ estimated differences. If furthermore the distribution of e_{ijk} does not depend on k and $\sigma_{ABC}^2 = 0$, then the result (9.2.18) below for the variance of the estimated contrast (9.2.17) is also valid.

the observed general mean is

(9.2.4)
$$y_{...} = m^{-2}\sum_i\sum\sum_k d_{ijk}y_{ijk},$$

and the total SS is

(9.2.5)
$$\text{SS}_{\text{tot}} = \sum_i\sum_j\sum_k d_{ijk}y_{ijk}^2,$$

where in every case y_{ijk} is to be replaced by (9.2.1). If we substitute (9.2.1) into (9.2.3) the structure of the observed means simplifies to

(9.2.6)
$$y_{..k} = \mu + \alpha_k^C + m^{-1}\sum_i\sum_j d_{ijk}(\alpha_{ij}^{AB} + \alpha_{ijk}^{ABC} + e_{ijk}),$$

with similar expressions for $y_{i..}$ and $y_{.j.}$, because of the relations

$$\sum_i d_{ijk} = \sum_j d_{ijk} = \sum_k d_{ijk} = 1$$

and the side conditions (9.2.2). We remark that the random variables g_k and e_k used in (5.1.15) consist of the following terms of (9.2.6):

$$g_k = m^{-1}\sum_i\sum_j d_{ijk}(\alpha_{ij}^{AB} + \alpha_{ijk}^{ABC}), \qquad e_k = m^{-1}\sum_i\sum_j d_{ijk}e_{ijk}.$$

The results of this section will depend on the first and second moments of the $\{d_{ijk}\}$. In deriving these moments it will be convenient to call the levels of A the rows, those of B the columns, and those of C the "numbers" (like the numbers in (5.1.1)). Now $E(d_{ijk}) = \Pr\{d_{ijk} = 1\}$; this is the probability of the number k appearing in the i,j cell, and hence

(9.2.7)
$$E(d_{ijk}) = m^{-1},$$

since the randomization is equivalent to randomizing on rows, columns, and numbers, and so all numbers k have the same probability of appearing in the i,j cell.

To derive the second moments we start with a device similar to (9.1.9),

(9.2.8)
$$E(d_{ijk}d_{i'j'k'}) = \Pr\{d_{ijk} = 1\}\text{P} = m^{-1}\text{P},$$

where P is the conditional probability

(9.2.9)
$$P = \Pr\{d_{i'j'k'} = 1 \,|\, d_{ijk} = 1\},$$

and to evaluate P we consider four cases, namely if (i) all, (ii) exactly two, (iii) exactly one, or (iv) none of the three conditions

(9.2.10)
$$i = i', \qquad j = j', \qquad k = k'$$

are satisfied. From the symmetry of the design in the three factors A, B, C it follows that, in calculating P, if exactly N of the conditions (9.2.10) are satisfied ($N = 0, 1, 2, 3$), it does not matter which N. In case (i), $P = \Pr\{d_{ijk}=1 | d_{ijk}=1\} = 1$. In case (ii) suppose that $k \neq k'$. Then we see from (9.2.9) that P is the conditional probability of the number k' falling in the i,j cell, given that k fell in the i,j cell, and hence $P = 0$. In case (iii) suppose that $i = i'$ and consider the numbers in the ith row only, where some permutation of the numbers $1, 2, \cdots, m$ appears. Then P is the conditional probability that the number k' appears in the j'th column, given that some other number k appeared in the jth column. Since the randomization generates a random permutation of the numbers, all $m!$ permutations of $1, 2, \cdots, m$ are equally likely in the ith row. Of these, $(m-1)!$ have k in the jth column, and of these permutations $(m-2)!$ have k' in the j'th column. Hence $P = (m-2)!/(m-1)! = (m-1)^{-1}$. In case (iv) consider the conditional probability for the number k to be in the i',j' cell, given that the number k is in the i,j cell. This conditional probability must be the same for all $k' \neq k$ because of the random permutation of the numbers, and hence is

$$
\begin{aligned}
&(m-1)^{-1}[1 - \Pr\{k \text{ in } i',j' \text{ cell} | k \text{ in } i,j \text{ cell}\}] \\
&\quad = (m-1)^{-1}[1 - \Pr\{d_{i'j'k} = 1 | d_{ijk}=1\}] = (m-1)^{-1}[1 - (m-1)^{-1}],
\end{aligned}
$$

the last equality following from the value of P in case (iii); and so in case (iv), $P = (m-1)^{-2}(m-2)$. Substituting these values of P in (9.2.8), we have

(9.2.11)

$$
E(d_{ijk}d_{i'j'k'}) = \begin{cases} m^{-1}, & \text{if all three} \\ 0, & \text{if exactly two} \\ m^{-1}(m-1)^{-1}, & \text{if exactly one} \\ m^{-1}(m-1)^{-2}(m-2), & \text{if none} \end{cases} \quad \text{of the conditions (9.2.10) are (or is) satisfied.}
$$

The SS's are defined and calculated as in sec. 5.1. Their structures in terms of the random variables $\{d_{ijk}\}$, the technical errors $\{e_{ijk}\}$, and the parameters μ and $\{\alpha_i\}$ would be given by substituting first (9.2.3) and (9.2.4) and then (9.2.1) into

$$SS_C = m \sum_k (y_{..k} - y_{...})^2,$$

with similar formulas for SS_A and SS_B, and by similarly treating, with the help of (9.2.5),

$$SS_e = SS_{\text{tot}} - my_{...}^2 - SS_A - SS_B - SS_C.$$

Their expectations are determined by these structures, and the first and second moments of the $\{d_{ijk}\}$ and $\{e_{ijk}\}$. The calculations are extremely tedious and we give the results[15] without proof:

$$E(\text{MS}_C) = \sigma_e^2 + (1-2m^{-1})\sigma_{ABC}^2 + \sigma_{AB}^2 + m\sigma_C^2, \tag{9.2.12}$$

$$E(\text{MS}_e) = \sigma_e^2 + (1-3m^{-1})\sigma_{ABC}^2 + \sigma_{AB}^2 + \sigma_{AC}^2 + \sigma_{BC}^2, \tag{9.2.13}$$

where

$$\sigma_C^2 = \frac{\sum_k (\alpha_k^C)^2}{m-1}, \qquad \sigma_{AB}^2 = \frac{\sum_i\sum_j (\alpha_{ij}^{AB})^2}{(m-1)^2}, \qquad \sigma_{ABC}^2 = \frac{\sum_i\sum_j\sum_k (\alpha_{ijk}^{ABC})^2}{(m-1)^3},$$

etc. The formulas for $E(\text{MS}_A)$ and $E(\text{MS}_B)$ may be obtained from (9.2.12) by permuting the symbols A, B, C. We note that under the hypothesis H_C: all $\alpha_k^C = 0$, or $\sigma_C^2 = 0$, $E(\text{MS}_e)$ would usually exceed $E(\text{MS}_C)$; more precisely, $E(\text{MS}_e) > E(\text{MS}_C)$ if

$$\sigma_{ABC}^2 < m(\sigma_{AC}^2 + \sigma_{BC}^2). \tag{9.2.14}$$

In the terminology introduced below (9.1.19), the design is biased for testing the hypothesis $\sigma_C^2 = 0$ (unless we assume that $\sigma_{AC}^2 = \sigma_{BC}^2 = \sigma_{ABC}^2 = 0$). If we recall that in case (iii), as discussed after (9.2.2), σ_{ABC}^2 is a measure of the magnitude of the unit–treatment interactions, we see that this design, unlike the randomized-blocks design is biased by the presence of unit–treatment interactions for testing the hypothesis $\sigma_C^2 = 0$, in the sense that σ_{ABC}^2 enters into the difference

$$E(\text{MS}_C) - E(\text{MS}_e) = m\sigma_C^2 - \sigma_{AC}^2 - \sigma_{BC}^2 + m^{-1}\sigma_{ABC}^2,$$

although only with the coefficient m^{-1}.

To get a rough idea of the effect of interactions on the F-test of a hypothesis when the σ^2 for the interactions (e.g., σ_{AB}^2) enters the numerator $E(\text{MS})$ as well as the denominator $E(\text{MS})$ we might observe the effect on a suitably defined noncentrality parameter. In the case of tests in the normal-theory fixed-effects models we found in sec. 2.6 and 1.6 that if MS_N and MS_D denote the numerator and denominator MS's of the F-statistic,

$$E(\text{MS}_N) = \sigma_e^2(1+q^{-1}\delta^2), \qquad E(\text{MS}_D) = \sigma_e^2,$$

where q is the number of d.f. of MS_N, and δ^2 is the noncentrality parameter. We may alternatively use (sec. 2.8) as noncentrality parameter $\phi^2 = (q+1)^{-1}\delta^2$, the power of the test being more nearly constant as q varies with ϕ^2 fixed than with δ^2 fixed. Then

$$\phi^2 = \frac{q}{q+1}\,\frac{E(\text{MS}_N) - E(\text{MS}_D)}{E(\text{MS}_D)}. \tag{9.2.15}$$

[15] These are special cases of formulas of Wilk and Kempthorne (1957), who treated the case where the factors may have their levels sampled from a population of levels.

We shall call the ϕ^2 defined by (9.2.15) the *generalized noncentrality parameter* in any situation where one would consider using a statistic that is the quotient of two mean squares, MS_N by MS_D. We may consider ϕ^2 a rough measure of the power of the test, decreasing ϕ^2 corresponding to decreasing power.

From (9.2.12) and (9.2.13) the generalized noncentrality parameter for the statistic $\mathrm{MS}_C/\mathrm{MS}_e$ is

$$(9.2.16)\quad \phi^2 = \frac{m-1}{m}\,\frac{m\sigma_C^2 - (\sigma_{AC}^2+\sigma_{BC}^2) + m^{-1}\sigma_{ABC}^2}{\sigma_e^2 + \sigma_{AB}^2 + (\sigma_{AC}^2+\sigma_{BC}^2) + (1-3m^{-1})\sigma_{ABC}^2}.$$

If we compare this with its value

$$\phi^2 = \frac{m-1}{m}\,\frac{m\sigma_C^2}{\sigma_e^2}$$

when there are no interactions present, we see that the presence of interactions whose σ^2, such as σ_{AC}^2, enters only $E(\mathrm{MS}_D)$ has a double effect in decreasing the power, decreasing the numerator of ϕ^2 as well as increasing the denominator, whereas the presence of those whose σ^2, such as σ_{AB}^2, enters both $E(\mathrm{MS}_N)$ and $E(\mathrm{MS}_D)$ with the same coefficient has only the latter effect. We can also draw conclusions of the following type: If $\sigma_{ABC}^2 = 0$ and if σ_{AB}^2, σ_{AC}^2, σ_{BC}^2 are of about the magnitude of σ_e^2 then the power is less than what it would be if there were no interactions present and the error variance σ_e^2 were roughly four times as large. Etc. If we are considering the effect of interactions in an application of type (iii) discussed at the beginning of this section, where σ_{AB}^2 represents the magnitude of unit errors not eliminated by the two-way grouping of the experimental units, we should compare (9.2.16) with

$$\phi^2 = \frac{m-1}{m}\,\frac{m\sigma_C^2}{\sigma_e^2 + \sigma_{AB}^2}.$$

From (9.2.6) and (9.2.7) we see that the observed treatment mean $y_{..k}$ is an unbiased estimate of the true treatment mean $\mu+\alpha_k^C$, no matter what interactions are present, and hence, if $\psi = \Sigma_k c_k \alpha_k^C$ $(\Sigma_k c_k = 0)$ is any contrast in the main effects of factor C, then

$$(9.2.17)\quad \hat{\psi} = \sum_k c_k y_{..k}$$

is an unbiased estimate of ψ. The calculation of Var $(\hat{\psi})$ from (9.2.6) and (9.2.11) is straightforward, but the resulting expression for Var $(\hat{\psi})$ looks complicated, and one would not know how to estimate it. It is possible to deduce some simple results about Var $(\hat{\psi})$ from (9.2.12): First suppose that $\sigma_{ABC}^2 = 0$. Then for the estimates $\{y_{..k}\}$ of $\{\mu + \alpha_k^C\}$ it may be argued from (9.2.6) that the m errors of estimate $\{y_{..k}-\mu-\alpha_k^C\}$ are symmetrically distributed, in particular they have equal variances and equal correlation coefficients. This permits the application of Lemma 1 at the end of this section to conclude that

$$\text{Var}(\hat{\psi}) = m^{-1}(\sigma_e^2+\sigma_{AB}^2)\sum_k c_k^2. \tag{9.2.18}$$

It is seen from (9.2.13) that $m^{-1}(\text{MS}_e)\Sigma_k c_k^2$ is an overestimate of Var $(\hat{\psi})$, that is, has positive bias, unless $\sigma_{AC}^2 = \sigma_{BC}^2 = 0$, in which case it is unbiased. In the general case where we do not assume that $\sigma_{ABC}^2 = 0$, if we are interested only in estimating the contrasts which are the differences $\{\alpha_k^C-\alpha_{k'}^C\}$, so that the estimates are $\{y_{..k}-y_{..k'}\}$, then the *average* value of the variance of these estimates, averaged over the $\frac{1}{2}m(m-1)$ differences, is[16]

$$2m^{-1}[\sigma_e^2 + (1-2m^{-1})\sigma_{ABC}^2 + \sigma_{AB}^2]; \tag{9.2.19}$$

this follows from Lemma 2 below. An estimate of this quantity would be useful only for "order-of-magnitude" considerations; the statistic $2m^{-1}\text{MS}_e$ will usually be an overestimate, according to the remark made above (9.2.14).

Two Lemmas

We conclude this section with the proof of the two lemmas we applied above. Both lemmas concern the variances of estimates of contrasts in parameters $\{\theta_1, \cdots, \theta_K\}$, say $\psi = \Sigma_k c_k\theta_k$, where $\Sigma_k c_k = 0$. Suppose that we have unbiased estimates $\{\hat{\theta}_k\}$,

$$\hat{\theta}_k = \theta_k + f_k,$$

so $E(f_k) = 0$. Then $\hat{\psi} = \Sigma_k c_k\hat{\theta}_k$ is an unbiased estimate of ψ. Define

$$Q = \sum_k(\hat{\theta}_k-\hat{\theta}_.)^2/(K-1), \tag{9.2.20}$$

and let

$$V = E(Q) - \sigma_\theta^2,$$

where the symbol σ_θ^2 denotes

$$\sigma_\theta^2 = \sum_k(\theta_k-\theta_.)^2/(K-1).$$

Lemma 1: If the $\{\hat{\theta}_k\}$ have equal variances and equal correlation coefficients then

$$\text{Var}(\hat{\psi}) = V\sum_k c_k^2.$$

[16] This result was given by Wilk and Kempthorne (1957).

Proof: We shall derive a formula for V in the general case where $E(f_k f_{k'}) = \rho_{kk'}\sigma_k\sigma_{k'}$, and then specialize it. If we substitute

$$\hat\theta_k - \hat\theta_. = (\theta_k - \theta_.) + (f_k - f_.)$$

in (9.2.20) and take expected values we get

$$E(Q) = \sigma_\theta^2 + (K-1)^{-1}E\left(\sum_k f_k^2 - Kf_.^2\right).$$

Now

$$E\left(\sum_k f_k^2\right) = \sum_k \sigma_k^2,$$

and

$$E(Kf_.^2) = K^{-1}E\left(\sum_k f_k \sum_{k'} f_{k'}\right) = K^{-1}\sum_k\sum_{k'}\rho_{kk'}\sigma_k\sigma_{k'},$$

and hence

$$V = (K-1)^{-1}\left(\sum_k \sigma_k^2 - K^{-1}\sum_k\sum_{k'}\rho_{kk'}\sigma_k\sigma_{k'}\right). \tag{9.2.21}$$

In the special case where all $\rho_{kk'} = \rho$ for $k \neq k'$, and all $\sigma_k^2 = \sigma_1^2$, the K terms for which $k = k'$ in the double sum in (9.2.21) have the value σ_1^2, and the remaining $K^2 - K$ terms have the value $\rho\sigma_1^2$. Hence

$$\begin{aligned} V &= (K-1)^{-1}\{K\sigma_1^2 - K^{-1}[K\sigma_1^2 + K(K-1)\rho\sigma_1^2]\} \\ &= \sigma_1^2(1-\rho). \end{aligned}$$

On the other hand, if in

$$\text{Var}(\hat\psi) = \sum_k\sum_{k'} c_k c_{k'}\rho_{kk'}\sigma_k\sigma_{k'}$$

we substitute $\rho_{kk'}\sigma_k\sigma_{k'} = \sigma_1^2[\rho + \delta_{kk'}(1-\rho)]$, we get

$$\begin{aligned} \text{Var}(\hat\psi) &= \sigma_1^2[\rho\sum_k\sum_{k'} c_k c_{k'} + (1-\rho)\sum_k\sum_{k'} c_k c_{k'}\delta_{kk'}] \\ &= \sigma_1^2(1-\rho)\sum_k c_k^2 = V\sum_k c_k^2. \end{aligned}$$

Lemma 2: The average value of the variance of the $\frac{1}{2}K(K-1)$ differences $\{\hat\theta_k - \hat\theta_{k'}\}$ is $2V$.

Proof: To calculate the average variance we may take $[K(K-1)]^{-1}$ times the sum of Var $(\hat\theta_k - \hat\theta_{k'})$ over all k,k' for which $k \neq k'$, where each term occurs twice; but this sum is the same as that over all k and k', since the terms for $k = k'$ are zero: The average variance is thus $[K(K-1)]^{-1}$ times

$$\begin{aligned} \sum_k\sum_{k'} \text{Var}(\hat\theta_k - \hat\theta_{k'}) &= \sum_k\sum_{k'}(\sigma_k^2 + \sigma_{k'}^2 - 2\rho_{kk'}\sigma_k\sigma_{k'}) \\ &= 2K\sum_k \sigma_k^2 - 2\sum_k\sum_{k'}\rho_{kk'}\sigma_k\sigma_{k'} = 2K(K-1)V, \end{aligned}$$

the last step following from (9.2.21). Thus the average variance is $2V$.

9.3. PERMUTATION TESTS

We now turn to the problem of hypothesis testing under the randomization models we have defined and used earlier in this chapter for estimation in the randomized-blocks and Latin-square designs. Under randomization models exact tests of certain hypotheses are possible, which are called permutation tests.[17] Permutation tests for a hypothesis exist whenever the joint distribution of the observations under the hypothesis has a certain kind of symmetry, namely, when there exists a set of permutations of the observations which leave the distribution the same (the distribution is invariant under a group of permutations). The tests are exact for a very broad kind of hypothesis in the sense that the validity of the calculated significance level depends only on the symmetry of the distribution and not on any further assumptions such as normality, homogeneity of variance, or independence. Such symmetry may be generated in three ways: (i) by the assumption of random sampling from one or more populations, as in the first example below, (ii) by an actual randomization in assigning treatments to experimental units, as in the randomized-blocks design, or (iii) by an actual randomization in choosing from the totality of the treatment combinations of a complete layout for several factors an incomplete layout of treatment combinations to be used in the experiment, as in the Latin-square design for case (i) of sec. 9.2.

We shall see that permutation tests are easy to define, but that the numerical calculations required to carry them out are usually hopelessly tedious. Their main interest for us is that the F-test derived for the corresponding hypothesis under a fixed-effects normal-theory model including assumptions of normality, independence, and equality of variance, or a slight modification of this test, can often be regarded as a good approximation to a permutation test, which is an exact test under a less restrictive model.

Before attempting to define permutation tests in general, we shall consider a numerical example which, while artificially simple, may serve to convey the intuitive idea. Suppose that it is desired to test the hypothesis H that a sample of three and a sample of four are independent random samples from the same population, and suppose the observations in the two samples, each arranged in increasing order, are

$$\{0, 3, 5\} \quad \text{and} \quad \{2, 3, 6, 9\}. \tag{9.3.1}$$

A permutation test is in practice usually based on some chosen statistic. Suppose we choose the statistic that would be appropriate under the usual

[17] Also called randomization tests. The idea was originated by R. A. Fisher (1925), sec. 24; (1935), sec. 21.

normal-theory assumptions for testing the above hypothesis H against two-sided alternatives of translation, i.e., against alternatives that the two samples are independent random samples from populations differing by a translation. We then choose the statistic

$$(9.3.2) \qquad |t| = c_1|\bar{x}-\bar{z}|S^{-1/2},$$

where large values of $|t|$ are significant, c_1 is a constant of no present interest, $\bar{x}$ and $\bar{z}$ are the sample means, S is the pooled SS

$$S = \sum_1^3(x_i-\bar{x})^2 + \sum_1^4(z_i-\bar{z})^2,$$

and $\{x_1, x_2, x_3\}$ and $\{z_1, z_2, z_3, z_4\}$ denote the samples. To facilitate treatment of the tie caused by the occurrence of the two 3's in (9.3.1), let us put subscripts 1 and 2 on the 3's, treating them as different in counting permutations or combinations involving the 3's, but assigning the value 3 to both in calculating the numerical value of the statistic $|t|$. The combined sample is then

$$(9.3.3) \qquad \{0, 2, 3_1, 3_2, 5, 6, 9\}.$$

The word "sample" in the examples referring to (9.3.1), but not elsewhere in this section, is to be understood as "sample arranged in increasing order." The number of ways into which this sample of seven can be combined into samples of three and four is

$$C_3^7 = 7!(4!\,3!)^{-1} = 35,$$

and under the hypothesis H stated above (9.3.1) all these samples are equally likely, in the sense that the conditional probability of the two samples being any particular combinations of three and four, given that the combined sample was (9.3.3), is 1/35. In principle the permutation test based on the statistic $|t|$ is made by choosing a significance level α, considering the 35 values of $|t|$, not all different, that would be calculated for the 35 combinations, noting for what proportion of the 35 combinations the value of $|t|$ is $\geqq$ the value of $|t|$ for the observed combination (9.3.1), and rejecting H if and only if this proportion is $\leqq \alpha$. To see that the resulting test controls the probability of type-I error (rejecting H when true) in the desired way, we note that if H is true, the conditional probability of rejecting H, given the ordered sample, is $\leqq \alpha$, and hence the unconditional probability of rejecting H, being the expectation, over all ordered samples, of the conditional probability, is also $\leqq \alpha$.

In practice several simplifications are possible in attaining the same result: First, it would be sufficient to calculate merely $d = |\bar{x}-\bar{z}|$ for the 35 combinations, and reject H for large values of this statistic instead of

$|t|$, since, given the combined sample, $|t|$ is a strictly increasing function of d: For any of the 35 combinations we have the identity

$$\text{SS}_A + \text{S} = A, \tag{9.3.4}$$

where SS_A is the between-groups SS for the two samples, S is the pooled error SS defined above, and A is the total SS about the grand mean, which is the same for all 35 combinations. We may calculate that

$$\text{SS}_A = c_2 d^2, \tag{9.3.5}$$

where $c_2 > 0$. Now $|t|$ is a strictly increasing function of t^2, and the formula

$$t^2 = c_1^2 d^2/(A - c_2 d^2),$$

obtained from (9.3.2), (9.3.4), and (9.3.5), shows that t^2 is a strictly increasing function of d.

Next, let $B = \Sigma_i x_i + \Sigma_j z_j$, which is the same for all the combinations. Then

$$\bar{x} - \bar{z} = m^{-1} \sum_{i=1}^{m} x_i - r^{-1} \sum_{j=1}^{r} z_j = m^{-1}\Sigma x_i - r^{-1}(B - \Sigma x_i)$$
$$= (m^{-1} + r^{-1})(\Sigma x_i - C),$$

where $C = mB/(m+r)$, and in the above example $m = 3$, $r = 4$, $B = 28$, $C = 12$. The second simplification is to base the test on the statistic $|\Sigma x_i - C|$. The 35 combinations and the corresponding values of this statistic are shown in Table 9.3.1. From the table we see that the value of the statistic for the observed combination is 4 and that 15/35 of the combinations give a value $\geqq 4$. The probability (conditional, given that (9.3.3) was observed) of obtaining a value of $|t|$ greater than that observed is thus 15/35, and the hypothesis would be rejected by the permutation test only at significance levels $\geqq 15/35$.

TABLE 9.3.1

VALUES OF THE STATISTIC $|\Sigma x_i - C|$, $C = 12$

First Sample	0	0	0	0	0	0	0	0	0	0	0	0	0	0	0	2	2	2
	2	2	2	2	2	3_1	3_1	3_1	3_1	3_2	3_2	3_2	5	5	6	3_1	3_1	3_1
	3_1	3_2	5	6	9	3_2	5	6	9	5	6	9	6	9	9	3_2	5	6
$\|\Sigma x_i - C\|$	7	7	5	4	1	6	4	3	0	4	3	0	1	2	3	4	2	1

First Sample	2	2	2	2	2	2	2	3_1	3_1	3_1	3_1	3_1	3_1	3_2	3_2	3_2	5
	3_1	3_2	3_2	3_2	5	5	6	3_2	3_2	3_2	5	5	6	5	5	6	6
	9	5	6	9	6	9	9	5	6	9	6	9	9	6	9	9	9
$\|\Sigma x_i - C\|$	2	2	1	2	1	4	5	1	0	3	2	5	6	2	5	6	8

The significance level should be chosen before the calculations are made, and then a third simplification is possible: Suppose for example we choose a significance level of 10 per cent. The hypothesis will be accepted if four of the 35 combinations give the statistic $|\Sigma x_i - C|$ a value $\geqq$ the observed value, hence if three other combinations besides the observed give such a value. We need now calculate only a few of the values in the table: We calculate first that the value of $|\Sigma x_i - C|$ for the observed combination is 4. We reflect that the value of $|\bar{x}-\bar{z}|$ will be largest if we make $\bar{x}$ as small as possible or as large as possible (in which case $\bar{z}$ is respectively as large or as small as possible). The smallest values for $\bar{x}$ will be obtained for the combinations $\{0, 2, 3_1\}$ and $\{0, 2, 3_2\}$, the largest for $\{5, 6, 9\}$. The values of the statistic are calculated for these three combinations, and since all are $\geqq 4$ the hypothesis is accepted.

Finally, we remark that it would not even be necessary to calculate the last three values if it could be seen by any easier method that they are $\geqq 4$. These simplifications apply to the exact calculation for a permutation test; approximation of permutation tests will be considered below.

It is usually easier to make the necessary calculations in cases where the permutation test accepts a hypothesis H, as above, then when it rejects: If the number of equally likely samples under H is N ($N = 35$, above), and if M is the smallest integer $> \alpha N - 1$, it suffices to accept H if it is possible to find by any method M samples other than the observed which give the statistic a value $\geqq$ its observed value, whereas if the test rejects H it may be difficult to show that all such combinations have been located without making calculations for many more.

The above hypothesis H could be tested against one-sided alternatives that the x-population differs from the z-population by a translation to the right by basing the permutation test on $\bar{x}-\bar{z}$ instead of on the statistic $|\bar{x}-\bar{z}|$ used above in the two-sided case. In these cases of alternatives of translation we might consider basing the permutation test on the statistic $\tilde{x}-\tilde{z}$ or $|\tilde{x}-\tilde{z}|$, where $\tilde{x}$ and $\tilde{z}$ are the sample medians. However, for the reason indicated above, we shall be mainly interested in permutation tests based on the same statistic that would be used in normal-theory tests. If the above hypothesis H were to be tested against alternatives that the populations differ in scale, not just in location,[18] we could base the permutation test on the statistic

$$\mathfrak{F} = c\left(\sum_i x_i^2 - m\bar{x}^2\right)\Big/\left(\sum_j z_j^2 - r\bar{z}^2\right).$$

This would be appropriate against alternatives that the x-population has a larger spread than the z-population, if we define permutation tests based

[18] As defined, say, by the population median.

on a statistic so that large values are significant.[19] If we wished to test H at the α level of significance against two-sided alternatives of difference of scale, a possibility would be to make a $\frac{1}{2}\alpha$-level permutation test based on $\mathfrak{F}$ and another on $1/\mathfrak{F}$, and reject H if either test rejects it. Just as the statistic could be expressed in terms of Σx_i alone in the test based on $|t|$, here the statistic could be expressed in terms of Σx_i and Σx_i^2.

The general idea of permutation tests can be distilled from the above example: Permutation tests can be constructed for testing a hypothesis H under which the distribution has the kind of symmetry mentioned above (invariance under a group of permutations). We shall now formulate this symmetry property in a way that is mathematically equivalent, less brief, but easier for application. Let the vector $\mathbf{y} = (y_1, \cdot \cdot \cdot, y_n)'$ of random variables denote the observation vector or sample, and let $\mathbf{y}_0 = (y_{10}, \cdot \cdot \cdot, y_{n0})'$ denote a possible value of $\mathbf{y}$. (It is necessary here to distinguish in the notation between the random sample $\mathbf{y}$ which can take on different possible values and one of the possible values $\mathbf{y}_0$ that $\mathbf{y}$ can take on. In applications of the definition we are formulating we may think of $\mathbf{y}_0$ as the possible value of $\mathbf{y}$ that was actually observed.) Suppose there exists a set G consisting of L permutations of the n elements of $\mathbf{y}$ with the following symmetry property: Let $S(\mathbf{y}_0)$ denote the set of L samples generated by applying the set G of permutations to $\mathbf{y}_0$. Then the required symmetry property is that, given the condition that $\mathbf{y}$ fell in the set $S(\mathbf{y}_0)$, the conditional probability under H that $\mathbf{y}$ took on any particular one of the L values in $S(\mathbf{y}_0)$ is $1/L$ for each of the L values in $S(\mathbf{y}_0)$; this symmetry property must hold for all values $\mathbf{y}_0$ that it is possible for $\mathbf{y}$ to take on. If $\mathbf{y}_0$ has some elements equal they must be distinguished in assigning the L equal probabilities, as was done in the above example by tagging the elements of the tie. A *permutation test* of the hypothesis H at significance level α is a rule[20] that defines for each possible set $S(\mathbf{y}_0)$ a

[19] If we wish to reject in some other range of a statistic (possibly, large absolute values, small algebraic values, or values near zero) we can always get the same result by redefining the statistic—for example, we used $|t|$ instead of t, above.

[20] For a more rigorous mathematical treatment of permutation tests see Scheffé (1943), Hoeffding (1952), or Lehmann (1959*a*). From a more advanced mathematical viewpoint than is adopted in this book certain technicalities would have to be considered which complicate the simple intuitive idea: Thus, the sets $S'(\mathbf{y}_0)$ have to be chosen so that their union for all $\mathbf{y}_0$ is a Borel set in the sample space. This will be the case if the test is based on a Borel-measurable statistic. Furthermore, there is some difficulty about the precise definition of conditional probabilities given that $\mathbf{y} \in S(\mathbf{y}_0)$, since the last event is one of probability zero if $\mathbf{y}$ has a continuous distribution. Finally, for any chosen α it is possible to make the probability of type-I error exactly equal to α instead of merely $\leqq \alpha$, with a resultant gain of power, by introducing into the decision to accept or reject H another randomization *after* the data have been obtained, for certain outcomes of the data. Although this is a useful computational artifice for comparing the powers of tests, I would find its employment in real applications reprehensible.

subset $S'(\mathbf{y}_0)$ of samples whose number is $\leqq \alpha L$, the hypothesis H to be rejected if and only if the observed sample $\mathbf{y}_0$ is in the subset $S'(\mathbf{y}_0)$. The subset $S'(\mathbf{y}_0)$ of "significant" samples is usually defined by means of a statistic as in the above examples.

Let us illustrate the general definition of permutation test in the case considered above: The vector $\mathbf{y}$ of the general definition is now $(x_1, \cdots, x_m, z_1, \cdots, z_r)'$, with $n = m+r$. The set G of permutations which has the required property is in this case the set of all permutations of the elements of $\mathbf{y}$, so the number L of permutations in G is $n!$ The value of any statistic that is symmetric in the x's and symmetric in the z's, like the examples t, $|t|$, F, $1/F$ above, will not be affected by the permutations in G which permute only the x's among themselves and only the z's among themselves; there are $m!r!$ such permutations in G, and instead of $L = n!$ equally likely permutations we may consider $N = n!(m!r!)^{-1}$ equally likely combinations,[21] as we did above. The application of the rest of the general definition to the above case should now be clear.

Permutation Test for Randomized Blocks

If we consider the very general model in which the observation on the ith treatment in the jth block has the structure

$$y_{ij} = \mu + \alpha_i + \beta_j + \gamma_{ij} + \tilde{e}_{ij} + e_{ij},$$

where the unit error $\tilde{e}_{ij}$ and technical error e_{ij} are

$$\tilde{e}_{ij} = \sum_\nu d_{ij\nu}\varepsilon_{ij\nu}, \qquad e_{ij} = \sum_\nu d_{ij\nu}e_{ij\nu},$$

and the random variables and constants (denoted respectively by Latin and Greek letters) are subject to the conditions stated in sec. 9.1, we find that the distribution does not have sufficient symmetry under the hypothesis H_A: all $\alpha_i = 0$ to permit a permutation test. We can however construct a permutation test of the hypothesis H that there is absolutely no difference between treatments in the sense that (i) $\sigma_A^2 = \sigma_{AB}^2 = \sigma_{AU}^2 = 0$, and (ii) the joint distribution of the technical errors on the IJ units does not depend on how the treatments are assigned to the units. We may then write under H

$$y_{ij} = \mu + \beta_j + \sum_\nu d_{ij\nu}(\xi_{j\nu} + u_{j\nu}), \tag{9.3.6}$$

where $u_{j\nu}$ is the technical error associated with the j,ν unit, formerly written $e_{ij\nu}$, which is now assumed not to depend on the treatment applied to the

[21] In the analysis of variance the calculations for the exact permutation test can be made for combinations instead of permutations in the case of the one-way layout.

unit. By definition, $E(u_{j\nu}) = 0$; but otherwise the joint distribution of the $\{u_{j\nu}\}$ is arbitrary.

Permutation tests will be based on the set G of permutations within blocks: A member of G is the result of making within each of the J blocks one of the $I!$ possible permutations of the observations in the block. The number of permutations in G is thus $L = (I!)^J$. Denote by $\mathbf{y}_0$ the observed sample of $\{y_{ij}\}$, and by $S(\mathbf{y}_0)$ the set of $(I!)^J$ samples obtained from $\mathbf{y}_0$ by applying the set G of permutations. It may be easier to see that, under H and the randomization incorporated into the experiment, all samples in $S(\mathbf{y}_0)$ have the same conditional probability, given that the observed sample was in $S(\mathbf{y}_0)$, if we rewrite (9.3.6) as

$$y_{ij} = \sum_{\nu} d_{ij\nu} z_{j\nu}, \tag{9.3.7}$$

where

$$z_{j\nu} = \mu + \beta_j + \xi_{j\nu} + u_{j\nu} \tag{9.3.8}$$

is the observation on the j,ν experimental unit; however, this may now be evident to the reader without following the small steps of our argument in the rest of this paragraph, in which case he may wish to hurdle them. We may think of starting with an arbitrary joint distribution of the $\{z_{j\nu}\}$; then under H

$$E(z_{j\nu}) = \mu + \beta_j + \xi_{j\nu}$$

by definition, and hence $E(u_{j\nu}) = 0$ from (9.3.8). Denote by $\mathbf{z}$ the vector $(z_{11}, \cdots, z_{1I}, z_{21}, \cdots, z_{2I}, \cdots, z_{J1}, \cdots, z_{JI})'$ and by $\mathbf{z}_0$ the value that $\mathbf{z}$ takes on in the experiment. A particular assignment of treatments to units in the randomization determines the $\{d_{ij\nu}\}$ and hence $\mathbf{y}$ in terms of $\mathbf{z}$ by (9.3.7), all such assignments having equal probability. Another way of saying this is that the elements of $\mathbf{y}$ are formed from the elements of $\mathbf{z}$ by one of the permutations in G and all the permutations have equal probability. It is clear now that, given $\mathbf{z} = \mathbf{z}_0$, all $\mathbf{y}$ in $S(\mathbf{z}_0)$ have equal probability, where $S(\mathbf{z}_0)$ denotes the set of samples generated by applying the permutations in G to the elements of $\mathbf{z}_0$. Since for any $\mathbf{z}$ in $S(\mathbf{z}_0)$, $S(\mathbf{z})$ equals $S(\mathbf{z}_0)$, we may now say that, given $\mathbf{z}$ is in $S(\mathbf{z}_0)$, all $\mathbf{y}$ in $S(\mathbf{z}_0)$ have equal probability. Finally, $S(\mathbf{z}_0)$ equals $S(\mathbf{y}_0)$, and $\mathbf{z}$ is in $S(\mathbf{z}_0)$ if and only if $\mathbf{y}$ is in $S(\mathbf{y}_0)$; so we get the desired result that, given that $\mathbf{y}$ is in $S(\mathbf{y}_0)$, all $\mathbf{y}$ in $S(\mathbf{y}_0)$ have the same conditional probability.

To carry out the exact permutation test of H based on the statistic

$$\mathscr{F} = c\text{SS}_A/\text{SS}_e,$$

where the SS's are those defined in sec. 4.2, we would note first that $\text{SS}_A + \text{SS}_e$ has a constant value in the set $S(\mathbf{y}_0)$ since

$$\text{SS}_A + \text{SS}_e = \text{SS}_{\text{tot}} - \text{SS}_B - IJy_{..}^2,$$

and the three terms on the right are constant, the sum of squares SS_B for blocks depending only on the block totals. It follows that $\mathscr{F}$ is a strictly increasing function of SS_A, and hence of the sum of squares of the treatment totals $\Sigma_i(\Sigma_j y_{ij})^2$. The permutation test based on this statistic would thus be equivalent to the permutation test based on $\mathscr{F}$. Since the value of the statistic is not affected by a permutation of the treatments, the $(I!)^J$ values are equal in sets of $(I!)$, that is, there are $(I!)^{J-1}$ values assumed with equal probability under H. We might calculate these by holding the observations in the first block fixed and making the $(I!)^{J-1}$ permutations of the observations within the other blocks. Usually this number of permutations is so large as to render the calculation impracticable. We therefore consider approximating the exact calculation.

We shall now consider the permutation test to be based on the statistic

$$U = SS_A/(SS_A + SS_e);$$

this is equivalent to basing it on $\mathscr{F}$. The reason for choosing U instead of SS_A is that it will be useful both for approximating the exact calculation and for comparing the permutation test with the normal-theory test: The normal-theory test based on the F-distribution of $\mathscr{F}$ is equivalent to one based on the β-distribution (defined below) of U, but is not equivalent to one based on the chi-square distribution of SS_A. The statistic U is more convenient than $\mathscr{F}$ because the proposed approximation and comparison utilize the moments of the statistic, and U, unlike $\mathscr{F}$, has a constant denominator under the permutations in G.

We shall assume at first that there are no technical errors, i.e., $\sigma_e^2 = 0$, hence all $u_{j\nu} = 0$. The conditions we have now assumed, namely

$$\sigma_A^2 = \sigma_{AB}^2 = \sigma_{AU}^2 = \sigma_e^2 = 0, \tag{9.3.9}$$

are a special case of the general conditions under which the formulas (9.1.19) for the E(MS)'s are valid.[22] Hence

$$E(SS_A) = (I-1)\sigma_U^2, \qquad E(SS_e) = (I-1)(J-1)\sigma_U^2,$$

where

$$\sigma_U^2 = J^{-1}(I-1)^{-1}\sum_j\sum_\nu \xi_{j\nu}^2.$$

The constant value of $SS_A + SS_e$ may be calculated as

$$SS_A + SS_e = E(SS_A + SS_e) = J(I-1)\sigma_U^2 = \sum_j\sum_\nu \xi_{j\nu}^2.$$

The expected value of U is thus

$$E(U) = (SS_A + SS_e)^{-1} E(SS_A) = J^{-1}. \tag{9.3.10}$$

[22] To derive (9.1.19) the general randomization model is specialized to the case of independent technical errors, and this is achieved here by the assumption $\sigma_e^2 = 0$.

The variance of U is tedious and not straightforward to calculate; we state the result without proof[23]:

(9.3.11) $$\text{Var}(U) = 2J^{-2}(I-1)^{-1}[1-J^{-2}\sigma_U^{-4}\sum_j \sigma_{U,j}^4],$$

where

$$\sigma_{U,j}^2 = (I-1)^{-1}\sum_\nu \xi_{j\nu}^2$$

may be called the "block variance" of the jth block. Under the assumptions (9.3.9) the block variances could be calculated without error from the observations $\{y_{ij}\}$, since from (9.3.8) with $u_{j\nu} = 0$, $\xi_{j\nu} = z_{j\nu} - z_{j.}$, and hence

$$\sigma_{U,j}^2 = (I-1)^{-1}\sum_\nu (z_{j\nu}-z_{j.})^2,$$

which is the same as

(9.3.12) $$\sigma_{U,j}^2 = (I-1)^{-1}\sum_i (y_{ij}-y_{.j})^2.$$

The result (9.3.11) may be expressed in terms of

$$V = \frac{\sum_j (\sigma_{U,j}^2 - \sigma_U^2)^2}{(J-1)(\sigma_U^2)^2},$$

the square of the coefficient of variation of the block variances, as[24]

(9.3.13) $$\text{Var}(U) = \frac{2(J-1)}{J^3(I-1)}\left(1 - \frac{V}{J}\right).$$

Under the assumptions (9.3.9) the statistic U has a discrete distribution with the range $0 \leqq U \leqq 1$. Suppose we approximate this by a (continuous) β-distribution with the same mean and variance. A random variable X is said to have the *β-distribution* with ν_1 and ν_2 d.f. (usually called the β-distribution with indices $\nu_1/2$ and $\nu_2/2$) if for $0 \leqq x \leqq 1$,

$$\Pr\{X \leqq x\} = c\int_0^x t^{(\nu_1-2)/2}(1-t)^{(\nu_2-2)/2}dt,$$

where

$$1/C = \int_0^1 t^{(\nu_1-2)/2}(1-t)^{(\nu_2-2)/2}dt.$$

The mean and variance of the β-variable are easily shown[25] to be

(9.3.14) $$E(X) = \frac{\nu_1}{\nu_1+\nu_2}, \qquad \text{Var}(X) = \frac{2\nu_1\nu_2}{(\nu_1+\nu_2)^2(\nu_1+\nu_2+2)}.$$

[23] Welch (1937), pp. 26–27; verified by Pitman (1937).

[24] This useful expression is given by Kempthorne (1952), p. 142.

[25] By calculating $E(X)$ and $E(X^2)$ by integration by parts, or by expressing $E(X^k)$ as a quotient of complete-beta functions, substituting for these in terms of gamma functions, and utilizing the recurrence formula $\Gamma(z+1) = z\Gamma(z)$ for the gamma functions.

If we equate these to the values of $E(U)$ and $\text{Var}(U)$, and solve for ν_1 and ν_2, we find

$$\nu_1 = \phi(I-1), \qquad \nu_2 = \phi(I-1)(J-1),$$

where

$$\phi = \frac{1}{1 - J^{-1}V} - \frac{2}{J(I-1)}. \tag{9.3.15}$$

Our approximation to the permutation test of H based on the statistic U consists in rejecting H if the value of U for the observed sample is $\geqq$ the upper α point of the β-distribution with the d.f. ν_1 and ν_2 found above.

If X is a β-variable with ν_1 and ν_2 d.f. then $\nu_2 X/[\nu_1(1-X)]$ is a strictly increasing function of X, and has the F-distribution with ν_1 and ν_2 d.f. Hence it is equivalent to reject H if $U \geqq$ the upper α point of the β-distribution with ν_1 and ν_2 d.f. and to reject H if $\mathscr{F} \geqq F_{\alpha;\nu_1,\nu_2}$. Our approximation to the permutation test of H based on the statistic $\mathscr{F}$ is thus equivalent to modifying the normal-theory test by multiplying the numbers of d.f. of the F-distribution by the factor ϕ. The factor ϕ may be calculated from the observations $\{y_{ij}\}$ by using (9.3.15) with

$$\frac{V}{J} = \frac{1}{J-1}\left[\frac{J\sum_j S_j^2}{\left(\sum_j S_j\right)^2} - 1\right], \tag{9.3.16}$$

where

$$S_j = \sum_i y_{ij}^2 - I^{-1}(\sum_i y_{ij})^2.$$

Ordinarily if the factor ϕ deviates from unity by a nonnegligible amount this deviation will be positive. The test is then more sensitive than the unadjusted normal-theory test: Inspection of the F-tables shows that for $\alpha \leqq 0.1$ and $\nu_2 > 2$, $F_{\alpha;\nu_1,\nu_2}$ is a decreasing function of ν_1 and of ν_2; thus values of the statistic $\mathscr{F}$ which miss significance with the unadjusted numbers of d.f. may achieve it with the adjusted numbers.

It has been shown[26] that the β-distribution fitted in this way by the first two moments of U also fits closely with respect to the third and fourth moments unless V/J is near 1.

We shall now justify the above approximation to the permutation test in the case where the technical errors $\{u_{j\nu}\}$ have an arbitrary distribution subject only to $E(u_{j\nu}) = 0$. Consider conditional distributions, given the $\{u_{j\nu}\}$: The terms $\{u_{j\nu}\}$ in (9.3.6) are then constants, and the equation (9.3.6) for the structure of the observations may be written

$$y_{ij} = \mu' + \beta_j' + \sum_\nu d_{ij\nu}\xi_{j\nu}',$$

[26] Pitman (1937), pp. 331–333; his $K = (1-J^{-1})(1-J^{-1}V)$.

where

$$\xi'_{j\nu} = \xi_{j\nu} + u_{j\nu} - u_{j.}, \qquad \beta'_j = \beta_j + u_{j.} - u_{..}, \qquad \mu' = \mu + u_{..},$$

and $\xi'_{j.} = 0$ for all j, $\beta'_. = 0$, so that the observations are distributed conditionally like those in (9.3.6) with the $\{\xi_{j\nu}\}$ replaced by the $\{\xi'_{j\nu}\}$, the $\{\beta_j\}$ by the $\{\beta'_j\}$, μ by μ', and the $\{u_{j\nu}\}$ in (9.3.6) replaced by zero. The present conditional distribution of U is thus the same as the distribution of U in the preceding case where the $\{u_{j\nu}\}$ are zero and with the $\{\xi_{j\nu}\}$ replaced by the $\{\xi'_{j\nu}\}$, the former distribution of U not depending on μ or the $\{\beta_j\}$. In particular, if in the definition of the block variances $\sigma^2_{U,j}$ we replace $\xi_{j\nu}$ by $\xi'_{j\nu}$, these "conditional" block variances are still given in terms of the observations $\{y_{ij}\}$ by (9.3.16). The approximation to the "conditional" permutation test is thus made by carrying out the normal-theory test with the numbers of d.f. adjusted by the factor ϕ calculated from (9.3.15) and (9.3.16); actually ϕ is now a random variable and only conditionally a constant. However, if the test conditionally controls the type-I error so that its probability is approximately α, the same will be true unconditionally.

It is a common belief among statisticians that for tests of means,[27] referring the usual statistic $\mathscr{F}$ to the F-tables in the way derived under normal theory gives a good approximation to the result of an exact permutation test based on the F-statistic under an appropriate randomization model. The author has had difficulty in trying to formulate clearly the sense in which this approximation is expected to hold. In the normal theory we reject the hypothesis at significance level α if $\mathscr{F} \geqq F_\alpha$ where F_α is a constant which does not depend on the observed sample $\mathbf{y}_0$. In the exact permutation test we may say that we reject if $\mathscr{F} \geqq F_\alpha(\mathbf{y}_0)$, where the value of $F_\alpha(\mathbf{y}_0)$ depends on the observed sample $\mathbf{y}_0$. We learned above how to approximate $F_\alpha(\mathbf{y}_0)$ in the case of randomized blocks by modifying the numbers of d.f. of the normal-theory F_α.

Most of the evidence in favor of this belief concerning the approximation of permutation tests by normal-theory tests is for cases where the observed sample $\mathbf{y}_0$ is subject to certain restrictions;[28] for example, it follows from the above discussion that the normal-theory test is a good approximation

[27] The distinction between tests of means and tests of variances will be clarified in sec. 10.2, the paragraph after that containing (10.2.8).

[28] However, Hoeffding (1952) proved that for randomized blocks, as J increases with fixed I, and under certain assumptions on the sequence of *distributions* of the observations, the random-variable "significance level" $F_\alpha(\mathbf{y})$ of the permutation test approaches a constant in probability. With this he was able to show that the permutation test has in a certain sense asymptotically the same power as the usual F-test against alternatives of the normal-theory model. Of course, what we would really like to know more about is the power of the usual F-test against the alternatives allowed by the randomization model.

to the permutation test if the function ϕ of $\mathbf{y}_0$ is near unity, which will usually be true if V/J is small compared with unity. The evidence is in general of three kinds:

First, *numerical* examples have been published[29] for particular $\mathbf{y}_0$, in which it transpires that $F_\alpha(\mathbf{y}_0)$ is close to the normal theory F_α.

Second, there are some asymptotic calculations. In the randomized-blocks design as the number J of blocks increases with the number I of treatments fixed, the limiting distribution of the statistic $\mathscr{F}$ under the normal-theory model is easily seen to be that of χ^2_{I-1}. It has been shown[30] that, as J increases with fixed I, if the sequence of $\{\mathbf{y}_0\}$ satisfies certain restrictions, then the permutation distribution of the statistic $\mathscr{F}$ has the same limiting form. A similar calculation[31] has been carried out for the one-way layout.

Third, there are moment calculations. These are usually made on the transform U of the statistic $\mathscr{F}$ as above, where $U = SS_H/(SS_H + SS_e)$, and SS_H and SS_e are respectively the numerator SS and the denominator SS of $\mathscr{F}$. We shall discuss here only the moment calculations for the randomized-blocks design. We found above for the permutation distribution of U, i.e., the discrete distribution which assigns equal probabilities to the L samples in $S(\mathbf{y}_0)$, that $E(U)$ and Var (U) are given by (9.3.10) and (9.3.13). Under normal theory $\mathscr{F} = MS_A/MS_e$ has the F-distribution with $(I-1)$ and $(I-1)(J-1)$ d.f. when H is true, and hence $U = SS_A/(SS_A+SS_e)$ is a β-variable with these numbers of d.f. From (9.3.14) it then follows that under normal theory

$$E(U) = J^{-1}, \qquad \text{Var}\,(U) = \frac{2(J-1)}{J^3(I-1)}\left[1 + \frac{2}{J(I-1)}\right]^{-1}.$$

Comparing with (9.3.10) and (9.3.13), we see that $E(U)$ is exactly the same in the permutation distribution as in the normal-theory distribution, but that Var (U) may differ somewhat, the ratio of Var (U) in the permutation distribution to Var (U) under normal theory being

$$\left(1 - \frac{V}{J}\right)\left[1 + \frac{2}{J(I-1)}\right].$$

[29] R. A. Fisher (1935), sec. 21; Welch (1937), p. 31; Pitman (1937), p. 334; Kempthorne (1952), p. 152; Eden and Yates (1933). The results of Eden and Yates can be regarded as a comparison with values in the F-tables of estimates of $F_\alpha(\mathbf{y}_0)$ obtained by empirical sampling of the permutation distribution of the statistic $\mathscr{F}$, for various levels α and a single set of "observations" $\mathbf{y}_0$ (not the actual observations but averages of sets of eight observations in a uniformity trial in randomized blocks; also, they use $z = \frac{1}{2}\log \mathscr{F}$ instead of $\mathscr{F}$). This paper is clarified by a discussion between Neyman and Yates (1935), pp. 164, 165.

[30] Wald and Wolfowitz (1944).

[31] Silvey (1954).

If V/J is small compared with unity this ratio will usually be close to unity; the ratio is exactly unity when the above correction factor ϕ for the numbers of d.f. is unity.

It is somewhat surprising[32] to find good agreement under H between the permutation distribution and the normal-theory distribution of U, or equivalently, of $\mathscr{F} = \text{MS}_A/\text{MS}_e$, since the joint distribution of MS_A and MS_e is very different in the two cases: In either case denote $E(\text{MS}_e)$ by σ^2. Then in the normal-theory distribution MS_A and MS_e are statistically independent with

$$E(\text{MS}_A) = E(\text{MS}_e) = \sigma^2,$$

$$\text{Var}\,(\text{MS}_A) = 2\nu_A^{-1}\sigma^4, \qquad \text{Var}\,(\text{MS}_e) = 2\nu_e^{-1}\sigma^4,$$

where $\nu_A = I-1$ and $\nu_e = (I-1)(J-1)$, whereas in the permutation distribution MS_A and MS_e are completely dependent, since $\nu_A\text{MS}_A + \nu_e\text{MS}_e$ = constant, with the following moments,

$$E(\text{MS}_A) = E(\text{MS}_e) = \sigma^2,$$

$$\text{(9.3.17)} \quad \text{Var}\,(\text{MS}_A) = 2\nu_A^{-1}\sigma^4(1-J^{-1})(1-J^{-1}V),$$

$$\text{Var}\,(\text{MS}_e) = 2\nu_e^{-1}\sigma^4 J^{-1}(1-J^{-1}V),$$

where (9.3.17) is easily derived from (9.3.13). Thus the means of MS_A and MS_e are the same in both distributions, Var (MS_A) differs a little, but Var (MS_e) differs enormously, by a factor of more than J.

It has been shown[33] that when $V = 0$ the permutation distribution of U agrees well with its normal-theory distribution with respect to the third and fourth moments.

Little is known about the power of F-tests against alternatives in the randomization models which are not normal-theory alternatives, beyond what is suggested by the formulas for the E(MS)'s.

Permutation Test for Latin Squares

With the model defined by (9.2.1) there exists no permutation test of the hypothesis H_C: $\sigma_C^2 = 0$, namely, that the main effects of factor C

[32] The phenomenon discussed below did not surprise the anonymous reader who reviewed the manuscript of this book for the publisher: He pointed out that the joint distribution of MS_A and MS_e in the permutation theory may be regarded as conditional, given certain conditions that imply $\nu_A\text{MS}_A + \nu_e\text{MS}_e$ = constant. It is easily calculated that in the normal theory $\mathscr{F} = \text{MS}_A/\text{MS}_e$ is statistically independent of $\text{SS}_A + \text{SS}_e$. This means that entirely within the framework of normal theory the conditional distribution of $\mathscr{F}$, given $\nu_A\text{MS}_A + \nu_e\text{MS}_e$ = constant, is the same as unconditionally, when MS_A and MS_e are independent.

[33] Pitman (1937), pp. 333–335.

are zero. We may however find a permutation test for the hypothesis H that factor C has absolutely no effect in the sense that $\sigma_C^2 = \sigma_{AC}^2 = \sigma_{BC}^2 = \sigma_{ABC}^2 = 0$ and that the technical errors $\{e_{ijk}\}$ have the following property: The m^2 variables $\{e_{ijk}\}$ occurring in an experiment are distributed like m^2 variables $\{e_{ij}\}$ whose joint distribution does not depend on how the levels of C are combined with the levels of A and B; in particular, in case (iii) discussed at the beginning of sec. 9.2, the joint distribution of the $\{e_{ij}\}$ for the m^2 experimental units does not depend on how the treatments are assigned to the units. The joint distribution of the $\{e_{ij}\}$ is subject to the conditions $E(e_{ij}) = 0$ but is otherwise arbitrary.[34] Under H, if an observation were made on the treatment combination where A is at the ith level, B at the jth, C at the kth, it would have the structure

$$y_{ijk} = \mu + \alpha_i^A + \alpha_j^B + \alpha_{ij}^{AB} + e_{ij}, \tag{9.3.18}$$

where

$$\alpha_.^A = \alpha_.^B = \alpha_{i.}^{AB} = \alpha_{.j}^{AB} = 0$$

for all i, j. The reader should recall that, in a situation of type (iii) discussed at the beginning of sec. 9.2 and following (9.2.2), the $\{\alpha_{ij}^{AB}\}$ are unit errors.

If the Latin square has been selected at random from only a single transformation set, the set G of permutations is that consisting of the $(m!)^3$ permutations of rows, columns, and numbers. If the transformation set was first chosen with certain probabilities from the totality of transformation sets, the permutation distribution of the statistic or its moments are determined by those for the transformation sets by weighting with the same probabilities.

The permutation test of H based on $\mathscr{F} = \text{MS}_C/\text{MS}_e$ is equivalent to that based on $U = \text{SS}_C/(\text{SS}_C + \text{SS}_e)$, which is a strictly increasing function of $\mathscr{F}$, and the denominator of which is constant in the permutation distribution, since the terms on the right side of

$$\text{SS}_C + \text{SS}_e = \text{SS}_{\text{tot}} - \text{SS}_A - \text{SS}_B - m^2 y_{...}^2,$$

are constant. To carry out the permutation test exactly it would be simplest to base it on the sum of squares for numbers (i.e., levels of factor C), namely on $\Sigma_k T_k^2$ where T_k is the total of the m observations where factor C is at level k. The number of different squares in a transformation set is $m!(m-1)!$ times the number of standard squares in the set. Since the statistic is invariant under the $m!$ permutation of the levels of C, the number of different values the statistic takes on with equal probability for a given transformation set is $(m-1)!$ times the number of standard

[34] The assumptions on the $\{e_{ijk}\}$ are much less restrictive than in sec. 9.2.

squares in the set. A systematic way of getting these values would be to arrange the observations in a square where the rows and columns correspond respectively to the levels of A and B, and then consider applying for each of the standard squares of the transformation set the $(m-1)!$ squares obtainable by permuting the rows of the standard square after the first.[35] Of course we would calculate the statistic first only for squares that look as though they might give the statistic a value $\geqq$ than that observed. For $m = 4$, there are two transformation sets, one with one standard square, the other with three, hence the statistic takes on 6 values in one set and 18 in the other. The 24 values will be equally likely if the transformation sets are chosen with probabilities proportional to the number of standard squares in the sets, or if one of the four standard squares was chosen with equal probabilities. Clearly the normal-theory test at the usual significance levels will be a wretched approximation to the permutation test if $m = 4$. For $m = 5$, the number of values for the F-statistic is $56 \times 4! = 1344$.

It can be shown from the formulas for the E(MS)'s under the randomization model of sec. 9.2 with $\sigma_e^2 = 0$ that $E(U)$ is the same in the permutation distribution as in the normal-theory distribution. Var (U) has been calculated[36] but the formula is so complicated to evaluate numerically that it is not suitable for practical application. The correction[37] to the numbers of d.f. of the normal-theory test to make it approximate the exact permutation test better is hence also not feasible to calculate, since it requires using the value of Var (U). Unfortunately the only evidence at present that indicates that the normal-theory test is not too bad an approximation to the permutation test if $m > 4$ seems to consist of the four numerical examples included in Table 9.3.1. The last column approximates the probability that $U > U_0$ in the permutation distribution if U_0 is the normal-theory 5 per cent point: this is the same as the probability in the permutation distribution that $\mathscr{F} > F_{0.05;m-1,(m-1)(m-2)}$;

[35] In sec. 5.1 it was stated that the different squares of a transformation set could be obtained from the set of standard squares in the transformation set by permuting columns, and rows after the first. Here we obtain them by permuting treatments, and rows after the first. These are equivalent ways of generating the totality of different squares in the transformation set, because of the symmetry of the definition of a transformation set in rows, columns, and numbers. The two methods in general associate the same square to different standard squares, but each associates $m!(m-1)!$ different squares with a standard square: they establish different equivalence relations among the squares of a transformation set.

[36] Welch (1937), p. 41; his u_{ij} corresponds to our α_{ij}^{AB} and equals $z_{ij}-z_{i.}-z_{.j}+z_{..}$, where z_{ij} is the observation in the ith row and jth column.

[37] This correction is given for the one-way layout, and thus also for the t-test for a difference of means, by Box and Andersen (1955), p. 13; a comma should be inserted before "$A =$" in their equations (35).

the approximation is made by fitting a β-distribution to the distribution of U by equating the first two moments (entries are omitted for $m = 4$ because the discrete distribution assigning equal probabilities to the 24 possible values of U would be poorly approximated in the upper tail by a continuous distribution). Thus the present unsatisfactory state of

TABLE* 9.3.1

COMPARISON OF PERMUTATION TEST WITH NORMAL-THEORY TEST FOR THREE ARTIFICIAL EXAMPLES AND THREE UNIFORMITY TRIALS

Example	m	Ratio of Var (U) under Permutation Distribution to Var (U) under Normal Theory	Approximate Probability in Permutation Distribution that $U > U_0$, where U_0 is 5 per cent Point for Normal Theory
I: Artificial example	4	0.64	—
II: Uniformity trial	4	0.47	—
III: Uniformity trial	5	0.75	0.029
IV: Uniformity trial	6†	0.72	0.027
V: Artificial example	6†	1.16	0.062
VI: Artificial example	6†	1.03	0.053

* From p. 45 of "On the z-test in randomized blocks and Latin squares" by B. L. Welch, *Biometrika*, Vol. 29 (1937). Reproduced with the kind permission of the author and editor.

† The examples for $m = 6$ were evaluated for only two of the 22 transformation sets, which give extreme values to a certain constant entering Var (U).

our knowledge suggests that if we suspect the normal-theory assumptions to be seriously violated we should avoid Latin squares of size $m < 4$, calculate the exact permutation test for $m = 4$, and use the normal-theory test for $m > 4$.

A Remark on Interval Estimation

Suppose the observations in the randomized-blocks design are distributed as under the hypothesis H tested by the permutation test except that we permit the presence of treatment main effects $\{\alpha_i\}$, i.e., y_{ij} is given by (9.3.6) plus a term $\alpha_i(\Sigma_i\alpha_i = 0)$, the assumptions on the other terms being those stated in connection with (9.3.6). Then the $\{y_{ij}-\alpha_i\}$ are distributed like the y_{ij} of (9.3.6) and hence, regardless of the true values of the $\{\alpha_i\}$,

$$(I-1)^{-1}J\sum_i(y_{i.}-y_{..}-\alpha_i)^2/\text{MS}_e \tag{9.3.19}$$

is distributed exactly like the statistic $\mathscr{F} = \text{MS}_A/\text{MS}_e$ under H. For a given α the S-method of multiple comparison is exact if the probability is α that (9.3.19) exceeds the upper α point of the F-distribution with $(I-1)$ and $(I-1)(J-1)$ d.f. The S-method will thus be a good approximation under this randomization model to the extent that the normal-theory test of H holds its nominal significance level under the randomization model. Encouraging evidence for this has been found above. However, the improvement in the approximation of the normal-theory test by adjusting the numbers of d.f. by the factor ϕ is not possible in the case of estimation because the calculation of ϕ from the observations depends on the "conditional" block variances $(I-1)^{-1}\Sigma_\nu(\xi_{j\nu}+u_{j\nu}-u_{j.})^2$ being known functions of the observations, which is not the case if the $\{\alpha_i\}$ are different from zero and unknown. The T-method cannot be justified by this argument since it is not related to the F-test like the S-method.

Similarly we may try to justify the S-method for Latin squares under a randomization model. The model would be that yielded by adding the term α_k^C to (9.3.18), where $\Sigma_k\alpha_k^C = 0$. Our reservations about normal-theory tests when the square is of size $m \leqq 4$ of course carry over to the S-method.

PROBLEMS

9.1. State precisely the structure assumed for the observations in a randomized-blocks design under the hypothesis of no treatment effects, for (a) the "normal-theory test" and (b) the permutation test. For the randomized-blocks data of Problem 4.5 calculate the adjustment of the numbers of d.f. that makes (a) a good approximation to (b).

9.2. The following Latin-square design* was used to measure the effect of four sizing treatments on warp breakage rate of cloth; letters refer to the four treatments, rows to four time periods, and columns to the four looms used in the experiment:

44 (D)	54 (A)	71 (C)	29 (B)
22 (C)	59 (B)	100 (D)	22 (A)
31 (A)	40 (C)	79 (B)	38 (D)
27 (B)	83 (D)	100 (A)	29 (C)

Use the randomization test at the 0.05 level for the hypothesis of no treatment effects. (The normal-theory test gives an $\mathscr{F}$ significant at the 0.01 level.)

9.3. Consider a contrast $\psi = \alpha_i - \alpha_{i'}$ which is a difference of main effects in a randomized-block design, its estimate $\hat{\psi} = y_{i.} - y_{i'.}$, the J estimates $\{\hat{\psi}_j = y_{ij} - y_{i'j}\}$ formed separately from the blocks, and the overestimate s^2/J of Var $(\hat{\psi})$, formed according to (9.1.14) from the $\{\hat{\psi}_j\}$. Show that the average value of the overestimate s^2/J averaged over the $\frac{1}{2}I(I-1)$ differences is $2\text{MS}_e/J$. [*Hint:* The algebraic manipulation in calculating the average is similar to that in the proof of Lemma 2 at the end of sec. 9.2.]

* From p. 66 of *Industrial Statistics* by H. A. Freeman, John Wiley, New York (1942). Reproduced with the kind permission of the author and the publisher.

9.4. Prove that if $\mathbf{M} = (\mu_{ij})$ is an $I\times\mathcal{J}$ matrix such that for some $J<\mathcal{J}$ every $I\times J$ submatrix of $\mathbf{M}$ has equal row sums (i.e., the row main effects are zero), then the rows of $\mathbf{M}$ are equal (i.e., every submatrix of $\mathbf{M}$ has zero interactions). [*Hint:* To prove that $\mu_{ij} = \mu_{i'j}$ for all i, i', j it suffices to prove that $\mu_{i1} = \mu_{i'1}$ for all i, i', since the columns of $\mathbf{M}$ may be permuted without affecting the above equality of row sums. Now consider the $J+1$ submatrices formed from the first $J+1$ columns of $\mathbf{M}$ by omitting the sth column ($s = 1, \cdots, J+1$). Express the above equality as $c_i-\mu_{is} = c_{i'}-\mu_{i's}$, where $c_i = \Sigma_{j=1}^{J+1}\mu_{ij}$, and sum from $s = 1$ to J.]

CHAPTER 10

The Effects of Departures from the Underlying Assumptions

10.1. INTRODUCTION

In this chapter we shall study the effects of violations of the following assumptions made elsewhere in various parts of this book: (i) normality of the errors, and also normality of the random effects in the models where these appear (Chs. 7 and 8), (ii) equality of variance of the errors, and (iii) statistical independence of the errors. A study of this kind cannot be exhaustive, for one reason, because assumptions like this can be violated in many more ways than they can be satisfied. Usually we shall treat the three kinds of violations one at a time. We shall not be able to treat all the basic designs under all the models we have considered. Some of our conclusions will have to be inductions from rather small numerical tables. However, we must tackle the important questions raised in this chapter even though the evidence is incomplete and we realize that standards of rigor possible in deducing a mathematical theory from certain assumptions generally cannot be maintained in deriving the consequences of departures from those assumptions.

In considering the effects of nonnormality we will find it convenient to use the measures[1] γ_1 of skewness and γ_2 of kurtosis of the distribution of a random variable x. If the mean and variance of the distribution are denoted by μ and σ^2, respectively, its skewness γ_1 is defined as

$$\gamma_1 = \sigma^{-3}E[(x-\mu)^3],$$

and its kurtosis γ_2 as

$$\gamma_2 = \sigma^{-4}E[(x-\mu)^4] - 3.$$

[1] Other commonly used measures are $\beta_1 = \gamma_1^2$ for the magnitude of the skewness, and $\beta_2 = \gamma_2+3$ for the kurtosis. To preserve the sign of the skewness in the β-system, γ_1 is usually denoted by $\sqrt{\beta_1}$, meaning $\pm\sqrt{\beta_1}$ with the sign of γ_1. We assume for all distributions whose skewness and kurtosis is considered in this chapter that the variance is positive and that the fourth moment (and hence the variance also) is finite.

These measures do not depend on the location or scale (measured respectively by μ and σ) of the distribution, i.e., if y is a linear function of x, $y = c(x-a)$, where a and c are constants with $c > 0$, then γ_1 and γ_2 are the same for y as for x. For a symmetrical distribution γ_1 is evidently zero. Positive values for γ_1 indicate that the distribution is "skewed to the right" so that the right tail is in a certain sense[2] heavier than the left. For every distribution $\gamma_2 \geqq -2$; for a normal distribution $\gamma_2 = 0$. If we imagine deforming a normal probability density curve, keeping it symmetric and unimodal, a deformation in which the tails are heavier and the central part more sharply peaked would have $\gamma_2 > 0$; one in which the tails are lighter and the central part is flatter, giving an effect of shoulders, would have[3] $\gamma_2 < 0$.

It is easy to show by mathematical induction that, for a linear combination

$$(10.1.1) \qquad v = \sum_{j=1}^{N} c_j x_j$$

of N independent random variables $\{x_j\}$, the γ_1 and γ_2 of v are related to those of the $\{x_j\}$ as follows: Suppose that x_j has variance σ_j^2, skewness $\gamma_{1,j}$, and kurtosis $\gamma_{2,j}$. Denote by λ_j the proportion of the variance of v contributed by the term $c_j x_j$,

$$\lambda_j = c_j^2 \sigma_j^2 / \sum_{j'=1}^{N} c_{j'}^2 \sigma_{j'}^2,$$

so $\Sigma_1^N \lambda_j = 1$. Then the γ_1 and γ_2 of v are

$$(10.1.2) \qquad \gamma_1 = \sum_{j=1}^{N} (\pm \lambda_j^{3/2}) \gamma_{1,j},$$

where the $\pm$ sign is the sign of c_j, and

$$(10.1.3) \qquad \gamma_2 = \sum_{j=1}^{N} \lambda_j^2 \gamma_{2,j}.$$

We remark that in the skewness of the difference of two independent

[2] Tautologically we might call $\int_{\mu}^{\pm\infty} (x-\mu)^3 \, dF(x)$ the heaviness of the tails, where $F(x)$ denotes the cumulative distribution function of x, and we use the limit $+\infty$ for the right tail, and $-\infty$ for the left tail.

[3] The reader will encounter in the literature the terms *leptokurtic* meaning $\gamma_2 > 0$, and *platykurtic* meaning $\gamma_2 < 0$, and may be aided and amused by the mnemonic of Student (1927, see figures in footnote, p. 160) that platykurtic curves are short in the tails like platypuses, and leptokurtic are heavy in the tails like kangaroos, noted for "lepping."

random variables of equal γ_1 and equal variances the γ_1's cancel. This follows from (10.1.2) with $N = 2$, $c_1 = -c_2 = 1$, $\sigma_1 = \sigma_2$, $\gamma_{1,1} = \gamma_{1,2}$; then $\gamma_1 = 0$. We remark also that when the distribution of the linear combination (10.1.1) approaches normality for large N the formulas (10.1.2) and (10.1.3) give some indication of how rapid the approach is, i.e., of how rapidly the γ_1 and γ_2 of (10.1.1) approach the value zero of a normal distribution; for example, if all $\{c_j^2\sigma_j^2\}$ are equal, and all $\{c_j\}$ positive, then $\lambda_j = N^{-1}$, and hence $\gamma_1 = N^{-1/2}\gamma_{1,.}$ and $\gamma_2 = N^{-1}\gamma_{2,.}$, where $\gamma_{m,.}$ denotes the average of $\{\gamma_{m,1}, \gamma_{m,2}, \cdots, \gamma_{m,N}\}$.

To give some idea of the range of (γ_1, γ_2) occurring in applications we mention that in nine observed distributions[4] of engineering data, each based on several hundred measurements at the Bell Telephone Laboratories, the estimated γ_1 varied from -0.7 to $+0.9$, and γ_2 from -0.4 to $+1.8$. Student (1927) found that the errors of routine chemical analysis usually have positive γ_2 and cited an extreme example of one distribution of 100 measurements with an estimated γ_2 of 7. Positive γ_2 of an error distribution could be caused by the occurrence of occasional gross errors or blunders. We remark that $\gamma_2 = -1.2$ for the uniform distribution, which is usually assumed appropriate in the analysis of round-off errors in numerical computations. Some further values[5] of (γ_1, γ_2), each estimated from a very large number of individuals (about 10,000 or more), are: (2.0, 6.3) for the (claimed) age of brides at marriage in Australia during the period 1907–14; (2.0, 5.3) for the same for the grooms; (0.3, -0.6) for the age of the mothers for births in Australia, 1922–26; (0.7, 0.6) for the same for the fathers; (0.5, 0.4) for barometric heights at Greenwich, 1848–1926; (-0.9, 1.8) for the lengths of beans (the seeds, not the pods) of a certain pure line; (-0.4, 0.6) for the same for the breadths.

The first four moments of a population of course do not determine even approximately the sampling distributions of all statistics of practical interest (e.g., the sample percentiles, the sample third and fourth moments). However, we shall find that the value of the γ_2, and to a lesser degree the value of the γ_1, of the errors (and of the random-effects factors, if any are present) are in the present state of our knowledge the most important indicators of the extent to which nonnormality affects the usual inferences made in the analysis of variance.

In discussing violations of the assumption of independence it will be convenient to consider a simple model of serial correlation of random variables $x_1, x_2, \cdots, x_I$, which may be appropriate if these are measurements which are successive in time or space. We shall say the $\{x_i\}$ have

[4] The distributions were collected by Shewhart, and the γ values published by E. S. Pearson (1931).

[5] From Pretorius (1930).

the *serial correlation coefficient*[6] ρ if the correlation coefficient of x_i and x_{i+1} is ρ for $i = 1, 2, \cdots, I-1$, and all the other correlation coefficients are zero. Not all values $-1 \leqq \rho \leqq 1$ are mathematically possible, because of the condition that the $I \times I$ matrix of correlation coefficients, like the covariance matrix, must be positive indefinite;[7] however, that all values in the interval $-\frac{1}{2} \leqq \rho \leqq \frac{1}{2}$ are possible is shown by the following artificial example: Let $x_i = z_i + cz_{i+1}$, where $z_1, \cdots, z_{I+1}$ are independent with equal variances, and c is a constant; we easily calculate $\rho = c/(1+c^2)$, which takes on all values from $-\frac{1}{2}$ to $+\frac{1}{2}$ as the constant c is varied from -1 to 1.

For a series of 100 routine analyses for each of five different chemical properties, made daily (during successive five-day weeks) on samples from the same batch of thoroughly mixed material, Student (1927) calculated the following correlation coefficients between successive analyses: 0.27, 0.31, 0.19, 0.09, 0.09; he also remarked that he had never observed such coefficients to be negative. In the case of successive yield measurements on a batch process where the vessel is not completely emptied of the product, and the measurement is made after emptying, a negative serial correlation coefficient could be caused by fluctuation of the amount of product left in the vessel.

10.2. SOME ELEMENTARY CALCULATIONS OF THE EFFECTS OF DEPARTURES

The calculations of this section are all made for the case where the number of d.f. for error is very large. This permits us to derive in an elementary way all the important conclusions which will later be substantiated in the general case where the number of d.f. for error is not necessarily large. Those violations of the assumptions that are here found to have a serious effect on the inferences may then be considered to invalidate them in the general case, but those that are found to be not serious need to be investigated further in the general case.

We begin with the case of a single sample: Suppose $\{y_1, y_2, \cdots, y_n\}$ is a random sample from a population with mean μ, variance σ^2, and kurtosis γ_2 (sec. 10.1). If the normality assumption held for the population we would have then the simplest example falling under the general

[6] This would usually be called the serial correlation coefficient with *lag* 1, the one with lag h being the correlation coefficient of x_i and x_{i+h}; we are assuming those with lag $h > 1$ to be zero.

[7] A necessary and sufficient condition for this matrix to be positive definite is $|\rho| < \{2 \cos [\pi/(I+1)]\}^{-1}$; this may be deduced from Grenander and Rosenblatt (1956), pp. 101–102.

theory of Ch. 2, and we can now examine the effect of violation of the normality assumption on inferences about the mean μ. Under normal theory these inferences are usually based on the central t-distribution of the random variable

$$t = n^{1/2}(y_{.} - \mu)/s, \tag{10.2.1}$$

where $y_{.}$ and $s^2 = MS_e$ are the sample mean and sample variance, if the inference is a confidence interval, or on the noncentral (we recall from App. IV that the noncentral distributions are defined to include the central) t-distribution of

$$t' = n^{1/2}(y_{.} - \mu_0)/s, \tag{10.2.2}$$

if the inference is a test of the hypothesis H: $\mu = \mu_0$. If equal tails of the t-distribution are used, the methods are respectively equivalent to the confidence "ellipsoid" (interval, in this case, centered at $y_{.}$) for μ based on t^2, which is distributed as central F under normal theory, and to the F-test based on t'^2, which is distributed as noncentral F under normal theory. Now if the number $n-1$ of d.f. for error is large, s may be replaced[8] by σ in considering the distribution of the ratio (10.2.1) or (10.2.2). Furthermore, by the central-limit theorem,[9] $n^{1/2}(y_{.} - \mu)/\sigma$ is $N(0, 1)$ for large n. Combining these two results we see that for large n the ratio (10.2.1) is $N(0, 1)$, and the ratio (10.2.2) is $N(\delta, 1)$, where δ is the noncentrality parameter

$$\delta = n^{1/2}(\mu - \mu_0)/\sigma.$$

For large n the distributions of t and t' are thus independent of the form of the population, and hence the inferences about the mean μ which are valid in the case of normality must be correct for large n regardless[10] of the form of the population.

[8] Since s converges in probability to σ, the limiting distribution of (10.2.1) or (10.2.2) remains the same if s is replaced by σ; see Cramér (1946), sec. 20.6. That s^2 converges in probability to σ^2, and hence s to σ, follows from Problem IV.3a. The arguments below where we replace a MS_e by σ_e^2 can be similarly rigorized. We remark here that if we say x is $N(a_n, b_n^2)$ for large n we may give this the following precise mathematical definition: We are really referring to a random variable $x = x_n$ whose distribution depends on n for $n = N, N+1, N+2, \cdots$, and some given N. The preceding statement means that, for the sequences $\{a_n\}$ and $\{b_n\}$ of constants, $\Pr\{(x_n - a_n)/b_n \le t\} \to (2\pi)^{-1/2}\int_{-\infty}^{t} \exp(-\frac{1}{2}z^2)\,dz$ as $n \to \infty$ for every fixed t.

[9] Cramér (1946), sec. 17.4.

[10] We are assuming finite γ_2, although the present conclusion can be justified merely with finite σ^2.

The story is very different for inferences about the variance σ^2. Inferences[11] about σ^2 derived under normal theory are usually based on the distribution of

$$(n-1)s^2/\sigma^2$$

being that of χ^2_{n-1}, or equivalently, on s^2/σ^2 being $\chi^2_{n-1}/(n-1)$. Under this distribution

$$E\left(\frac{s^2}{\sigma^2}\right) = 1, \tag{10.2.3}$$

$$\text{Var}\left(\frac{s^2}{\sigma^2}\right) = \frac{2}{n-1}. \tag{10.2.4}$$

Already we can see that for any n something is very wrong if γ_2 differs much from zero, since, while (10.2.3) is correct, instead of (10.2.4) we have

$$\text{Var}\left(\frac{s^2}{\sigma^2}\right) = \frac{2}{n-1} + \frac{\gamma_2}{n} \tag{10.2.5}$$

from (3.8.1). The ratio of (10.2.5) to (10.2.4) is

$$1 + \frac{n-1}{2n}\gamma_2. \tag{10.2.6}$$

For large n, the ratio approaches $1 + \frac{1}{2}\gamma_2$, and s^2 is normal.[12] From this it follows that

$$\left(\frac{n-1}{2}\right)^{-1/2}\left(\frac{s^2}{\sigma^2} - 1\right)$$

is actually $N(0, 1 + \frac{1}{2}\gamma_2)$ for large n, and not in general $N(0, 1)$ as under normal theory, where $\gamma_2 = 0$. This obviously causes a serious error in any confidence coefficient, significance level, or power calculated under normal theory if γ_2 differs much from 0. For example, if for large n we use a normal-theory confidence interval for σ^2 with nominal confidence coefficient $1-\alpha$ the probability of the interval not covering σ^2 is

$$2(2\pi)^{-1/2}\int_{z_{\alpha/2}/B}^{\infty} \exp\left(-\tfrac{1}{2}t^2\right) dt \tag{10.2.7}$$

instead of

$$2(2\pi)^{-1/2}\int_{z_{\alpha/2}}^{\infty} \exp\left(-\tfrac{1}{2}t^2\right) dt = \alpha, \tag{10.2.8}$$

[11] Power calculations for a test of the hypothesis $\sigma^2 = \sigma_0^2$ are based on the distribution of s^2/σ_0^2, but this is the constant σ^2/σ_0^2 times the s^2/σ^2 considered here.

[12] Cramér (1946), sec. 28.3. If we rigorize this calculation a question arises about the propriety of assuming that the variance of the limit distribution is the limit of the variance; however, Cramér's theorem directly gives the desired result with $n-1$ replaced by n in the next displayed expression.

where $z_{\alpha/2}$ is the upper $\alpha/2$ point of $N(0, 1)$, defined by (10.2.8), and

$$B = (1 + \tfrac{1}{2}\gamma_2)^{1/2}.$$

Some values are shown in Table 10.2.1 for the case $\alpha = 0.05$. Table 10.2.1 of course gives also the actual probability of type-I error for the usual two-tailed test of the hypothesis $\sigma^2 = \sigma_0^2$ at the nominal 5 per cent significance level.

TABLE 10.2.1

EFFECT OF NONNORMALITY ON TRUE PROBABILITY OF NOMINAL 95 PER CENT CONFIDENCE INTERVAL FOR σ^2 NOT COVERING TRUE σ^2 FOR LARGE n

γ_2	-1.5	-1	-0.5	0	0.5	1	2	4	7
Probability	$9 \cdot 10^{-5}$	0.006	0.024	0.050	0.080	0.11	0.17	0.26	0.36

We shall call inferences involving only fixed effects in the model equation of an analysis of variance "inferences about means," and inferences involving only random effects "inferences about variances." Examples of inferences about means are tests of hypotheses about fixed main effects or interactions, confidence ellipsoids for estimable functions, and the S- and T-methods for multiple comparison. Examples of inferences about variances are confidence intervals for the error variance σ_e^2, for a variance component σ_A^2, where A is a random-effects factor, or for a ratio of variance components, and tests for the equality of variances. The conclusion just found in the single-sample case will be found in every case examined, and for basically the same reason. The conclusion is that the effect of violation of the normality assumption is slight on inferences about means but dangerous on inferences about variances.[13] The reason is that in both cases the inferences about the parameters (means or variances) are based on the distribution of deviations of estimates from the parameters in some sense (the deviation is the difference for μ, the ratio for σ^2, in this example), but, whereas in the case of means the extent of the observed deviation is measured against an internal estimate of error calculated from the observations, and valid without normality assumptions, in the case of variances it is measured against the per cent points of a theoretical distribution, derived under the assumption that the effects characterized by the variances are normal, the theoretical distribution having the correct location, and at least for large n the correct shape, but the wrong spread if the γ_2 of the effects differs from zero. Another way of putting this is to say that in the case of means the criterion used is "Studentized" for the second moments of the estimates, and in the case of variances it isn't.

[13] This effect was noted by E. S. Pearson (1931); the reason, by Box (1953).

To see the possible effect of violation of the independence assumption let us modify the above example so that the observations $\{y_i\}$ are serially correlated with the serial correlation coefficient ρ, as defined at the end of sec. 10.1. To simplify the discussion we shall assume the n observations jointly normal, so that the $\{y_i\}$ have a multivariate normal distribution with $E(y_i) = \mu$, $\text{Var}(y_i) = \sigma^2$, the correlation coefficient of y_i and y_{i+1} is ρ for $i = 1, \cdots, n-1$, and all other correlation coefficients are zero. Then the sample mean $y_.$ is normal, and we easily calculate for $y_.$ and the sample variance s^2 that

$$E(y_.) = \mu,$$

$$\text{(10.2.9)} \qquad \text{Var}(y_.) = \frac{\sigma^2}{n}\left[1 + 2\rho\left(1 - \frac{1}{n}\right)\right],$$

$$E(s^2) = \sigma^2\left(1 - \frac{2\rho}{n}\right).$$

The last expression may be obtained by first taking expectations in

$$\sum_{i=1}^{n}(y_i - y_.)^2 = \sum_{i=1}^{n} y_i^2 - n^{-1}\sum_{i=1}^{n}\sum_{i'=1}^{n} y_i y_{i'},$$

for which purpose we may pretend that $E(y_i) = 0$, so $E(y_i^2) = \sigma^2$, and then noting that expectations in the double sum are zero except for the n terms with $i' = i$, where they are σ^2, and the $2(n-1)$ terms where i and i' differ by unity, where they are $\rho\sigma^2$.

We have already noted that the usual normal-theory calculations of significance levels and confidence coefficients for inferences about μ are based on the ratio (10.2.1) having the t-distribution with $n-1$ d.f., and hence being $N(0, 1)$ for large n. We shall now drop the term in $1/n^2$ from (10.2.9). For large n, s^2 may be replaced by its expected value σ^2 and hence s by σ in considering the distribution of a quotient like (10.2.1); employing the formulas (10.2.9), we then find that (10.2.1) is $N(0, 1+2\rho)$ for large n, instead of $N(0, 1)$. The probability of a confidence interval with nominal confidence coefficient $1-\alpha$ not covering μ, or the probability of a type-I error with a two-tailed test of μ at the nominal α level of significance, is then given by (10.2.7) witn

$$B = (1 + 2\rho)^{1/2}.$$

As ρ approaches $-\frac{1}{2}$, this approaches zero. As ρ approaches $\frac{1}{2}$, this approaches 0.17 if $\alpha = 0.05$. Some other values are shown in Table 10.2.2. Clearly the effect of serial correlation on inferences about means can be serious.

TABLE 10.2.2

EFFECT OF SERIAL CORRELATION ON TRUE PROBABILITY OF NOMINAL 95 PER CENT CONFIDENCE INTERVAL FOR μ NOT COVERING TRUE μ FOR LARGE n

ρ	-0.4	-0.3	-0.2	-0.1	0	0.1	0.2	0.3	0.4
Probability	$1\cdot10^{-5}$	0.002	0.011	0.028	0.050	0.074	0.098	0.12	0.14

The simplest example where we can see the effect of violation of the assumption of equality of error variances is that of two samples, i.e., the one-way layout with two groups of size J_1 and J_2, respectively. Suppose that, for $i = 1, 2$, $\{y_{i1}, y_{i2}, \cdots, y_{iJ_i}\}$ is a random sample from a population with mean μ_i, variance σ_i^2, and kurtosis $\gamma_{2,i}$, the two samples being independent. Let $y_{i.}$ and s_i^2 $(i = 1, 2)$ denote the sample means and sample variances. We shall assume that J_1 and J_2, and hence $n = J_1+J_2$, are large, and when we say "for large n" we shall mean "for large J_1 and J_2." We denote by

$$R = J_1/J_2, \qquad \theta = \sigma_1^2/\sigma_2^2$$

the ratios of the sample sizes and of the variances. The usual normal-theory confidence intervals for the difference of means

$$\Delta = \mu_1 - \mu_2$$

or significance level calculations for tests of the hypothesis $\Delta = \Delta_0$ are based on the ratio

$$\frac{y_{1.} - y_{2.} - \Delta}{[(J_1^{-1} + J_2^{-1})\text{MS}_e]^{1/2}} \tag{10.2.10}$$

having the t-distribution with $n-2$ d.f. if $\sigma_1^2 = \sigma_2^2$, or equivalently, being $N(0, 1)$ for large n; here MS_e is formed from the pooled SS for error,

$$\text{MS}_e = \frac{(J_1-1)s_1^2 + (J_2-1)s_2^2}{J_1 + J_2 - 2}. \tag{10.2.11}$$

Considering the distribution of the quotient (10.2.10), we see that, for large n, $y_{i.}$ is $N(\mu_i, \sigma_i^2/J_i)$, hence the numerator is $N(0, J_1^{-1}\sigma_1^2+J_2^{-1}\sigma_2^2)$, while in the denominator we may replace[14] s_1^2 by σ_1^2 and s_2^2 by σ_2^2, where they enter MS_e through (10.2.11). Also putting $J_1 = R(1+R)^{-1}n$, $J_2 = (1+R)^{-1}n$, $\sigma_1^2 = \theta\sigma_2^2$, and neglecting terms in $1/n$, we find that the quotient (10.2.10) is $N(0, (\theta+R)(R\theta+1)^{-1})$ for large n. Hence if the sample sizes are equal ($R = 1$), the quotient is $N(0, 1)$ for large n as in the normal theory assuming equal variances, even though the two populations are nonnormal or have unequal variances. This means that our

[14] The statistic MS_e converges in probability to $(1+R)^{-1}(R\theta+1)\sigma_2^2$ as $n \to \infty$ with fixed R and fixed θ.

confidence coefficients and significance levels, calculated under the assumptions of normality and equal variance, are valid for large n when these assumptions are violated, in the case of the one-way layout with two large *equal* groups.

The effect of unequal variances ($\theta \neq 1$) on the true probabilities corresponding to nominal confidence coefficients of $1-\alpha$ and nominal type-I error probabilities of α is that instead of getting the probability α of error we get (10.2.7) for large n with

$$B = \left(\frac{\theta+R}{R\theta+1}\right)^{1/2}.$$

Some values are shown in Table 10.2.3 for the case $\alpha = 0.05$. The limiting values $\theta = 0, \infty$ are of course impossible to attain but give bounds for the behavior of the entries in any row; similarly for $R = \infty$, and any column.

TABLE 10.2.3

EFFECT OF UNEQUAL ERROR VARIANCE AND UNEQUAL GROUP SIZE ON TRUE PROBABILITY OF 95 PER CENT CONFIDENCE INTERVAL FOR $\mu_1-\mu_2$ NOT COVERING TRUE VALUE FOR LARGE n

θ = ratio of variances, R = ratio of group sizes

	θ						
R	0*	$\frac{1}{5}$	$\frac{1}{2}$	1	2	5	∞*
1	0.050	0.050	0.050	0.050	0.050	0.050	0.050
2	0.17	0.12	0.080	0.050	0.029	0.014	0.006
5	0.38	0.22	0.12	0.050	0.014	0.002	$1\cdot 10^{-5}$
∞*	1.00	0.38	0.17	0.050	0.006	$1\cdot 10^{-5}$	0

* Unattainable limiting cases to show bounds.

The power of the test of the hypothesis $\Delta = \Delta_0$ depends on the distribution of the ratio (10.2.10) with Δ replaced by Δ_0. Proceeding as before, we find that this distribution is normal for large n with mean

$$\left(\frac{1}{J_1} + \frac{1}{J_2}\right)^{-1/2} \frac{\Delta - \Delta_0}{\sqrt{(\sigma^2)_{.w}}}, \tag{10.2.12}$$

where $(\sigma^2)_{.w}$ is the weighted mean of the $\{\sigma_i^2\}$ with weights J_i,

$$(\sigma^2)_{.w} = (J_1\sigma_1^2 + J_2\sigma_2^2)/(J_1 + J_2),$$

and variance

$$(\theta + R)/(R\theta + 1). \tag{10.2.13}$$

Under the assumption of equality of variance, $\sigma_1^2 = \sigma_2^2 = \sigma^2$, the mean (10.2.12) becomes

$$\delta = \left(\frac{1}{J_1} + \frac{1}{J_2}\right)^{-1/2} \frac{\Delta - \Delta_0}{\sqrt{\sigma^2}}, \tag{10.2.14}$$

and the variance (10.2.13) becomes unity. If we calculate as though the σ_1^2 and σ_2^2 were equal with a common value σ^2, then in order to make (10.2.14) coincide with (10.2.12) we must replace σ^2 in (10.2.14) by $(\sigma^2)_{.w}$. However, if the assumption of equality of variance is violated ($\theta \neq 1$) the resulting power calculation will be correct if and only if the group sizes are equal ($R = 1$), for otherwise the variance (10.2.13) differs from the value of unity required to make the calculation correct.

The case of two equal groups is exceedingly well behaved with regard to violation of the equality-of-variance assumption, since it shows no effect at all for large n. However when we now consider the case of I groups in the one-way layout under the fixed-effects model we shall find that violating the equality-of-variance assumption when $I > 2$ has some effect even when group sizes are equal, although it then appears to be slight.

Suppose then that for each $i = 1, \cdots, I$, $\{y_{i1}, \cdots, y_{iJ_i}\}$ is a random sample from a population with mean μ_i, variance σ_i^2, and kurtosis $\gamma_{2,i}$, and that the I samples are independent. The calculations of Ch. 3 for the significance level of the F-test of the hypothesis $\mu_1 = \mu_2 = \cdots = \mu_I$, for confidence ellipsoids for $I-1$ independent contrasts in the $\{\mu_i\}$, and for the S-method of multiple comparison of the contrasts in the $\{\mu_i\}$, all made under the assumptions of normality of the $\{y_{ij}\}$ and equality of the $\{\sigma_i^2\}$, are based on the ratio

$$(I-1)^{-1}\sum_i J_i(v_i-\bar{v})^2/\text{MS}_e \tag{10.2.15}$$

having the F-distribution with $I-1$ and $\Sigma_i(J_i-1)$ d.f., where

$$v_i = y_{i.} - \mu_i,$$

$$\bar{v} = \sum_i J_i v_i \Big/ \sum_i J_i,$$

$$\text{MS}_e = \sum_i (J_i-1)s_i^2 \Big/ \sum_i (J_i-1),$$

and $y_{i.}$ and s_i^2 are the sample mean and sample variance of the ith group.

We suppose now that the $\{J_i\}$, and hence also $n = \Sigma_i J_i$, are large,[15] and this is what we shall mean when we say "for large n." In treating

[15] The asymptotic calculations we make are for the case that $n \to \infty$ with the ratios $\{J_i/n\}$ all fixed.

the distribution of (10.2.15) we may then replace s_i^2 in MS_e by σ_i^2, or, since n is large, MS_e by

$$(\sigma^2)_{.w} = \sum_i J_i \sigma_i^2 / \sum_i J_i,$$

the weighted average of the $\{\sigma_i^2\}$ with weights $\{J_i\}$. We may utilize the central-limit theorem on the $\{v_i\}$, with the result that (10.2.15) is distributed for large n like

$$(I-1)^{-1} \sum_i J_i (v_i - \bar{v})^2 / (\sigma^2)_{.w}, \tag{10.2.16}$$

where the $\{v_i\}$ are independent and v_i is $N(0, \sigma_i^2/J_i)$. We may therefore apply the lemma at the end of sec. 7.6 to calculate the mean and variance of

$$S = \sum_i J_i (v_i - \bar{v})^2.$$

After a little manipulation the results of this calculation may be expressed as follows: For large n the expected value of the random variable (10.2.16) is

$$(I-1)^{-1}\{[I(\sigma^2)_{.u}/(\sigma^2)_{.w}] - 1\}, \tag{10.2.17}$$

where $(\sigma^2)_{.u}$ is the unweighted average of the $\{\sigma_i^2\}$, and its variance is

$$(I-1)^{-2}\{2I(1 + V_u)[(\sigma^2)_{.u}/(\sigma^2)_{.w}]^2 - 4V_w - 2\}, \tag{10.2.18}$$

where V_u and V_w are respectively the squares of unweighted and weighted coefficients of variation of the $\{\sigma_i^2\}$; i.e.,

$$V_u = [(\sigma^2)_{.u}]^{-2} \sum_i [\sigma_i^2 - (\sigma^2)_{.u}]^2 / I,$$

$$V_w = [(\sigma^2)_{.w}]^{-2} \sum_i J_i [\sigma_i^2 - (\sigma^2)_{.w}]^2 / \sum_i J_i.$$

Under the assumptions of normality and equal variances, (10.2.15) is distributed as $(I-1)^{-1}\chi_{I-1}^2$ for large n and hence has expected value unity and variance $2/(I-1)$. From (10.2.17), we see that (10.2.16) will have the same expected value (unity) in general if and only if the weighted average $(\sigma^2)_{.w}$ and the unweighted average $(\sigma^2)_{.u}$ are equal. This will be true for all $\{\sigma_i^2\}$ if and only if all $\{J_i\}$ are equal. If all $\{J_i\}$ are equal, $V_w = V_u$ and the variance (10.2.18) becomes

$$\frac{2}{I-1}\left(1 + V_u \frac{I-2}{I-1}\right). \tag{10.2.19}$$

If $I = 2$, or if the $\{\sigma_i^2\}$ are equal so that $V_u = 0$, (10.2.16) then has the correct variance $2/(I-1)$, but otherwise its variance is too large by the factor in parentheses in (10.2.19). Some idea of the magnitude of this factor may be obtained from its numerical values for some particular sets of $\{\sigma_{ij}^2\}$, which may be read from the first, second, and last columns of Table 10.4.2 below. For large n, (10.2.16) is a quadratic form in the I central normal variables $\{v_i\}$ and hence is distributed as a linear combination of independent chi-square variables (Problem V.2). It is obviously nonnegative, and for this case it has been shown[16] that a very close approximation to its distribution is that of $c\chi_\nu^2$, where the constant c and the number ν of d.f. are chosen to give the correct first two moments. From this and (10.2.19) it follows that the true error probabilities of type-I errors, confidence ellipsoids, and the S-method of multiple comparison are larger (by how much could be calculated by the approximation just stated) than the nominal α in the case of equal group sizes and unequal variances in the one-way layout.[17]

Power calculations do not appear to be so elementary, but the Rule for E(MS)'s at the beginning of sec. 10.4 suggests that in the case of equal $\{J_i\}$ the power is approximately that which would be calculated under the assumption of equal variance if the σ^2 in Rule 1 of sec. 2.6 is replaced by the average $(\sigma^2)_{.u}$.

Our last example will exhibit the effects of nonnormality in a random-effects model. The simplest case is the one-way layout with I groups of J observations each, considered in sec. 7.2. In that case the model equation is

$$y_{ij} = \mu + a_i + e_{ij}, \tag{10.2.20}$$

where μ is the general mean, the $\{a_i\}$ are main effects for the factor A, the $\{e_{ij}\}$ are errors, and it is assumed that the $\{a_i\}$ and $\{e_{ij}\}$ are completely independent. Let us assume that the $\{a_i\}$ are a random sample from a population of effects with zero mean, variance σ_A^2, and kurtosis $\gamma_{2,A}$, while the $\{e_{ij}\}$ are a random sample from a population with zero mean, variance σ_e^2, and kurtosis $\gamma_{2,e}$. The MS's are

$$\text{MS}_e = I^{-1}(J-1)^{-1}\sum_i\sum_j(y_{ij}-y_{i.})^2 = I^{-1}(J-1)^{-1}\sum_i\sum_j(e_{ij}-e_{i.})^2$$

and

$$\text{MS}_A = (I-1)^{-1}J\sum_i(y_{i.}-y_{..})^2 = (I-1)^{-1}J\sum_i(v_i-v_.)^2,$$

where

$$v_i = a_i + e_{i.}. \tag{10.2.21}$$

[16] Box (1954*a*), Table I, p. 294.

[17] This is also true for the two-way layout with large equal numbers of observations in the cells (Problem 10.7), but not in general for the two-way layout with one observation per cell (sec. 10.4), where the denominator of $\mathcal{F}$ is a residual MS instead of a within-cell MS.

We recall that under normal theory MS_e and MS_A are independent, MS_e is $\sigma_e^2\chi_{\nu_e}^2/\nu_e$, where $\nu_e = I(J-1)$, MS_A is $(\sigma_e^2+J\sigma_A^2)\chi_{I-1}^2/(I-1)$, and hence, with

$$\theta = \sigma_A^2/\sigma_e^2,$$

$$(1+J\theta)^{-1}\mathrm{MS}_A/\mathrm{MS}_e \tag{10.2.22}$$

is F_{I-1,ν_e}. On this distribution are based the usual calculations of confidence intervals for θ and the probabilities of both kinds of error in testing the hypothesis that $\sigma_A^2 = 0$, or, more generally, that $\theta \leqq$ a given constant.

In studying the distribution of (10.2.22) for large J we shall find very different results in the two cases $\sigma_A^2 = 0$ and $\sigma_A^2 > 0$: If $\sigma_A^2 = 0$ the distribution of (10.2.22) must be identical with that of the statistic $\mathscr{F}$ for testing the hypothesis of no difference of means in the fixed-effects model considered above for the case where the $\{J_i\}$ are all equal to J and the $\{\sigma_i^2\}$ are all equal to σ_e^2, since the structure (10.2.20) of the observations is then the same. But we have seen that the statistic is then distributed for large J as under the normality assumption. This means that the probability of type-I errors in testing the hypothesis $\sigma_A^2 = 0$ is not affected for large J by nonnormality.

If $\sigma_A^2 > 0$ we shall again replace MS_e by σ_e^2 in considering the distribution of (10.2.22) and consequently consider the distribution of

$$(1+J\theta)^{-1}\mathrm{MS}_A/\sigma_e^2, \tag{10.2.23}$$

which may be written[18] s_v^2/σ_v^2, where

$$s_v^2 = (I-1)^{-1}\sum_i(v_i-v_.)^2,$$

and

$$\sigma_v^2 = \sigma_A^2 + J^{-1}\sigma_e^2.$$

Now s_v^2 is distributed like the sample variance of a random sample of I from a population of v's, and hence, as in (10.2.5),

$$E\left(\frac{s_v^2}{\sigma_v^2}\right) = 1, \qquad \mathrm{Var}\left(\frac{s_v^2}{\sigma_v^2}\right) = \frac{2}{I-1} + \frac{\gamma_{2,v}}{I}.$$

From (10.1.3) with $N = J+1$, $x_j = e_{ij}$, and $c_j = J^{-1}$ for $j = 1, \cdots, J$; $x_N = a_i$, and $c_N = 1$, we find

$$\gamma_{2,v} = \left(\frac{\theta}{\theta+J^{-1}}\right)^2\gamma_{2,A} + \left(\frac{1}{\theta+J^{-1}}\right)^2\frac{\gamma_{2,e}}{J^3}.$$

[18] The ensuing calculation could be simplified for large J by replacing $e_{i.}$ by 0, since it converges to zero in probability, and dropping terms in J^{-1}. I prefer the present calculation since it gives the variance of MS_A correctly for all J.

The mean of (10.2.23) is thus the same as under normal theory, but its variance is in general different, depending on the kurtoses $\gamma_{2,A}$ and $\gamma_{2,e}$; for large J it is

$$\frac{2}{I-1}\left\{1+\frac{I-1}{2I}\gamma_{2,A}\right\}.$$

Our conclusions for large J in this example are then that, if the kurtosis $\gamma_{2,A}$ of the random effects of the factor A is zero, nonnormality in these effects or in the errors does not invalidate the inferences about $\theta = \sigma_A^2/\sigma_e^2$, but that if $\gamma_{2,A} \neq 0$ confidence coefficients and the probabilities of both kinds of errors are affected, except the probability of type-I errors in testing the hypothesis $\sigma_A^2 = 0$. The direction of the effect is such that for confidence coefficients $1-\alpha$ and significance levels α the true α will be less than the nominal α if $\gamma_{2,A} < 0$, greater if $\gamma_{2,A} > 0$, and the magnitude of the effect[19] increases with the magnitude of $\gamma_{2,A}$.

Our conclusions from the examples of this section may be briefly summarized as follows: (i) Nonnormality has little effect on inferences about means but serious effects on inferences about variances of random effects whose kurtosis γ_2 differs from zero. (ii) Inequality of variances in the cells of a layout has little effect on inferences about means if the cell numbers are equal, serious effects with unequal cell numbers. (iii) The effect of correlation in the observations can be very serious on inferences about means.

10.3. MORE ON THE EFFECTS OF NONNORMALITY

In this section it will be assumed, unless otherwise stated, that all errors have the same skewness $\gamma_{1,e}$ and the same kurtosis $\gamma_{2,e}$. We shall begin with some consideration of the effect of nonnormality on point estimates: We remind the reader that in earlier chapters we have noted that nonnormality does not bias our point estimates of estimable functions (sec. 1.4) or our point estimates of variance components. However, nonnormality will in general invalidate the normal-theory formulas for the variance of point estimates of variance components, since these are based on theoretical distributions which distribute the mean squares for random effects as constants times chi-square variables, and such distributions may be bad approximations in the presence of nonzero kurtosis γ_2, as we have seen in the above examples.

The variances and covariances of point estimates in random-effects

[19] This could be approximated from the relation that (10.2.23) is approximately $C\chi_\nu^2$, where C and ν are evaluated by use of the first two moments of (10.2.23) calculated above.

models have been calculated[20] without the normality assumption for the following cases: one-way layout with possibly unequal groups, two-way and higher-way layouts with equal numbers of observations in the cells, the Latin square, and balanced incomplete blocks; in all cases all interactions are assumed to be zero.[21] To indicate the kind of formulas obtained we shall state the results for the case of the one-way layout with equal groups, under the same assumptions as in the last example of sec. 10.2. If $\hat{\sigma}_A^2$ and $\hat{\sigma}_e^2$ denote the estimated components of variance,

$$\hat{\sigma}_A^2 = J^{-1}(\text{MS}_A - \text{MS}_e), \qquad \hat{\sigma}_e^2 = \text{MS}_e,$$

then Var $(\hat{\sigma}_A^2)$ is given by the formula resulting from the addition of the term $I^{-1}\gamma_{2,A}\sigma_A^4$ to the normal-theory formula (7.2.16), Var $(\hat{\sigma}_e^2)$ by the addition of $I^{-1}J^{-1}\gamma_{2,e}\sigma_e^4$ to (7.2.15), and Cov $(\hat{\sigma}_A^2, \hat{\sigma}_e^2)$ by (7.2.16*a*) without change.

In general the variances of the estimated variance components depend on the variances σ^2 and kurtoses γ_2 of the various effects and the errors. That nonnormality of the errors is less perturbing in the one-, two-, and higher-way layouts when the cell numbers are equal is suggested by the result that in these cases the variances of the estimated variance components other than $\hat{\sigma}_e^2$, the estimated error component, are not affected by the kurtosis $\gamma_{2,e}$ of the errors.

Most studies of the effect of nonnormality on tests of hypotheses have dealt only with type-I errors. There have been many investigations of the effect of nonnormality on *t*-tests, from which we now summarize some results. From many sampling experiments with nonnormal distributions it has been concluded[22] that tests based on $|t|$ or t^2 for inferences about the mean of a single population, which are equivalent to *F*-tests with 1 d.f. in the numerator, are not sensitive to the skewness γ_1 of the errors, but that this is not true of a one-tailed use of t; also that although the kurtosis γ_2 of the errors has some effect on the distribution of t, it is in general slight. These empirical conclusions have been substantiated by approximate moment calculations.[23] Such moment calculations have also indicated that even in the one-tailed use of t, if it is for the difference of two means (one-way layout with two groups), the effect of nonzero γ_1 is

[20] By Tukey (1956, 1957*a*); a few of the results were published earlier by Hammersley (1949). Tukey's results are given in a more general form than here, covering the case where the effects are sampled from finite populations.

[21] Tukey (1956), when he did not assume the interactions zero, assumed them completely independent, which is unrealistic in the case of nonnormality, as we saw in sec. 7.4. Hooke (1956) obtained some results for the more realistic treatment of interactions, but they look discouragingly complicated.

[22] E. S. Pearson (1929, 1931).

[23] Bartlett (1935) and Geary (1936, 1947).

small if the groups are of equal size and have equal γ_1 and equal variances, but that if γ_1 is different for the two groups, or the groups are of different sizes, or have different variances, this will tend to increase the effect of nonzero γ_1; these results are not surprising in view of the two remarks we made concerning formula (10.1.2) and the skewness of differences and averages.

For the one-way layout with more than two equal groups sampling experiments[24] have again indicated that nonnormality of the errors has small effect on the F-test of the hypothesis that the means are equal. In these experiments the observed distribution of the beta transform of the statistic $\mathscr{F}$ (sec. 9.3) was studied. The extent of these sampling experiments is shown in Table 10.3.1. It would be inconvenient to reproduce here the seven observed distributions. The row labeled P at the bottom of the table gives the probability that the conventional chi-square statistic for the fit of the observed distribution to the theoretical β-distribution

TABLE* 10.3.1

EXTENT OF SAMPLING EXPERIMENTS ON THE EFFECT OF NONNORMALITY OF THE ERRORS ON THE DISTRIBUTION OF THE STATISTIC $\mathscr{F}$ FOR TESTING EQUALITY OF MEANS IN THE ONE-WAY LAYOUT

I = no. of groups	5	5	5	10	10	10	10	10	10
J = group size	5	5	5	5	5	5	5	4	10
No. of samples of IJ	200	200	200	100	100	100	100	50	50
γ_1 of errors	0	0.5	1.0	0	0	0	1.0	0	0.5
γ_2 of errors	4.1	0.7	0.8	−0.5	1.1	4.1	3.8	−1.2	0.7
P	0.5	0.97	0.5	0.7	0.7	0.2	0.07	0.9	0.3

* Contents quoted from "The analysis of variance in cases of nonnormal variation" by E. S. Pearson, *Biometrika*, Vol. 23 (1931) with the kind permission of the author and editor.

derived under the normality assumption would exceed the observed value: Of course it makes no sense to consider a chi-square test of a null hypothesis that we know to be false; rather, we may think of P being a conventional measure on a scale familiar to statisticians, where values of less than 0.05 are considered to indicate a bad fit.

An ingenious method[25] of assessing the effects of nonnormality by means of an easily understood approximation in the form of a correction

[24] E. S. Pearson (1931).

[25] Invented by Box and Andersen (1955). They used it also for the effects of inequality of variance, but carried this through only for the case of the two-way layout with one observation per cell. Comparison with the exact results by Box for this case that we will tabulate in Table 10.4.3 showed very close agreement.

to the numbers of d.f. may be applied in cases where the permutation test based on the statistic $\mathscr{F}$ has been worked out—at least to the extent of the first two moments of the beta transform U of the statistic (sec. 9.3). Suppose that we have formulas for $E_p(U)$ and $E_p(U^2)$, where E_p denotes the expectation calculated under the permutation distribution. Then under any distribution $\mathscr{P}$ of the sample $\mathbf{y}$ we may regard these as conditional[26] expectations, as noted in sec. 9.3, and express the unconditional expectations as

$$E(U^i) = E[E_p(U^i)] \qquad (i = 1, 2), \tag{10.3.1}$$

where E denotes the expectation calculated under $\mathscr{P}$. If we succeed in calculating or approximating the right side of (10.3.1) we may then fit a β-distribution to U by the first two moments, and this approximation of what will usually be a continuous distribution of U would be expected to be even better than the similar approximation of the discrete permutation distribution of U used in sec. 9.3. By comparing the numbers of d.f. of this approximating β-distribution of U with the numbers of d.f. of U or $\mathscr{F}$ under normality and equal variance, we can express the effect on the distribution of $\mathscr{F}$ as a correction on its numbers of d.f., as we did in sec. 9.3.

In a two-way layout[27] with factors A and B and one observation per cell, if we assume that the errors are independently and identically distributed, and that there are no interactions, we may then test the hypothesis H_A that the factor A has no effects, by using the permutation test for randomized blocks based on the statistic $\mathscr{F}$ (which we studied in sec. 9.3), the validity of the permutation test now depending, not on a randomization "within blocks" defined by the levels of factor B, but on the symmetry of the distribution function under H_A generated by the assumption of independently and identically distributed errors: In sec. 9.3 we found that in the permutation distribution of U, $E_p(U)$ is a constant not depending on the observations, and $\text{Var}_p(U)$ is a linear function of the statistic V defined by (9.3.16), which may be regarded as the squared sample coefficient of variation of the "block variances"; so $E_p(U^2)$ is also a linear function of V. Hence it follows for the unconditional moments of U, calculated by (10.3.1), that $E(U)$ is the constant $E_p(U)$, while $E(U^2)$ is given by replacing V in $E_p(U^2)$ by $E(V)$ calculated under the distribution $\mathscr{P}$ of the sample. Since the numbers of d.f. of the approximating

[26] Given that $\mathbf{y}$ fell in $S(\mathbf{y}_0)$ in the notation of sec. 9.3.

[27] I do not make this discussion for the randomized-blocks case, as do Box and Andersen (1955), because the assumption of independently and identically distributed errors looks very unrealistic in the light of our model of sec. 9.1.

β-distribution are functions of the first two moments only, it follows that the numbers of d.f. under $\mathscr{P}$ can be obtained by replacing V by $E(V)$ in the formulas for the numbers of d.f. in the permutation test. Applying this device to (9.3.15) we see that the numbers of d.f. under $\mathscr{P}$ are given by multiplying the normal-theory numbers of d.f. by

$$\frac{1}{1-J^{-1}E(V)} - \frac{2}{J(I-1)}.$$

To order J^{-1} this is

$$1+J^{-1}[E(V) - 2(I-1)^{-1}].$$

It may be shown[28] that

$$E(V) = I^{-1}\gamma_2 + 2(I-1)^{-1}$$

plus terms of order J^{-1} and higher, where γ_2 denotes the kurtosis of the error distribution. Thus to terms of order J^{-1} the numbers of d.f. in the normal-theory F-distribution of the statistic $\mathscr{F}$ must be multiplied by

$$1 + (IJ)^{-1}\gamma_2 \tag{10.3.2}$$

to approximate its distribution under $\mathscr{P}$. It is seen that the normal-theory distribution is rapidly approached as $n = IJ$ increases, and we have an easily understood measure of the deviation from the normal-theory distribution.

A similar treatment of the one-way layout with I groups of J each gives the same factor (10.3.2), to order J^{-1}, for correcting the numbers of d.f. of the distribution of the statistic $\mathscr{F}$ for testing the hypothesis that there is no difference between the group means, where γ_2 is the kurtosis of the errors, which are again assumed to be identically and independently distributed. The numerical values in Table 10.3.2 were obtained by calculating the factor to order[29] J^{-2} and applying it to the case of five groups of 5 each and nominal significance level 5 per cent. The Pearson populations and Edgeworth populations[30] are certain kinds of populations whose distribution is completely determined by their first four moments. The alternate calculation for the Edgeworth population was made by an

[28] From the first two terms in formula (40) of Box and Andersen (1955).

[29] It should not be forgotten that, if the correction factor for the numbers of d.f. is improved by calculating more accurately the expectations (10.3.1), we are still approximating the exact distribution of the statistic $\mathscr{F}$ by fitting to the distribution of its beta transform by only two moments and the range.

[30] Defined by Pearson curves and Edgeworth series; see Cramér (1946), secs. 19.4 and 17.7.

entirely different method.[31] The deviations from the nominal value are in no case of any practical importance.

TABLE* 10.3.2

PROBABILITIES OF TYPE-I ERROR WITH THE F-TEST FOR TESTING THE EQUALITY OF MEANS OF FIVE GROUPS OF FIVE EACH AT THE NOMINAL 5 PER CENT SIGNIFICANCE LEVEL, APPROXIMATED BY A CORRECTION ON THE NUMBERS OF D.F.; γ_1 AND γ_2 ARE THE SKEWNESS AND KURTOSIS OF THE POPULATION OF ERRORS

γ_1^2	Population	γ_2 = −1	−0.5	0	0.5	1
0	Pearson	0.053	0.051	0.050	0.048	‡
	Edgeworth	0.053	0.051	0.050	0.049	0.048
	Edgeworth†	0.052	0.051	0.050	0.049	0.048
0.5	Pearson	0.052	0.051	0.050	0.049	‡
	Edgeworth	0.053	0.051	0.050	0.049	0.048
	Edgeworth†	0.053	0.052	0.050	0.049	0.048
1	Pearson	0.052	0.050	0.049	0.048	0.048
	Edgeworth	0.053	0.052	0.050	0.050	0.049
	Edgeworth†	0.053	0.052	0.051	0.050	0.049

* From p. 14 of "Permutation theory in the derivation of robust criteria and the study of departures from assumption" by G. E. P. Box and S. L. Andersen, *J. of the Royal Stat. Soc.*, series B, Vol. 17 (1955). Reproduced with the kind permission of the authors and the editor.

† Probability approximated by a different method.

‡ The approximation here was unsatisfactory.

We have already remarked that the effect of nonnormality on the probability of type-II errors has not received much attention. Our elementary treatment in sec. 10.2 suggests that for inferences about means the power calculated under normal theory should not be affected much by nonnormality of the errors. A sampling investigation in the case of the two-tailed t-test for a single mean and sample size 5 or 10 verified that there was little effect on the power caused by nonnormality.[32]

[31] Gayen (1950) approximated the probability in the upper tail of the distribution of $\mathscr{F}$ as a linear combination of incomplete-beta functions, the coefficients depending on I, J, and the γ_1 and γ_2 of the errors. I am troubled by the meaning of sampling from an Edgeworth population, which is defined by a probability density that becomes negative in a certain part of its domain when the population is nonnormal.

[32] E. S. Pearson (1929), sec. 5. He found little effect with symmetrical populations but somewhat more effect with skew populations. However, for a t-test for the difference of means, which is the simplest example of the one-way layout, our two remarks

The question of whether F-tests preserve against nonnormal alternatives the power calculated under normal theory should not be confused with that of their efficiency against such alternatives relative to other kinds of tests. We shall touch on the latter question in sec. 10.6.

10.4. MORE ON THE EFFECTS OF INEQUALITY OF VARIANCE

The following rule enables us to extend all formulas we have obtained for E(MS)'s under the assumption of equal error variances to cases where this assumption is violated, provided the errors are independent (of each other and of other random effects, if any, in the model):

Rule: If Q is a quadratic form in $\{y_i\}$ and the $\{y_i\}$ have the structure $y_i = m_i + e_i$, where the $\{m_i\}$ are constants or random variables, the $\{e_i\}$ are independent, and also independent of the $\{m_i\}$ if the $\{m_i\}$ are random variables, and $E(e_i) = 0$, Var $(e_i) = \sigma_i^2$, then the formula for $E(Q)$ may be obtained from that for the special case where all σ_i^2 have a common value σ_e^2 by replacing σ_e^2 in $E(Q)$ by a weighted average $\Sigma_i w_i \sigma_i^2 / \Sigma_i w_i$. The weight $w_{i'}$ may be evaluated as the value of Q obtained by setting $y_{i'} = 1$ and all other $y_i = 0$. In particular if Q is a MS, $\Sigma_i w_i$ calculated in this way will always equal unity, and if furthermore Q is a MS for a layout with equal numbers in the cells, then σ_e^2 is replaced by the unweighted average of the $\{\sigma_i^2\}$ (so that if all observations in the same cell have the same variance, σ_e^2 is replaced by the unweighted average of the cell variances).

Proof: Suppose that

$$Q = \sum_i \sum_j a_{ij} y_i y_j. \tag{10.4.1}$$

Substituting $y_i = m_i + e_i$, we get

$$Q = \sum_i \sum_j a_{ij} m_i m_j + 2 \sum_i \sum_j a_{ij} m_i e_j + \sum_i \sum_j a_{ij} e_i e_j,$$

and taking expectations,

$$E(Q) = E\left(\sum_i \sum_j a_{ij} m_i m_j \right) + \sum_i w_i \sigma_i^2, \tag{10.4.2}$$

where

$$w_i = a_{ii}. \tag{10.4.3}$$

---

above concerning formula (10.1.2) would lead us to expect that the skewness would have much less effect on the t-test for a difference, especially in the case of equal group sizes, than on the test for a single mean. In testing the hypothesis $\mu = \mu_0$ the power turns out to be greater or smaller than the normal-theory power depending on whether the sign of $\mu - \mu_0$ is the same or opposite to that of the γ_1 of the errors.

In particular, if all $\sigma_i^2 = \sigma_e^2$,

$$E(Q) = E\left(\sum_i \sum_j a_{ij} m_i m_j\right) + \sigma_e^2 \sum_i w_i. \tag{10.4.4}$$

Comparing the value (10.4.2) of $E(Q)$ in the general case with its value (10.4.4) in the special case, we verify the rule. The value stated for $w_{i'}$ follows from (10.4.3) and (10.4.1). If Q is a MS and all $\sigma_i^2 = \sigma_e^2$ we know from earlier chapters that the last term in (10.4.4) is σ_e^2, and hence $\Sigma_i w_i = 1$. Finally, in order to establish that the $\{w_i\}$ are all equal in the case of layouts with equal numbers in the cells we must convince ourselves that if one y_i equals 1 and all others zero then the value of the MS is the same no matter which y_i equals 1; this is perhaps most easily seen by thinking of how computations like those of the following illustration are affected by the rules given in sec. 8.2 for the form of the SS suitable for numerical computation.

Let us illustrate the application of the rule in the example of the one-way layout with unequal numbers: With our usual notation for this example, the σ_e^2 appearing in $E(\text{MS}_A)$ is to be replaced by $\Sigma_i \Sigma_j w_{ij} \sigma_{ij}^2$, where $\sigma_{ij}^2 = \text{Var}(y_{ij})$, and the value of $w_{i'j'}$ is obtained by putting $y_{i'j'} = 1$ and all other $y_{ij} = 0$ in

$$\text{MS}_A = (I-1)^{-1}\left(\sum_i J_i^{-1} T_i^2 - n^{-1} T^2\right),$$

where T_i is the total of the ith group, T is the grand total, and $n = \Sigma_i J_i$; namely

$$w_{i'j'} = (I-1)^{-1}(J_{i'}^{-1} - n^{-1}).$$

Hence σ_e^2 is replaced in $E(\text{MS}_A)$ by

$$(I-1)^{-1} \sum_i \sum_j (J_i^{-1} - n^{-1}) \sigma_{ij}^2. \tag{10.4.5}$$

Similarly, σ_e^2 in $E(\text{MS}_e)$ is to be replaced by $\Sigma_i \Sigma_j v_{ij} \sigma_{ij}^2$ where $v_{i'j'}$ is the value obtained by putting $y_{i'j'} = 1$ and all other $y_{ij} = 0$ in

$$\text{MS}_e = \nu_e^{-1}\left(\sum_i \sum_j y_{ij}^2 - \sum_i J_i^{-1} T_i^2\right),$$

with $\nu_e = \Sigma_i (J_i - 1)$; namely

$$v_{i'j'} = \nu_e^{-1}(1 - J_{i'}^{-1}).$$

Hence σ_e^2 is replaced in $E(\text{MS}_e)$ by

$$\nu_e^{-1} \sum_i \sum_j (1 - J_i^{-1}) \sigma_{ij}^2. \tag{10.4.6}$$

In the case where all the errors in the ith group have the same variance $\sigma_{ij}^2 = \sigma_i^2$, (10.4.5) becomes

$$(I-1)^{-1}[I(\sigma^2)_{.u} - (\sigma^2)_{.w}],$$

where $(\sigma^2)_{.u}$ is the unweighted average of the $\{\sigma_i^2\}$ and $(\sigma^2)_{.w}$ is the weighted average with weights $\{J_i\}$, while (10.4.6) becomes the weighted average with weights $\{J_i-1\}$. We check that if the $\{J_i\}$ are all equal, these various averages are equal, and the results agree with those obtained by applying the last part of the rule for equal cell numbers.

The above rule has the following implication concerning an advantage of equal cell numbers in complete layouts when the equality-of-variance assumption is violated: Recall the definition (sec. 9.1) of a design being unbiased for testing the hypothesis $\sigma_x^2 = 0$, where σ_x^2 denotes any σ^2 employed in the E(MS) formulas, namely, that there exist two MS's whose E(MS)'s differ by $c\sigma_x^2$ with c a known nonzero constant. Since with independent errors and equal cell numbers σ_e^2 is replaced in every E(MS) by the same average of the variances, the consequence is that if a design with equal cell numbers is unbiased for testing $\sigma_x^2 = 0$ under the assumption of equal variance it continues to be unbiased when the assumption is violated. That this is not true in general of designs with unequal numbers is seen from the calculation carried out above for the one-way layout.

TABLE* 10.4.1

EFFECT OF INEQUALITY OF POPULATION VARIANCES ON PROBABILITY OF TYPE-I ERROR WITH TWO-TAILED t-TEST FOR EQUALITY OF MEANS AT NOMINAL 5 PER CENT SIGNIFICANCE LEVEL. J_1 AND J_2 ARE SAMPLE SIZES, $\theta = \sigma_1^2/\sigma_2^2$, AND σ_1^2 AND σ_2^2 ARE POPULATION VARIANCES

(J_1, J_2)	θ								
	0	0.1	0.2	0.5	1	2	5	10	∞
(15, 5)	0.32	0.23	0.18	0.098	0.050	0.025	0.008	0.005	0.002
(5, 3)	0.22	0.14	0.10	0.072	0.050	0.038	0.031	0.030	0.031
(7, 7)	0.072	0.070	0.063	0.058	0.050	0.051	0.058	0.063	0.072

* From p. 12 of "Contributions to the theory of Student's t-test as applied to the problem of two samples" by P. L. Hsu, *Stat. Research Memoirs*, Vol. 2 (1938*a*). Values are here reproduced with the kind permission of the editors.

The rest of this section will deal with tests for means in fixed-effects models when the errors are independently normal. Some exact results are then available for the probability of type-I errors. For the two-tailed t-test for the equality of two means at the nominal 5 per cent level, Table 10.4.1 gives the probabilities[33] of type-I errors when the population

[33] Calculated by P. L. Hsu (1938*a*).

variances are in the ratio $\theta = \sigma_1^2/\sigma_2^2$ for sample sizes $(J_1, J_2) = (15, 5)$, (5, 3), and (7, 7). Of course the table gives also the probability of a confidence interval with nominal confidence coefficient 95 per cent failing to cover the true difference of means. Our large sample conclusions in

TABLE* 10.4.2

EFFECT OF INEQUALITY OF VARIANCES ON PROBABILITY OF TYPE-I ERROR WITH F-TEST FOR EQUALITY OF MEANS IN ONE-WAY LAYOUT AT NOMINAL 5 PER CENT LEVEL

n = total number of observations, V_u = squared coefficient of variation of group variances

No. of Groups, I	Ratio of Group Variances $\{\sigma_i^2\}$	Group Sizes $\{J_i\}$	n	Probability of Type-I Error	V_u	$1+\frac{I-2}{I-1}V_u$
3	1:2:3	5, 5, 5	15	0.056	0.25	1.12
		3, 9, 3	15	0.056		
		7, 5, 3	15	0.092		
		3, 5, 7	15	0.040		
3	1:1:3	5, 5, 5	15	0.059	0.48	1.24
		7, 5, 3	15	0.11		
		9, 5, 1	15	0.17		
		1, 5, 9	15	0.013		
5	1:1:1:1:3	5, 5, 5, 5, 5	25	0.074	0.40	1.31
		9, 5, 5, 5, 1	25	0.14		
		1, 5, 5, 5, 9	25	0.025		
7	1:1: · · · :1:7	3, 3, · · ·, 3, 3	21	0.12	1.49	2.24

* From p. 299 of "Some theorems on quadratic forms applied in the study of analysis of variance problems, I. Effect of inequality of variance in the one-way classification" by G. E. P. Box, *Ann. Math. Stat.*, Vol. 25 (1954*a*). Here reproduced with the kind permission of the author and the editor.

sec. 10.2, that the effect is small with equal group sizes but may be severe with unequal sizes, is borne out also for these small samples. However, our large sample conclusion for the one-way classification in the case of I equal groups, that inequality of group variances causes the true probability of type-I error to *exceed* the nominal probability when $I > 2$ is seen to hold here[34] even in the case $I = 2$; this conclusion for $I > 2$ will be substantiated for small equal group sizes in the case we consider next.

For the one-way layout the probabilities shown in Table 10.4.2 have been calculated from theorems giving under certain conditions exact numerical values for the distribution of the ratio of two quadratic forms in central normal variables.[35] Let us consider first the four lines of the table for equal group sizes, immediately below the horizontal ruled lines. Formula (10.2.19) suggests that the deviation of the probability from the nominal value of 5 per cent should always be positive if the group variances $\{\sigma_i^2\}$ are unequal and should increase with the value of the factor $1 + (I-2)(I-1)^{-1}V_u$, where V_u is the squared coefficient of variation of the group variances; this is seen to be the case. If we let θ denote the ratio of the maximum to the minimum of the $\{\sigma_i^2\}$, it can be shown that, for fixed θ, V_u is maximum when the $\{\sigma_i^2\}$ are in the ratio $1:1:\cdots:1:\theta$. The last line of the table, where $\theta = 7$, was included to show how bad the deviation might be in a very unfavorable case not likely to be approached ordinarily in practice. For two of the other three lines, if not for all three, where the group sizes are equal, the deviations can be considered bearable. However, this cannot be said for six of the eight lines where the group sizes are unequal.

Some probabilities[36] for the two-way layout with one observation per cell are listed in Table 10.4.3. They are for the case where the variance of the error in the i,j cell has the same value $\sigma_{ij}^2 = \sigma_i^2$ throughout the ith row. Under the hypothesis H_A of no row effects the distribution of MS_A is then the same as in the one-way layout with equal numbers just considered, although MS_e is defined differently, to eliminate the column effects. It turns out that MS_A and MS_e are independent (MS_B and MS_e are not), and this permits the use of chi-square approximations of the kind mentioned after (10.2.19) to give a central-F approximation. As in the corresponding one-way classification we note that the true probabilities of type-I error in testing H_A always exceed the nominal probability. The situation for the test of the hypothesis H_B of no column effects does not correspond to the one-way classification considered above, since here the

[34] P. L. Hsu (1938*a*) proved the general result that for any $J_1 = J_2$ and any α the probability of type-I error with the t-test for the equality of two means is a strictly decreasing function of θ as θ increases from 0 to 1, and strictly increasing as θ increases from 1 to ∞.

[35] Box (1954*a*).

[36] Obtained by Box (1954*b*). I have interchanged his rows and columns to conform better to the notation of this chapter. Both the present results and those of sec. 10.5 for the two-way layout were obtained by Box by applying his general theorems to the case where our columns (his rows) are independently distributed according to the same I-variate normal distribution and then specializing to these two cases.

variances differ within columns, and there it was constant within groups. We note that the true probabilities are all less than the nominal. For neither test are the deviations[37] alarming.

TABLE* 10.4.3

EFFECT OF INEQUALITY OF VARIANCE ON PROBABILITY OF TYPE-I ERROR WITH F-TESTS FOR EQUALITY OF ROW MEANS AND OF COLUMN MEANS IN TWO-WAY LAYOUT WITH ONE OBSERVATION PER CELL AT NOMINAL 5 PER CENT LEVEL

I = number of rows, J = number of columns, σ_i^2 = variance in ith row

I	J	Ratio of Row Variances $\{\sigma_i^2\}$	Probability of Type-I Error Row Test	Column Test
3	11	1:2:3	0.055†	0.042
3	5	1:2:3	0.056†	0.043
3	11	1:1:3	0.059†	0.038
3	5	1:1:3	0.060	0.039
5	3	1:1:1:1:3	0.068	0.045
11	3	1:1: · · · :1:3	0.071†	0.049

* From p. 493 of "Some theorems on quadratic forms applied in the study of analysis of variance problems, II. Effects of inequality of variance and of correlation between errors in the two-way classification" by G. E. P. Box, *Ann. Math. Stat.*, Vol. 25 (1954*b*). Values are here reproduced with the kind permission of the author and the editor.

† Approximate; see discussion in text.

Most of what is known[38] about the effect of inequality of variance on the power of the F-test is contained in Figs. 10.4.1 and 10.4.2 which graph the power of the F-test of equality of means at the nominal 5 per cent level in a one-way layout with four groups. Each of the six graphs is plotted from the four points indicated, the ordinates of which were obtained by approximations based on moment calculations. In every case the group variances are in the ratio $\sigma_1^2:\sigma_2^2:\sigma_3^2:\sigma_4^2 = 1:1:1:3$ and the group sizes are equal for the three groups with equal variances, the group sizes for the three graphs in each figure being $(J_1, J_2, J_3, J_4) = (7, 7, 7, 19)$, $(10, 10, 10, 10)$, and $(12, 12, 12, 4)$, so that the total number

[37] Box (1954*b*), p. 493, remarks that "Comparison of the last four lines of the table suggests that the between-rows discrepancy is worse when the number of rows exceeds the number of columns, while the between-columns discrepancy is worse when the number of columns exceeds the number of rows."

[38] The numerical calculations and their graphs reproduced in Figs. 10.4.1 and 10.4.2 are by G. Horsnell (1953), based on methods of David and Johnson (1951). Some exact results for the effect of inequality of variance on the power of the t-test for the difference of two means were obtained by P. L. Hsu (1938*a*), but it is difficult to assess their practical import.

of observations is always 40. The true means $\{\mu_i\}$ consist of three equal means and one different; in Fig. 10.4.1 the divergent mean is in a group with low variance, $\mu_1 \neq \mu_2 = \mu_3 = \mu_4$; in Fig. 10.4.2 the divergent mean is in the group with high variance, so that $\mu_1 = \mu_2 = \mu_3 \neq \mu_4$. If the variances $\{\sigma_i^2\}$ were all equal to σ^2 the noncentrality parameter ϕ would be defined as (sec. 2.8)

$$\phi = [(\nu_1+1)^{-1} \sum_i J_i(\mu_i - \mu)^2/\sigma^2]^{1/2}, \tag{10.4.7}$$

where $\nu_1 = I-1 = 3$, and $\mu = \Sigma_i J_i \mu_i / \Sigma_i J_i$. The abscissa on the figures is the same as (10.4.7) with σ^2 replaced by $(\sigma^2)_{.w} = \Sigma_i J_i \sigma_i^2 / \Sigma_i J_i$. The graph for the case of equal variances is included as the dashed line in each figure.

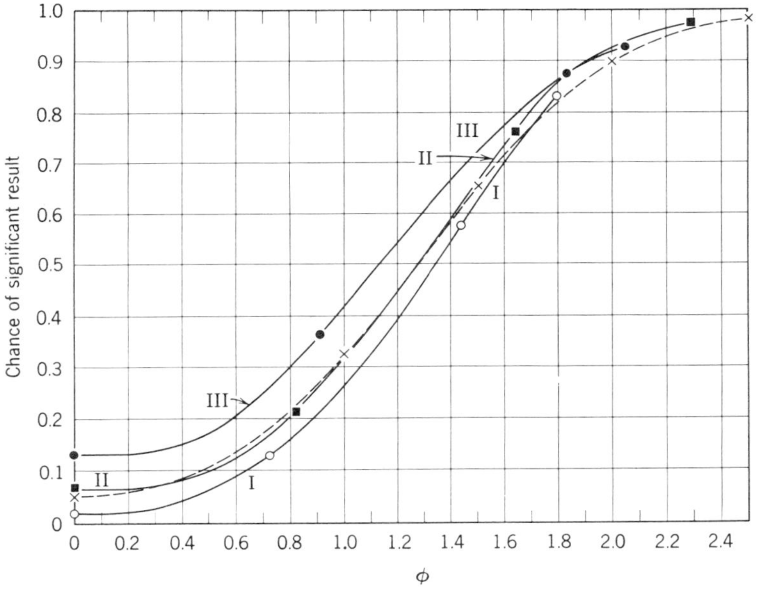

FIG. 10.4.1. Divergent mean in a group with low variance
$\sigma_1^2 : \sigma_2^2 : \sigma_3^2 : \sigma_4^2 = 1:1:1:3$

Key:	J_1	J_2	J_3	J_4
I	7	7	7	19
II	10	10	10	10
III	12	12	12	4
— — —	Equal variances			

(From p. 134 of "The effect of unequal group variances on the *F*-test for homogeneity of group means" by G. Horsnell, *Biometrika*, Vol. 40 (1953). Here reproduced with the kind permission of the author and the editor.)

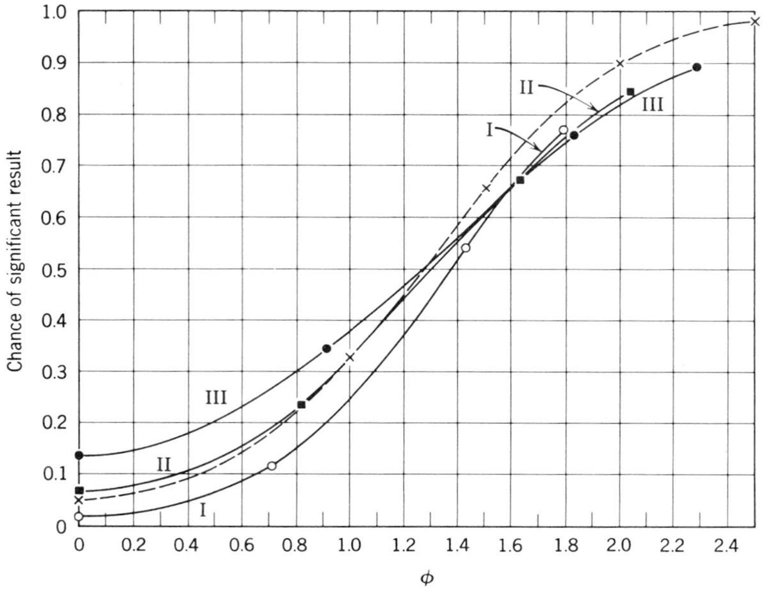

FIG. 10.4.2. Divergent mean in the group with high variance $\sigma_1^2:\sigma_2^2:\sigma_3^2:\sigma_4^2 = 1:1:1:3$

		J_1	J_2	J_3	J_4
Key:	I	7	7	7	19
	II	10	10	10	10
	III	12	12	12	4
	— — —	Equal variances			

(From p. 134 of "The effect of unequal group variances on the *F*-test for homogeneity of group means" by G. Horsnell, *Biometrika*, Vol. 40 (1953). Here reproduced with the kind permission of the author and the editor.)

The reader should draw his own conclusions from studying these figures. Perhaps they will agree with those of the original author[39] and the present author, that equal group sizes should be used unless one is sure for certain groups that the variance is larger, in which case there is no harm in allocating more of the measurements to those groups. However, for a two-way or higher layout one would usually consider choosing only equal numbers or "proportional frequencies" (end of sec. 4.4) in any case, because of the computational difficulties otherwise.

[39] G. Horsnell (1953).

10.5. MORE ON THE EFFECTS OF STATISTICAL DEPENDENCE

The reader is reminded that in Ch. 9 we studied to what extent the inferences derived under the Ω-assumptions of Ch. 2 are valid under the randomization models for the incomplete blocks and Latin-square designs, where dependence of the unit errors is caused by the randomization. Besides the results of Ch. 9 and the elementary example with serial correlation in sec. 10.2 we have only one further example to offer: This concerns the effect of a certain kind of serial correlation in a two-way layout with one observation per cell.

Suppose in a two-way layout, where the rows correspond to a factor A and the columns to a factor B, that the observations within a column are serially correlated but that the columns of observations are independent of each other. This could happen if the levels of factor A were equally spaced intervals in time or space, for example[40] if the rows are the 24 hours of the day, the columns are the 12 months of the year, and the entry is the frequency of rain at a certain place during a 10-year period. The entries in successive rows and within the same column would be positively correlated, since the probability of rain during any hour period is greater or smaller depending on whether there was rain during the preceding hour period, but any similar effect between columns would be much smaller and probably negligible.

From the previously mentioned general results for quadratic forms in normal variables one can derive the effect on the tests if the observations are serially correlated[41] within columns with correlation coefficient ρ as defined in sec. 10.1. More precisely, the model is that the errors have a joint normal distribution with zero means, equal variances, and all correlation coefficients zero except for pairs of successive observations in the same column, for which the correlation coefficient is ρ. Table 10.5.1 gives the probabilities of type-I errors under this model in the case of a 5×5 layout and a nominal 5 per cent level. The approximation used for some of the row tests was of the same kind that we indicated in connection with Table 10.4.3. The conclusion is clear: Serial correlation within

[40] Given by R. A. Fisher (1925); his assessments of the general nature of the effects of the correlation on the two tests for main effects were stated without proof, and were validated by the theoretical investigation of Box (1954*b*).

[41] The above rainfall example would be better fitted by a model with *circular* correlation within columns in which last and first observations (for the periods of the 24th and 1st hours) have the same correlation as pairs of successive observations. Furthermore, we would not expect the correlation coefficients with lags 2, 3, and possibly more, to be exactly zero.

columns has little effect on the row test but a powerful effect on the column test, positive serial correlation (the more frequent kind, see sec. 10.1) increasing the probability of type-I error.

TABLE* 10.5.1

EFFECT OF SERIAL CORRELATION COEFFICIENT ρ WITHIN COLUMNS IN A 5×5 LAYOUT WITH ONE OBSERVATION PER CELL ON PROBABILITY OF TYPE-I ERROR WITH F-TESTS FOR EQUALITY OF ROW MEANS AND OF COLUMN MEANS AT NOMINAL 5 PER CENT LEVEL

ρ	-0.4	-0.2	0	0.2	0.4
Row test	0.059†	0.053†	0.050†	0.054†	0.064
Column test	0.0003	0.010	0.050	0.13	0.25

* From p. 497 of "Some theorems on quadratic forms applied in the study of analysis of variance problems, II. Effects of inequality of variance and of correlation between errors in the two-way classification" by G. E. P. Box, *Ann. Math. Stat.*, Vol. 25 (1954*b*). Here reproduced with the kind permission of the author and the editor.

† Approximate; see text.

10.6. CONCLUSIONS

The further evidence we have examined since stating the conclusions at the end of sec. 10.2 about the effects of the three kinds of departures we have considered does not necessitate our revising those conclusions.

Among the underlying assumptions made in deriving statistical methods are usually some that are apt to be violated in applications and are introduced only to ease the mathematics of the derivation, for instance, the assumption of normality wherever we have made it in this book. Statistical methods have been called *robust*[42] if the inferences are not seriously invalidated by the violation of such assumptions. Optimality of statistical methods is usually proved under such assumptions. Robustness of the methods may then be judged to be of greater practical importance.[43] In the case of tests, optimality guarantees the best possible power in some sense against the alternatives possible under the assumptions Ω of the derivation. If the optimum test is not robust and a robust test exists we would surely prefer it if its power under Ω is not much inferior to that of the optimum test. It is clear how the notion of robustness applies to the probability of type-I errors and confidence coefficients,

[42] By Box (1953).

[43] When I became aware that the nominal probability of type-I error for the standard test of the equality of variances of two populations is invalidated by nonnormality to the same order of magnitude as found in Table 10.2.1, I found little consolation in the optimum properties someone once established for that test (Scheffé, 1942).

but it may be more difficult to extend it to the probability of type-II errors: for example, our conclusion that the power of the F-test of the equality of means in the one-way layout with equal groups is not much affected by inequality of variance depended on finding a suitable way of defining the noncentrality parameter in the case of inequality of variance. In any event, robustness for type-I errors is not a sufficient recommendation for a test; the power must also be considered against some alternatives of interest.

The robustness, concerning type-I errors, of F-tests about means extends to the corresponding S-methods of multiple comparison. To every such test there corresponds an S-method, as we saw in sec. 3.5, and the distribution theory on which the S-method rests is the same as that under the null hypothesis of the corresponding test, as indicated in the remark on interval estimation at the end of Ch. 9. This result deserves some elaboration: We illustrate it with the comparison of means in a one-way layout with four equal groups with means $\{\mu_i\}$ and variances $\{\sigma_i^2\}$. The S-method will share with the F-test its robustness against nonnormality and inequality of variance; this is not evident for the T-method of multiple comparison. Robustness against nonnormality would hold also for interval estimates of individual contrasts based on the t-distribution. However, the individual interval estimates based on t could be badly invalidated by inequality of variance; consider for example the interval estimates of $\mu_1-\mu_2$ and $\mu_3-\mu_4$ made in the usual way if actually $\sigma_1^2=\sigma_2^2$ are smaller and $\sigma_3^2=\sigma_4^2$ are larger than the average of the $\{\sigma_i^2\}$. Then Var $(\hat{\mu}_1-\hat{\mu}_2)$ will be overestimated, Var $(\hat{\mu}_3-\hat{\mu}_4)$ will be underestimated, and hence the probability of failing to cover the true value will be less than the nominal probability for the first interval and greater for the second, and the discrepancies would obviously be large if the inequality of variance were large. We may imagine that the S-method in offering statements about all the contrasts somehow averages out the variation in probability encountered with individual contrasts.

Returning to the consideration of the power of tests robust with regard to type-I errors, it appears[44] that there probably exist tests which have the robustness of the F-tests concerning type-I errors, a little less power against normal alternatives, but much greater power against "most" nonnormal alternatives. At present such tests have not been developed for the relatively complicated hypotheses usually considered in the analysis of variance, and even if they were, the methods of estimation with which one would usually want to follow them up when they rejected, analogous to the S-method after an F-test, while then possible in principle, would

[44] See for example Hodges and Lehmann (1956); the Wilcoxon test seems to be of this nature when it is applicable.

seem hopelessly complicated to carry out in any but the very simplest cases, such as that of comparing the means of two populations which are assumed identical except for a translation.

Robustness against nonnormality of the standard methods of inference in the case of means, and its lack in the case of variances, has the following practical consequence:[45] Since the analysis-of-variance methods for means are derived from assumptions that include the equality of variance, it has sometimes been recommended that this assumption first be subjected to a statistical test. The standard test[46] for equality of variance tends to mask differences of variances when they exist if the kurtosis $\gamma_2 < 0$ and to find differences when none exist if $\gamma_2 > 0$; for some populations with $\gamma_2 > 0$ the test has a sensitivity to nonnormality comparable to that of standard tests of nonnormality. If the variances are equal but the data are nonnormal with $\gamma_2 > 0$, the preliminary test is then likely to reject the hypothesis of equality of variance and the user will accordingly refrain from applying the analysis of variance of the means in a case where it is dependable. Furthermore, if the situation is a layout where equal numbers in the cells can be chosen there is no need to worry about possible inequality of variance in the first place unless it is extreme.

Such robustness of the analysis of variance of means as we found against inequality of variance in the case of a layout with equal cell numbers suggests the following quick approximate analysis to replace the tedious exact calculations in the case of unequal cell numbers, or as a quick preliminary analysis, after which we may dispense with the exact analysis if the results are sufficiently conclusive: Form the usual error MS by pooling the SS's within cells about the cell means, and denote this by MS_e' and its number of d.f. by ν_e'. Imagine a layout with one "observation" per cell, which we will call L, in which the "observation" is the observed cell mean. Analyze the layout L as though the "observations" had equal variance, say σ_L^2, except that whenever an error MS is needed to estimate σ_L^2 it is estimated as $c\,\text{MS}_e'$ with ν_e' d.f., where c is the average of the reciprocals of the cell numbers for the original observations. If any of these cell numbers are zero the method cannot be applied.

To see the justification of this approximate analysis, suppose that the original observations are independently normal, that there are n_p observations in the pth cell, with sample mean $\bar{y}_p$, sample variance s_p^2, and population variance σ_p^2 $(p = 1, \cdots, N)$. Then the variance of the "observation" $\bar{y}_p$ in the pth cell of L is σ_p^2/n_p. We have seen that with unequal cell variances and equal cell numbers, as in L, the F-test and S-method of multiple comparison behave approximately the same as though the

[45] These considerations are due to Box (1953).

[46] Bartlett's (1937) test.

variances were equal to the average cell variance, which is in this case

$$N^{-1}\sum_{p=1}^{N} n_p^{-1}\sigma_p^2. \tag{10.6.1}$$

We might estimate (10.6.1) as

$$N^{-1}\sum_{p} n_p^{-1}s_p^2, \tag{10.6.2}$$

but if we assume all $\{\sigma_p^2\}$ equal to σ_0^2 (as we usually do with the exact method for unequal cell numbers), so that (10.6.1) becomes $\sigma_0^2 N^{-1}\Sigma_p n_p^{-1}$, we would use the pooled estimate MS_e' of σ_0^2 and be led to estimate (10.6.1) as $c\ MS_e'$ with $c = N^{-1}\Sigma_p n_p^{-1}$, as suggested above. Obviously the $\{\bar{y}_p\}$ and the $\{s_p^2\}$ are completely independent, and so either estimate of (10.6.1) is independent of the "observations" in L and hence of any estimates or MS's calculated from L. The analysis suggested above corresponds to the exact analysis assuming all $\{\sigma_p^2\}$ equal, but is in general somewhat less efficient.[47] The estimate (10.6.2) should be used only if considerable inequality of the $\{\sigma_p^2\}$ is believed to exist; to assign a number of d.f. to (10.6.2) we may use the method at the end of sec. 7.5.

Now let us briefly consider what if anything may be done to alleviate those effects of departures from the underlying assumptions which are serious.

A device to avoid the serious effects of nonnormality with inferences about variances was used in sec. 3.8 on the comparison of several variances, namely analysis of variance applied to the sample variances. This device will not be possible for the estimates of variance components other than the error component σ_e^2. If we had some idea of the magnitude of the kurtosis γ_2 of the effects measured by the variance component we might use this in assessing the error of the point estimate, but obviously we don't ordinarily have any such idea. The situation is not very hopeful, and normal-theory inferences about variance components must be accepted as being much less reliable than those about means. This conclusion is reinforced by the consideration that models with variance components have, even without the normality assumption, a rather tenuous relation to those frequent applications where nothing is done to insure the random sampling of the effects which is assumed in the model.

Ordinarily in analyzing means we should not bother to attempt to transform data to reduce nonnormality. An exception is rank data, for example from experiments in which judges assign integers $1, 2, \cdots, m$ to m objects submitted to them for rating the order of some qualitative

[47] In the case of the one-way layout with two groups it is identical with the exact analysis; in any case it utilizes the sufficient statistics $\{\bar{y}_p\}$ and $\{s_p^2\}$.

characteristic. By use of tables[48] of the expected values of the ordered elements of a random sample of m from $N(0, 1)$ such data may be transformed to behave less nonnormally.

We have seen that the simplest safeguard against bad effects from inequality of variance in complete layouts is the use of equal cell numbers. If the ratios of the variances are known we can use the weighted analysis indicated in sec. 1.5, where there is some mention of the effects of using wrong weights, and in sec. 3.8, where there is an example after (3.8.3). If extreme inequality of variance is suspected and there are several observations in each of the cells of a layout, or at each abscissa in curve fitting, we may consider performing a weighted analysis with the weights inversely proportional to estimated variances.[49] For layouts higher than one-way this is computationally awkward, for even with equal cell numbers, orthogonality and the resulting simplicity of computation are lost in a weighted analysis. Furthermore, there is little theoretical knowledge of the effect of replacing unknown constant weights by random variables.[50] Transformations to reduce inequality of variance will be treated in sec. 10.7.

Where a correlation is suspected to which might be applied the simple model we have used of serial correlation with a single coefficient ρ, one might try to estimate ρ from the data and to approximate[51] the effect in the inferences of a true ρ equal to the estimated ρ. Some kinds of correlation can be brought under existing multivariate theory, like the correlation within columns in the mixed model for the two-way layout in sec. 8.1. In general, of the three kinds of possible departures from assumptions we have considered, those caused by lack of independence are the most formidable to cope with.

10.7. TRANSFORMATIONS OF THE OBSERVATIONS

Transformations are used sometimes to reduce interactions (end of sec. 4.1) or to reduce nonnormality (see above), but most frequently to reduce inequality of variance.[52] Most of the commonly used transformations

[48] Fisher and Yates (1943), Table XXI.

[49] This is standard procedure in fitting curves for quantal response to dosage by the probit method; see Fisher and Yates (1943), introduction to Tables X and XI.

[50] For the one-way layout with I groups there are some exact results for $I = 2$ by P. L. Hsu (1938*a*), and approximate results for $I > 2$ by James (1951) and Welch (1951), which see for further references.

[51] Box's (1954*b*) exact results are available for the two-way layout.

[52] It is curious that if the transformation (10.7.1), derived as one to stabilize variance, is applied to the sample correlation coefficient r (when it is called Fisher's transformation) it makes the distribution behave much more like a normal one; see Cramér (1946), sec. 29.7.

are special cases or modifications[53] of the following general transformation: Suppose that the mean, in general unknown, of a random variable y is denoted by μ and that the standard deviation of y can be expressed as a function of μ only, which is either completely known or known up to an unknown constant multiplier, say $\sigma_y = \phi(\mu)$. This will generally be the case if the distribution of y depends on a single parameter (which may however be a function of other parameters of interest in a problem). For example, for the binomial distribution of the number y of successes in n trials with constant probability p, $E(y) = np$, $\sigma_y = [np(1-p)]^{1/2}$, and so y has the required property with $\phi(\mu) = [\mu(1-n^{-1}\mu)]^{1/2}$. Consider a transformation $z = f(y)$, which we will try to determine so that the standard deviation of z is equal, at least approximately, to a predetermined constant σ_z. By the approximate formula $\sigma_z = \sigma_y f'(\mu)$, obtained by approximating z as a linear function of y in the neighborhood of $y = \mu$, we get $f'(\mu) = \sigma_z/\phi(\mu)$ and, integrating this, and switching notation,

$$f(y) = \sigma_z \int \frac{dy}{\phi(y)}. \tag{10.7.1}$$

Thus in the above example we have

$$f(y) = \sigma_z \int [y(1-n^{-1}y)]^{-1/2} dy = 2n^{1/2}\sigma_z \arcsin (y/n)^{1/2} + C,$$

and we take $C = 0$. If we choose $\sigma_z = (4n)^{-1/2}$, the transformation becomes the "angular transformation" $z = \arcsin (y/n)^{1/2}$, where y/n is the observed proportion of successes and the arcsine is in radians; if the arcsine is in degrees, $\sigma_z = 28.6n^{-1/2}$.

The most common transformation is to take logarithms. "Logging" the observations is appropriate for equalizing their variances in cases where their per cent error is constant; putting $\sigma_y = c\mu$ in (10.7.1) generates this transformation.

It is important to remember that transformations transform the mean as well as the variance, so that for the transformed variable approximately $E(z) = f(\mu)$, and this may help or hinder the analysis: As an example where it helps, suppose that in a pilot-plant experiment to convert a certain chemical substance, which we shall call the "reactant," to a desired product by means of a catalyst we intend to vary the following four factors: (i) the kind of catalyst, (ii) the amount of the reactant, (iii) the contact time of the reactant with the catalyst, and (iv) the temperature of the reaction (controlled by a water jacket or a heating coil). Denote by y the amount of reactant converted to the product in a single run of the pilot plant. If the set of factor levels in the experiment produces a large

[53] Freeman and Tukey (1950).

range of y in a percentage sense, and if we expect the per cent error rather than the absolute error of y to be more nearly constant, this would suggest using the transformation $f(y) = \log y$ to stabilize the variance. If we then consider the effect on the mean we might decide that in this case it is beneficial, because the amount of reactant converted might be better expressed as a product of four functions for the four factors, rather than a sum; i.e., $E(\log y)$ might be more nearly linear than $E(y)$ in four functions expressing the effects of the factors. We shall return to this example below.

In testing the hypothesis of equal group means in a one-way layout, transformation would seem to cause no difficulty since for 1 : 1 transformations the original means are equal if and only if their transforms are equal. However, if after a transformation the hypothesis is rejected by an F-test and we wish to follow up with multiple comparison or other forms of estimation on the means, the original scale of means may give us more meaningful statements than the transformed scale, for example if y were the observed proportion of diseased animals cured by one of several drugs being compared in an experiment: Then if μ_1 and μ_2 are the population proportions for the first and second drugs in the experiment it is easier to convey and appreciate the statement that $\mu_1 - \mu_2$ lies in a certain interval, than a similar statement about $\arcsin \mu_1^{1/2} - \arcsin \mu_2^{1/2}$.

With two-way and higher layouts not only is there the question of the meaningfulness of comparison or estimation on different scales, but the meaning of the usual hypotheses tested also depends on the transformation: We have already seen this to be true of the hypotheses of no interactions (sec. 4.1), but in the presence of interactions (and interactions will in general be present on at least one of the two scales if the transformation is nonlinear) it is also true of the hypotheses about main effects: Consider the example[54] of a two-way layout where one factor is kind of insecticide, the other a method of spraying, the observation y_{ij} is the proportion of bugs killed when the ith insecticide is sprayed by the jth method, and $\mu_{ij} = E(y_{ij})$. If we remember that the main effect for the ith insecticide is a certain average over the methods of spraying we see that if the hypothesis of no difference of main effects is true for one of the sets $\{\mu_{ij}\}$ or $\{\arcsin \mu_{ij}^{1/2}\}$, it will in general be false for the other set. Furthermore, if we consider the possible economic interpretation of the parameters we might agree that if the different methods of spraying are known to be used in practice with frequencies $\{\pi_1, \pi_2, \cdots\}$ then a reasonable measure of the effectiveness of the ith insecticide is $\Sigma_j \pi_j \mu_{ij}$, whereas it would be hard to justify $\Sigma_j \pi_j \arcsin \mu_{ij}^{1/2}$ as a meaningful measure.

[54] This example and its implications were suggested by Professor William Kruskal.

The belief is sometimes encountered that if a transformation equalizes the variances it is also likely to reduce nonnormality and nonadditivity. That additivity and equality of variance may be irreconcilable desiderata is shown by the following example in which we can achieve the one only at the cost of destroying the other: Suppose there are $I+J+1$ independent Poisson variables, $\{u_i\}$ with means $\{\alpha_i\}$, $\{v_j\}$ with means $\{\beta_j\}$, and w with mean γ, and the observation y_{ij} in the i, j cell of a two-way layout has the structure u_i+v_j+w. Then $E(y_{ij}) = \alpha_i+\beta_j+\gamma$ and we have additivity, but also $\text{Var}(y_{ij}) = \alpha_i+\beta_j+\gamma$, and so we do not have equality of variance unless all $\{\alpha_i\}$ are equal and all $\{\beta_j\}$ are equal. The transformation (10.7.1) to equalize the variance becomes $z = y^{1/2}$ for a Poisson distribution, and this destroys the additivity. It is not obvious whether y or $y^{1/2}$ is more nearly normal for a Poisson variable, but in the present context it hardly matters.

If our quantitative knowledge of the phenomena investigated in an experiment is not purely empirical but is derived in part from some theoretical principles, this theoretical knowledge will often suggest useful and meaningful transformations. Let us illustrate with the example of chemical conversion introduced above. If A denotes the original amount of the reactant, and if it is possible to get the amount y converted to the desired product to be very close to A, then it is physically impossible for y to fluctuate very much since y is jammed under the ceiling of A. In that case the error σ_y could be expected to become small as y approaches A, and then the transformation $f(y) = \log y$ would not equalize the variance.

To stabilize the variance one must counter the jamming effect mentioned above by a stretch of the scale of y near $y = A$, similar to the stretch the arcsine transformation gives the scale near its ends. Such a new scale may also be intuitively more desirable in that equal increases on different parts of the scale more nearly correspond to equal difficulty of attaining the increase (it is usually harder to raise y from $0.98A$ to $0.99A$ than from $0.90A$ to $0.91A$). Now suppose we know, or are told, that the reaction is expected to behave approximately like a "first-order reaction," for which

$$\log \frac{A}{A-y} = kt, \tag{10.7.2}$$

where t is the contact time. This suggests using the transformation $f(y)$ defined by the left member of (10.7.2), which stretches the scale in the desired way. The transformed quantity is physically meaningful since it is closely related to a theoretical rate constant k in (10.7.2), which is regarded as a rather basic physical parameter. Furthermore, this transformation will make the dependence on the contact time, one of the factors in the experiment, more nearly linear, the more nearly that

(10.7.2) is valid. We remark also that further theoretical knowledge of the process might suggest the nature of the interaction between the factors of temperature and contact time, and such knowledge should aid in designing the experiment.

We conclude by relating the theoretical equation (10.7.2) to the logarithmic transformation considered above. If we write (10.7.2) in the form

$$E(y) = A(1 - e^{-kt}), \tag{10.7.3}$$

where we have replaced y in (10.7.2) by $E(y)$, since (10.7.2) really refers to y stripped of its error, and if we note that $k = k(i, T)$, where i labels the catalyst and T is the temperature, we see that $E(y)$ factors into a function of A (namely, A) times a function of i, t, and T. If furthermore kt is sufficiently small compared with unity, then $E(y) = Akt$ approximately, $E(y)$ does not get near the point A of the y-scale where the jamming occurs, we get a further factorization of $E(y)$ into a function of A times a function of t times a function of i and T, and logging the y-measurements is hence a useful transformation, since it makes the effects of the factors corresponding to A and t additive and also brings in the rate constant $k(i, T)$ additively; at least this would be true if the theoretical relation (10.7.2) held exactly; if the relation holds approximately we may still expect the analysis of $\log y$ to be somewhat simpler than that of y.

PROBLEMS

10.1. Perform the approximate F-tests described in connection with (10.6.1) on the data of Problem 4.8, and compare the three resulting F-ratios with the exact values obtained in Problem 4.8.

10.2. For the (γ_1, γ_2) mentioned in sec. 10.1 can you give some verbal "explanation" of why the ages of brides and grooms have a large positive γ_1? Why the ages of fathers and mothers for births have smaller γ_1? Why the γ_2 for fathers is larger than that for mothers?

10.3. If J_1 and J_2 are the sample sizes in a nominal α-level t-test of the difference between two means, show that as J_1 and $J_2 \to \infty$ in any way with fixed α the probability of a type-II error approaches zero. Do not assume normality or equality of variances. [*Hint:* Using the formulas (10.2.12) and (10.2.13) for the asymptotic mean and variance of (10.2.10) prove that the variance stays between θ and $1/\theta$, and the mean becomes infinite.]

10.4. Suppose that in the t-test for the difference of two means the usual estimate of the standard deviation of $y_{1.} - y_{2.}$ is replaced by $(J_1^{-1}s_1^2 + J_2^{-1}s_2^2)^{1/2}$, in the notation above (10.2.10). Show that as J_1 and $J_2 \to \infty$ with fixed ratio $R = J_1/J_2$ this test has the correct significance level no matter what the value of $\theta = \sigma_1^2/\sigma_2^2$.

10.5. In Problem 10.4 it did not matter what number ν of d.f. was assigned to the approximate t-distribution as long as $\nu \to \infty$ with J_1 and J_2. What value of ν would you employ if you used this test with small samples? [*Hint:* Approximate $J_1^{-1}s_1^2+J_2^{-1}s_2^2$ as $c\chi_\nu^2$, and calculate ν by equating the first two moments as at the end of sec. 7.5, assuming normality but not equal variances. In the resulting formula for ν the $\{\sigma_i^2\}$ will have to be replaced by the $\{s_i^2\}$.]

10.6. In sec. 10.2 we remarked that s^2 is normal for large n, and in Problem IV.2 that χ_ν^2 is normal for large ν. It follows that for large n, s^2/σ^2 is $c\chi_\nu^2$ for some c,ν. (*a*) Argue that for small n a chi-square approximation to s^2/σ^2 should be better than a normal approximation. (*b*) Show that if for any n the c and ν of the chi-square approximation are calculated by fitting the first two moments then $c = 1/\nu$ and

$$\nu = (n-1)\left(1+\frac{n-1}{2n}\gamma_2\right)^{-1}.$$

(*c*) Suppose that s_1^2 and s_2^2 are independent sample variances, s_i^2 being calculated from a random sample of n_i from a population with $\sigma^2 = \sigma_i^2$ and $\gamma_2 = \gamma_{2i}$. Normal-theory tests and confidence intervals for comparing σ_1^2 and σ_2^2 are based on the ratio $(s_1^2/s_2^2)/(\sigma_1^2/\sigma_2^2)$. Approximate this as F_{ν_1,ν_2}. What correction factors to the normal-theory ν_1 and ν_2 are necessary?

10.7. For the two-way layout with K observations per cell and K large, assume $\{y_{ijk}\}$ independent with means $\{\mu+\alpha_i+\beta_j+\gamma_{ij}\}$, where $\alpha_. = \beta_. = \gamma_{i.} = \gamma_{.j} = 0$ for all i,j, and variances $\{\sigma_{ij}^2\}$, and show the true probability of type-I error in testing the hypothesis that all $\alpha_i = 0$ always exceeds the nominal significance level. [*Hint:* The calculations made for (10.2.16) can be shown to cover this case if the $\{J_i\}$ and $\{\sigma_i^2\}$ of that case are suitably identified.]

10.8. That inequality of variance may destroy the statistical independence of estimates and mean squares may be seen from the following simple example: Suppose that $\{x_i\}$ are independently normal with zero means and variances $\{\sigma_i^2\}$. A sufficient condition for the mean $x_.$ or the mean square $Ix_.^2$ to be independent of the error mean square $\Sigma_i(x_i-x_.)^2$ is that all $\{x_i-x_.\}$ have zero correlation with $x_.$: Show that this is the case only if all the $\{\sigma_i^2\}$ are equal.

10.9. Apply the transformation (10.7.1) to the cases where y is (*a*) a Poisson variable, (*b*) a sample variance s^2 (assume γ_2 known), (*c*) the sample correlation coefficient r. In (*c*) assume that the bivariate population is normal and use the approximations* $E(r) \doteqdot \rho$, $\text{Var}(r) \doteqdot n^{-1}(1-\rho^2)^2$, where ρ is the population correlation coefficient.

* See, for example, Cramér (1946), sec. 27.8.

APPENDIX I

Vector Algebra

This appendix is intended to give an introduction to vector algebra, and the next one, to matrix algebra, sufficient for the understanding of this book. We shall limit ourselves to finite-dimensional vector spaces. A mathematically more attractive approach is the abstract[1] one, but the present one is better adapted to our specific needs. The reader desiring a treatment similar to the present one but more detailed is referred to the book of Murdoch (1957).

Definition 1: A vector is an ordered n-tuple

$$\begin{pmatrix} x_1 \\ x_2 \\ \cdot \\ \cdot \\ \cdot \\ x_n \end{pmatrix} \tag{I.1}$$

of real numbers.

Here "n-tuple" means "set of n," thus a 2-tuple is a pair, a 3-tuple is a triple, a 4-tuple a quadruple, etc. "Ordered" means for example that

$$\begin{pmatrix} 2 \\ 1 \\ 4 \end{pmatrix} \text{ is different from } \begin{pmatrix} 1 \\ 2 \\ 4 \end{pmatrix}.$$

The real numbers are the numbers used in analytic geometry and calculus, such as 2, -1, 1.67, π, e, $\sqrt{7}$. The reader is familiar with picturing the

[1] The abstract approach is developed for example in the book by Birkhoff and MacLane (1953).

real numbers on an axis of a graph. Examples of nonreal (imaginary) numbers are $\sqrt{-1}$ and $2+3\sqrt{-1}$.

The n numbers in Definition 1 are usually written in a row as $(x_1, x_2, \cdots, x_n)$, but it will be more convenient for us to write them in a column as in (I.1) for reasons that will emerge in App. II. For any given n the set of all vectors is denoted by V_n.

For $n = 1, 2$, or 3, the vector (I.1) might be pictured as the point P with coordinates $(x_1, x_2, \cdots, x_n)$ in an n-dimensional Euclidean space, but it is better to let its geometrical picture be the directed line segment from the origin O to the point P; thus in V_3 we have Fig. I.1. (Sometimes it is more convenient to permit a nonunique picture consisting of any translation of the directed line segment OP—which keeps it parallel to OP.) This permits using geometrical language in discussing vectors in V_1, V_2, or V_3. By extension we will use the same language in V_n for $n > 3$. For instance, in V_3, the number x_i, called the ith component of the vector, is the algebraic length of the projection of the vector on the ith of the orthogonal axes, and we shall see that this is true in V_n as well.

Notation: We write simply **x** for the vector (I.1). We write $\mathbf{x} \in V_n$ to mean that **x** is a vector in V_n.

We now introduce the fundamental operations on vectors.

Definition 2: We define the *sum* $\mathbf{x}+\mathbf{y}$ of the vectors

$$\mathbf{x} = \begin{pmatrix} x_1 \\ x_2 \\ \cdot \\ \cdot \\ \cdot \\ x_n \end{pmatrix} \quad \text{and} \quad \mathbf{y} = \begin{pmatrix} y_1 \\ y_2 \\ \cdot \\ \cdot \\ \cdot \\ y_n \end{pmatrix} \quad \text{to be the vector} \quad \begin{pmatrix} x_1 + y_1 \\ x_2 + y_2 \\ \cdot \\ \cdot \\ \cdot \\ x_n + y_n \end{pmatrix}.$$

In other words, vectors are added by addition of their components.

Geometrically (and we now "draw" the n-dimensional space as we would draw the three-dimensional one) we obtain the picture shown in Fig. I.2. This picture is intended to show that the vector sum of **x** and **y** is obtained as the diagonal through the origin of the parallelogram constructed in the "plane" determined by **x** and **y**, and having **x**, **y** for adjacent sides. Another geometrical interpretation is evident from Fig. I.2: If we move the vectors parallel to themselves so that the tail of **y** coincides with the tip of **x**, then the sum is the vector drawn from the tail of **x** to the tip of **y**; this "tail-to-tip" interpretation gives a simpler picture for the sum of more than two vectors. With this picture a vector should not be considered as "attached" to the origin, but as free to move in the space, provided it is kept always parallel to itself. Such displacements do not modify the values of the components.

In elementary physics, vectors are used to represent forces. The direction of the vector indicates the direction of the force, and its "length"

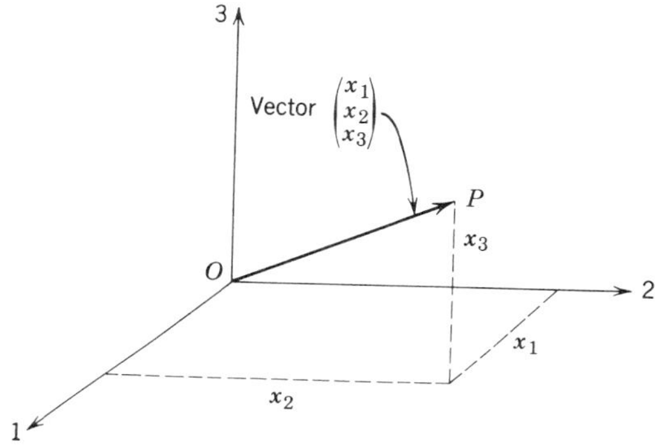

FIG. I.1

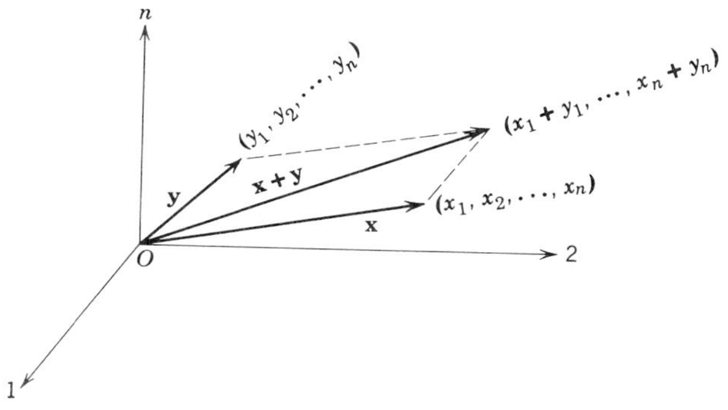

FIG. I.2

measures the magnitude of the force. Addition of vectors as we defined it is equivalent to addition of the corresponding forces: If the vectors $\mathbf{x}$, $\mathbf{y}$ represent two forces, the vector $\mathbf{x}+\mathbf{y}$ represents their resultant.

It is pretty obvious that addition of vectors is associative and commutative, i.e., if $\mathbf{x}$, $\mathbf{y}$, $\mathbf{z}$, are vectors, then

$$(\mathbf{x}+\mathbf{y}) + \mathbf{z} = \mathbf{x} + (\mathbf{y}+\mathbf{z}) = \mathbf{x} + \mathbf{y} + \mathbf{z} \quad \text{and} \quad \mathbf{x} + \mathbf{y} = \mathbf{y} + \mathbf{x}.$$

Definition 3: The *product* $c\mathbf{x}$ of a vector

$$\mathbf{x} = \begin{pmatrix} x_1 \\ x_2 \\ \cdot \\ \cdot \\ \cdot \\ x_n \end{pmatrix} \text{ by a scalar } c \text{ (i.e., a real number) is the vector } \begin{pmatrix} cx_1 \\ cx_2 \\ \cdot \\ \cdot \\ \cdot \\ cx_n \end{pmatrix}.$$

The geometrical interpretation is indicated in Fig. I.3. This operation is called *scalar multiplication*. It evidently has the properties that, if $\mathbf{x}$ and $\mathbf{y}$ are vectors, and c and d are scalars, then

$$1\mathbf{x} = \mathbf{x}, \qquad c(\mathbf{x}+\mathbf{y}) = c\mathbf{x} + c\mathbf{y},$$
$$(c+d)\mathbf{x} = c\mathbf{x} + d\mathbf{x}, \qquad c(d\mathbf{x}) = d(c\mathbf{x}) = cd\mathbf{x}.$$

The two vector operations which have been defined permit us to speak of linear combinations of vectors. We say that the vector $\mathbf{z}$ is a *linear combination* of the vectors $\boldsymbol{\alpha}_1, \boldsymbol{\alpha}_2, \cdots, \boldsymbol{\alpha}_r$ with coefficients $c_1, c_2, \cdots, c_r$ (scalars) if

$$\mathbf{z} = c_1\boldsymbol{\alpha}_1 + c_2\boldsymbol{\alpha}_2 + \cdots + c_r\boldsymbol{\alpha}_r.$$

This notation is abbreviated to

$$\mathbf{z} = \sum_{i=1}^{r} c_i\boldsymbol{\alpha}_i,$$

or $\mathbf{z} = \Sigma_1^r c_i\boldsymbol{\alpha}_i$, or, if the range of i clear, $\mathbf{z} = \Sigma_i c_i\boldsymbol{\alpha}_i$, or even $\mathbf{z} = \Sigma c_i\boldsymbol{\alpha}_i$. The reader not at home with this notation should convince himself that $\Sigma_{i=1}^r c_i\boldsymbol{\alpha}_i = \Sigma_{j=1}^r c_j\boldsymbol{\alpha}_j$, analogous to $\int_1^3 f(x)\,dx = \int_1^3 f(y)\,dy$.

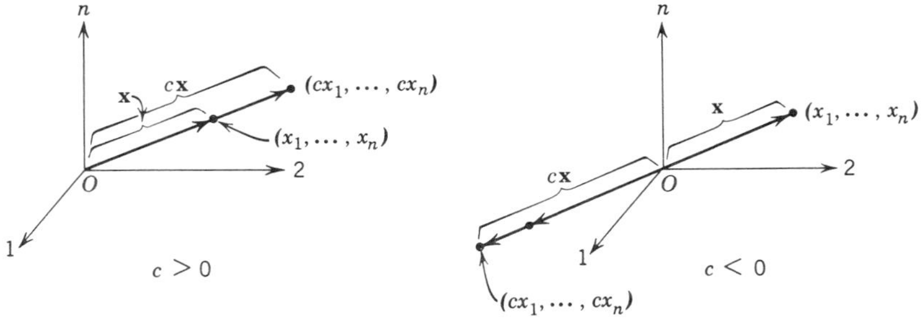

FIG. I.3

We introduce a third fundamental operation, which for any pair of vectors yields a scalar.

Definition 4: The *scalar product* (or *inner product*) of two vectors

$$\mathbf{x} = \begin{pmatrix} x_1 \\ x_2 \\ \cdot \\ \cdot \\ \cdot \\ x_n \end{pmatrix} \quad \text{and} \quad \mathbf{y} = \begin{pmatrix} y_1 \\ y_2 \\ \cdot \\ \cdot \\ \cdot \\ y_n \end{pmatrix} \quad \text{is the scalar } \Sigma_{i=1}^n x_i y_i.$$

Notation: The definition shows that the scalar product is commutative. For reasons that will become clear in App. II we write it

$$\mathbf{x}'\mathbf{y} = \mathbf{y}'\mathbf{x} = \sum_{i=1}^{n} x_i y_i.$$

If c is a scalar and $\mathbf{x}$, $\mathbf{y}$, $\mathbf{z}$ are vectors, the following properties are easily verified:

$$\mathbf{x}'(\mathbf{y}+\mathbf{z}) = \mathbf{x}'\mathbf{y} + \mathbf{x}'\mathbf{z}, \qquad \mathbf{x}'(c\mathbf{y}) = (c\mathbf{x})'\mathbf{y} = c(\mathbf{x}'\mathbf{y}), \qquad \mathbf{x}'\mathbf{x} \geqq 0;$$

in fact, $\mathbf{x}'\mathbf{x} = 0$ if and only if $\mathbf{x}$ is the zero vector, i.e.,

$$\mathbf{x} = \begin{pmatrix} 0 \\ 0 \\ \cdot \\ \cdot \\ \cdot \\ 0 \end{pmatrix}.$$ It is important to distinguish the zero vector $\mathbf{0} = \begin{pmatrix} 0 \\ 0 \\ \cdot \\ \cdot \\ \cdot \\ 0 \end{pmatrix}$

from the scalar zero. We note that the two zeros are connected by the relation $0\mathbf{x} = \mathbf{0}$ for every $\mathbf{x} \in V_n$. The zero vector $\mathbf{0}$ has the property analogous to that of the scalar zero 0 that $x + 0 = x$ for every scalar x, namely $\mathbf{x} + \mathbf{0} = \mathbf{x}$ for every $\mathbf{x} \in V_n$.

In one, two, or three dimensions, we have respectively $\mathbf{x}'\mathbf{x} = x_1^2$, $\mathbf{x}'\mathbf{x} = x_1^2+x_2^2$, $\mathbf{x}'\mathbf{x} = x_1^2+x_2^2+x_3^2$. Thus in those three cases $(\mathbf{x}'\mathbf{x})^{1/2}$ is the length of the vector $\mathbf{x}$. It is natural to generalize to

Definition 5: The *norm* (or *length*) of the vector (I.1) written $\|\mathbf{x}\|$, is defined to be

(I.2) $$\|\mathbf{x}\| = (\mathbf{x}'\mathbf{x})^{1/2}.$$

From (I.2) it follows that $\|c\mathbf{x}\| = |c|\ \|\mathbf{x}\|$.

In V_2 or V_3, the scalar product has an important geometrical significance. Any pair **x**, **y** of vectors determines a triangle as in Fig. I.4, and the "law of cosines" states that

$$\|\mathbf{x}-\mathbf{y}\|^2 = \|\mathbf{x}\|^2 + \|\mathbf{y}\|^2 - 2\|\mathbf{x}\|\,\|\mathbf{y}\| \cos \alpha,$$

where α is the angle between **x** and **y**. Substituting $\|\mathbf{x}-\mathbf{y}\|^2 = (\mathbf{x}-\mathbf{y})'(\mathbf{x}-\mathbf{y}) = \mathbf{x}'\mathbf{x} - \mathbf{y}'\mathbf{x} - \mathbf{x}'\mathbf{y} + \mathbf{y}'\mathbf{y}$, $\|\mathbf{x}\|^2 = \mathbf{x}'\mathbf{x}$, $\|\mathbf{y}\|^2 = \mathbf{y}'\mathbf{y}$, and $\mathbf{y}'\mathbf{x} = \mathbf{x}'\mathbf{y}$, we obtain

$$\mathbf{x}'\mathbf{y} = \|\mathbf{x}\|\,\|\mathbf{y}\| \cos \alpha.$$

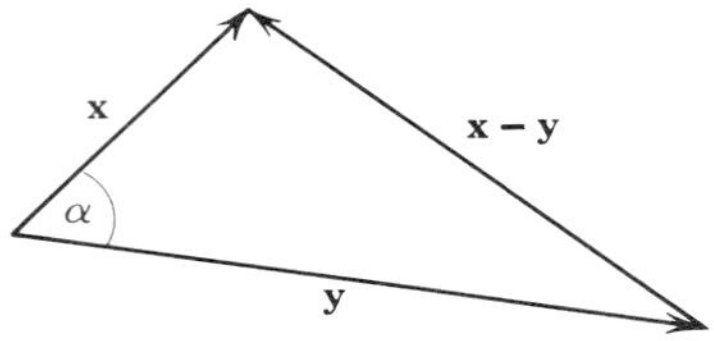

FIG. I.4

Now, $\|\mathbf{x}\| \cos \alpha$ is the algebraic length of the projection of **x** on **y** (*algebraic* length because it is considered negative if $90° < \alpha \leq 180°$). Thus the scalar product $\mathbf{x}'\mathbf{y}$ is equal to the algebraic length of the projection of one of the vectors **x**, **y** on the other, times the norm of the other. In particular, if **y** is a unit vector, i.e., if $\|\mathbf{y}\| = 1$, then $\mathbf{x}'\mathbf{y}$ is the algebraic length of the projection of **x** on **y**, and so the vector $(\mathbf{x}'\mathbf{y})\mathbf{y}$ is the projection of **x** on **y**; see Fig. I.5. We generalize this to V_n. Noticing that if **y** is any nonzero vector in V_n then $\|\mathbf{y}\|^{-1}\mathbf{y}$ is a unit vector in the same direction, we frame

Definition 6: If **x** and **y** are in V_n and $\mathbf{y} \neq \mathbf{0}$, the vector $\|\mathbf{y}\|^{-2}(\mathbf{x}'\mathbf{y})\mathbf{y}$ is called the projection of **x** on **y**.

More insight into the notion of projection will be obtained when we consider a more general definition (Definition 15) below, of which this is

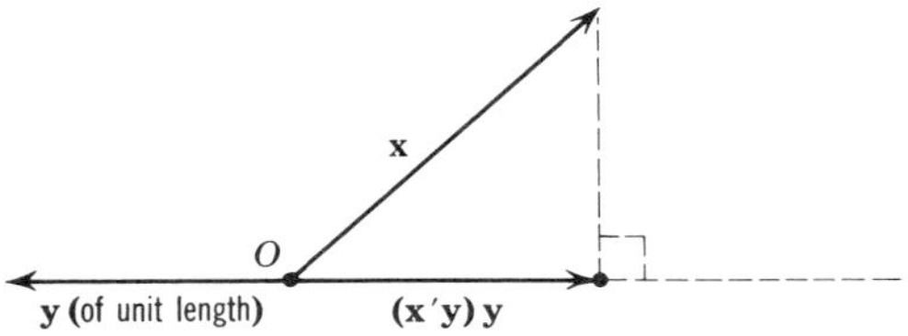

FIG. I.5

a special case. In terms of the scalar product we also define the important notion of orthogonality of vectors. In V_3, the two nonzero vectors $\mathbf{x}$, $\mathbf{y}$ of Fig. I.4 are said to be orthogonal if and only if $\cos \alpha = 0$. We will agree to say that $\mathbf{x}$ and $\mathbf{y}$ are also orthogonal if $\mathbf{x}$ or $\mathbf{y}$ is the zero vector. Then $\mathbf{x}$ and $\mathbf{y}$ are orthogonal if and only if $\mathbf{x}'\mathbf{y} = 0$. Similarly we formulate

Definition 7: Two vectors $\mathbf{x}$, $\mathbf{y}$ in V_n are said to be *orthogonal* if and only if $\mathbf{x}'\mathbf{y} = 0$.

It follows from Definition 6 that two nonzero vectors in V_n are orthogonal if and only if the projection of either one on the other is $\mathbf{0}$.

Notation: We write $\mathbf{x} \perp \mathbf{y}$ to mean that the vectors $\mathbf{x}$ and $\mathbf{y}$ are orthogonal.

Definition 8: Let $\{\boldsymbol{\alpha}_1, \boldsymbol{\alpha}_2, \cdots, \boldsymbol{\alpha}_s\}$ be a set of s vectors in V_n. The *vector space V spanned by* $\{\boldsymbol{\alpha}_1, \boldsymbol{\alpha}_2, \cdots, \boldsymbol{\alpha}_s\}$ is the set of all vectors that are linear combinations of $\boldsymbol{\alpha}_1, \boldsymbol{\alpha}_2, \cdots, \boldsymbol{\alpha}_s$ and $\mathbf{0}$.

The notation $\{\boldsymbol{\alpha}_1, \boldsymbol{\alpha}_2, \cdots, \boldsymbol{\alpha}_s\}$ is understood to mean $\{\boldsymbol{\alpha}_1, \boldsymbol{\alpha}_2\}$ if $s = 2$, $\{\boldsymbol{\alpha}_1\}$ if $s = 1$, and the *empty set* if $s = 0$. The reason for including $\mathbf{0}$ with $\boldsymbol{\alpha}_1, \boldsymbol{\alpha}_2, \cdots, \boldsymbol{\alpha}_s$ in Definition 8 is that it is convenient (for example in Lemma 2 below) to allow the empty set and have this span the vector space which consists solely of $\mathbf{0}$.

The V of Definition 8 is also called a *linear subspace of* V_n. It is called a "linear" space because it is easily verified that if $\mathbf{x}$ and $\mathbf{y}$ are in V, so is $a\mathbf{x}+b\mathbf{y}$ for any scalars a and b. Furthermore, V is called a *subspace* of V_n, and we write $V \subset V_n$, because $\mathbf{x} \in V$ implies $\mathbf{x} \in V_n$. (The symbols $\in$ and $\subset$ are both inclusion symbols; $\in$ means "is a member of the set," $\subset$ means "is a subset of the set"; usually either may be read "in" or "is in.")

Consider two nonzero vectors $\mathbf{u}$ and $\mathbf{v}$ in V_3 which are nonparallel. If V is the set of all linear combinations of $\mathbf{u}$ and $\mathbf{v}$, then V is a plane. If $\mathbf{w}$ is any vector in V, the linear space spanned by the augmented set $\{\mathbf{u}, \mathbf{v}, \mathbf{w}\}$ is again V. It appears, therefore, desirable to characterize minimal sets that span a linear space.

Definition 9: A set of vectors $\{\boldsymbol{\beta}_1, \boldsymbol{\beta}_2, \cdots, \boldsymbol{\beta}_r\}$ in V_n is said to be *linearly dependent* if there exists a set of scalars $\{c_1, c_2, \cdots, c_r\}$ not all zero such that $c_1\boldsymbol{\beta}_1+c_2\boldsymbol{\beta}_2+\cdots+c_r\boldsymbol{\beta}_r = \mathbf{0}$. If no such scalars exist, the set of vectors $\{\boldsymbol{\beta}_1, \boldsymbol{\beta}_2, \cdots, \boldsymbol{\beta}_r\}$ is called *linearly independent*. The empty set is agreed to be a linearly independent set of vectors.

Thus for $r > 1$ the set of vectors $\{\boldsymbol{\beta}_1, \boldsymbol{\beta}_2, \cdots, \boldsymbol{\beta}_r\}$ is linearly independent if and only if none of the vectors is a linear combination of the others; for $r = 1$ the set $\{\boldsymbol{\beta}_1\}$ is linearly independent if and only if $\boldsymbol{\beta}_1 \neq \mathbf{0}$; for

$r = 0$ the (empty) set is always linearly independent. Thus in every case the set is linearly independent if and only if none of the vectors lies in the vector space spanned by the others. It follows that a linearly independent set cannot contain the zero vector, and that if a set of vectors is linearly independent, then so is every subset. In the previous example, the set $\{\mathbf{u}, \mathbf{v}\}$ is linearly independent, but the set $\{\mathbf{u}, \mathbf{v}, \mathbf{w}\}$ is linearly dependent. We will find the following lemma useful:

Lemma 1: If $\{\boldsymbol{\alpha}_1, \boldsymbol{\alpha}_2, \cdots, \boldsymbol{\alpha}_r\}$ is a linearly independent set of vectors and $\boldsymbol{\beta}$ is a nonzero vector which is not a linear combination of them, then the set $\{\boldsymbol{\alpha}_1, \cdots, \boldsymbol{\alpha}_r, \boldsymbol{\beta}\}$ is linearly independent.

Proof: If the set $\{\boldsymbol{\alpha}_1, \cdots, \boldsymbol{\alpha}_r\}$ is empty, the lemma is obvious; so suppose that it is nonempty and the lemma is false. Then there exist constants $c_1, \cdots, c_r, c$, not all zero, such that $c_1\boldsymbol{\alpha}_1 + \cdots + c_r\boldsymbol{\alpha}_r + c\boldsymbol{\beta} = \mathbf{0}$. If c were zero, at least one of the $\{c_i\}$ would have to be different from zero, and the $\{\boldsymbol{\alpha}_i\}$ would satisfy $c_1\boldsymbol{\alpha}_1 + \cdots + c_r\boldsymbol{\alpha}_r = \mathbf{0}$, in violation of the linear independence of the $\{\boldsymbol{\alpha}_i\}$. The contradiction is that $\boldsymbol{\beta} = -\Sigma_1^r(c_i/c)\boldsymbol{\alpha}_i$, a linear combination of the $\{\boldsymbol{\alpha}_i\}$.

Remark: The reason for imposing the restriction that $\boldsymbol{\beta}$ is nonzero is to exclude the case where $\{\boldsymbol{\alpha}_1, \cdots, \boldsymbol{\alpha}_r\}$ is the empty set and $\boldsymbol{\beta} = \mathbf{0}$.

Definition 10: A *basis* for a vector space V is a set of linearly independent vectors that span V.

Lemma 2: Every vector space has a basis.

Proof: By Definition 8 there exists a set $\{\boldsymbol{\alpha}_1, \cdots, \boldsymbol{\alpha}_s\}$ which spans V, and if this set is empty it is a basis for V, which then consists just of $\mathbf{0}$. If $\{\boldsymbol{\alpha}_1, \cdots, \boldsymbol{\alpha}_s\}$ is not empty but all $\boldsymbol{\alpha}_i = \mathbf{0}$, we get the same V and the same basis. Now suppose that there is at least one $\boldsymbol{\alpha}_i \neq \mathbf{0}$. We then form a subset of $\{\boldsymbol{\alpha}_1, \cdots, \boldsymbol{\alpha}_s\}$ by discarding all the $\boldsymbol{\alpha}_i$ which are equal to $\mathbf{0}$, keeping the first $\boldsymbol{\alpha}_i \neq \mathbf{0}$, and then successively considering each of the remaining $\boldsymbol{\alpha}_i$ in order of occurrence and discarding or keeping it according as it is or is not a linear combination of the ones kept before considering this one. From Lemma 1 it follows that this subset is linearly independent. It is evident that discarding the $\boldsymbol{\alpha}_i$ according to the above process does not diminish the space spanned by the remaining $\boldsymbol{\alpha}_i$: For, suppose that $\boldsymbol{\beta}_1, \cdots, \boldsymbol{\beta}_q$ denote the $\{\boldsymbol{\alpha}_i\}$ which are discarded and $\boldsymbol{\gamma}_1, \cdots, \boldsymbol{\gamma}_r$ the $\{\boldsymbol{\alpha}_i\}$ which are kept $(q+r = s)$. Then every $\mathbf{x} \in V$ is a linear combination of $\{\boldsymbol{\alpha}_1, \cdots, \boldsymbol{\alpha}_s\} = \{\boldsymbol{\beta}_1, \cdots, \boldsymbol{\beta}_q, \boldsymbol{\gamma}_1, \cdots, \boldsymbol{\gamma}_r\}$, say $\mathbf{x} = \Sigma_1^q a_i\boldsymbol{\beta}_i + \Sigma_1^r b_j\boldsymbol{\gamma}_j$. But each $\boldsymbol{\beta}_i$ is a linear combination of the $\boldsymbol{\gamma}_j$, say $\boldsymbol{\beta}_i = \Sigma_j c_{ij}\boldsymbol{\gamma}_j$, and hence $\mathbf{x}$ is a linear combination of the $\boldsymbol{\gamma}_j$ alone, $\mathbf{x} = \Sigma_1^r d_j\boldsymbol{\gamma}_j$, where $d_j = b_j + \Sigma_i a_i c_{ij}$. The subset thus obtained is hence a basis for V.

Definition 11: The *dimension* of a vector space V is the number of vectors in any basis for V.

It follows that the dimension of the vector space consisting solely of $\mathbf{0}$ is 0. That the dimension of any vector space is uniquely determined by this definition no matter what basis is used follows from the *basis theorem*, which we consider in a moment. We shall then show that the definition also assigns the dimension n to the space V_n which we are used to calling "n-dimensional."

Theorem 1 (Basis theorem): Any two bases for a vector space contain the same number of vectors.

Proof: If the vector space V is $\{\mathbf{0}\}$ it has the unique basis consisting of the empty set and the theorem is true. Otherwise let $\{\boldsymbol{\alpha}_1, \cdots, \boldsymbol{\alpha}_r\}$ and $\{\boldsymbol{\beta}_1, \cdots, \boldsymbol{\beta}_s\}$ be two bases for V, and suppose that $r < s$. Now $\boldsymbol{\alpha}_1, \cdots, \boldsymbol{\alpha}_r$ span V and are linearly independent, since they constitute a basis. A fortiori $\boldsymbol{\alpha}_1, \cdots, \boldsymbol{\alpha}_r, \boldsymbol{\beta}_1$ span V. They are however linearly dependent, since $\boldsymbol{\beta}_1$, being in V, is a linear combination of the basis vectors $\boldsymbol{\alpha}_1, \cdots, \boldsymbol{\alpha}_r$. Thus there exist $a_{11}, \cdots, a_{1r}$ such that $\boldsymbol{\beta}_1 = a_{11}\boldsymbol{\alpha}_1 + \cdots + a_{1r}\boldsymbol{\alpha}_r$. Not all $a_{1i} = 0$; else $\boldsymbol{\beta}_1 = \mathbf{0}$; but $\boldsymbol{\beta}_1$ is a basis vector. Without loss of generality we can assume that $a_{11} \neq 0$. Then $\boldsymbol{\alpha}_1$ is a linear combination of $\boldsymbol{\alpha}_2, \cdots, \boldsymbol{\alpha}_r, \boldsymbol{\beta}_1$, which therefore must also span V. A fortiori $\boldsymbol{\alpha}_2, \cdots, \boldsymbol{\alpha}_r, \boldsymbol{\beta}_1, \boldsymbol{\beta}_2$ span V. They are however linearly dependent since $\boldsymbol{\beta}_2$ is a linear combination of $\boldsymbol{\alpha}_2, \cdots, \boldsymbol{\alpha}_r, \boldsymbol{\beta}_1$, say $\boldsymbol{\beta}_2 = a_{22}\boldsymbol{\alpha}_2 + \cdots + a_{2r}\boldsymbol{\alpha}_r + b_{21}\boldsymbol{\beta}_1$. If all a_{2i} $(i = 2, \cdots, r)$ were 0, then $\boldsymbol{\beta}_1, \cdots, \boldsymbol{\beta}_s$ would not be linearly independent, contrary to the assumption. Thus at least one of the a_{2i} is $\neq 0$, and without loss of generality we can assume that $a_{22} \neq 0$. Then $\boldsymbol{\alpha}_2$ is a linear combination of $\boldsymbol{\alpha}_3, \cdots, \boldsymbol{\alpha}_r, \boldsymbol{\beta}_1, \boldsymbol{\beta}_2$, which therefore must span V. . . .. Repeating the same argument, we finally obtain that $\boldsymbol{\beta}_1, \boldsymbol{\beta}_2, \cdots, \boldsymbol{\beta}_r$ span V. It follows that $\boldsymbol{\beta}_{r+1}$ is a linear combination of $\boldsymbol{\beta}_1, \cdots, \boldsymbol{\beta}_r$, contrary to the assumption that $\boldsymbol{\beta}_1, \cdots, \boldsymbol{\beta}_s$ form a basis for V. Therefore, we cannot have $r < s$. Similarly, we cannot have $s < r$. Thus $r = s$.

Example: Consider again the vector space V_n. Let $\mathscr{R}$ denote the set of vectors

$$\boldsymbol{\rho}_1 = \begin{pmatrix} 1 \\ 0 \\ 0 \\ \cdot \\ \cdot \\ \cdot \\ 0 \end{pmatrix}, \quad \boldsymbol{\rho}_2 = \begin{pmatrix} 0 \\ 1 \\ 0 \\ \cdot \\ \cdot \\ \cdot \\ 0 \end{pmatrix}, \quad \cdots, \quad \boldsymbol{\rho}_n = \begin{pmatrix} 0 \\ 0 \\ \cdot \\ \cdot \\ \cdot \\ 0 \\ 1 \end{pmatrix}.$$

The set $\mathscr{R}$ is linearly independent

because $\Sigma_{i=1}^{n} c_i \boldsymbol{\rho}_i = \begin{pmatrix} c_1 \\ c_2 \\ \cdot \\ \cdot \\ \cdot \\ c_n \end{pmatrix} = \mathbf{0}$ implies that $c_1 = c_2 = \cdots = c_n = 0$. Furthermore, the vector space spanned by $\mathscr{R}$ is V_n because any vector (I.1) in V_n can be written $\mathbf{x} = \Sigma_{i=1}^{n} x_i \boldsymbol{\rho}_i$. The set $\mathscr{R} = \{\boldsymbol{\rho}_1, \boldsymbol{\rho}_2, \cdots, \boldsymbol{\rho}_n\}$ is therefore a basis for V_n, and the theorem above yields as an immediate

Corollary 1: Any basis for V_n contains exactly n vectors, i.e., the dimension of V_n is n.

Notation: We write $V_r \subset V_n$ to denote that V_r is an r-dimensional vector space contained in V_n. (We always assume $n > 0$.) This notation is ambiguous in that heretofore V_r would have denoted the set of all r-tuples of real numbers, but now it also denotes an r-dimensional vector space of n-tuples—but this will cause no harm.[2]

Lemma 3: If $V_r \subset V_n$, $r > 0$, V_r is spanned by $\{\boldsymbol{\alpha}_1, \cdots, \boldsymbol{\alpha}_s\}$, and $\mathbf{x} \in V_r$, then the coefficients in

$$\mathbf{x} = a_1\boldsymbol{\alpha}_1 + a_2\boldsymbol{\alpha}_2 + \cdots + a_s\boldsymbol{\alpha}_s$$

are unique if and only if $\{\boldsymbol{\alpha}_1, \cdots, \boldsymbol{\alpha}_s\}$ are linearly independent, i.e., if and only if $s = r$ and $\{\boldsymbol{\alpha}_1, \cdots, \boldsymbol{\alpha}_r\}$ is a basis for V_r.

Proof: Suppose that $\mathbf{x} = \Sigma_1^s a_i \boldsymbol{\alpha}_i$, and suppose that $\{b_i\}$ is any other set of coefficients such that $\mathbf{x} = \Sigma_1^s b_i \boldsymbol{\alpha}_i$. Then $\Sigma_1^s (b_i - a_i)\boldsymbol{\alpha}_i = \mathbf{0}$. If the $\{\boldsymbol{\alpha}_i\}$ are linearly independent then all $b_i - a_i = 0$, and hence the a_i are unique. If the $\{\boldsymbol{\alpha}_i\}$ are linearly dependent there exist $\{c_i\}$ not all zero such that $\Sigma c_i \boldsymbol{\alpha}_i = \mathbf{0}$. Given the $\{a_i\}$ in $\mathbf{x} = \Sigma_1^s a_i \boldsymbol{\alpha}_i$, take $b_i = a_i + c_i$; then $\mathbf{x} = \Sigma_1^s b_i \boldsymbol{\alpha}_i$, and not all $b_i - a_i = 0$.

Definition 12: If $\{\boldsymbol{\alpha}_1, \boldsymbol{\alpha}_2, \cdots, \boldsymbol{\alpha}_n\}$ is a basis for V_n and $\mathbf{x}$ is any vector in V_n, the coefficient a_i ($i = 1, \cdots, n$) of $\boldsymbol{\alpha}_i$ in the (unique) linear representation $\mathbf{x} = \Sigma_{i=1}^{n} a_i \boldsymbol{\alpha}_i$ of $\mathbf{x}$ in terms of vectors of the basis is called the ith *coordinate* of $\mathbf{x}$ with respect to that basis.

With the above definition we see that the ith coordinate of the point $(x_1, x_2, \cdots, x_n)$ in V_n is the same as the ith coordinate of the vector (I.1) with respect to the basis $\mathscr{R}$ introduced in the above example.

[2] The two V_r's are *isomorphic*; see for example, Birkhoff and MacLane (1953), Ch. 7, Theorem 5.

Definition 13: A basis $\{\boldsymbol{\alpha}_1, \cdots, \boldsymbol{\alpha}_r\}$ for $V_r \subset V_n$ is called *orthonormal* if the r vectors $\boldsymbol{\alpha}_i$ are pairwise orthogonal and have unit norm.

Using the "Kronecker delta," $\delta_{ij} = 1$ if $i = j$, $\delta_{ij} = 0$ if $i \neq j$, we can write that $\{\boldsymbol{\alpha}_1, \cdots, \boldsymbol{\alpha}_r\}$ is an orthonormal basis if $\boldsymbol{\alpha}_i'\boldsymbol{\alpha}_j = \delta_{ij}$ $(i, j = 1, \cdots, r)$.

The simplest example of an orthonormal basis is the basis $\mathscr{R}$ for V_n introduced above, consisting of a unit vector in the positive direction on each coordinate axis. In this case we see that the algebraic length of the projection of the vector $\mathbf{x}$ on the ith coordinate axis is $\boldsymbol{\rho}_i'\mathbf{x} = x_i$ as mentioned earlier. More generally, if $\{\boldsymbol{\alpha}_1, \cdots, \boldsymbol{\alpha}_n\}$ is any orthonormal basis for V_n then for any vector $\mathbf{x}$ in V_n the ith coordinate of $\mathbf{x}$ relative to the basis is $\boldsymbol{\alpha}_i'\mathbf{x}$, the projection of $\mathbf{x}$ on the unit vector $\boldsymbol{\alpha}_i$; this follows since there exist unique coordinates $a_1, \cdots, a_n$ such that $\mathbf{x} = \Sigma a_j\boldsymbol{\alpha}_j$, hence

$$\boldsymbol{\alpha}_i'\mathbf{x} = \boldsymbol{\alpha}_i'\sum_j a_j\boldsymbol{\alpha}_j = \sum_j a_j\boldsymbol{\alpha}_i'\boldsymbol{\alpha}_j = \sum_j a_j\delta_{ij} = a_i.$$

Lemma 4: If the vectors $\boldsymbol{\alpha}_1, \boldsymbol{\alpha}_2, \cdots, \boldsymbol{\alpha}_r$ are pairwise orthogonal and nonzero they are linearly independent.

Proof: We shall prove that if $\mathbf{0} = c_1\boldsymbol{\alpha}_1 + c_2\boldsymbol{\alpha}_2 + \cdots + c_r\boldsymbol{\alpha}_r$ then for every $i = 1, 2, \cdots, r$, the scalar $c_i = 0$: Scalar-multiply both sides of the equation by $\boldsymbol{\alpha}_i'$ on the left to get $0 = \boldsymbol{\alpha}_i'\Sigma_j c_j\boldsymbol{\alpha}_j = \Sigma_j c_j\boldsymbol{\alpha}_i'\boldsymbol{\alpha}_j = c_i\|\boldsymbol{\alpha}_i\|^2$. Since $\boldsymbol{\alpha}_i \neq \mathbf{0}$, $c_i = 0$.

Lemma 5: Any linearly independent set of r vectors in $V_r \subset V_n$ is a basis for V_r.

Proof: Suppose the contrary, that $\{\boldsymbol{\alpha}_1, \cdots, \boldsymbol{\alpha}_r\}$ is a linearly independent set with $\boldsymbol{\alpha}_i \in V_r$ but is not a basis for V_r. By Lemma 2, V_r has a basis $\{\boldsymbol{\beta}_1, \cdots, \boldsymbol{\beta}_r\}$. Apply to the set $\{\boldsymbol{\alpha}_1, \cdots, \boldsymbol{\alpha}_r, \boldsymbol{\beta}_1, \cdots, \boldsymbol{\beta}_r\}$ of $2r$ vectors spanning V_r the process used to prove Lemma 2 to get a basis for V_r. This process would not discard any of $\boldsymbol{\alpha}_1, \cdots, \boldsymbol{\alpha}_r$ because they are linearly independent, nor discard all of $\boldsymbol{\beta}_1, \cdots, \boldsymbol{\beta}_r$, else $\{\boldsymbol{\alpha}_1, \cdots, \boldsymbol{\alpha}_r\}$ would be a basis. The resulting basis would have more than r elements, contradicting the basis theorem.

Corollary 2: Any $r+1$ vectors in $V_r \subset V_n$ are linearly dependent.

From Lemmas 4 and 5 follows also

Corollary 3: Any set of r orthogonal nonzero vectors in $V_r \subset V_n$ is a basis for V_r.

Orthonormal bases often lead to particularly simple proofs of results. We will show that it is always possible to "orthonormalize" an arbitrary basis for V_r. In V_2, the way to proceed is obvious. If, as pictured in

Fig. I.6, $\{\boldsymbol{\alpha}_1, \boldsymbol{\alpha}_2\}$ is an arbitrary basis and if $\boldsymbol{\pi}$ is the projection of $\boldsymbol{\alpha}_2$ on $\boldsymbol{\alpha}_1$ given by Definition 6, then the two vectors $\boldsymbol{\alpha}_1, \boldsymbol{\alpha}_2-\boldsymbol{\pi}$ are orthogonal and

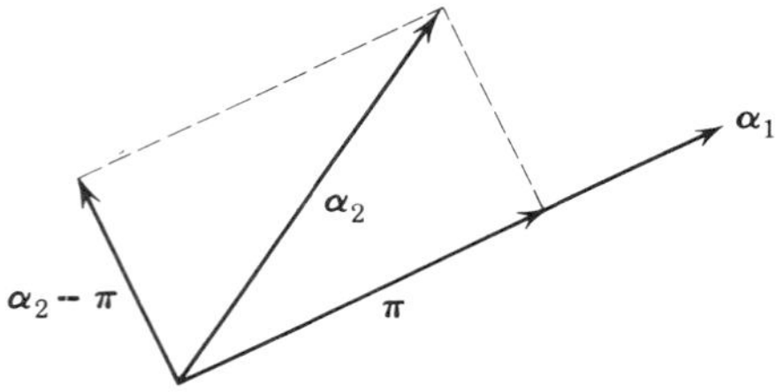

FIG. I.6

therefore $\|\boldsymbol{\alpha}_1\|^{-1}\boldsymbol{\alpha}_1$ and $\|\boldsymbol{\alpha}_2-\boldsymbol{\pi}\|^{-1}(\boldsymbol{\alpha}_2-\boldsymbol{\pi})$ constitute an orthonormal basis for V_2. The method (Schmidt process) can be generalized and leads to

Lemma 6: Given an arbitrary basis $\{\boldsymbol{\alpha}_1, \boldsymbol{\alpha}_2, \cdots, \boldsymbol{\alpha}_r\}$ for V_r, there exists an orthonormal basis $\{\boldsymbol{\gamma}_1, \boldsymbol{\gamma}_2, \cdots \boldsymbol{\gamma}_r\}$ for V_r such that each $\boldsymbol{\gamma}_i$ is a linear combination of $\boldsymbol{\alpha}_1, \boldsymbol{\alpha}_2, \cdots, \boldsymbol{\alpha}_i$.

Proof: Let $\boldsymbol{\beta}_1 = \boldsymbol{\alpha}_1$. Let $\boldsymbol{\beta}_2 = \boldsymbol{\alpha}_2 - c_{21}\boldsymbol{\beta}_1$, where c_{21} is to be determined so that $\boldsymbol{\beta}_1'\boldsymbol{\beta}_2 = 0$. This gives $c_{21}\boldsymbol{\beta}_1'\boldsymbol{\beta}_1 = \boldsymbol{\beta}_1'\boldsymbol{\alpha}_2$, $c_{21} = \boldsymbol{\beta}_1'\boldsymbol{\alpha}_2/\boldsymbol{\beta}_1'\boldsymbol{\beta}_1$. We have to show that $\boldsymbol{\beta}_2 \neq \mathbf{0}$: This would imply that $c_{21}\boldsymbol{\alpha}_1-\boldsymbol{\alpha}_2 = \mathbf{0}$, in contradiction with the linear independence of $\{\boldsymbol{\alpha}_1, \cdots, \boldsymbol{\alpha}_r\}$. Now, let $\boldsymbol{\beta}_3 = \boldsymbol{\alpha}_3-c_{31}\boldsymbol{\beta}_1-c_{32}\boldsymbol{\beta}_2$, with the conditions $\boldsymbol{\beta}_1'\boldsymbol{\beta}_3 = 0$, $\boldsymbol{\beta}_2'\boldsymbol{\beta}_3 = 0$. Taking into account that $\boldsymbol{\beta}_1'\boldsymbol{\beta}_2 = 0$, these give $c_{31} = \boldsymbol{\beta}_1'\boldsymbol{\alpha}_3/\boldsymbol{\beta}_1'\boldsymbol{\beta}_1$ and $c_{32} = \boldsymbol{\beta}_2'\boldsymbol{\alpha}_3/\boldsymbol{\beta}_2'\boldsymbol{\beta}_2$. Again, $\boldsymbol{\beta}_3$ cannot be $\mathbf{0}$, because if it were we would have $\boldsymbol{\alpha}_3-c_{31}\boldsymbol{\alpha}_1-c_{32}(\boldsymbol{\alpha}_2-c_{21}\boldsymbol{\alpha}_1) = \mathbf{0}$, which would make the $\{\boldsymbol{\alpha}_j\}$ linearly dependent. Proceeding stepwise in the same fashion, we will obtain a set of r vectors $\{\boldsymbol{\beta}_1, \boldsymbol{\beta}_2, \cdots, \boldsymbol{\beta}_r\}$, where $\boldsymbol{\beta}_i = \boldsymbol{\alpha}_i - \Sigma_{j=1}^{i-1}c_{ij}\boldsymbol{\beta}_j$, and $c_{ij} = \boldsymbol{\beta}_j'\boldsymbol{\alpha}_i/\boldsymbol{\beta}_j'\boldsymbol{\beta}_j$. These vectors are pairwise orthogonal by construction. They are nonzero, since $\boldsymbol{\beta}_i$ is of the form $\boldsymbol{\alpha}_i-\Sigma_{j=1}^{i-1}d_{ij}\boldsymbol{\alpha}_j$, and thus if $\boldsymbol{\beta}_i = \mathbf{0}$ the $\boldsymbol{\alpha}$'s would be linearly dependent. They therefore constitute a basis for V_r by Corollary 3. Then $\{\boldsymbol{\gamma}_1, \boldsymbol{\gamma}_2, \cdots, \boldsymbol{\gamma}_r\}$, with $\boldsymbol{\gamma}_i = \boldsymbol{\beta}_i/\|\boldsymbol{\beta}_i\|$, is the desired orthonormal basis for V_r.

Lemma 7: If $\{\boldsymbol{\alpha}_1, \boldsymbol{\alpha}_2, \cdots, \boldsymbol{\alpha}_r\}$ is an orthonormal basis for $V_r \subset V_n$ it is always possible to extend it to an orthonormal basis $\{\boldsymbol{\alpha}_1, \cdots, \boldsymbol{\alpha}_r, \boldsymbol{\alpha}_{r+1}, \cdots, \boldsymbol{\alpha}_n\}$ for V_n.

Proof: Reasoning as in Lemma 5 we see that it is possible to find vectors $\boldsymbol{\beta}_{r+1}, \cdots, \boldsymbol{\beta}_n$ such that $\{\boldsymbol{\alpha}_1, \cdots, \boldsymbol{\alpha}_r, \boldsymbol{\beta}_{r+1}, \cdots, \boldsymbol{\beta}_n\}$ is a basis for V_n. Then by Lemma 6 we can orthonormalize it and obtain a basis $\{\boldsymbol{\gamma}_1, \cdots, \boldsymbol{\gamma}_r, \boldsymbol{\gamma}_{r+1}, \cdots, \boldsymbol{\gamma}_n\}$. But, $\boldsymbol{\alpha}_1, \cdots, \boldsymbol{\alpha}_r$ being already pairwise orthogonal, we easily see that all coefficients $c_{21}, c_{31}, c_{32}, \cdots, c_{r1}, \cdots, c_{r,r-1}$ will be zero, so that $\boldsymbol{\gamma}_i = \boldsymbol{\alpha}_i$ for $i = 1, \cdots, r$. Thus we have extended $\{\boldsymbol{\alpha}_1, \cdots, \boldsymbol{\alpha}_r\}$ to an orthonormal basis $\{\boldsymbol{\alpha}_1, \cdots, \boldsymbol{\alpha}_r, \boldsymbol{\gamma}_{r+1}, \cdots, \boldsymbol{\gamma}_n\}$ for V_n.

Definition 14: Let $V_r \subset V_n$, and $\mathbf{x} \in V_n$. Then $\mathbf{x}$ is said to be *orthogonal to* V_r (we write $\mathbf{x} \perp V_r$) if and only if $\mathbf{x}$ is orthogonal to every vector in V_r.

Lemma 8: If $\{\boldsymbol{\alpha}_1, \boldsymbol{\alpha}_2, \cdots, \boldsymbol{\alpha}_s\}$ spans $V_r \subset V_n$, then a vector $\mathbf{x} \in V_n$ is orthogonal to V_r if and only if $\mathbf{x}$ is orthogonal to each $\boldsymbol{\alpha}_i$ $(i = 1, \cdots, s)$.

Proof: If $\mathbf{x} \perp V_r$, then $\mathbf{x} \perp \boldsymbol{\alpha}_i$ by definition. If $\mathbf{x} \perp \boldsymbol{\alpha}_i$ for $i = 1, \cdots, s$, and if $\mathbf{y} \in V_r$, we can write $\mathbf{y} = \Sigma_1^s b_i \boldsymbol{\alpha}_i$. Then $\mathbf{x}'\mathbf{y} = \Sigma_1^s b_i \mathbf{x}'\boldsymbol{\alpha}_i = 0$. Thus $\mathbf{x} \perp \mathbf{y}$, for all $\mathbf{y} \in V_r$.

Lemma 9: Let $V_r \subset V_n$ and $\mathbf{x} \in V_n$. There exist vectors $\mathbf{y}$ and $\mathbf{z}$ such that $\mathbf{x} = \mathbf{y}+\mathbf{z}$, $\mathbf{y} \in V_r$, $\mathbf{z} \perp V_r$. This decomposition is unique; i.e., if we also have $\mathbf{x} = \mathbf{y}^*+\mathbf{z}^*$, $\mathbf{y}^* \in V_r$, and $\mathbf{z}^* \perp V_r$, then necessarily $\mathbf{y} = \mathbf{y}^*$ and $\mathbf{z} = \mathbf{z}^*$.

Proof: Let $\{\boldsymbol{\alpha}_1, \boldsymbol{\alpha}_2, \cdots, \boldsymbol{\alpha}_r\}$ be an orthonormal basis for V_r. Let $\mathbf{y} = \Sigma_{i=1}^r c_i \boldsymbol{\alpha}_i$, where $c_i = \mathbf{x}'\boldsymbol{\alpha}_i$. Clearly $\mathbf{y} \in V_r$. Let $\mathbf{z} = \mathbf{x}-\mathbf{y} = \mathbf{x} - \Sigma_1^r c_i \boldsymbol{\alpha}_i$. We have, for $k = 1, \cdots, r$, $\mathbf{z}'\boldsymbol{\alpha}_k = \mathbf{x}'\boldsymbol{\alpha}_k - \Sigma_{i=1}^r c_i \boldsymbol{\alpha}_i'\boldsymbol{\alpha}_k = c_k - \Sigma_{i=1}^r c_i \delta_{ik} = c_k - c_k = 0$. Thus by Lemma 8 we have $\mathbf{z} \perp V_r$, and so the vectors $\mathbf{y}$ and $\mathbf{z}$ we have constructed satisfy the conditions of Lemma 9. Assume now that we also have $\mathbf{x} = \mathbf{y}^*+\mathbf{z}^*$ where $\mathbf{y}^* \in V_r$ and $\mathbf{z}^* \perp V_r$. Then $(\mathbf{y}^*-\mathbf{y})+(\mathbf{z}^*-\mathbf{z}) = \mathbf{x}-\mathbf{x} = \mathbf{0}$. But $\tilde{\mathbf{y}} = (\mathbf{y}^*-\mathbf{y}) \in V_r$, $\tilde{\mathbf{z}} = (\mathbf{z}^*-\mathbf{z}) \perp V_r$ on the one hand, and $\tilde{\mathbf{z}} = -\tilde{\mathbf{y}} \in V_r$ on the other. Hence, $\tilde{\mathbf{z}}$ must be orthogonal to itself, which is possible only if $\tilde{\mathbf{z}} = \mathbf{0}$. Then $\tilde{\mathbf{y}}+\tilde{\mathbf{z}} = \mathbf{0}$ implies $\tilde{\mathbf{y}} = \mathbf{0}$. Thus $\mathbf{y} = \mathbf{y}^*$ and $\mathbf{z} = \mathbf{z}^*$.

The proof of the uniqueness of $\mathbf{y}$ implies that $\mathbf{y}$ is in fact independent of the particular orthonormal basis $\{\boldsymbol{\alpha}_1, \cdots, \boldsymbol{\alpha}_r\}$ used in its definition. It is natural at this point to introduce the following

Definition 15: Given a vector $\mathbf{x} \in V_n$, the vector $\mathbf{y} \in V_r$ defined in Lemma 9, which is such that $(\mathbf{x}-\mathbf{y}) \perp V_r$, is called the *projection* of $\mathbf{x}$ on V_r.

In the case $r = 1$, this definition reduces to Definition 6 (where we speak of the projection of $\mathbf{x}$ on a nonzero vector instead of its projection on the V_1 spanned by the vector).

Theorem 2: Given a fixed $V_r \subset V_n$, a fixed vector $\mathbf{x} \in V_n$, and a variable vector $\mathbf{y} \in V_r$, then $\|\mathbf{x}-\mathbf{y}\|$ has a minimum value. This minimum is attained if and only if $\mathbf{y}$ is the projection of $\mathbf{x}$ on V_r.

Proof: Let $\mathbf{y}^*$ be the projection of $\mathbf{x}$ on V_r. Write $\mathbf{x}-\mathbf{y} = (\mathbf{x}-\mathbf{y}^*) + (\mathbf{y}^*-\mathbf{y})$. Then

$$\|\mathbf{x}-\mathbf{y}\|^2 = (\mathbf{x}-\mathbf{y}^*)'(\mathbf{x}-\mathbf{y}^*) + (\mathbf{y}^*-\mathbf{y})'(\mathbf{y}^*-\mathbf{y}) + (\mathbf{x}-\mathbf{y}^*)'(\mathbf{y}^*-\mathbf{y}) + (\mathbf{y}^*-\mathbf{y})'(\mathbf{x}-\mathbf{y}^*).$$

But $(\mathbf{x}-\mathbf{y}^*) \perp V_r$, while $\mathbf{y}^*$, $\mathbf{y}$, and hence $\mathbf{y}^*-\mathbf{y}$ are all in V_r. Thus the last two terms in the above equation are zero, and

$$\|\mathbf{x}-\mathbf{y}\|^2 = \|\mathbf{x}-\mathbf{y}^*\|^2 + \|\mathbf{y}^*-\mathbf{y}\|^2.$$

When $\mathbf{y}$ varies in V_r the first of the two terms on the right is fixed while the second is variable with value $\geqq 0$, and $=0$ if and only if $\mathbf{y} = \mathbf{y}^*$. Thus $\|\mathbf{x}-\mathbf{y}\|^2$ attains its minimum if and only if $\mathbf{y} = \mathbf{y}^*$. This proof may be pictured by Fig. I.7.

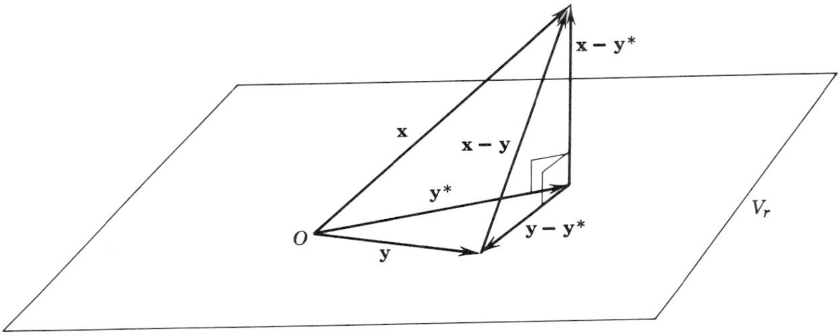

FIG. I.7

The following theorem, which seems geometrically obvious, is sometimes useful for proving the existence of a solution of a set of linear equations (see for example the end of sec. 1.4).

Theorem 3: If $V_r \subset V_n$, $\mathbf{x} \in V_n$, and $\mathbf{x}$ is orthogonal to every $\mathbf{y}$ orthogonal to V_r, then $\mathbf{x} \in V_r$.

Proof: Let $\mathbf{x} = \mathbf{z} + \mathbf{w}$, where $\mathbf{z} \in V_r$, $\mathbf{w} \perp V_r$, and take the scalar product with $\mathbf{w}$ to get $\mathbf{w}'\mathbf{x} = \mathbf{w}'\mathbf{z} + \mathbf{w}'\mathbf{w}$. But $\mathbf{w}'\mathbf{z} = 0$, and by the hypothesis of the theorem $\mathbf{w}'\mathbf{x} = 0$; thus $\mathbf{w}'\mathbf{w} = 0$, $\mathbf{w} = \mathbf{0}$, and hence $\mathbf{x} = \mathbf{z} \in V_r$.

Implicit in Theorem 3 is the notion of orthocomplement, which we will also find useful (secs. 2.4 and 2.5), and which we define by

Definition 16: If V_s is an s-dimensional subspace of the r-dimensional subspace $V_r \subset V_n$, then the totality of vectors in V_r which are orthogonal to V_s is called the *orthocomplement* of V_s in V_r.

Example: In the three-dimensional Euclidean space of points with coordinates (x_1, x_2, x_3) the orthocomplement of the (x_1,x_2)-plane is the x_3-axis.

Lemma 10: The orthocomplement of V_s in V_r of Definition 16 is an $(r-s)$-dimensional subspace of V_r.

Proof: Let $\{\gamma_1, \cdots, \gamma_s\}$ be an orthonormal basis for V_s, and complete it to an orthonormal basis $\{\gamma_1, \cdots, \gamma_s, \cdots, \gamma_r\}$ for V_r. Then any $\mathbf{x} \in V_r$ is of the form $\Sigma_1^r c_i \gamma_i$, and is $\perp V_s$ if and only if $c_1 = c_2 = \cdots = c_s = 0$, i.e., if and only if it is in the $(r-s)$-dimensional space of vectors of the form $\Sigma_{s+1}^r c_i \gamma_i$.

If V_{r-s} denotes the orthocomplement of V_s in V_r, then we see that the orthocomplement of V_{r-s} in V_r is V_s, since it is the totality of vectors of the form $\Sigma_1^s c_i \gamma_i$ in the above proof. Theorem 3 may be regarded as the special case of this remark when $r = n$.

PROBLEMS

I.1. (*a*) Let V be the vector space spanned by $\boldsymbol{\alpha}_1, \boldsymbol{\alpha}_2, \boldsymbol{\alpha}_3, \boldsymbol{\alpha}_4$, where

$$\boldsymbol{\alpha}_1 = \begin{pmatrix} 2 \\ 1 \\ 0 \\ -1 \\ 1 \end{pmatrix}, \quad \boldsymbol{\alpha}_2 = \begin{pmatrix} 4 \\ 4 \\ 4 \\ 0 \\ 2 \end{pmatrix}, \quad \boldsymbol{\alpha}_3 = \begin{pmatrix} 2 \\ 0 \\ -2 \\ -2 \\ 1 \end{pmatrix}, \quad \boldsymbol{\alpha}_4 = \begin{pmatrix} 1 \\ 3 \\ -1 \\ 5 \\ 0 \end{pmatrix}.$$

Apply the Schmidt process to $\boldsymbol{\alpha}_1, \boldsymbol{\alpha}_2, \boldsymbol{\alpha}_3, \boldsymbol{\alpha}_4$ in this order, in order to obtain an orthonormal basis for V. (*b*) Are the following vectors in the orthocomplement of V?

$$\begin{pmatrix} 1 \\ -5 \\ 1 \\ 3 \\ 6 \end{pmatrix}, \quad \begin{pmatrix} -1 \\ 1 \\ -3 \\ 5 \\ 6 \end{pmatrix}.$$

(*c*) Decompose the vector

$$\mathbf{x} = \begin{pmatrix} 16 \\ -7 \\ 12 \\ 1 \\ -10 \end{pmatrix}$$

into a sum $\mathbf{x} = \mathbf{y} + \mathbf{z}$ such that $\mathbf{y} \in V$ and $\mathbf{z} \perp V$. Check that

$$\|\text{projection of } \mathbf{x} \text{ on } V\|^2 = \sum_{i=1}^{r} \|\text{projection of } \mathbf{x} \text{ on } \gamma_i\|^2,$$

where $\{\gamma_1, \cdots, \gamma_r\}$ is the orthonormal basis obtained in (*a*). Compute the minimum value of $\|\mathbf{x} - \mathbf{w}\|^2$, where $\mathbf{w}$ is a variable vector in V.

I.2. Let **x**, **y** be any two vectors in V_n. Prove that $|\mathbf{x}'\mathbf{y}|^2 \leqq \|\mathbf{x}\|^2\|\mathbf{y}\|^2$ (Schwartz inequality). [*Hint:* Consider the second-order polynomial in c, $\|c\mathbf{x}+\mathbf{y}\|^2$, and express in terms of its discriminant the fact that it is never negative.]

I.3. Consider the solution of m linear equations $\Sigma_{j=1}^n a_{ij}x_j = c_j$ $(i = 1, \cdots, m)$ in n unknowns $\{x_1, \cdots, x_n\}$. (*a*) Define vectors $\{\boldsymbol{\alpha}_j\}$ and **c** such that the m equations may be expressed as a single vector equation $\Sigma_{j=1}^n x_j\boldsymbol{\alpha}_j = \mathbf{c}$. (*b*) Verify that a solution exists if and only if **c** is a linear combination of the $\{\boldsymbol{\alpha}_j\}$. (*c*) If a solution exists, under what conditions is it unique?

I.4. Prove that if V and W are vector spaces in V_n then the orthocomplement of the union of V and W is the intersection of the orthocomplements of V and W. (The *union* of two sets A and B is the totality of elements belonging to A or B; the *intersection* of A and B is the totality of elements belonging to both A and B.)

APPENDIX II

Matrix Algebra

In this appendix we shall state some of the theorems without proof.[1] Throughout, all variables and constants are understood to be *real* numbers.

Consider a linear transformation from n variables $x_1, \cdots, x_n$ to m variables $y_1, \cdots, y_m$ (or from a vector $\mathbf{x}$ with n components to a vector $\mathbf{y}$ with m components),

$$\text{(II.1)} \qquad \begin{aligned} y_1 &= a_{11}x_1 + a_{12}x_2 + \cdots + a_{1n}x_n, \\ y_2 &= a_{21}x_1 + a_{22}x_2 + \cdots + a_{2n}x_n, \\ &\cdot \quad \cdot \quad \cdot \quad \cdots \quad \cdot \\ y_m &= a_{m1}x_1 + a_{m2}x_2 + \cdots + a_{mn}x_n. \end{aligned}$$

The rectangular array of coefficients of the transformation is called the *matrix* of the linear transformation; we write it

$$\text{(II.2)} \qquad \mathbf{A}^{m\times n} = \begin{pmatrix} a_{11} & a_{12} & \cdots & a_{1n} \\ a_{21} & a_{22} & \cdots & a_{2n} \\ \cdot & \cdot & \cdots & \cdot \\ a_{m1} & a_{m2} & \cdots & a_{mn} \end{pmatrix},$$

where the superscript $m\times n$ (read "m by n") on $\mathbf{A}$ indicates that $\mathbf{A}$ has m rows and n columns. We shall often omit the superscript when it is clear what the size of a matrix is. We also abbreviate the notation (II.2) as $\mathbf{A}^{m\times n} = (a_{ij})$; this means that the i,j element (element in ith row and jth column) of $\mathbf{A}$ is a_{ij}. No confusion will be caused by ignoring the

[1] The missing proofs may be found in Birkhoff and MacLane (1953) or Murdoch (1957).

distinction between a 1×1 matrix and the real number which is its only element; in this case we may then write $\mathbf{A}^{1 \times 1} = (a_{ij}) = a_{11}$.

Definition 1: The *transpose* of the $m \times n$ matrix $\mathbf{A}$ in (II.2), written $\mathbf{A}'$, is the $n \times m$ matrix

$$\mathbf{A}' = \begin{pmatrix} a_{11} & a_{21} & \cdots & a_{m1} \\ a_{12} & a_{22} & \cdots & a_{m2} \\ \cdot & \cdot & \cdots & \cdot \\ a_{1n} & a_{2n} & \cdots & a_{mn} \end{pmatrix}$$

obtained by interchanging the rows and columns of $\mathbf{A}$.

Thus, if $\mathbf{A}' = (a'_{ij})$, then $a'_{ij} = a_{ji}$.

We now introduce three operations on matrices.

Definition 2: The *sum* of two matrices $\mathbf{A}^{m \times n} = (a_{ij})$ and $\mathbf{B}^{m \times n} = (b_{ij})$ is the $m \times n$ matrix $\mathbf{A}+\mathbf{B} = (a_{ij}+b_{ij})$, obtained by adding corresponding elements.

The sum of two matrices is defined only for matrices of the same size. Clearly, addition of matrices is commutative and associative:

$$\mathbf{A} + \mathbf{B} = \mathbf{B} + \mathbf{A},$$

$$(\mathbf{A}+\mathbf{B}) + \mathbf{C} = \mathbf{A} + (\mathbf{B}+\mathbf{C}) = \mathbf{A} + \mathbf{B} + \mathbf{C},$$

and the transpose of a sum is the sum of the transposes,

$$(\mathbf{A}+\mathbf{B})' = \mathbf{A}' + \mathbf{B}'.$$

Definition 3: The product of an $m \times n$ matrix $\mathbf{A}$ by a scalar (real number) c is the $m \times n$ matrix $c\mathbf{A} = (ca_{ij})$, obtained by multiplying each element by c.

This multiplication has the same properties as scalar multiplication of a vector and is also called *scalar multiplication*. We note that

$$(c\mathbf{A})' = c\mathbf{A}'.$$

To motivate the definition of matrix multiplication we consider two successive linear transformations, a transformation from $w_1, \cdots, w_r$ to $x_1, \cdots, x_n$ followed by a transformation from $x_1, \cdots, x_n$ to $y_1, \cdots, y_m$. Suppose that the transformation from the x's to the y's is (II.1) with matrix $\mathbf{A}^{m \times n}$ in (II.2), and the transformation from the w's to the x's is

$$x_j = \sum_{k=1}^{r} b_{jk} w_k \qquad (j = 1, \cdots, n) \tag{II.3}$$

with matrix

$$\mathbf{B}^{n\times r} = \begin{pmatrix} b_{11} & b_{12} & \cdots & b_{1r} \\ b_{21} & b_{22} & \cdots & b_{2r} \\ \cdot & \cdot & \cdots & \cdot \\ b_{n1} & b_{n2} & \cdots & b_{nr} \end{pmatrix}.$$

The resulting transformation, obtained by substituting (II.3) into

$$y_i = \sum_{j=1}^{n} a_{ij}x_j \qquad (i = 1, \cdots, m),$$

is

(II.4)
$$y_i = \sum_{j=1}^{n} a_{ij} \sum_{k=1}^{r} b_{jk}w_k,$$

or

$$y_i = \sum_{k=1}^{r} c_{ik}w_k$$

with

(II.5)
$$c_{ik} = \sum_{j=1}^{n} a_{ij}b_{jk}.$$

Its matrix is

$$\mathbf{C}^{m\times r} = \begin{pmatrix} c_{11} & c_{12} & \cdots & c_{1r} \\ c_{21} & c_{22} & \cdots & c_{2r} \\ \cdot & \cdot & \cdots & \cdot \\ c_{m1} & c_{m2} & \cdots & c_{mr} \end{pmatrix}.$$

We shall define matrix multiplication of **A** and **B** so that **AB** = **C**. In other words:

Definition 4: The *matrix product* of $\mathbf{A}^{m\times n} = (a_{ij})$ by $\mathbf{B}^{n\times r} = (b_{jk})$ is defined to be $\mathbf{C}^{m\times r} = (c_{ik})$ with c_{ik} given by (II.5), and is written $\mathbf{C} = \mathbf{AB}$.

Verbalizing this definition we say that the element in the ith row and kth column of **AB** is obtained by multiplying pairwise the elements in the ith row of **A** by the corresponding elements in the kth column of **B** and summing (as in the formation of a scalar product of vectors).

We note that matrix multiplication is defined only when the first matrix has as many columns as the second has rows, and that if these matrices are $m\times n$ and $n\times r$ then the product is $m\times r$. We remark that matrix multiplication is *not commutative*: If **A** is $m\times n$ and **B** is $n\times r$, then **AB** is defined, and **BA** is not, unless $m = r$, and even then $\mathbf{AB} \neq \mathbf{BA}$ in general, as may be seen from the example

$$\mathbf{A} = \begin{pmatrix} 0 & 0 \\ 0 & 1 \end{pmatrix}, \qquad \mathbf{B} = \begin{pmatrix} 1 & 1 \\ 0 & 0 \end{pmatrix},$$

where

(II.6)
$$\mathbf{AB} = \begin{pmatrix} 0 & 0 \\ 0 & 0 \end{pmatrix},$$

but

$$\mathbf{BA} = \begin{pmatrix} 0 & 1 \\ 0 & 0 \end{pmatrix}.$$

A matrix is called a *zero matrix* and written $\mathbf{0}^{m\times n}$ or $\mathbf{0}$ if every element is 0. The result of multiplying any matrix on the left or right by a zero matrix (with the necessary number of rows or columns) is another zero matrix. While this is as in the familiar real-number system, it is important to note that the following is not: We may have $\mathbf{AB} = \mathbf{0}$ while $\mathbf{A} \neq \mathbf{0}$ and $\mathbf{B} \neq \mathbf{0}$; an example is (II.6).

A vector $\mathbf{x}$ with n components may be regarded as an $n\times 1$ matrix (or "column matrix"). Instead of writing $\mathbf{x}$ in the form (I.1) we may now use the typographically more convenient notation $\mathbf{x} = (x_1, x_2, \cdots, x_n)'$. We now note that in App. I the three vector operations $\mathbf{x}+\mathbf{y}$, $c\mathbf{x}$, and $\mathbf{x}'\mathbf{y}$ were defined to be consistent with the matrix operations of addition, scalar multiplication, and matrix multiplication, respectively; thus

$$\mathbf{x}'\mathbf{y} = \begin{pmatrix} x_1 \\ x_2 \\ \cdot \\ \cdot \\ \cdot \\ x_n \end{pmatrix}' \begin{pmatrix} y_1 \\ y_2 \\ \cdot \\ \cdot \\ \cdot \\ y_n \end{pmatrix} = (x_1, x_2, \cdots, x_n) \begin{pmatrix} y_1 \\ y_2 \\ \cdot \\ \cdot \\ \cdot \\ y_n \end{pmatrix} = \sum_{j=1}^{n} x_j y_j.$$

Since the columns of a matrix may be regarded as vectors or column matrices we can form linear combinations of them; a similar statement holds for the rows as row matrices. An *interpretation* of matrix multiplication that we will often find useful is that the kth column of $\mathbf{AB}$ is a linear combination of the columns of $\mathbf{A}$, the coefficients in the linear combination being the elements of the kth column of $\mathbf{B}$. This follows from Definition 4 according to which the kth column of $\mathbf{AB}$ is

(II.7)
$$\begin{pmatrix} \sum_{j=1}^n a_{1j} b_{jk} \\ \sum_{j=1}^n a_{2j} b_{jk} \\ \cdot \\ \cdot \\ \cdot \\ \sum_{j=1}^n a_{mj} b_{jk} \end{pmatrix} = \sum_{j=1}^{n} b_{jk} \begin{pmatrix} a_{1j} \\ a_{2j} \\ \cdot \\ \cdot \\ \cdot \\ a_{mj} \end{pmatrix},$$

the linear combination of column matrices on the right side to be evaluated according to Definitions 3 and 2. Similarly we may conclude that the ith row of **AB** is a linear combination of the rows of **B** with coefficients from the ith row of **A**. A mnemonic for remembering from which factor the columns of the product come is to reflect that if **A** is $m \times n$ and **B** is $n \times r$ then **AB** is $m \times r$, hence the columns of **AB** have m elements and must therefore come from the columns of **A**; analogously for the rows.

Properties of Matrix Multiplication

If $\mathbf{A}^{m\times n}$, $\mathbf{B}^{n\times r}$, $\mathbf{C}^{n\times r}$, and $\mathbf{D}^{r\times p}$ are matrices and c is a scalar, then the following laws are easily verified from the definitions of the three matrix operations:

$$\mathbf{A}(\mathbf{B}+\mathbf{C}) = \mathbf{AB} + \mathbf{AC},$$

$$(\mathbf{B}+\mathbf{C})\mathbf{D} = \mathbf{BD} + \mathbf{CD},$$

$$(c\mathbf{A})\mathbf{B} = \mathbf{A}(c\mathbf{B}) = c(\mathbf{AB}) = c\mathbf{AB},$$

$$(\mathbf{AB})' = \mathbf{B}'\mathbf{A}',$$

$$(\mathbf{AB})\mathbf{D} = \mathbf{A}(\mathbf{BD}) = \mathbf{ABD}.$$

The last law implies that parentheses may be inserted or dropped at will in a product of any finite number of matrices.

Matrix notation permits very simple handling of linear transformations. Thus if we denote by **w**, **x**, **y** the vectors

$$\mathbf{w}^{r\times 1} = \begin{pmatrix} w_1 \\ w_2 \\ \cdot \\ \cdot \\ \cdot \\ w_r \end{pmatrix}, \quad \mathbf{x}^{n\times 1} = \begin{pmatrix} x_1 \\ x_2 \\ \cdot \\ \cdot \\ \cdot \\ x_n \end{pmatrix}, \quad \mathbf{y}^{m\times 1} = \begin{pmatrix} y_1 \\ y_2 \\ \cdot \\ \cdot \\ \cdot \\ y_m \end{pmatrix},$$

then the transformations (II.1) and (II.3) may be written

$$\mathbf{y} = \mathbf{Ax}, \tag{II.8}$$

$$\mathbf{x} = \mathbf{Bw}, \tag{II.9}$$

and the resulting transformation (II.4) is obtained by substituting (II.9) into (II.8) to get $\mathbf{y} = \mathbf{ABw}$.

Determinants

We shall skip the definition of the determinant $|\mathbf{A}|$ of a square matrix **A**: Its value (a scalar) may be calculated by successive application of the remark after Theorem 3: Each application reduces the order of the determinants by 1, and finally determinants of order 1 are equal to the single element they contain. We say that $|\mathbf{A}|$ and $\mathbf{A}$ are of *order n* if $\mathbf{A}$ is $n\times n$. We list some definitions and theorems concerning determinants:

Theorem 1: If $\mathbf{A}$ is $n\times n$, $|\mathbf{A}'| = |\mathbf{A}|$.

Theorem 2: If $\mathbf{A}$ and $\mathbf{B}$ are $n\times n$, then $|\mathbf{AB}| = |\mathbf{A}|\cdot|\mathbf{B}|$.

Definition 5: The *cofactor* A_{ij} of the element a_{ij} in the nth ($n > 1$) order matrix $\mathbf{A}$ is $(-1)^{i+j}$ times the $(n-1)$th-order determinant obtained by deleting the ith row and jth column of $\mathbf{A}$ and taking the determinant of the remaining matrix.

For example, the cofactor of a_{23} in $\mathbf{A}^{3\times 3} = (a_{ij})$ is

$$A_{23} = -\begin{vmatrix} a_{11} & a_{12} \\ a_{31} & a_{32} \end{vmatrix}.$$

Theorem 3: If $\mathbf{A}$ is $n\times n$,

$$\text{(II.10)} \qquad \sum_{j=1}^{n} a_{kj}A_{ij} = \delta_{ki}|\mathbf{A}|, \qquad \sum_{i=1}^{n} a_{ik}A_{ij} = \delta_{kj}|\mathbf{A}|.$$

When $k = i$ in particular we obtain $|\mathbf{A}| = \Sigma_{j=1}^{n} a_{ij}A_{ij}$, a formula useful for evaluating a determinant. For example,

$$\begin{vmatrix} a_{11} & a_{12} & a_{13} \\ a_{21} & a_{22} & a_{23} \\ a_{31} & a_{32} & a_{33} \end{vmatrix} = -a_{21}\begin{vmatrix} a_{12} & a_{13} \\ a_{32} & a_{33} \end{vmatrix} + a_{22}\begin{vmatrix} a_{11} & a_{13} \\ a_{31} & a_{33} \end{vmatrix} - a_{23}\begin{vmatrix} a_{11} & a_{12} \\ a_{31} & a_{32} \end{vmatrix}.$$

By successive applications of this rule we see that if for a matrix $\mathbf{A}^{n\times n} = (a_{ij})$, $a_{ij} = 0$ for $i \neq j$ (in which case $\mathbf{A}$ is called a *diagonal matrix*), then $|\mathbf{A}|$ is the product $a_{11}a_{22}\cdots a_{nn}$ of the diagonal elements of $\mathbf{A}$.

Definition 6: A square matrix $\mathbf{A}$ is called *singular* if $|\mathbf{A}| = 0$; it is called *nonsingular* if $|\mathbf{A}| \neq 0$.

Inverse of a Matrix

The matrix of the identity transformation $\mathbf{y}^{n\times 1} = \mathbf{x}^{n\times 1}$ is

$$\text{(II.11)} \qquad \mathbf{I}^{n\times n} = \begin{pmatrix} 1 & 0 & \cdots & 0 \\ 0 & 1 & \cdots & 0 \\ \cdot & \cdot & \cdots & \cdot \\ 0 & 0 & \cdots & 1 \end{pmatrix} = (\delta_{ij}),$$

for this transformation is the special case of $\mathbf{y} = \mathbf{Ax}$ for $\mathbf{A} = \mathbf{I}$.

Definition 7: $\mathbf{I}^{n\times n}$ defined in (II.11) is called the *identity matrix* of order n.

Whatever the matrices $\mathbf{A}^{m\times n}$ and $\mathbf{B}^{n\times r}$, we have $\mathbf{IB} = \mathbf{B}$, $\mathbf{AI} = \mathbf{A}$ (this may be seen most quickly from the rules stated in connection with (II.7)). We also note that $|\mathbf{I}| = 1$.

Definition 8: If for a matrix $\mathbf{A}^{n\times n}$ there exists a matrix $\mathbf{B}^{n\times n}$ such that $\mathbf{BA} = \mathbf{AB} = \mathbf{I}$ then $\mathbf{B}$ is called the *inverse* of $\mathbf{A}$ and written $\mathbf{B} = \mathbf{A}^{-1}$.

We note that the inverse is defined only for square matrices.

Theorem 4: The matrix $\mathbf{A}^{n\times n}$ has an inverse if and only if $\mathbf{A}$ is nonsingular. Then $\mathbf{A}^{-1}$ is unique, and if $n > 1$ the i,j element of $\mathbf{A}^{-1}$ is $A_{ji}/|\mathbf{A}|$. (Note the permutation of the subscripts!)

Proof: If $\mathbf{A}$ has an inverse $\mathbf{A}^{-1}$ then $1 = |\mathbf{I}| = |\mathbf{AA}^{-1}| = |\mathbf{A}| \cdot |\mathbf{A}^{-1}|$, thus $|\mathbf{A}| \neq 0$. Conversely, suppose that $|\mathbf{A}| \neq 0$. If $\mathbf{A}$ is 1×1, $\mathbf{A} = (a)$, $a \neq 0$, and the 1×1 matrix $\mathbf{B} = (a^{-1})$ satisfies $\mathbf{BA} = \mathbf{AB} = \mathbf{I}$ and is thus an inverse of $\mathbf{A}$. If $n > 1$ then $b_{ij} = A_{ji}/|\mathbf{A}|$ is well defined, and if $\mathbf{B} = (b_{ij})$ the i,j element of $\mathbf{BA}$ is $\Sigma_k b_{ik}a_{kj} = |\mathbf{A}|^{-1}\Sigma_k a_{kj}A_{ki} = \delta_{ij}$ by Theorem 3. Thus $\mathbf{BA} = \mathbf{I}$, and similarly we find that $\mathbf{AB} = \mathbf{I}$. Therefore, $\mathbf{B}$ is an inverse of $\mathbf{A}$. Suppose that $\mathbf{C}$ were another. Then $\mathbf{BAC} = (\mathbf{BA})\mathbf{C} = \mathbf{IC} = \mathbf{C}$ and $\mathbf{BAC} = \mathbf{B}(\mathbf{AC}) = \mathbf{BI} = \mathbf{B}$ imply $\mathbf{B} = \mathbf{C}$. Thus $\mathbf{A}^{-1} = \mathbf{B}$ is unique.

Remark: To verify that a matrix $\mathbf{M}^{n\times n}$ is the inverse of a matrix $\mathbf{A}^{n\times n}$ it is sufficient to check that $\mathbf{MA} = \mathbf{I}$ (or else to check that $\mathbf{AM} = \mathbf{I}$): By taking determinants on both sides of $\mathbf{MA} = \mathbf{I}$ we see that $|\mathbf{A}| \neq 0$, and hence by Theorem 4, $\mathbf{A}^{-1}$ exists. Multiplying $\mathbf{MA} = \mathbf{I}$ on the right by $\mathbf{A}^{-1}$ we get $\mathbf{M} = \mathbf{A}^{-1}$.

Lemma 1: If $\mathbf{A}^{n\times n}$ is nonsingular, then the inverse of the transpose is the transpose of the inverse, i.e., $(\mathbf{A}')^{-1} = (\mathbf{A}^{-1})'$.

Proof: By Theorem 1, $\mathbf{A}'$ is also nonsingular; hence by Theorem 4 it has an inverse which we write $\mathbf{B} = (\mathbf{A}')^{-1}$. Take transposes in $\mathbf{A}'\mathbf{B} = \mathbf{I}$ to get $\mathbf{B}'\mathbf{A} = \mathbf{I}' = \mathbf{I}$. Then from the above remark $\mathbf{B}' = \mathbf{A}^{-1}$, or $\mathbf{B} = (\mathbf{A}^{-1})'$, i.e., $(\mathbf{A}')^{-1} = (\mathbf{A}^{-1})'$.

Lemma 2: If $\mathbf{A}^{n\times n}$ and $\mathbf{B}^{n\times n}$ are nonsingular matrices, so is their product $\mathbf{AB}$, and we have $(\mathbf{AB})^{-1} = \mathbf{B}^{-1}\mathbf{A}^{-1}$.

Proof: If $|\mathbf{A}| \neq 0$, $|\mathbf{B}| \neq 0$, then $|\mathbf{AB}| = |\mathbf{A}| \cdot |\mathbf{B}| \neq 0$; so $\mathbf{AB}$ is also nonsingular. Furthermore, $(\mathbf{B}^{-1}\mathbf{A}^{-1})(\mathbf{AB}) = \mathbf{B}^{-1}(\mathbf{A}^{-1}\mathbf{A})\mathbf{B} = \mathbf{B}^{-1}\mathbf{IB} = \mathbf{B}^{-1}\mathbf{B} = \mathbf{I}$ shows that $(\mathbf{AB})^{-1} = \mathbf{B}^{-1}\mathbf{A}^{-1}$.

Rank of a Matrix

As already mentioned, in a matrix $\mathbf{A}^{n\times m} = (\boldsymbol{\alpha}_1, \boldsymbol{\alpha}_2, \cdots, \boldsymbol{\alpha}_m)$ the columns $\boldsymbol{\alpha}_1, \cdots, \boldsymbol{\alpha}_m$ may be considered to be vectors in V_n.

Definition 9: The rank of $\mathbf{A} = (\boldsymbol{\alpha}_1, \boldsymbol{\alpha}_2, \cdots, \boldsymbol{\alpha}_m)$ is the maximum number of linearly independent vectors in the set $\{\boldsymbol{\alpha}_1, \cdots, \boldsymbol{\alpha}_m\}$; i.e., it is the dimension of the vector space spanned by the columns of **A**.

Example: Suppose the determinant $|\mathbf{A}^{n \times n}|$ is nonzero. Then there exists $\mathbf{B}^{n \times n}$ such that $\mathbf{AB} = \mathbf{I}$. Writing $\mathbf{A} = (\boldsymbol{\alpha}_1, \cdots, \boldsymbol{\alpha}_m)$ and $\mathbf{I} = (\boldsymbol{\rho}_1, \cdots, \boldsymbol{\rho}_n)$, where the vectors $\boldsymbol{\rho}$ are those of the basis $\mathscr{R}$ introduced after Theorem 1 in App. I, and using the interpretation of matrix multiplication above (II.7), we see that the $\boldsymbol{\rho}$'s, which span V_n, are linear combinations of the $\boldsymbol{\alpha}$'s, which therefore must also span V_n, and thus constitute a basis for V_n. Therefore, rank $\mathbf{A} = n$.

This example suggests that there is an intimate relationship between rank and determinants. In fact, the following result can be established: (Here *submatrix* of **A** means a matrix obtained by deleting any number of rows and columns from **A**.)

Theorem 5: Consider all square submatrices of $\mathbf{A}^{m \times n}$ which are nonsingular: The rank of **A** equals the maximum order of these nonsingular submatrices. (Roughly speaking, rank = maximum order of nonzero subdeterminants.)

Corollary 1: The maximum number of linearly independent columns of $\mathbf{A}^{m \times n}$ (i.e., rank **A**) equals the maximum number of linearly independent rows.

Proof utilizes Theorem 1.

Corollary 2: Rank $\mathbf{A}^{m \times n} \leq \min(m, n)$.

Lemma 3: Rank $\mathbf{AB} \leq \min(\text{rank } \mathbf{A}, \text{rank } \mathbf{B})$.

Proof: Since by the rule above (II.7) the columns of **AB** are linear combinations of the columns of **A**, the number of linearly independent columns in **AB** cannot exceed the number in **A**; hence rank $\mathbf{AB} \leq$ rank **A**. Arguing similarly about rows we get rank $\mathbf{AB} \leq$ rank **B**.

Lemma 4: If **A** is $m \times n$ and if $\mathbf{P}^{m \times m}$ and $\mathbf{Q}^{n \times n}$ are nonsingular, then rank $\mathbf{PAQ}$ = rank **A**, that is, the rank of a matrix is not altered by multiplying it on the right or left by a nonsingular matrix.

Proof: By Lemma 3, rank $\mathbf{PA} \leq$ rank **A**. Let $\mathbf{PA} = \mathbf{B}$. Since $\mathbf{P}^{-1}$ exists, $\mathbf{A} = \mathbf{P}^{-1}\mathbf{B}$. Hence, again by Lemma 3, rank $\mathbf{A} \leq$ rank $\mathbf{B}$ = rank **PA**. We now have rank $\mathbf{PA} \leq$ rank $\mathbf{A} \leq$ rank **PA**; hence rank **PA** = rank **A**. Similarly we find that if **C** is $m \times n$ then rank **CQ** = rank **C**. Now apply this to $(\mathbf{PA})\mathbf{Q}$ with $\mathbf{PA} = \mathbf{C}$ to get rank $(\mathbf{PA})\mathbf{Q}$ = rank **PA** = rank **A**.

Quadratic Forms

Definition 10: A *quadratic form* in the n variables $x_1, x_2, \cdots, x_n$ is a function of the form

$$Q = \sum_{i=1}^{n} \sum_{j=1}^{n} a_{ij} x_i x_j,$$

where the $\{a_{ij}\}$ are constants.

We remind the reader of our blanket assumption that all variables and constants are real.

We may write the quadratic form in matrix notation by defining the vector $\mathbf{x} = (x_1, x_2, \cdots, x_n)'$ and the matrix $\mathbf{A} = (a_{ij})$, so that

$$Q = \mathbf{x}'\mathbf{A}\mathbf{x}.$$

$\mathbf{A}$ is called the *matrix of the quadratic form* Q. We shall always assume that the matrix $\mathbf{A}$ of a quadratic form is *symmetric* (i.e., $\mathbf{A}' = \mathbf{A}$) because of

Lemma 5: Without loss of generality we may assume the matrix of a quadratic form to be *symmetric*.

Proof: Since Q is a 1×1 matrix, $Q' = Q$; therefore, $\mathbf{x}'\mathbf{A}'\mathbf{x} = \mathbf{x}'\mathbf{A}\mathbf{x}$, or $\mathbf{Q} = \frac{1}{2}\mathbf{x}'\mathbf{A}\mathbf{x} + \frac{1}{2}\mathbf{x}'\mathbf{A}'\mathbf{x} = \mathbf{x}'\mathbf{B}\mathbf{x}$, where $\mathbf{B} = \frac{1}{2}(\mathbf{A}+\mathbf{A}')$. Thus if we replace the matrix of the quadratic form Q by $\mathbf{B}$ we get the same Q and $\mathbf{B}' = \mathbf{B}$.

Example: If $Q = 5x_1^2 + 12x_1x_2 + 7x_2^2$, write

$$Q = 5x_1^2 + 6x_1x_2 + 6x_2x_1 + 7x_2^2 = (x_1, x_2)\begin{pmatrix} 5 & 6 \\ 6 & 7 \end{pmatrix}\begin{pmatrix} x_1 \\ x_2 \end{pmatrix},$$

and the matrix of the form is symmetric.

In many situations where a quadratic form Q is met, one is interested in making a nonsingular linear transformation of the variables such that the form of Q is particularly simple when expressed in the new variables. We shall see that it is always possible to eliminate the cross-product terms in Q, so that in the new variables $y_1, y_2, \cdots, y_n$ the form of Q is simply $Q = \Sigma_1^n \lambda_i y_i^2$, and that this can be done with an especially simple kind of linear transformation called orthogonal. The resulting coefficients $\{\lambda_i\}$ are then of great interest.

Suppose then we transform from $\mathbf{x} = (x_1, \cdots, x_n)'$ to $\mathbf{y} = (y_1, \cdots, y_n)'$ by a nonsingular linear transformation $\mathbf{y} = \mathbf{P}^{-1}\mathbf{x}$, so

$$\mathbf{x} = \mathbf{P}\mathbf{y}.$$

Then $\mathbf{x}' = \mathbf{y}'\mathbf{P}'$, hence $Q = \mathbf{x}'\mathbf{A}\mathbf{x} = \mathbf{y}'\mathbf{P}'\mathbf{A}\mathbf{P}\mathbf{y}$, and the matrix of the same quadratic form when expressed in terms of $\mathbf{y}$ is $\mathbf{P}'\mathbf{A}\mathbf{P}$. The matrix formulation of the transformation problem is thus, given a symmetric matrix $\mathbf{A}$, to find a nonsingular $\mathbf{P}$ such that $\mathbf{P}'\mathbf{A}\mathbf{P}$ has a simple form.

Orthogonal Matrices and Transformations

Definition 11: The matrix $\mathbf{P}^{n\times n}$ is called an *orthogonal matrix* if $\mathbf{P}'\mathbf{P} = \mathbf{I}$; the transformation $\mathbf{x} = \mathbf{P}\mathbf{y}$ is then called an *orthogonal transformation*.

We remark that the condition $\mathbf{P}'\mathbf{P} = \mathbf{I}$ is equivalent to $\mathbf{P}\mathbf{P}' = \mathbf{I}$ since both are equivalent to $\mathbf{P}^{-1} = \mathbf{P}'$. Furthermore, the inverse of an orthogonal matrix or transformation is also orthogonal, since $(\mathbf{P}\mathbf{P}')^{-1} = \mathbf{I}$ implies $(\mathbf{P}^{-1})'\mathbf{P}^{-1} = \mathbf{I}$ by Lemmas 1 and 2.

The following properties of orthogonal matrices and transformations follow immediately from the definition:

Lemma 6: The matrix $\mathbf{P}^{n\times n}$ is orthogonal if and only if its columns (or the transposes of its rows) constitute an orthonormal basis for V_n.

Lemma 7: Inner products $\mathbf{x}'\mathbf{z}$ are invariant if both $\mathbf{x}$ and $\mathbf{z}$ are subjected to the same orthogonal transformation.

Proof: If $\mathbf{P}'\mathbf{P} = \mathbf{I}$, $\mathbf{x} = \mathbf{P}\mathbf{x}^*$, $\mathbf{z} = \mathbf{P}\mathbf{z}^*$, then $\mathbf{x}'\mathbf{z} = \mathbf{x}^{*\prime}\mathbf{P}'\mathbf{P}\mathbf{z}^* = \mathbf{x}^{*\prime}\mathbf{z}^*$.

It follows that the length of any vector $\mathbf{x}$ is invariant under orthogonal transformation. We may regard the orthogonal transformation as one on the *points* of the n-dimensional Euclidean space, the points being located by vectors $\mathbf{x}$ drawn from the origin. Then the distance between any two points is invariant, since if the points are located by $\mathbf{x}$ and $\mathbf{y}$ their distance is $\|\mathbf{x}-\mathbf{y}\|$ (see Fig. I.4), and if we write $\mathbf{z} = \mathbf{x}-\mathbf{y}$, $\mathbf{z}^* = \mathbf{x}^*-\mathbf{y}^*$, where $\mathbf{x} = \mathbf{P}\mathbf{x}^*$, $\mathbf{y} = \mathbf{P}\mathbf{y}^*$, and $\mathbf{P}$ is the matrix of the orthogonal transformation, then $\mathbf{z} = \mathbf{P}\mathbf{z}^*$, and hence $\mathbf{z}^{*\prime}\mathbf{z}^* = \mathbf{z}'\mathbf{z}$ or $\|\mathbf{x}^*-\mathbf{y}^*\|^2 = \|\mathbf{x}-\mathbf{y}\|^2$. Since under an orthogonal transformation the distances of all pairs of points of any configuration are preserved and the origin is a fixed point, it is clear to our geometric intuition that the transformation must consist of a rotation about the origin—except for some possible reflections in planes, in addition. This qualification is illustrated by the orthogonal transformation with matrix

$$\begin{pmatrix} -1 & 0 & 0 \\ 0 & 1 & 0 \\ 0 & 0 & 1 \end{pmatrix},$$

which is a reflection in the x_2, x_3-plane.

The following lemma illustrates one way in which orthogonal transformations arise:

Lemma 8: If $\{\boldsymbol{\alpha}_1, \boldsymbol{\alpha}_2, \cdots, \boldsymbol{\alpha}_n\}$ and $\{\boldsymbol{\beta}_1, \boldsymbol{\beta}_2, \cdots, \boldsymbol{\beta}_n\}$ are orthonormal bases in V_n, if $a_1, \cdots, a_n$ and $b_1, \cdots, b_n$ are the coordinates of the vector $\mathbf{x}$ relative to these bases, then the coordinates are related by an orthogonal transformation, i.e., if $\mathbf{a} = (a_1, \cdots, a_n)'$, $\mathbf{b} = (b_1, \cdots, b_n)'$, then there exists an orthogonal $\mathbf{P}$ such that $\mathbf{b} = \mathbf{P}\mathbf{a}$.

Proof: Let $\mathbf{A}$ and $\mathbf{B}$ be the matrices whose columns are the vectors of the bases, $\mathbf{A} = (\boldsymbol{\alpha}_1, \cdots, \boldsymbol{\alpha}_n)$, $\mathbf{B} = (\boldsymbol{\beta}_1, \cdots, \boldsymbol{\beta}_n)$, so $\mathbf{A}$ and $\mathbf{B}$ are each orthogonal. Then the relations $\mathbf{x} = \Sigma_1^n a_j \boldsymbol{\alpha}_j = \Sigma_1^n b_j \boldsymbol{\beta}_j$ may be written $\mathbf{x} = \mathbf{Aa} = \mathbf{Bb}$ by the interpretation of matrix multiplication above (II.7). It follows that $\mathbf{b} = \mathbf{Pa}$ with $\mathbf{P} = \mathbf{B}^{-1}\mathbf{A}$ and $\mathbf{P}'\mathbf{P} = \mathbf{I}$.

The Principal-Axis Theorem

Definition 12: A square matrix (a_{ij}) is called a *diagonal matrix* if $a_{ij} = 0$ for $i \neq j$ (i.e., all "off-diagonal" elements are 0); a quadratic form is called a *diagonal quadratic form* if its matrix is diagonal (i.e., no cross-product terms are present in the form).

Theorem 6 (Principal-Axis Theorem): For every quadratic form $Q = \mathbf{x}'\mathbf{A}\mathbf{x}$ in n variables there exists an orthogonal transformation $\mathbf{x} = \mathbf{Py}$ which reduces Q to a diagonal quadratic form $Q = \lambda_1 y_1^2 + \lambda_2 y_2^2 + \cdots + \lambda_n y_n^2$.

We shall skip the proof. The reason for calling this the "principal-axis theorem" is that $\mathbf{x}'\mathbf{A}\mathbf{x} =$ constant may be regarded as a central quadric surface in n-dimensional Euclidean space (for example, in three-dimensional space, an ellipsoid, or hyperboloid of one or two sheets, or degenerate form of one of these) and the transformation can be interpreted as a change of coordinate axes to the principal axes of the quadric: this makes the cross-product terms vanish. The equivalent matrix formulation of the theorem is evidently

Theorem 6′: If $\mathbf{A}^{n \times n}$ is symmetric,[2] there exists an orthogonal $\mathbf{P}^{n \times n}$ such that $\mathbf{P}'\mathbf{A}\mathbf{P}$ is diagonal, $\mathbf{P}'\mathbf{A}\mathbf{P} = (\lambda_i \delta_{ij})$.

We shall now show that, regardless of what orthogonal $\mathbf{P}$ is used to reduce the matrix $\mathbf{A}$ in Theorem 6′, the elements $\{\lambda_i\}$ of the diagonal matrix obtained are always the same except for order, and we shall see how the $\{\lambda_i\}$ may be calculated.

Definition 13: The *characteristic polynomial* of a matrix $\mathbf{A}^{n \times n}$ is defined to be the determinant $|\mathbf{A} - \lambda \mathbf{I}|$, a polynomial in λ of degree n.

Examples: (1) If $\mathbf{A} = \begin{pmatrix} a_{11} & a_{12} \\ a_{21} & a_{22} \end{pmatrix}$, then its characteristic polynomial is

$$\begin{vmatrix} a_{11}-\lambda & a_{12} \\ a_{21} & a_{22}-\lambda \end{vmatrix} \text{ or } \lambda^2 - \lambda(a_{11}+a_{22}) + (a_{11}a_{22} - a_{12}a_{21}).$$

(2) If $\mathbf{A}^{n \times n}$ is diagonal, with diagonal elements $a_{11}, \cdots, a_{nn}$, so that $\mathbf{A} = (a_{ii}\delta_{ij})$ then its characteristic polynomial is $(a_{11}-\lambda)(a_{22}-\lambda) \cdots (a_{nn}-\lambda)$.

Lemma 9: If $\mathbf{P}$ is orthogonal the characteristic polynomial of $\mathbf{A}$ is

[2] We recall that all matrices are assumed real.

invariant under the transformation $\mathbf{P}'\mathbf{A}\mathbf{P}$ (i.e., $\mathbf{P}'\mathbf{A}\mathbf{P}$ has the same characteristic polynomial as $\mathbf{A}$).

Proof: If $\mathbf{A}^* = \mathbf{P}'\mathbf{A}\mathbf{P}$, its characteristic polynomial is $|\mathbf{A}^*-\lambda\mathbf{I}| = |\mathbf{P}'\mathbf{A}\mathbf{P}-\lambda\mathbf{P}'\mathbf{I}\mathbf{P}| = |\mathbf{P}'(\mathbf{A}-\lambda\mathbf{I})\mathbf{P}| = |\mathbf{P}'| \cdot |\mathbf{A}-\lambda\mathbf{I}| \cdot |\mathbf{P}| = |\mathbf{A}-\lambda\mathbf{I}|$, since $\mathbf{P}'\mathbf{P} = \mathbf{I}$ implies $|\mathbf{P}'| \cdot |\mathbf{P}| = 1$.

Definition 14: The roots of its characteristic polynomial are called the *characteristic roots* of $\mathbf{A}^{n\times n}$.

Since the characteristic polynomial is invariant under the transformation $\mathbf{P}'\mathbf{A}\mathbf{P}$ if $\mathbf{P}$ is orthogonal, so are the characteristic roots. We now see that the numbers $\{\lambda_i\}$ in the principal-axis theorem are the characteristic roots of the matrix of the quadratic form, and may thus be calculated by solving a polynomial equation of the nth degree. Theorem 6′ implies that the characteristic roots $\{\lambda_i\}$ are all real. We note that the rank of $\mathbf{A}$ equals the number of nonzero characteristic roots: this follows from Example 2, Theorem 6′, and Lemma 4.

Definition 15: The symmetric matrix $\mathbf{A}^{n\times n}$ and the quadratic form $\mathbf{x}'\mathbf{A}\mathbf{x}$ are called *positive definite* if[3] all the characteristic roots of $\mathbf{A}$ are >0, *positive indefinite* (or *positive semidefinite*—note that this includes positive definite) if all the characteristic roots are $\geqq 0$.

The reason for this terminology is indicated by

Lemma 10: The quadratic form $Q = \mathbf{x}'\mathbf{A}\mathbf{x}$ is positive definite if and only if $Q > 0$ for all $\mathbf{x} \neq \mathbf{0}$; it is positive indefinite if and only if $Q \geqq 0$ for all $\mathbf{x}$.

Proof: Let $\mathbf{x} = \mathbf{P}\mathbf{y}$ be the orthogonal transformation of Theorem 6, so $Q = \Sigma_1^n \lambda_i y_i^2$. If Q is positive definite and $\mathbf{x} \neq \mathbf{0}$ then $\mathbf{y} \neq \mathbf{0}$ (else $\mathbf{x} = \mathbf{P}\mathbf{y} = \mathbf{0}$), thus some $y_i \neq 0$, and all $\lambda_i > 0$, hence $Q > 0$. Conversely, suppose that $Q > 0$ for all $\mathbf{x} \neq \mathbf{0}$, and Q were not positive definite. Then there exists a $\lambda_k \leqq 0$. Consider the particular vector $\mathbf{y}$ whose kth component is 1, the others 0: There is an $\mathbf{x} = \mathbf{x}_0$ which gives this $\mathbf{y}$, namely $\mathbf{x}_0 = \mathbf{P}\mathbf{y}$, $\mathbf{x}_0 \neq \mathbf{0}$ (else $\mathbf{y} = \mathbf{P}^{-1}\mathbf{x}_0 = \mathbf{0}$), and for $\mathbf{x} = \mathbf{x}_0$, $Q = \lambda_k \leqq 0$, contradicting the assumption that $Q > 0$ for all $\mathbf{x} \neq \mathbf{0}$. This proves the first part of the lemma; the second part may be proved similarly.

From Theorems 6 and 6′ we may deduce the following results about reduction of quadratic forms and matrices under nonsingular (in general not orthogonal) transformations:

[3] It is not necessary to calculate the characteristic roots of $\mathbf{A}$ to determine whether $\mathbf{A}$ is positive definite: A submatrix of $\mathbf{A}$ whose diagonal elements are diagonal elements of $\mathbf{A}$ is called a *principal minor*. A set of *leading principal* minors is a set of n principal minors of orders $1, 2, \cdots, n$, each of which is a submatrix of the next, except the last (which is $\mathbf{A}$). A necessary and sufficient condition that $\mathbf{A}$ be positive definite is that the determinants of any set of leading principal minors be all positive. For a proof see Seelye (1958).

Lemma 11: For every quadratic form $Q = \mathbf{x}'\mathbf{A}\mathbf{x}$ in n variables there exists a nonsingular transformation $\mathbf{x} = \mathbf{P}\mathbf{y}$ which reduces Q to the form $Q = \Sigma_1^n \delta_i y_i^2$, where $\delta_i = 1, -1$, or 0.

Lemma 11′: For every symmetric $\mathbf{A}^{n\times n}$ there exists a nonsingular $\mathbf{P}$ such that $\mathbf{P}'\mathbf{A}\mathbf{P}$ has the diagonal form $(\delta_i\delta_{ij})$ where $\delta_i = 1, -1$, or 0.

Proof: By Theorem 6′ there exists an orthogonal $\mathbf{T}$ such that $\mathbf{T}'\mathbf{A}\mathbf{T} = (\lambda_i\delta_{ij})$. Let $\mathbf{S}$ be the nonsingular diagonal matrix whose ith diagonal element is $(\lambda_i)^{-1/2}$ if $\lambda_i > 0$, $(-\lambda_i)^{-1/2}$ if $\lambda_i < 0$, 1 if $\lambda_i = 0$. Then $\mathbf{S}'(\lambda_i\delta_{ij})\mathbf{S}$ is of the required form, and equals $\mathbf{P}'\mathbf{A}\mathbf{P}$ with $\mathbf{P} = \mathbf{T}\mathbf{S}$.

Although this proof utilizes the $\{\lambda_i\}$ whose calculation involves the solution of a polynomial equation of degree n, it may be shown by a different proof that a transforming matrix $\mathbf{P}$ giving $\mathbf{P}'\mathbf{A}\mathbf{P}$ the same form may be calculated by using only rational operations (addition, subtraction, multiplication, and division) and square-root extractions on the elements of $\mathbf{A}$. It is evident from Lemma 4 that, no matter what nonsingular $\mathbf{P}$ is used to make the reduction, the total number of $\{\delta_i\}$ equal to 1 or -1 must always be the same (namely, rank $\mathbf{A}$); however, it may also be shown that the number of $\{\delta_i\}$ equal to 1 (and hence the number equal to -1) is also invariant. This is sometimes called the "law of inertia." With the transformation $\mathbf{P}$ used in the above proof of Lemmas 11 and 11′, the $\{\delta_i\}$ obtained will evidently be all 1 if Q and $\mathbf{A}$ are positive definite, all 1 or 0 with the number of 1's equal to rank $\mathbf{A}$ if Q and $\mathbf{A}$ are positive indefinite. But this must be true for *any* nonsingular $\mathbf{P}$, from the law of inertia, or by direct calculations as in the proof of Lemma 10.

Corollary 3: If $Q = \mathbf{x}'\mathbf{A}\mathbf{x}$ is a quadratic form in n variables $x_1, \cdots, x_n$ and Q is of rank r (i.e., rank $\mathbf{A} = r$) then there exist r linear forms in (i.e., r linear combinations of) the n variables $x_1, \cdots, x_n$, say $z_1, \cdots, z_r$, such that $Q = \Sigma_1^r \delta_i z_i^2$, and each $\delta_i = 1$ or -1.

Proof: Employ the nonsingular $\mathbf{P}$ of Lemma 11 to define n linear combinations $y_1, \cdots, y_n$ of the $\{x_j\}$ by $\mathbf{y} = (y_1, \cdots, y_n)' = \mathbf{P}^{-1}\mathbf{x}$, so that $Q = \Sigma_1^n \delta_i y_i^2$, where r of the δ_i are ± 1, the rest 0. Now take the $\{z_i\}$ to be the r of the forms $\{y_i\}$ for which $\delta_i \neq 0$.

We include here another result that we shall need often in the text:

Theorem 7: For any (real) $\mathbf{A}$, the matrix $\mathbf{A}\mathbf{A}'$ is symmetric, positive indefinite, and of the same rank as $\mathbf{A}$.

Proof: (*a*) Symmetry is established by taking the transpose of $\mathbf{A}\mathbf{A}'$ to find $\mathbf{A}\mathbf{A}'$.

(*b*) Let $\mathbf{B} = \mathbf{A}\mathbf{A}'$. By the principal-axis theorem there exists an orthogonal $\mathbf{P}$ such that $\mathbf{P}'\mathbf{B}\mathbf{P} = (\lambda_i\delta_{ij})$ and $\{\lambda_i\}$ are the characteristic roots of **B**. Now $\mathbf{P}'\mathbf{B}\mathbf{P} = \mathbf{P}'\mathbf{A}\mathbf{A}'\mathbf{P} = \mathbf{C}'\mathbf{C}$, where $\mathbf{C} = \mathbf{A}'\mathbf{P}$. Let $\boldsymbol{\gamma}_j$ be the jth column of **C**. Then the i,j element of $\mathbf{C}'\mathbf{C}$ is

$$\boldsymbol{\gamma}_i'\boldsymbol{\gamma}_j = \lambda_i\delta_{ij}. \tag{II.12}$$

In particular

$$\lambda_i = \boldsymbol{\gamma}_i'\boldsymbol{\gamma}_i \tag{II.13}$$

is $\geqq 0$, so **B** is positive indefinite.

(*c*) Since $\mathbf{C} = \mathbf{A}'\mathbf{P}$, Lemma 4 implies that rank **A** = rank **C** = (maximum number of linearly independent $\boldsymbol{\gamma}_j$) = number of nonzero $\boldsymbol{\gamma}_j$, since the $\{\boldsymbol{\gamma}_j\}$ are orthogonal by (II.12). Hence by (II.13), rank **A** = number of nonzero λ_i = rank **B**.

Partitioned Matrices

These are used in this book only at the end of sec. 1.4, so a reader skipping the proofs there may also omit the rest of this appendix. Suppose

TABLE II.1

PARTITIONING A MATRIX

	n_1 columns	n_2 columns			n_ν columns
m_1 rows	$\mathbf{A}_{11}$	$\mathbf{A}_{12}$			$\mathbf{A}_{1\nu}$
m_2 rows	$\mathbf{A}_{21}$	$\mathbf{A}_{22}$			$\mathbf{A}_{2\nu}$
m_μ rows	$\mathbf{A}_{\mu 1}$	$\mathbf{A}_{\mu 2}$			$\mathbf{A}_{\mu\nu}$

that the $m \times n$ matrix $\mathbf{A}$ is divided into $\mu\nu$ submatrices $\mathbf{A}_{ij}$ by means of horizontal and vertical dividing lines as indicated in Table II.1, so that $\mathbf{A}_{ij}$ is $m_i \times n_j$, and $m_i > 0$, $\Sigma_1^\mu m_i = m$, $n_i > 0$, $\Sigma_1^\nu n_i = n$. Suppose further that the $n \times r$ matrix $\mathbf{B}$ is likewise partitioned, with the restriction that the partitioning of the n rows of $\mathbf{B}$ is the same as that of the n columns of $\mathbf{A}$; so the rows are partitioned into sets of $n_1, n_2, \cdots, n_\nu$. Let the columns of $\mathbf{B}$ be partitioned into sets of $r_1, r_2, \cdots, r_\rho$, where $r_k > 0$ and $\Sigma_1^\rho r_k = r$. This partitions $\mathbf{B}$ in $\nu\rho$ submatrices $\mathbf{B}_{jk}$, where $\mathbf{B}_{jk}$ is $n_j \times r_k$. Now form the product $\mathbf{C} = \mathbf{AB}$ and partition its m rows the same as the m rows of $\mathbf{A}$ and its r columns the same as the r columns of $\mathbf{B}$, so that there are $\mu\rho$ submatrices $\mathbf{C}_{ik}$, where $\mathbf{C}_{ik}$ is $m_i \times r_k$. We now have

$$\begin{pmatrix} \mathbf{A}_{11} & \mathbf{A}_{12} & \cdots & \mathbf{A}_{1\nu} \\ \mathbf{A}_{21} & \mathbf{A}_{22} & \cdots & \mathbf{A}_{2\nu} \\ \cdot & \cdot & \cdots & \cdot \\ \mathbf{A}_{\mu 1} & \mathbf{A}_{\mu 2} & \cdots & \mathbf{A}_{\mu\nu} \end{pmatrix}^{m\times n} \begin{pmatrix} \mathbf{B}_{11} & \mathbf{B}_{12} & \cdots & \mathbf{B}_{1r} \\ \mathbf{B}_{21} & \mathbf{B}_{22} & \cdots & \mathbf{B}_{2r} \\ \cdot & \cdot & \cdots & \cdot \\ \mathbf{B}_{\nu 1} & \mathbf{B}_{\nu 2} & \cdots & \mathbf{B}_{\nu r} \end{pmatrix}^{n\times r}$$

$$= \begin{pmatrix} \mathbf{C}_{11} & \mathbf{C}_{12} & \cdots & \mathbf{C}_{1\rho} \\ \mathbf{C}_{21} & \mathbf{C}_{22} & \cdots & \mathbf{C}_{2\rho} \\ \cdot & \cdot & \cdots & \cdot \\ \mathbf{C}_{\mu 1} & \mathbf{C}_{\mu 2} & \cdots & \mathbf{C}_{\mu\rho} \end{pmatrix}^{m\times r}.$$

It may be verified that the submatrices in the product satisfy the relation

$$\text{(II.14)} \qquad \mathbf{C}_{ik} = \sum_j \mathbf{A}_{ij}\mathbf{B}_{jk},$$

as in ordinary matrix multiplication where all the "submatrices" are 1×1. Briefly: we can multiply partitioned matrices as though the submatrices were real numbers, except that we must of course observe the two rules that (1) the partitioning of the rows of the second factor is the same as the partitioning of the columns of the first, and (2) the order of factors must be preserved in the terms on the right of (II.14).

PROBLEMS

II.1. Let

$$\mathbf{A} = \begin{pmatrix} 2 & 0 & -1 & 4 \\ 3 & \frac{1}{3} & 0 & -2 \\ 5 & 0 & 3 & 1 \\ -2 & 4 & 0 & 3 \\ 1 & 0 & 3 & 4 \end{pmatrix}, \qquad \mathbf{B} = \begin{pmatrix} 0 & 1 & 2 & 3 \\ 3 & 0 & -4 & 4 \\ 0 & 1 & 1 & 2 \\ -1 & 3 & -2 & 0 \\ 0 & 0 & 3 & 1 \end{pmatrix},$$

$$\mathbf{C} = \begin{pmatrix} 1 & -3 & 0 & 3 & -2 \\ 3 & 2 & 1 & 2 & -1 \\ 0 & 1 & -1 & 0 & 0 \\ 4 & -2 & 0 & 0 & 5 \end{pmatrix},$$

Compute successively $(3\mathbf{A}+\mathbf{B})$, $(3\mathbf{A}+\mathbf{B})\mathbf{C}$, $\mathbf{C}'(3\mathbf{A}+\mathbf{B})'$, and $\mathbf{C}'(3\mathbf{A}'+\mathbf{B}')$.

II.2. Let

$$\mathbf{D} = \begin{pmatrix} 1 & \frac{1}{2} & -\frac{1}{2} \\ -\frac{1}{2} & 0 & 1 \\ \frac{3}{2} & \frac{1}{2} & \frac{1}{2} \end{pmatrix} \quad \text{and} \quad \mathbf{E} = \begin{pmatrix} -\frac{1}{7} & -\frac{26}{7} & 3 \\ -1 & 2 & 0 \\ \frac{3}{7} & \frac{29}{7} & -2 \end{pmatrix}.$$

Compute $\mathbf{D}^{-1}$ and $\mathbf{E}^{-1}$. Check that $\mathbf{D}\mathbf{D}^{-1} = \mathbf{I}$ and $\mathbf{E}^{-1}\mathbf{E} = \mathbf{I}$. Compute $(\mathbf{ED})^{-1}$ by inversion of $(\mathbf{ED})$. Then check that $(\mathbf{ED})^{-1} = \mathbf{D}^{-1}\mathbf{E}^{-1}$, while $(\mathbf{ED})^{-1} \neq \mathbf{E}^{-1}\mathbf{D}^{-1}$. Also check that $|\mathbf{ED}| = |\mathbf{E}|\cdot|\mathbf{D}|$, and $|\mathbf{E}^{-1}| = |\mathbf{E}|^{-1}$.

II.3. In evaluating a determinant, direct application of the formula following (II.10) becomes tedious when the order n is >4. The following property is then useful: The value of the determinant $|\mathbf{A}|$ is unchanged if to any row of $\mathbf{A}$ is added a linear combination of the other rows of $\mathbf{A}$, or to any column a linear combination of the other columns. By this device one can quickly get all but one element zero in some row or column, and application of the above-mentioned formula then gives a constant times a single determinant of order $n-1$, and the process is then repeated. If there is an element 1 or -1, this may be exploited; if not, we may first try to produce an element ± 1 by the above device. For example, in

$$\begin{vmatrix} 2 & 3 & -4 & 5 \\ 5 & -2 & 3 & 2 \\ 4 & 1 & 3 & -4 \\ 3 & 4 & -4 & 6 \end{vmatrix}$$

we may exploit the 1 in the second column and third row to make the other elements of the column zero by multiplying the third row by -3 and adding to the first row, by 2 and adding to the second row, and by -4 and adding to the fourth row, leaving the third row as it; this gives

$$\begin{vmatrix} -10 & 0 & -13 & 17 \\ 13 & 0 & 9 & -6 \\ 4 & 1 & 3 & -4 \\ -13 & 0 & -16 & 22 \end{vmatrix} = - \begin{vmatrix} -10 & -13 & 17 \\ 13 & 9 & -6 \\ -13 & -16 & 22 \end{vmatrix},$$

where the last step follows from application of the above-mentioned formula to the elements of the second column.

Compute

$$\begin{vmatrix} 4 & 6 & 8 & 1 & 9 \\ -2 & -3 & 4 & 2 & 1 \\ 5 & 4 & 3 & 1 & 2 \\ 6 & 7 & 4 & 3 & 1 \\ 8 & 9 & 4 & 5 & 7 \end{vmatrix}.$$

II.4. Consider the $n\times n$ matrix all of whose diagonal elements are a and all of whose off-diagonal elements are b. Prove that the determinant of this matrix is $[a+(n-1)b](a-b)^{n-1}$. [*Hint:* Subtract the first row from each of the others, next add to the first column the sum of the other columns, then expand by the elements of the first column.]

II.5. Prove that for a symmetric matrix $\mathbf{A}$ the characteristic values of $\mathbf{A}^2$ are the squares of those of $\mathbf{A}$. [*Hint:* Write $\mathbf{A}$ in the form $\mathbf{P\Lambda P'}$, where $\mathbf{\Lambda}$ is a diagonal matrix $(\lambda_i\delta_{ij})$ and $\mathbf{P'P} = \mathbf{I}$, and form $\mathbf{A}^2$.]

II.6. Prove that for a symmetric matrix $\mathbf{A}$, $\mathbf{A} = \mathbf{A}^2$ if and only if every characteristic value of $\mathbf{A}$ is 0 or 1.

II.7. In this book we usually need to know of the principal-axis transformation only that it exists. In the sequence of Problems II.7 to II.10 we discover how it could actually be computed. Let $\mathbf{A}^{n\times n}$ be a symmetric matrix and $\{\lambda_1, \cdots, \lambda_n\}$ its characteristic values. A vector $\boldsymbol{\rho} \neq \mathbf{0}$, satisfying $(\mathbf{A}-\lambda_j\mathbf{I})\boldsymbol{\rho} = \mathbf{0}$ is called a *characteristic vector* of $\mathbf{A}$ corresponding to the characteristic value λ_j. Prove that characteristic vectors corresponding to different characteristic values are orthogonal. [*Hint:* Multiply $(\mathbf{A}-\lambda_j\mathbf{I})\boldsymbol{\rho}_j = \mathbf{0}$ and $(\mathbf{A}-\lambda_{j'}\mathbf{I})\boldsymbol{\rho}_{j'} = \mathbf{0}$ on the left by $\boldsymbol{\rho}_{j'}'$ and $\boldsymbol{\rho}_j'$, respectively, and subtract.]

II.8. Let $\mathbf{A}^{n\times n}$ be symmetric, let $\mathbf{P}^{n\times n}$ be an orthogonal matrix such that $\mathbf{P'AP}$ is a diagonal matrix with diagonal elements $\{\lambda_1, \cdots, \lambda_n\}$, and let $\{\boldsymbol{\rho}_1, \cdots, \boldsymbol{\rho}_n\}$ denote the columns of $\mathbf{P}$. Show that $\boldsymbol{\rho} = \boldsymbol{\rho}_j$ is a characteristic vector of $\mathbf{A}$ corresponding to the characteristic value λ_j. [*Hint:* Consider corresponding columns in $\mathbf{AP} = \mathbf{P\Lambda}$, where $\mathbf{\Lambda}$ is defined in Problem II.5.]

II.9. Let $Q = \mathbf{x'Ax}$, where

$$\mathbf{A} = \begin{pmatrix} -\frac{4}{5} & \sqrt{6} & \frac{2}{\sqrt{30}} \\ \frac{-34}{5\sqrt{6}} & \frac{49}{30} & \frac{\sqrt{5}}{3} \\ \frac{2}{\sqrt{30}} & \frac{2}{\sqrt{5}} & \frac{1}{6} \end{pmatrix}.$$

Find a symmetric matrix $\mathbf{B}$ such that $\mathbf{Q} = \mathbf{x'Bx}$. Find the characteristic roots $\{\lambda_1, \lambda_2, \lambda_3\}$ of $\mathbf{B}$. Calculate a set of characteristic vectors $\{\boldsymbol{\rho}_1, \boldsymbol{\rho}_2, \boldsymbol{\rho}_3\}$, which satisfy $(\mathbf{B}-\lambda_j\mathbf{I})\boldsymbol{\rho}_j = \mathbf{0}$, and normalize them (i.e., make $\boldsymbol{\rho}_j'\boldsymbol{\rho}_j = 1$). For the matrix $\mathbf{P}$ whose columns are $\{\boldsymbol{\rho}_1, \boldsymbol{\rho}_2, \boldsymbol{\rho}_3\}$, check that $\mathbf{P}$ is orthogonal and that $\mathbf{P'BP} = (\lambda_i\delta_{ij})$.

II.10. The roots of a polynomial $f(x) = a_0x^n+a_1x^{n-1}+\cdots+a_n$, or more generally a differentiable function, can be found numerically by succession approximation as follows (Newton–Raphson method): Find two values of x, say x' and x'' (for example, two successive integers) such that $f(x')$ and $f(x'')$ have opposite signs. Obtain a first approximation x_0 to the root (or one of the roots) λ of $f(x)$ between x' and x'' by linear interpolation between the two points $(x', f(x'))$ and $(x'', f(x''))$, or by graphing several points including these two. From x_0 obtain successively $x_1, x_2, x_3, \cdots$ by the recurrence formula

$$x_r = x_{r-1} - [f(x_{r-1})/f'(x_r)],$$

$r = 1, 2, \cdots$, where $f'(x)$ is the derivative of $f(x)$, $f'(x) = na_0x^{n-1}+(n-1)a_1x^{n-2}+\cdots+a_{n-1}$ in the polynomial case. Then x_r converges toward λ as r increases, except perhaps if the initial approximation x_0 was not good enough. This may be shown with the help of a diagram indicating the graph of $f(x)$ near $x = \lambda$, and a series of triangles for $r = 1, 2, \cdots$, the rth triangle having as sides the tangent at x_{r-1}, the ordinate at x_{r-1}, and the x-axis, so that the intercept of the tangent on the x-axis is x_r. Using this method, find to three decimals the characteristic roots of the matrix

$$\mathbf{A} = \begin{pmatrix} 3 & 2 & 1 \\ 2 & -1 & 5 \\ 1 & 5 & 4 \end{pmatrix}.$$

[*Hint:* The roots are located near -4, 2, and 8.]

II.11. In Problem I.3 call the $m \times n$ matrix (a_{ij}) the *coefficient matrix* and the $m \times (n+1)$ matrix obtained by bordering (a_{ij}) on the right with the $\{c_i\}$, the *augmented matrix*. Formulate the conditions in (*b*) and (*c*) of Problem I.3 in terms of the ranks of the coefficient matrix and the augmented matrix.

II.12. Show that matrices of the form

$$\mathbf{Z} = \begin{pmatrix} x & y \\ -y & x \end{pmatrix}$$

behave like complex numbers $z = x+iy$ under addition and multiplication.

II.13. Let us call an $m \times n$ matrix "of type T" if all its diagonal elements are equal and all its off-diagonal elements are equal. Prove that multiplication of matrices of type T is commutative and that products and differences of matrices of type T are of the same type.

II.14. Prove that if a matrix is of type T (Problem II.13) with diagonal elements equal to a and off-diagonal elements equal to b, if $\Delta = (a-b)[a+(n-1)b]$, and $\Delta \neq 0$, then the matrix has an inverse of type T, and derive the inverse. [*Hint:* Assume at first that the inverse is of type T with x in the diagonal and y off. Get two equations for x and y, and solve. Verify directly that the matrix of type T thus determined is actually the inverse.]

II.15. Prove that if a $(2n) \times (2n)$ matrix is of the composite type

$$\mathbf{A} = \begin{pmatrix} \mathbf{U} & \mathbf{V} \\ \mathbf{V} & \mathbf{W} \end{pmatrix},$$

where $\mathbf{U}$, $\mathbf{V}$, $\mathbf{W}$ are of type T (Problem II.13), and if $\mathbf{M} = \mathbf{UW}-\mathbf{V}^2$ is nonsingular, then $\mathbf{A}$ has an inverse and it is

$$\mathbf{A}^{-1} = \begin{pmatrix} \mathbf{X} & \mathbf{Y} \\ \mathbf{Y} & \mathbf{Z} \end{pmatrix},$$

where $\mathbf{X} = \mathbf{W}\mathbf{M}^{-1}$, $\mathbf{Y} = -\mathbf{V}\mathbf{M}^{-1}$, $\mathbf{Z} = \mathbf{U}\mathbf{M}^{-1}$, and all the matrices $\mathbf{M}$, $\mathbf{M}^{-1}$, $\mathbf{X}$, $\mathbf{Y}$, $\mathbf{Z}$ are of type T.

II.16. Prove that if $\mathbf{A}$ and $\mathbf{B}$ are any matrices which are respectively $m \times n$ and $n \times m$, then the set of nonzero characteristic roots is the same for $\mathbf{AB}$ and $\mathbf{BA}$. [*Hint:*[4] Prove the identity

$$\begin{pmatrix} \lambda\mathbf{I}_m - \mathbf{AB} & \mathbf{A} \\ \mathbf{0} & \lambda\mathbf{I}_n \end{pmatrix} \begin{pmatrix} \mathbf{I}_m & \mathbf{0} \\ \mathbf{B} & \mathbf{I}_n \end{pmatrix} = \begin{pmatrix} \mathbf{I}_m & \mathbf{0} \\ \mathbf{B} & \mathbf{I}_n \end{pmatrix} \begin{pmatrix} \lambda\mathbf{I}_m & \mathbf{A} \\ \mathbf{0} & \lambda\mathbf{I}_n - \mathbf{BA} \end{pmatrix}$$

in the composite matrices, where $\mathbf{I}_r$ denotes the identity matrix of order r, and take determinants to get $|\lambda\mathbf{I}_m - \mathbf{AB}|\lambda^n = |\lambda\mathbf{I}_n - \mathbf{BA}|\lambda^m$.]

II.17. Prove that if $\mathbf{A}$ and $\mathbf{B}$ are square matrices and $\mathbf{A}$ is nonsingular then

$$\begin{vmatrix} \mathbf{A} & \mathbf{U} \\ \mathbf{V} & \mathbf{B} \end{vmatrix} = |\mathbf{A}| \cdot |\mathbf{B} - \mathbf{V}\mathbf{A}^{-1}\mathbf{U}|,$$

and if $\mathbf{B}$ is also nonsingular the last determinant may be written

$$|\mathbf{B} - \mathbf{V}\mathbf{A}^{-1}\mathbf{U}| = |\mathbf{B}| \cdot |\mathbf{I} - \mathbf{V}\mathbf{A}^{-1}\mathbf{U}\mathbf{B}^{-1}|.$$

[*Hint:* Prove the identity

$$\begin{pmatrix} \mathbf{I}_m & \mathbf{0} \\ -\mathbf{V}\mathbf{A}^{-1} & \mathbf{I}_n \end{pmatrix} \begin{pmatrix} \mathbf{A} & \mathbf{U} \\ \mathbf{V} & \mathbf{B} \end{pmatrix} \begin{pmatrix} \mathbf{I}_m - \mathbf{A}^{-1} & \mathbf{U} \\ \mathbf{0} & \mathbf{I}_n \end{pmatrix} = \begin{pmatrix} \mathbf{A} & \mathbf{0} \\ \mathbf{0} & \mathbf{B} - \mathbf{V}\mathbf{A}^{-1}\mathbf{U} \end{pmatrix},$$

where $\mathbf{I}_m$ and $\mathbf{I}_n$ are the identity matrices of the orders of $\mathbf{A}$ and $\mathbf{B}$, respectively, and take determinants.]

[4] The proofs suggested in Problems II.16 and II.17 are from p. 802 of "Orthogonal and oblique projectors and the characteristics of pairs of vector spaces" by S. N. Afriat, *Proc. Cambridge Philos. Soc.*, Vol. 53, 1957.

APPENDIX III

Ellipsoids and Their Planes of Support[1]

The analytic geometry of this section will be developed with matrix notation. We shall be dealing with certain sets of points in an n-dimensional Euclidean space. It will be convenient to say "the point $\mathbf{x}$" to mean "the point located by the vector $\mathbf{x}$ drawn from the origin," similarly for "the point $\mathbf{a}$," etc. We define a (solid n-dimensional) *sphere* with center at the point $\mathbf{a}$ and radius r to be the set of all points whose distance from $\mathbf{a}$ is $\leqq r$, in other words the set of points $\mathbf{x}$ satisfying $\|\mathbf{x}-\mathbf{a}\| \leqq r$, or

$$\|\mathbf{x}-\mathbf{a}\|^2 \leqq r^2.$$

The case where $\mathbf{a} = \mathbf{0}$ and $r = 1$, namely

$$\mathbf{x}'\mathbf{x} \leqq 1, \tag{III.1}$$

we shall call the *unit sphere at the origin.*

We shall define an *ellipsoid in canonical position* to be the result of applying to the unit sphere at the origin a uniform stretch along each axis: A set of points has received a *uniform stretch by a factor c_i* ($c_i > 0$) *in the x_i-direction* if each point is displaced on a line parallel to the x_i-axis so that its ith coordinate x_i is multiplied by c_i. The old x_i is then equal to the new x_i divided by c_i; therefore, if the point set to receive the stretch is defined[2] by an equation or inequality, the stretch may be accomplished by replacing x_i by x_i/c_i in the equation or inequality. We thus see that if stretches are made along several axes it does not matter in what order

[1] The planes of support of the ellipsoid are its tangent planes; however, it is the support and not the tangent property that is relevant to our statistical applications in secs. 3.5 and 8.1. When the planes of support do not coincide with the tangent planes, as for example with the convex polyhedron in sec. 3.7, where the tangent planes are those in which the faces lie, we need the planes of support.

[2] In principle, any point set S can be defined by an equation or inequality: Define the function $g_S(\mathbf{x})$ to be 1 for $\mathbf{x} \in S$, 0 otherwise; $g_S(\mathbf{x})$ is called the characteristic function of S. Then S is defined by the equation $g_S(\mathbf{x}) = 1$ or by the inequality $g_S(\mathbf{x}) > 0$.

the stretches are made. A stretch with factor $c_i < 1$ is actually a compression. If the unit sphere at the origin (III.1) is stretched by factors $c_1, c_2, \cdots, c_n$ along the axes, the resulting ellipsoid in canonical position then satisfies

$$\text{(III.2)} \qquad \sum_{i=1}^{n} x_i^2/c_i^2 \leqq 1.$$

The numbers $\{c_i\}$ are called the *semi-axes* of the ellipsoid: If we think of inscribing the sphere (III.1) in the cube defined by $|x_i| \leqq 1$ ($i = 1, \cdots, n$), the sphere and cube will stretch into the ellipsoid (III.2) and the box defined by $|x_i| \leqq c_i$, and the ellipsoid will be inscribed in the box; all this is evident from our geometrical intuition of the stretch, and could also be shown rigorously from the results below about planes of support of the sphere and ellipsoid. The ellipsoid in canonical position is symmetrical in all coordinate planes, since replacing x_i by $-x_i$ does not affect (III.2); we may therefore call the origin its *center.*

We shall define an *ellipsoid* as any point set which may be brought by translation and subsequent orthogonal transformation to coincide with an ellipsoid in canonical position. A *translation* of a set by the vector $\mathbf{a}$ consists in displacing each point $\mathbf{x}$ of the set so that it goes into $\mathbf{x}+\mathbf{a}$. This means that a set defined by an equation or inequality can be translated by the vector $\mathbf{a}$ by substituting $\mathbf{x}-\mathbf{a}$ for $\mathbf{x}$ or x_i-a_i for x_i, where $\mathbf{a} = (a_1, \cdots, a_n)'$. We recall that the geometric meaning of an orthogonal transformation is a rotation plus possibly some reflections in coordinate planes. The *center* of the ellipsoid just defined is the point into which the center of the ellipsoid in canonical position goes.

The main result we shall want is that, if $\mathbf{M}$ is a symmetric positive definite matrix, the inequality

$$\text{(III.3)} \qquad (\mathbf{x}-\mathbf{a})' \mathbf{M}(\mathbf{x}-\mathbf{a}) \leqq 1$$

defines an ellipsoid with center at $\mathbf{a}$. First we translate the set defined by (III.3) by the vector $-\mathbf{a}$, by substituting $\mathbf{x}+\mathbf{a}$ for $\mathbf{x}$, to get

$$\text{(III.4)} \qquad \mathbf{x}'\mathbf{M}\mathbf{x} \leqq 1.$$

From Theorem 6 of App. II we know that there exists an orthogonal transformation which reduces (III.4) to the form $\Sigma_1^n \lambda_i x_i^2 \leqq 1$, where the $\{\lambda_i\}$ are the characteristic values of $\mathbf{M}$ (we have denoted the coordinates of points in the transformed set also by $\{x_i\}$ instead of $\{y_i\}$ as in Theorem 6). But this is of the form (III.2), which defines an ellipsoid in canonical position, with semi-axes $\{c_i = \lambda_i^{-1/2}\}$; the $\{\lambda_i\}$ are positive because we assumed $\mathbf{M}$ to be positive definite. Thus, (III.3) defines an ellipsoid with center at $\mathbf{a}$.

We shall also need some results about the planes of support of the ellipsoid (III.3). For any vector $\mathbf{h} \neq \mathbf{0}$, we define the *plane through* $\mathbf{0}$ *orthogonal to* $\mathbf{h}$ to be the set of points whose location vectors $\mathbf{x}$ drawn from $\mathbf{0}$ are orthogonal to $\mathbf{h}$. These vectors $\mathbf{x}$ constitute an $(n-1)$-dimensional vector space. Points $\mathbf{x}$ are on the plane if and only if $\mathbf{h}'\mathbf{x} = 0$. We define a *plane* in general to be any set of points that can be brought by a translation to become a plane through $\mathbf{0}$. Hence if $\mathbf{h} \neq \mathbf{0}$

$$\text{(III.5)} \qquad \mathbf{h}'(\mathbf{x}-\mathbf{x}_0) = 0$$

defines a plane through the point $\mathbf{x}_0$ and orthogonal to $\mathbf{h}$. Under translation by any vector $\mathbf{a}$, or linear transformation with any nonsingular matrix $\mathbf{P}^{n\times n}$, a plane goes into another plane: To translate (III.5) by the vector $\mathbf{a}$ we substitute $\mathbf{x}-\mathbf{a}$ for $\mathbf{x}$ and get another equation of the form (III.5) with $\mathbf{x}_0$ replaced by $\mathbf{x}_0+\mathbf{a}$. A linear transformation with the nonsingular matrix $\mathbf{P}$ may be achieved by replacing $\mathbf{x}$ by $\mathbf{P}^{-1}\mathbf{x}$; this again gives an equation of the form (III.5) with $\mathbf{h}'$ and $\mathbf{x}_0$ replaced by $\mathbf{h}'\mathbf{P}^{-1}$ and $\mathbf{P}\mathbf{x}_0$, respectively.

The plane through $\mathbf{x}_0$ orthogonal to $\mathbf{h} \neq \mathbf{0}$ divides the n-dimensional space of points $\mathbf{x}$ into three parts according as the linear function

$$f(\mathbf{x}) = \mathbf{h}'(\mathbf{x}-\mathbf{x}_0)$$

is 0, >0, or <0. We shall say that two points $\mathbf{x}_{(1)}$ and $\mathbf{x}_{(2)}$ are *on the same side of the plane* if

$$\text{(III.6)} \qquad f(\mathbf{x}_{(1)})f(\mathbf{x}_{(2)}) \geqq 0;$$

in particular this will be true if $\mathbf{x}_{(1)}$ or $\mathbf{x}_{(2)}$ is on the plane. Suppose we subject the plane and the points $\mathbf{x}_{(1)}$ and $\mathbf{x}_{(2)}$ to a linear transformation with nonsingular matrix $\mathbf{P}$. The equation of the transformed plane is then $\tilde{f}(\mathbf{x}) = 0$, where $\tilde{f}(\mathbf{x}) = \mathbf{h}'(\mathbf{P}^{-1}\mathbf{x}-\mathbf{x}_0)$, and the transformed points are $\mathbf{P}\mathbf{x}_{(1)}$ and $\mathbf{P}\mathbf{x}_{(2)}$, respectively. We see that $\tilde{f}(\mathbf{P}\mathbf{x}_{(1)})\tilde{f}(\mathbf{P}\mathbf{x}_{(2)}) = f(\mathbf{x}_{(1)})f(\mathbf{x}_{(2)})$, from which we conclude that the relationship of two points being on the same side of a plane is unaltered by nonsingular linear transformation. Similarly we can show it is invariant under translation.

We may define a *plane of support to the ellipsoid* (III.3) as a plane that has at least one point in common[3] with the ellipsoid and such that the ellipsoid is entirely on one side of the plane. The sphere (III.1) is a special case of (III.3) for which it is easy to treat the planes of support. Our intuition suggests that through every point $\mathbf{x}_0$ on the surface of the sphere (i.e., through every point $\mathbf{x}_0$ for which $\mathbf{x}_0'\mathbf{x}_0 = 1$) there exists a plane of support, namely the plane through $\mathbf{x}_0$ orthogonal to the vector

[3] We would have to complicate this a little more had we not made the ellipsoid *closed*, for example if our definitions had led to (III.3) with "$\leqq$" replaced by "$<$."

from $\mathbf{0}$ to $\mathbf{x}_0$ (i.e., orthogonal to the vector $\mathbf{x}_0$). The equation of this plane is $\mathbf{x}_0'(\mathbf{x}-\mathbf{x}_0) = 0$, or

$$\mathbf{x}_0'\mathbf{x} = 1. \tag{III.7}$$

To prove formally that (III.7) is a plane of support we note first that it has the point $\mathbf{x}_0$ in common with the sphere. Second, to show that any two points $\mathbf{x}_{(1)}$ and $\mathbf{x}_{(2)}$ in the sphere are on the same side of the plane, take $f(\mathbf{x}) = \mathbf{x}_0'\mathbf{x}-1$. For $i = 1, 2$, the absolute value of $\mathbf{x}_0'\mathbf{x}_{(i)}$, the length of the projection of $\mathbf{x}_{(i)}$ on $\mathbf{x}_0$, must be $\leqq \|\mathbf{x}_{(i)}\| \leqq 1$, so $f(\mathbf{x}_{(i)}) \leqq 0$, and hence (III.6) is satisfied.

Replacing $\mathbf{x}_0$ by $-\mathbf{x}_0$ in (III.7) we see that $-\mathbf{x}_0'\mathbf{x} = 1$ is a plane of support to the sphere through the point $-\mathbf{x}_0$, and hence parallel planes of support at $\mathbf{x}_0$ and $-\mathbf{x}_0$ are given by

$$\mathbf{x}_0'\mathbf{x} = \pm 1, \tag{III.8}$$

one plane for $+1$, the other for -1.

Suppose we transform the sphere (III.1) into the ellipsoid (III.4) by first stretching it to (III.2), and then making an orthogonal transformation to (III.4). The stretch is accomplished by a substitution $\mathbf{Cx}$ for $\mathbf{x}$ in (III.1) where $\mathbf{C}^{n\times n} = (\delta_{ij}/c_i)$ is nonsingular; we then substitute $\mathbf{P}^{-1}\mathbf{x}$ for $\mathbf{x}$ in (III.2) where $\mathbf{P}$ is orthogonal. But this is the same as substituting $\mathbf{Qx}$ for $\mathbf{x}$ where $\mathbf{Q} = \mathbf{CP}^{-1}$, and hence $\mathbf{Q}$ is nonsingular. We have already remarked that under nonsingular linear transformation a plane goes into a plane, and hence a point common to the sphere and to a plane of support of the sphere goes into a point common to their transforms. Also, by a remark above, since the sphere is on one side of its plane of support, the same is true of their transforms. It follows that the transform of a plane of support of the sphere (III.1) is a plane of support of the ellipsoid (III.4). By a similar argument concerning a translation by the vector $\mathbf{a}$, it follows that a plane of support of the ellipsoid (III.4) goes into a plane of support of the ellipsoid (III.3) under the translation.

Under the substitution of $\mathbf{Qx}$ for $\mathbf{x}$, $\mathbf{x}'\mathbf{x}$ goes into $\mathbf{x}'\mathbf{Q}'\mathbf{Qx} = \mathbf{x}'\mathbf{Mx}$; hence $\mathbf{Q}'\mathbf{Q} = \mathbf{M}$. If we substitute $\mathbf{Qx}$ for $\mathbf{x}$ in (III.8) we get

$$\mathbf{x}_0'\mathbf{Qx} = \pm 1 \tag{III.9}$$

as the equations of parallel planes of support of the ellipsoid (III.4), a pair for every $\mathbf{x}_0$ with $\|\mathbf{x}_0\| = 1$. We will now show that for every vector $\mathbf{h} \neq \mathbf{0}$ there exist two planes of support (III.9) orthogonal to $\mathbf{h}$, and we will find their equations. The planes (III.9) are evidently orthogonal to $(\mathbf{x}_0'\mathbf{Q})'$ and hence will be orthogonal to $\mathbf{h}$ if and only if $\mathbf{x}_0'\mathbf{Q} = c\mathbf{h}'$, where c is a scalar, i.e., if and only if $\mathbf{x}_0' = c\mathbf{h}'\mathbf{Q}^{-1}$. But $\mathbf{x}_0'\mathbf{x}_0 = 1$, hence c must satisfy $c\mathbf{h}'\mathbf{Q}^{-1}\mathbf{Q}'^{-1}\mathbf{h}c = 1$, or $c^2\mathbf{h}'\mathbf{M}^{-1}\mathbf{h} = 1$. Hence the planes (III.9) will be orthogonal to $\mathbf{h}$ if and only if $\mathbf{x}_0'\mathbf{Q} = \pm(\mathbf{h}'\mathbf{M}^{-1}\mathbf{h})^{-1/2}\mathbf{h}'$. Substituting this in (III.9) we get $\mathbf{h}'\mathbf{x} = \pm(\mathbf{h}'\mathbf{M}^{-1}\mathbf{h})^{1/2}$, and, translating by the vector $\mathbf{a}$, we get

$$\text{(III.10)} \qquad \mathbf{h}'(\mathbf{x}-\mathbf{a}) = \pm(\mathbf{h}'\mathbf{M}^{-1}\mathbf{h})^{1/2}$$

as the equations of the two planes of support orthogonal to $\mathbf{h}$.

Finally, we shall need the inequality (III.11) below which defines the strip between the two planes (III.10): Perhaps (III.11) is geometrically obvious to the reader (he might for instance note that the inequality $|\mathbf{x}_0'\mathbf{x}| \leqq 1$ says that the projection of $\mathbf{x}$ on $\mathbf{x}_0$ is of length $\leqq 1$, and interpret (III.11) as its transform); if not, he may be interested in the following formal derivation. We define the set of points *between* the planes (III.10) to be the set of those points which are on the same side of both planes as the ellipsoid. This is the same as the set of points $\mathbf{x}$ which are on the same side of both planes as the center $\mathbf{a}$ of the ellipsoid. Using the condition (III.6) to determine the latter set of points, we let

$$f_\pm(\mathbf{x}) = \mathbf{h}'(\mathbf{x}-\mathbf{a}) \pm c_h,$$

where $c_h = -(\mathbf{h}'\mathbf{M}^{-1}\mathbf{h})^{1/2}$, the $\pm$ signs having the same value as in (III.10). For either plane, both $\mathbf{x}$ and $\mathbf{a}$ will be on the same side if and only if

$$f_\pm(\mathbf{x})f_\pm(\mathbf{a}) \geqq 0,$$

or

$$\pm c_h\mathbf{h}'(\mathbf{x}-\mathbf{a}) + c_h^2 \geqq 0,$$

or

$$\pm\mathbf{h}'(\mathbf{x}-\mathbf{a}) \leqq -c_h.$$

The two conditions (for + and −) will both be satisfied if and only if

$$c_h \leqq \mathbf{h}'(\mathbf{x}-\mathbf{a}) \leqq -c_h,$$

or

$$\text{(III.11)} \qquad |\mathbf{h}'(\mathbf{x}-\mathbf{a})| \leqq (\mathbf{h}'\mathbf{M}^{-1}\mathbf{h})^{1/2}.$$

This is the desired condition defining the strip between the two planes of support orthogonal to $\mathbf{h}$.

PROBLEM

III.1. Consider the ellipsoid (III.3) determined by its center $\mathbf{a} = (3, -1, 2, 0)'$ and by the symmetric positive definite matrix

$$\mathbf{M} = \frac{1}{36}\begin{pmatrix} 12 & -2 & 2 & -2 \\ -2 & 7 & 1 & 0 \\ 2 & 1 & 15 & -4 \\ -2 & 0 & -4 & 8 \end{pmatrix}.$$

(*a*) Find the equations of the planes of support orthogonal to the vector $\mathbf{h} = (-1, 0, 3, 1)'$. (*b*) What is the distance between those two planes? (*c*) Does the point $\mathbf{x} = (14, -25, 7, 12)'$ lie between them? [*Hint:* For computing $\mathbf{M}^{-1}$: $\mathbf{M} = \mathbf{P}'\mathbf{\Lambda P}$ where

$$\mathbf{P} = \frac{1}{\sqrt{6}}\begin{pmatrix} 1 & 2 & 0 & 1 \\ -2 & 1 & 1 & 0 \\ 0 & -1 & 1 & 2 \\ 1 & 0 & 2 & -1 \end{pmatrix}, \qquad \mathbf{\Lambda} = \frac{1}{6}\begin{pmatrix} 1 & 0 & 0 & 0 \\ 0 & 2 & 0 & 0 \\ 0 & 0 & 1 & 0 \\ 0 & 0 & 0 & 3 \end{pmatrix},$$

and $\mathbf{P}'\mathbf{P} = \mathbf{I}$. One could also evaluate the quadratic form $\mathbf{h}'\mathbf{M}^{-1}\mathbf{h}$ without calculating $\mathbf{M}^{-1}$ by use of (V.2).]

APPENDIX IV

Noncentral χ^2, F, and t

Noncentral Chi-Square

If a random variable x has a normal distribution with mean ξ and variance σ^2 we denote this by writing "x is $N(\xi, \sigma^2)$."

Definition 1: If $x_1, x_2, \cdots, x_\nu$ are independently distributed and x_i is $N(\xi_i, 1)$ then the random variable $U = \Sigma_1^\nu x_i^2$ is called a *noncentral chi-square variable with* ν d.f. (*degrees of freedom*). We call $\delta = (\Sigma_1^\nu \xi_i^2)^{1/2}$ the *noncentrality parameter* of the distribution.

Remarks: We note that the ordinary or *central* chi-square distribution is the special case of the noncentral distribution when the noncentrality parameter $\delta = 0$. (We are thus involved in the somewhat unfortunate terminology that if we say "chi-square variable" without specifying it to be central it is understood to be central, but if we say "noncentral chi-square variable" it may be central.) The same goes for the F- and t-distributions. In the literature the noncentrality parameter is usually defined differently, some authors using $\lambda = \delta^2$, while others use $\lambda = \frac{1}{2}\delta^2$, both using the same symbol λ.

Notation: We use the symbol $\chi'^2_{\nu,\delta}$ for a noncentral chi-square variable with ν d.f. and noncentrality parameter δ, and χ^2_ν for $\chi'^2_{\nu,0}$. The notation $\chi'^2_{\nu,\delta}$ is justified by the fact that the distribution depends on $\xi_1, \xi_2, \cdots, \xi_\nu$ only through the function δ. Geometrically, this is quite obvious: The cumulative distribution function of U is given by $F(u) = \Pr\{U \leq u\}$. This is the probability that the point $P = (x_1, \cdots, x_\nu)$ does not fall outside the ν-dimensional sphere with center at the origin and radius $u^{1/2}$. The distribution of $(x_1, \cdots, x_\nu)$ is spherically symmetric about the point $Q = (\xi_1, \cdots, \xi_\nu)$. Hence $F(u)$ will not be modified when the point Q moves on the sphere of constant radius δ centered at the origin.

Analytically, one would reach the same conclusion by considering an othogonal transformation to new independent variables $y_1, \cdots, y_\nu$ made in such a way that the point Q comes to lie on the y_1-axis. Then

$U = y_1^2 + \Sigma_2^\nu y_i^2$, where y_1 is $N(\delta, 1)$, while y_i is $N(0, 1)$ for $i = 2, \cdots, \nu$, and $y_1, y_2, \cdots, y_\nu$ are independent. Hence

$$U = \chi'^2_{\nu,\delta} = (v + \delta)^2 + \chi^2_{\nu-1}, \tag{IV.1}$$

where v is independent of $\chi^2_{\nu-1}$ and is $N(0, 1)$. From this relation, the probability density of $\chi'^2_{\nu,\delta}$ could be derived by well-known methods. This density does not have any simple closed form.[1]

Mean and Variance of $\chi'^2_{\nu,\delta}$

We shall employ the following elementary formulas: If x is $N(0, 1)$,

$$\text{Var}(x) = E(x^2) = 1,$$

$$\text{Var}(x^2) = E(x^4) - [E(x^2)]^2 = 3 - 1 = 2,$$

$$\text{Cov}(x, x^2) = E\{[x - E(x)][x^2 - E(x^2)]\} = E[x(x^2 - 1)] = E(x^3) - E(x) = 0.$$

In obtaining (IV.1) we had

$$\chi'^2_{\nu,\delta} = (v+\delta)^2 + \sum_{i=2}^{\nu} y_i^2,$$

where $v, y_2, \cdots, y_\nu$ are all independently normal with zero mean and unit variance. Hence

$$E(\chi'^2_{\nu,\delta}) = E\left(v^2 + 2\delta v + \delta^2 + \sum_{i=2}^{\nu} y_i^2\right) = (1+\delta^2) + (\nu - 1),$$

$$E(\chi'^2_{\nu,\delta}) = \nu + \delta^2, \tag{IV.2}$$

and

$$\text{Var}(\chi'^2_{\nu,\delta}) = \text{Var}[(v+\delta)^2] + \sum_{i=2}^{\nu} \text{Var}(y_i^2) = \text{Var}(v^2 + 2\delta v) + 2(\nu - 1).$$

But

$$\text{Var}(v^2 + 2\delta v) = \text{Var}(v^2) + \text{Var}(2\delta v) + 2\,\text{Cov}(v^2, 2\delta v) = 2 + (2\delta)^2 + 0,$$

so that finally

$$\text{Var}(\chi'^2_{\nu,\delta}) = 2\nu + 4\delta^2. \tag{IV.3}$$

Addition of Noncentral Chi-Square Variables

Consider independent random variables U_1 and U_2 with noncentral chi-square distributions,

$$U_1 \text{ is } \chi'^2_{\nu_1,\delta_1}, \qquad U_2 \text{ is } \chi'^2_{\nu_2,\delta_2}.$$

[1] See, e.g., Patnaik (1949).

Then it follows immediately from Definition 1 that U_1+U_2 has again a noncentral chi-square distribution,

(IV.4) $$U_1 + U_2 \text{ is } \chi'^2_{\nu,\delta}, \text{ with } \nu = \nu_1 + \nu_2,\ \delta = (\delta_1^2 + \delta_2^2)^{1/2}.$$

Noncentral F

Definition 2: If U_1 and U_2 are independent random variables and U_1 is $\chi'^2_{\nu_1,\delta}$, U_2 is $\chi^2_{\nu_2}$, the distribution of the quotient $(U_1/\nu_1)/(U_2/\nu_2)$ is called a *noncentral F-distribution with ν_1 and ν_2 d.f. and noncentrality parameter δ.*

Notation: We write $F'_{\nu_1,\nu_2;\delta}$ for a noncentral F-variable with ν_1 and ν_2 d.f. and noncentrality parameter δ.

The probability density of noncentral F can be derived by standard methods, but like that of noncentral χ^2 it does not have any simple closed form.[2]

Noncentral t

Definition 3: If x and U are independent random variables and x is $N(\delta, 1)$, U is χ^2_ν, the distribution of the quotient $x/(U/\nu)^{1/2}$ is called the *noncentral t-distribution with ν d.f. and noncentrality parameter δ.*

Notation: We denote by $t'_{\nu,\delta}$ a noncentral t-variable with ν d.f. and noncentrality parameter δ.

Evidently

$$t'^2_{\nu,\delta} = F'_{1,\nu;\delta}.$$

Tables and charts of the distribution of noncentral χ^2, F, and t are discussed in sec. 2.8; tables of the distribution and per cent points of central χ^2, F, and t in sec. 2.2.

Approximation[3] to Noncentral χ^2 and F

It is sometimes useful to approximate χ'^2 by χ^2, or F' by F. A possible way is to approximate $\chi'^2_{\nu,\delta}$ as $c\chi^2_{\tilde{\nu}}$, where the constants c and $\tilde{\nu}$ are determined by equating the means and variances of the two distributions,

(IV.5) $$c\tilde{\nu} = \nu + \delta^2, \qquad c^2\tilde{\nu} = \nu + 2\delta^2,$$

from (IV.2) and (IV.3). The number $\tilde{\nu}$ of d.f. is then generally no longer an integer. The approximation of $\chi'^2_{\nu_1,\delta}$ as $c\chi^2_{\tilde{\nu}_1}$ immediately gives an approximation of $F'_{\nu_1,\nu_2;\delta}$ as $c\nu_1^{-1}\tilde{\nu}_1 F_{\tilde{\nu}_1,\nu_2}$, where c and $\tilde{\nu}_1$ are determined by (IV.5) with ν and $\tilde{\nu}$ replaced by ν_1 and $\tilde{\nu}_1$. The method can be extended

[2] See, e.g., Patnaik (1949).

[3] These approximations were given by Patnaik (1949). Numerical evidence on the adequacy of the approximation to noncentral χ^2 at its upper 0.05 point may be found in Tukey (1957*b*).

in an obvious way to the approximation of a linear combination of independent noncentral chi-square variables with positive coefficients.

PROBLEMS

IV.1. Occasionally (e.g., in sec. 4.8) one encounters a doubly noncentral F-variable, which may be denoted by $F''_{\nu_1,\nu_2;\delta_1,\delta_2}$, and whose distribution is that of the ratio of two independent noncentral χ^2-variables with respective numbers of d.f. ν_1 and ν_2, and noncentrality parameters δ_1 and δ_2. By using (IV.5) show how to approximate the doubly noncentral F-variable by a central F-variable, i.e., $F''_{\nu_1,\nu_2;\delta_1,\delta_2}$ by $cF_{\tilde{\nu}_1,\tilde{\nu}_2}$, and determine the constant c and numbers of d.f. $\tilde{\nu}_1$, $\tilde{\nu}_2$ in terms of ν_1, ν_2, δ_1, δ_2.

IV.2. Prove that, for large ν, χ^2_ν is $N(\nu, 2\nu)$. [*Hint:* Central-limit theorem.]

IV.3. If for a sequence $\{X_\nu\}$ of random variables ($\nu = 1, 2, \cdots$) there exists a constant θ such that for every $\varepsilon > 0$, $\Pr\{|X_\nu - \theta| < \varepsilon\} \to 1$ as $\nu \to \infty$, X_ν is said to converge in probability to θ. (*a*) Prove that a sufficient condition for this is that $E(X_\nu) \to \theta$, $\text{Var}(X_\nu) \to 0$. [*Hint:* Use the Tchebycheff inequality.] (*b*) Hence show that χ^2_ν/ν converges in probability to 1. (*c*) Prove (*b*) from the result of Problem IV.2.

IV.4. Derive the following approximation for the cumulative distribution function of noncentral t in terms of that of $N(0, 1)$:

$$\Pr\{t'_{\nu,\delta} \leq x\} = \Pr\left\{z \leq (x - \delta)\left(1 + \frac{x^2}{2\nu}\right)^{-1/2}\right\},$$

where z is $N(0, 1)$. [*Hint:* Suppose v and s^2 are independent, v is $N(0, 1)$, and s^2 is χ^2_ν/ν. Then s is approximately $N(1, (2\nu)^{-1})$, and $(v+\delta)/s \leq x$ if and only if $v - xs + \delta \leq 0$.]

APPENDIX V

The Multivariate Normal Distribution

A multivariate normal distribution in m variables might be defined as the joint distribution of m linear (nonhomogeneous) functions of $x_1, x_2, \cdots, x_n$, where the $\{x_i\}$ are independently $N(0, 1)$, that is, the joint distribution of $y_1, y_2, \cdots, y_m$, where $\mathbf{y} = \mathbf{A}\mathbf{x}+\boldsymbol{\eta}$, $\mathbf{x} = (x_1, \cdots, x_n)'$, $\mathbf{y} = (y_1, \cdots, y_m)'$, $\mathbf{A}^{m\times n}$ is a constant matrix, and $\boldsymbol{\eta}^{m\times 1}$ is a constant vector. We will also say the random vector $\mathbf{y}$ has a multivariate normal distribution. It follows from the end of sec. 1.2, since $\mathbf{x}$ has the mean $E(\mathbf{x}) = \mathbf{0}$, and the covariance matrix $\boldsymbol{\Sigma}_x = \mathbf{I}$, that the mean and covariance matrix of $\mathbf{y}$ are $E(\mathbf{y}) = \boldsymbol{\eta}$ and $\boldsymbol{\Sigma}_y = \mathbf{A}\mathbf{A}'$.

It may be shown[1] that the multivariate normal distribution of $\mathbf{y}$ is completely determined by the first two moments $E(\mathbf{y})$ and $\boldsymbol{\Sigma}_y$. That $\boldsymbol{\Sigma}_y$ is a symmetric positive indefinite matrix follows from Theorem 7 of App. II; however, this is true for *any* covariance matrix. It may also be shown that if $\boldsymbol{\Sigma}_y$ is nonsingular then $y_1, \cdots, y_m$ have the joint probability density $(2\pi)^{-m/2}|\boldsymbol{\Sigma}_y|^{-1/2}e^{-Q/2}$, where Q is the quadratic form

$$(\mathbf{y}-\boldsymbol{\eta})'\boldsymbol{\Sigma}_y^{-1}(\mathbf{y}-\boldsymbol{\eta}). \tag{V.1}$$

From this it follows that if $y_1, \cdots, y_m$ have a multivariate normal distribution and $\boldsymbol{\Sigma}_y$ is nonsingular then $y_1, \cdots, y_m$ are statistically independent if and only if $\boldsymbol{\Sigma}_y$ is diagonal, since in that case and only in that case will the joint density factor into the product of the marginal densities. If $\boldsymbol{\Sigma}_y$ is singular $\mathbf{y}$ is said to have a *degenerate* distribution, and if rank $\boldsymbol{\Sigma}_y = t$ the total "mass" of the distribution lies in a t-dimensional subspace of the m-dimensional space of $\mathbf{y}$ but not in any lower-dimensional subspace (t = rank $\mathbf{A}$ in the above representation).

We use the notation "$\mathbf{y}^{m\times 1}$ is $N(\boldsymbol{\eta}, \boldsymbol{\Sigma}_y)$" to mean that $\mathbf{y}$ is multivariate

[1] For proofs of the statements in this paragraph the reader may refer to Cramér's book (1946), Chs. 22, 24.

normal with the indicated moments. With this notation we would say that the above $\mathbf{x}^{n\times 1}$ is $N(\mathbf{0}, \mathbf{I})$. It follows from the above definition of the multivariate normal distribution that under any linear transformation a multivariate normal $\mathbf{y}$ remains multivariate normal: More precisely, if $\mathbf{y}^{m\times 1}$ is $N(\boldsymbol{\eta}, \boldsymbol{\Sigma}_y)$ and $\mathbf{w}^{q\times 1} = \mathbf{B}\mathbf{y}$, where $\mathbf{B}^{q\times m}$ is a constant matrix, then $\mathbf{w}$ is $N(\mathbf{B}\boldsymbol{\eta}, \mathbf{B}\boldsymbol{\Sigma}_y\mathbf{B}')$.

Theorem: If $\mathbf{y}^{m\times 1}$ is $N(\boldsymbol{\eta}, \boldsymbol{\Sigma}_y)$ and $\boldsymbol{\Sigma}_y$ is nonsingular then the quadratic form (V.1) has the chi-square distribution with m d.f.

Proof: By the discussion following Lemma 11 of App. II there exists a nonsingular $m\times m$ matrix $\mathbf{P}$ such that $\mathbf{P}\boldsymbol{\Sigma}_y\mathbf{P}' = \mathbf{I}$, hence $\boldsymbol{\Sigma}_y = (\mathbf{P}'\mathbf{P})^{-1}$. Define the random vector $\mathbf{w}^{m\times 1} = \mathbf{P}(\mathbf{y}-\boldsymbol{\eta})$. Then $\mathbf{w}$ is $N(\mathbf{0}, \boldsymbol{\Sigma}_w)$ with $\boldsymbol{\Sigma}_w = \mathbf{P}\boldsymbol{\Sigma}_y\mathbf{P}' = \mathbf{I}$. Therefore $\mathbf{w}'\mathbf{w} = \Sigma_1^m w_i^2$ is χ^2_m by App. IV. But

$$\mathbf{w}'\mathbf{w} = (\mathbf{y}-\boldsymbol{\eta})'\,\mathbf{P}'\mathbf{P}(\mathbf{y}-\boldsymbol{\eta}) = (\mathbf{y}-\boldsymbol{\eta})'\boldsymbol{\Sigma}_y^{-1}(\mathbf{y}-\boldsymbol{\eta}).$$

Remark: In numerical computations of the value of a quadratic form whose matrix is the inverse of a given matrix, one can avoid actually inverting the matrix by calculating merely two determinants and using the relation[2]

$$\text{(V.2)} \qquad \mathbf{z}'\mathbf{A}^{-1}\mathbf{z} = \frac{|\mathbf{A}+\mathbf{z}\mathbf{z}'|}{|\mathbf{A}|} - 1.$$

This may be useful in applications of the above theorem.

Hotelling's T^2

Suppose we have a random sample of J from an m-variate normal population, i.e., the J vectors $\mathbf{x}^{(j)} = (x_{1j}, \cdots, x_{mj})'$, $j = 1, \cdots, J$, are independently distributed like $\mathbf{x} = (x_1, \cdots, x_m)'$, where $\mathbf{x}$ is $N(\boldsymbol{\xi}, \boldsymbol{\Sigma})$. Then the vector of sample means $\bar{\mathbf{x}} = \Sigma_j \mathbf{x}^{(j)}/J$, whose ith component is $\bar{x}_i = \Sigma_j x_{ij}/J$, is $N(\boldsymbol{\xi}, J^{-1}\boldsymbol{\Sigma})$. An unbiased estimate of the (i, i') element of $\boldsymbol{\Sigma}$, that is, of Cov $(x_i, x_{i'})$, is the sample covariance

$$\text{(V.3)} \qquad s_{ii'} = (J-1)^{-1}\sum_j (x_{ij}-\bar{x}_i)(x_{i'j}-\bar{x}_{i'}),$$

and so an unbiased estimate of $\boldsymbol{\Sigma}$ is the matrix $\mathbf{S}$ whose (i, i') element is (V.3). It may be shown that $\mathbf{S}$ is statistically independent of $\bar{\mathbf{x}}$, a fact familiar to the reader in the case $m = 1$.

From the above theorem we know that, if $\boldsymbol{\Sigma}$ is nonsingular,

$$\text{(V.4)} \qquad J(\bar{\mathbf{x}}-\boldsymbol{\xi})'\boldsymbol{\Sigma}^{-1}(\bar{\mathbf{x}}-\boldsymbol{\xi})$$

is a chi-square variable with m d.f. The random variable obtained by

[2] Wilks (1932), pp. 487–488: equivalence of his (36) and equation below (40) gives our (V.2).

replacing $\boldsymbol{\Sigma}$ in (V.4) by the estimate $\mathbf{S}$ is called *Hotelling's* T^2. It may be shown[3] that T^2 is distributed like a constant times an F-variable with m and $J-m$ d.f.,

$$T^2 = J(\bar{\mathbf{x}}-\boldsymbol{\xi})' \mathbf{S}^{-1}(\bar{\mathbf{x}}-\boldsymbol{\xi}) \quad \text{is} \quad \frac{(J-1)m}{J-m} F_{m,J-m}. \tag{V.5}$$

In applications of (V.5) the above *Remark* may be useful.

If we use the T^2-statistic to test the hypothesis H: $\boldsymbol{\xi} = \mathbf{0}$, then to calculate the power of the test we need the distribution of $J\bar{\mathbf{x}}' \mathbf{S}^{-1}\bar{\mathbf{x}}$ under the alternatives to H. It may be shown[4] that if $\mathbf{x}$ is $N(\boldsymbol{\xi}, \boldsymbol{\Sigma})$ then

$$J\bar{\mathbf{x}}' \mathbf{S}^{-1}\bar{\mathbf{x}} \quad \text{is} \quad \frac{(J-1)m}{J-m} F'_{m,J-m;\delta}, \tag{V.6}$$

where the noncentrality parameter δ of the noncentral F-distribution (App. IV) is given by

$$\delta^2 = J\boldsymbol{\xi}'\boldsymbol{\Sigma}^{-1}\boldsymbol{\xi}. \tag{V.7}$$

Again, in the case $m = 1$, where Hotelling's T^2 becomes Student's t^2, the reader is familiar with (V.5) and can easily verify (V.6) and (V.7).

PROBLEMS

V.1. If $\{x_1, \cdots, x_n\}$ are independently normal with equal variance and $\{y_1, \cdots, y_n\}$ are linear functions of the $\{x_j\}$, prove that the $\{y_i\}$ are independent with equal variance if and only if the matrix of the transformation is a scalar times an orthogonal matrix.

V.2. Prove that, if Q is a quadratic form in random variables whose joint distribution is nondegenerate multivariate normal, then Q is distributed as a linear combination of independent noncentral χ^2-variables. [*Hint:* Let $Q = \mathbf{y}'\mathbf{A}\mathbf{y}$, where $\mathbf{y}$ is $N(\boldsymbol{\eta}, \boldsymbol{\Sigma}_y)$. There exist nonsingular $\mathbf{P}$ and orthogonal $\mathbf{T}$ such that $\mathbf{P}\boldsymbol{\Sigma}_y\mathbf{P}' = \mathbf{I}$ and $\mathbf{T}(\mathbf{P}'^{-1}\mathbf{A}\mathbf{P}^{-1})\mathbf{T}' = (\lambda_i\delta_{ij})$. Let $\mathbf{z} = \mathbf{T}\mathbf{P}\mathbf{y}$, so $\boldsymbol{\Sigma}_z = \mathbf{I}$. In $Q = \Sigma_i\lambda_i z_i^2$ collect terms on equal λ_i.]

[3] A direct proof, showing that T^2 is a constant times a quotient of independent chi-square variables, was given by Wijsman (1957).

[4] See Wijsman (1957). This result was originally given by Hsu (1938*c*).

APPENDIX VI

Cochran's Theorem

When in various cases of the analysis of variance the total SS is partitioned into a sum of other SS's, the joint distribution of the constituent SS's is often deduced from Cochran's theorem. Although we did not use this method in this book, the theorem is of sufficient importance in the literature of the analysis of variance to justify its inclusion here. We shall derive Cochran's theorem as an easy corollary of the following

Theorem 1: Suppose

$$\sum_{i=1}^{n} y_i^2 = Q_1 + \cdots + Q_s,$$

where Q_j $(j = 1, \cdots, s)$ is a quadratic form in the variables $y_1, \cdots, y_n$ of rank n_j. A necessary and sufficient condition that there exist an orthogonal transformation $\mathbf{z} = \mathbf{Ay}$ from the vector $\mathbf{y} = (y_1, \cdots, y_n)'$ to a vector $\mathbf{z} = (z_1, \cdots, z_n)'$ such that

(VI.1)
$$\begin{aligned} Q_1 &= \sum_{1}^{n_1} z_i^2, \\ Q_2 &= \sum_{n_1+1}^{n_1+n_2} z_i^2, \\ &\cdots\cdots\cdots \\ Q_s &= \sum_{n_1+\cdots+n_{s-1}+1}^{n_1+\cdots+n_s} z_i^2 \end{aligned}$$

is

$$n_1 + \cdots + n_s = n.$$

Proof:[1] (1) *Necessity:* If such an orthogonal transformation exists, then

$$\sum_{i=1}^{n} y_i^2 = \sum_{i=1}^{n_1+\cdots+n_s} z_i^2.$$

[1] Essentially follows Cochran (1934).

The left-hand member is a quadratic form of rank n, the right-hand member a quadratic form of rank $n_1+\cdots+n_s$. It follows from Lemma 4 of App. II that $n_1+\cdots+n_s = n$.

(2) *Sufficiency:* Since rank $Q_j = n_j$ it follows from Corollary 3 of App. II that there exist n_j linear forms $\{z_i\}$ in $y_1, \cdots, y_n$ such that $Q_j = \Sigma_i \delta_i z_i^2$, where each $\delta_i = +1$ or -1. In Q_1 we shall take the subscripts i on the z_i's equal to $1, 2, \cdots, n_1$; in Q_2, equal to $n_1+1, \cdots, n_1+n_2$, etc. If $\Sigma_1^s n_j = n$, there are then n linear forms z_i, so that in matrix notation $\mathbf{z}^{n\times 1} = \mathbf{A}^{n\times n}\mathbf{y}^{n\times 1}$. Introducing the $n\times n$ diagonal matrix $\mathbf{D}$ with diagonal elements $\delta_1, \cdots, \delta_n$, we can write

$$\sum_{j=1}^{s} Q_j = \sum_{i=1}^{n} \delta_i z_i^2 = \mathbf{z}'\mathbf{D}\mathbf{z} = \mathbf{y}'\mathbf{A}'\mathbf{D}\mathbf{A}\mathbf{y}.$$

But we have also

$$\sum_{i=1}^{s} Q_j = \sum_{i=1}^{n} y_i^2 = \mathbf{y}'\mathbf{y}.$$

Since the symmetric matrix of a quadratic form is unique we deduce that $\mathbf{A}'\mathbf{D}\mathbf{A} = \mathbf{I}$, hence $\mathbf{A}$ is nonsingular. We can now argue that $\mathbf{D} = \mathbf{I}$. In fact, suppose that $\delta_k = -1$. Then, for the y's corresponding to the particular values $z_i = 0$ for $i \neq k$, $z_k = 1$, through $\mathbf{y} = \mathbf{A}^{-1}\mathbf{z}$, we would have $\Sigma_1^n y_i^2 = \Sigma_1^n \delta_i z_i^2 = \delta_k = -1$, which is impossible. Therefore, $\mathbf{D} = \mathbf{I}$ and $\mathbf{A}'\mathbf{A} = \mathbf{I}$, which establishes that the transformation $\mathbf{z} = \mathbf{A}\mathbf{y}$ is orthogonal.

It is somewhat remarkable that the condition $\Sigma_1^s n_i = n$ also insures that each of the quadratic forms Q_j is positive, and has all its characteristic values equal to 0 or 1 (since it is equivalent under orthogonal transformation to a sum of n_j of the $\{z_i^2\}$).

Theorem 2 (Cochran's Theorem):[2] Let the random variables y_i $(i = 1, \cdots, n)$ be independently $N(\eta_i, 1)$, and let $Q_1, Q_2, \cdots, Q_s$ be quadratic forms in the $\{y_i\}$ such that

$$\sum_{i=1}^{n} y_i^2 = Q_1 + \cdots + Q_s.$$

Let $n_j = \text{rank } Q_j$. Then $Q_1, \cdots, Q_s$ will have independent noncentral chi-square distributions with $n_1, \cdots, n_s$ d.f., respectively, if and only if $\Sigma_1^s n_j = n$. Then, if δ_j denotes the noncentrality parameter of Q_j, the value of δ_j^2 may be obtained by replacing y_i by η_i in Q_j.

Proof: (1) If the $\{Q_j\}$ are independently noncentral chi-square with $\{n_j\}$ d.f., respectively, then from (IV.4) it follows that $\Sigma_1^s Q_j$ is noncentral

[2] Central case given by Cochran (1934); non-central by Madow (1940).

chi-square with $\Sigma_1^s n_j$ d.f. But $\Sigma_1^s Q_j = \Sigma_1^n y_i^2$, which is noncentral chi-square with n d.f. Therefore $\Sigma_1^s n_j = n$.

(2) Assume that $\Sigma_1^s n_j = n$. Then, under the orthogonal transformation $\mathbf{z} = \mathbf{A}\mathbf{y}$ of Theorem 1, the random variables $\{z_i\}$ are again independently normally distributed. From the relations (VI.1) it immediately follows that the $\{Q_j\}$ have independent noncentral chi-square distributions with $\{n_j\}$ d.f., respectively. The value of δ_j^2 may be obtained by applying Rule 1 of sec. 2.6.

Cochran's theorem is often applied in the central case, i.e., in the case where $E(y_i) = 0$ $(i = 1, \cdot \cdot \cdot, n)$. Then the $\{Q_j\}$ have central chi-square distributions.

The following corollary to the above two theorems is useful for establishing orthogonality relations if one uses the Cochran-theorem approach, mentioned at the beginning of this appendix, to the distribution theory of the constituent SS's in a partition of the total SS:

Corollary[3] 1: Suppose that the above assumptions of Theorem 2 are satisfied and $\Sigma_1^s n_j = n$. Suppose further than each Q_j is written in any way[4] as a sum of squares of linear forms $\{L_{jt}\}$ in the $\{y_i\}$,

$$Q_j = \sum_{t=1}^{\tau_j} L_{jt}^2 \qquad (j = 1, \cdot \cdot \cdot, s).$$

Then, for $j \neq j'$ and any t,t' $(t = 1, \cdot \cdot \cdot, \tau_j;\ t' = 1, \cdot \cdot \cdot, \tau_{j'})$, the forms L_{jt} and $L_{j't'}$ are orthogonal.

Proof: Since the $\{L_{jt}\}$ are linear forms in the $\{y_i\}$ they are also linear forms in the $\{z_i\}$ used in the proof of Theorem 2. Denote by S_j the set of n_j values of i for which z_i enters into Q_j, i.e.,

$$Q_j = \sum_{i \in S} z_i^2 \qquad (j = 1, \cdot \cdot \cdot, s). \tag{VI.2}$$

Then L_{jt} can involve with nonzero coefficient only the $\{z_i\}$ with $i \in S_j$. For, suppose the contrary: for some $t \leqq \tau_j$, and some $k \notin S_j$, $L_{jt} = \Sigma_1^n c_i z_i$, and $c_k \neq 0$. Then for the particular values $\{z_i = \delta_{ik}\}$ we have the contradiction that $Q_j = 0$ from (VI.2), whereas

$$Q_j = L_{j1}^2 + \cdot \cdot \cdot + L_{j\tau_j}^2 \geqq L_{jt}^2 = c_k^2 > 0.$$

Since L_{jt} and $L_{j't'}$ are linear combinations of the sets of $\{z_i\}$ with i in S_j and $S_{j'}$, respectively, and the two sets are statistically independent, therefore L_{jt} and $L_{j't'}$ are statistically independent, and hence orthogonal as linear forms in the $\{y_i\}$.

[3] Pointed out to me by Professor Robert Wijsman.

[4] It follows from Lemma 2 below that $\tau_j \geq n_j$.

To illustrate the use of Cochran's theorem consider the two-way layout with one observation per cell under the assumptions of sec. 4.2. There we arrived at the SS's

$$SS_A = J\sum_i (y_{i.} - y_{..})^2,$$

$$SS_B = I\sum_j (y_{.j} - y_{..})^2,$$

$$SS_e = \sum_j \sum_j (y_{ij} - y_{i.} - y_{.j} + y_{..})^2$$

by applying general theory. We might also reach them by asking what is an intuitively appealing statistic to measure the differences among rows, etc.; SS_e might also be suggested by the derivation indicated below for the identity

$$\Sigma\Sigma y_{ij}^2 = IJy_{..}^2 + SS_A + SS_B + SS_e. \tag{VI.3}$$

While this identity also fell out of our general theory, it can be verified directly, by preserving the parentheses in, squaring, and summing the expression

$$y_{ij} = y_{..} + (y_{i.} - y_{..}) + (y_{.j} - y_{..}) + (y_{ij} - y_{i.} - y_{.j} + y_{..}),$$

and this is the way the distribution theory of the sums of squares would be approached via Cochran's theorem. In order to apply the theorem to (VI.3) we need the ranks of the four quadratic forms on the right. Again, while these ranks are byproducts of our general theory (namely the numbers of d.f. for each of the corresponding noncentral χ^2-distributions), they would now have to be obtained directly, but this is not difficult. We may utilize the following:

Lemma 1: The rank of a sum of quadratic forms is $\leq$ the sum of their ranks.

Proof: It will suffice to show that if $\mathbf{A}_1$ and $\mathbf{A}_2$ are matrices of the same size with rank $\mathbf{A}_i = \mathbf{r}_i$ then rank $(\mathbf{A}_1 + \mathbf{A}_2) \leq r_1 + r_2$. For each $\mathbf{A}_i$ pick a basis of r_i vectors for the vector space spanned by the columns. Then since the columns of $\mathbf{A}_1 + \mathbf{A}_2$ are the sums of the corresponding columns of $\mathbf{A}_1$ and $\mathbf{A}_2$ they are linear combinations of the $r_1 + r_2$ vectors in the two bases; hence the number of linearly independent columns in $\mathbf{A}_1 + \mathbf{A}_2$ cannot exceed $r_1 + r_2$.

Corollary 2: If $\Sigma_1^n y_i^2 = Q_1 + \cdots + Q_s$, where Q_j is a quadratic form of rank $\leq m_j$ $(j = 1, \cdots, s)$, and if $m_1 + \cdots + m_s = n$, then rank $Q_j = m_j$.

Lemma 2: If Q is a quadratic form in $x_1, \cdots, x_n$ which can be expressed as a quadratic form in $z_1, \cdots, z_p$, where the z_i are linear forms in $x_1, \cdots, x_n$, then rank $Q \leq p$.

Proof: Suppose that $Q = \mathbf{x}'\mathbf{A}^{n\times n}\mathbf{x} = \mathbf{z}'\mathbf{B}^{p\times p}\mathbf{z}$ and $\mathbf{z} = \mathbf{C}^{p\times n}\mathbf{x}$, where $\mathbf{A}$ and $\mathbf{B}$ are symmetric. Then $Q = \mathbf{x}'\mathbf{C}'\mathbf{B}\mathbf{C}\mathbf{x}$ implies that $\mathbf{A} = \mathbf{C}'\mathbf{B}\mathbf{C}$, rank Q = rank $\mathbf{A} \leqq$ rank $\mathbf{C}$ by Lemma 3 of App. II, and rank $\mathbf{C} \leqq p$ since $\mathbf{C}$ is $p\times n$.

Now SS_A is a quadratic form in the $\{z_i = y_{i.}-y_{..}\}$, which are linear forms in the observations $\{y_{ij}\}$; in fact, $SS_A = J\Sigma_1^I z_i^2$. Since $\Sigma_1^I z_i = 0$ we may substitute $z_I = -\Sigma_1^{I-1} z_i$ in SS_A to express it as a quadratic form in the $I-1$ linear forms $\{z_1, \cdots, z_{I-1}\}$. Hence rank $SS_A \leqq I-1$ by Lemma 2; likewise rank $SS_B \leqq J-1$. Similarly SS_e is of rank $\leqq(I-1)(J-1)$, for it is a quadratic form in the IJ linear forms $\{z_{ij} = y_{ij}-y_{i.}-y_{.j}+y_{..}\}$, where $\Sigma_i z_{ij} = 0$, $\Sigma_j z_{ij} = 0$, and can hence be expressed as a quadratic form in the $(I-1)(J-1)$ of the $\{z_{ij}\}$ with $i < I$, $j < J$, since $z_{iJ} = -\Sigma_{j=1}^{J-1} z_{ij}$ for $i < I$, $z_{Ij} = -\Sigma_{i=1}^{I-1} z_{ij}$ for $j < J$, and $z_{IJ} = -\Sigma_{j=1}^{J-1} z_{Ij} = \Sigma_{j=1}^{J-1}\Sigma_{i=1}^{I-1} z_{ij}$. Finally, the rank of $IJy_{..}^2$ is $\leqq 1$ by Lemma 2. It now follows by Corollary 2 that rank $(IJy_{..}^2) = 1$, rank $SS_A = I-1$, rank $SS_B = J-1$, and rank $SS_e = (I-1)(J-1)$.

We rewrite (VI.3) as

$$\sum_i \sum_j (y_{ij}/\sigma)^2 = \sigma^{-2} IJy_{..}^2 + \sigma^{-2} SS_A + \sigma^{-2} SS_B + \sigma^{-2} SS_e,$$

and apply Theorem 2, which tells us that the four sums of squares on the right have independent noncentral χ^2 distributions with the familiar numbers of d.f. and the familiar values of the noncentrality parameters. Furthermore, Corollary 1 yields the familiar orthogonality property that any linear form in any of the four SS's (such as $y_{i.}-y_{..}$) is orthogonal to any linear form in any of the other SS's (such as $y_{i'j}-y_{i'.}-y_{.j}+y_{..}$).

PROBLEMS

VI.1. Work Problem 4.13 by applying Cochran's theorem.

VI.2. Derive the joint-distribution theory for the sums of squares employed with the Latin square, and the orthogonality relations of the corresponding sets of linear forms, by the approach via Cochran's theorem illustrated above for the two-way layout.

UPPER α POINT* OF F WITH ν_1 AND ν_2 D.F.

$\alpha = 0.10$

ν_2 \ ν_1	1	2	3	4	5	6	7	8	9
1	39.9	49.5	53.6	55.8	57.2	58.2	58.9	59.4	59.9
2	8.53	9.00	9.16	9.24	9.29	9.33	9.35	9.37	9.38
3	5.54	5.46	5.39	5.34	5.31	5.28	5.27	5.25	5.24
4	4.54	4.32	4.19	4.11	4.05	4.01	3.98	3.95	3.94
5	4.06	3.78	3.62	3.52	3.45	3.40	3.37	3.34	3.32
6	3.78	3.46	3.29	3.18	3.11	3.05	3.01	2.98	2.96
7	3.59	3.26	3.07	2.96	2.88	2.83	2.78	2.75	2.72
8	3.46	3.11	2.92	2.81	2.73	2.67	2.62	2.59	2.56
9	3.36	3.01	2.81	2.69	2.61	2.55	2.51	2.47	2.44
10	3.29	2.92	2.73	2.61	2.52	2.46	2.41	2.38	2.35
11	3.23	2.86	2.66	2.54	2.45	2.39	2.34	2.30	2.27
12	3.18	2.81	2.61	2.48	2.39	2.33	2.28	2.24	2.21
13	3.14	2.76	2.56	2.43	2.35	2.28	2.23	2.20	2.16
14	3.10	2.73	2.52	2.39	2.31	2.24	2.19	2.15	2.12
15	3.07	2.70	2.49	2.36	2.27	2.21	2.16	2.12	2.09
16	3.05	2.67	2.46	2.33	2.24	2.18	2.13	2.09	2.06
17	3.03	2.64	2.44	2.31	2.22	2.15	2.10	2.06	2.03
18	3.01	2.62	2.42	2.29	2.20	2.13	2.08	2.04	2.00
19	2.99	2.61	2.40	2.27	2.18	2.11	2.06	2.02	1.98
20	2.97	2.59	2.38	2.25	2.16	2.09	2.04	2.00	1.96
21	2.96	2.57	2.36	2.23	2.14	2.08	2.02	1.98	1.95
22	2.95	2.56	2.35	2.22	2.13	2.06	2.01	1.97	1.93
23	2.94	2.55	2.34	2.21	2.11	2.05	1.99	1.95	1.92
24	2.93	2.54	2.33	2.19	2.10	2.04	1.98	1.94	1.91
25	2.92	2.53	2.32	2.18	2.09	2.02	1.97	1.93	1.89
26	2.91	2.52	2.31	2.17	2.08	2.01	1.96	1.92	1.88
27	2.90	2.51	2.30	2.17	2.07	2.00	1.95	1.91	1.87
28	2.89	2.50	2.29	2.16	2.06	2.00	1.94	1.90	1.87
29	2.89	2.50	2.28	2.15	2.06	1.99	1.93	1.89	1.86
30	2.88	2.49	2.28	2.14	2.05	1.98	1.93	1.88	1.85
40	2.84	2.44	2.23	2.09	2.00	1.93	1.87	1.83	1.79
60	2.79	2.39	2.18	2.04	1.95	1.87	1.82	1.77	1.74
120	2.75	2.35	2.13	1.99	1.90	1.82	1.77	1.72	1.68
∞	2.71	2.30	2.08	1.94	1.85	1.77	1.72	1.67	1.63

* Rounded off to three significant figures from tables of M. Merrington and C. M. Thompson in *Biometrika*, Vol. 33, pp. 78–87, 1943. Reproduced with the kind permission of the authors and the editor.

UPPER α POINT* OF F WITH ν_1 AND ν_2 D.F.

$\alpha = 0.10$

ν_2 \ ν_1	10	12	15	20	24	30	40	60	120	∞
1	60.2	60.7	61.2	61.7	62.0	62.3	62.5	62.8	63.1	63.3
2	9.39	9.41	9.42	9.44	9.45	9.46	9.47	9.47	9.48	9.49
3	5.23	5.22	5.20	5.18	5.18	5.17	5.16	5.15	5.14	5.13
4	3.92	3.90	3.87	3.84	3.83	3.82	3.80	3.79	3.78	3.76
5	3.30	3.27	3.24	3.21	3.19	3.17	3.16	3.14	3.12	3.10
6	2.94	2.90	2.87	2.84	2.82	2.80	2.78	2.76	2.74	2.72
7	2.70	2.67	2.63	2.59	2.58	2.56	2.54	2.51	2.49	2.47
8	2.54	2.50	2.46	2.42	2.40	2.38	2.36	2.34	2.32	2.29
9	2.42	2.38	2.34	2.30	2.28	2.25	2.23	2.21	2.18	2.16
10	2.32	2.28	2.24	2.20	2.18	2.16	2.13	2.11	2.08	2.06
11	2.25	2.21	2.17	2.12	2.10	2.08	2.05	2.03	2.00	1.97
12	2.19	2.15	2.10	2.06	2.04	2.01	1.99	1.96	1.93	1.90
13	2.14	2.10	2.05	2.01	1.98	1.96	1.93	1.90	1.88	1.85
14	2.10	2.05	2.01	1.96	1.94	1.91	1.89	1.86	1.83	1.80
15	2.06	2.02	1.97	1.92	1.90	1.87	1.85	1.82	1.79	1.76
16	2.03	1.99	1.94	1.89	1.87	1.84	1.81	1.78	1.75	1.72
17	2.00	1.96	1.91	1.86	1.84	1.81	1.78	1.75	1.72	1.69
18	1.98	1.93	1.89	1.84	1.81	1.78	1.75	1.72	1.69	1.66
19	1.96	1.91	1.86	1.81	1.79	1.76	1.73	1.70	1.67	1.63
20	1.94	1.89	1.84	1.79	1.77	1.74	1.71	1.68	1.64	1.61
21	1.92	1.88	1.83	1.78	1.75	1.72	1.69	1.66	1.62	1.59
22	1.90	1.86	1.81	1.76	1.73	1.70	1.67	1.64	1.60	1.57
23	1.89	1.84	1.80	1.74	1.72	1.69	1.66	1.62	1.59	1.55
24	1.88	1.83	1.78	1.73	1.70	1.67	1.64	1.61	1.57	1.53
25	1.87	1.82	1.77	1.72	1.69	1.66	1.63	1.59	1.56	1.52
26	1.86	1.81	1.76	1.71	1.68	1.65	1.61	1.58	1.54	1.50
27	1.85	1.80	1.75	1.70	1.67	1.64	1.60	1.57	1.53	1.49
28	1.84	1.79	1.74	1.69	1.66	1.63	1.59	1.56	1.52	1.48
29	1.83	1.78	1.73	1.68	1.65	1.62	1.58	1.55	1.51	1.47
30	1.82	1.77	1.72	1.67	1.64	1.61	1.57	1.54	1.50	1.46
40	1.76	1.71	1.66	1.61	1.57	1.54	1.51	1.47	1.42	1.38
60	1.71	1.66	1.60	1.54	1.51	1.48	1.44	1.40	1.35	1.29
120	1.65	1.60	1.55	1.48	1.45	1.41	1.37	1.32	1.26	1.19
∞	1.60	1.55	1.49	1.42	1.38	1.34	1.30	1.24	1.17	1.00

* Rounded off to three significant figures from tables of M. Merrington and C. M. Thompson in *Biometrika*, Vol. 33, pp. 78–87, 1943. Reproduced with the kind permission of the authors and the editor.

UPPER α POINT* OF F WITH ν_1 AND ν_2 D.F.

$\alpha = 0.05$

ν_2 \ ν_1	1	2	3	4	5	6	7	8	9
1	161	200	216	225	230	234	237	239	241
2	18.5	19.0	19.2	19.2	19.3	19.3	19.4	19.4	19.4
3	10.1	9.55	9.28	9.12	9.01	8.94	8.89	8.85	8.81
4	7.71	6.94	6.59	6.39	6.26	6.16	6.09	6.04	6.00
5	6.61	5.79	5.41	5.19	5.05	4.95	4.88	4.82	4.77
6	5.99	5.14	4.76	4.53	4.39	4.28	4.21	4.15	4.10
7	5.59	4.74	4.35	4.12	3.97	3.87	3.79	3.73	3.68
8	5.32	4.46	4.07	3.84	3.69	3.58	3.50	3.44	3.39
9	5.12	4.26	3.86	3.63	3.48	3.37	3.29	3.23	3.18
10	4.96	4.10	3.71	3.48	3.33	3.22	3.14	3.07	3.02
11	4.84	3.98	3.59	3.36	3.20	3.09	3.01	2.95	2.90
12	4.75	3.89	3.49	3.26	3.11	3.00	2.91	2.85	2.80
13	4.67	3.81	3.41	3.18	3.03	2.92	2.83	2.77	2.71
14	4.60	3.74	3.34	3.11	2.96	2.85	2.76	2.70	2.65
15	4.54	3.68	3.29	3.06	2.90	2.79	2.71	2.64	2.59
16	4.49	3.63	3.24	3.01	2.85	2.74	2.66	2.59	2.54
17	4.45	3.59	3.20	2.96	2.81	2.70	2.61	2.55	2.49
18	4.41	3.55	3.16	2.93	2.77	2.66	2.58	2.51	2.46
19	4.38	3.52	3.13	2.90	2.74	2.63	2.54	2.48	2.42
20	4.35	3 49	3.10	2.87	2.71	2.60	2.51	2.45	2.39
21	4.32	3.47	3.07	2.84	2.68	2.57	2.49	2.42	2.37
22	4.30	3.44	3.05	2.82	2.66	2.55	2.46	2.40	2.34
23	4.28	3.42	3.03	2.80	2.64	2.53	2.44	2.37	2.32
24	4.26	3.40	3.01	2.78	2.62	2.51	2.42	2.36	2.30
25	4.24	3.39	2.99	2.76	2.60	2.49	2.40	2.34	2.28
26	4.23	3.37	2.98	2.74	2.59	2.47	2.39	2.32	2.27
27	4.21	3.35	2.96	2.73	2.57	2.46	2.37	2.31	2.25
28	4.20	3.34	2.95	2.71	2.56	2.45	2.36	2.29	2.24
29	4.18	3.33	2.93	2.70	2.55	2.43	2.35	2.28	2.22
30	4.17	3.32	2.92	2.69	2.53	2.42	2.33	2.27	2.21
40	4.08	3.23	2.84	2.61	2.45	2.34	2.25	2.18	2.12
60	4.00	3.15	2.76	2.53	2.37	2.25	2.17	2.10	2.04
120	3.92	3.07	2.68	2.45	2.29	2.17	2.09	2.02	1.96
∞	3.84	3.00	2.60	2.37	2.21	2.10	2.01	1.94	1.88

* Rounded off to three significant figures from tables of M. Merrington and C. M. Thompson in *Biometrika*, Vol. 33, pp. 78–87, 1943. Reproduced with the kind permission of the authors and the editor.

UPPER α POINT* OF F WITH ν_1 AND ν_2 D.F.

$\alpha = 0.05$

ν_2 \ ν_1	10	12	15	20	24	30	40	60	120	∞
1	242	244	246	248	249	250	251	252	253	254
2	19.4	19.4	19.4	19.4	19.5	19.5	19.5	19.5	19.5	19.5
3	8.79	8.74	8.70	8.66	8.64	8.62	8.59	8.57	8.55	8.53
4	5.96	5.91	5.86	5.80	5.77	5.75	5.72	5.69	5.66	5.63
5	4.74	4.68	4.62	4.56	4.53	4.50	4.46	4.43	4.40	4.36
6	4.06	4.00	3.94	3.87	3.84	3.81	3.77	3.74	3.70	3.67
7	3.64	3.57	3.51	3.44	3.41	3.38	3.34	3.30	3.27	3.23
8	3.35	3.28	3.22	3.15	3.12	3.08	3.04	3.01	2.97	2.93
9	3.14	3.07	3.01	2.94	2.90	2.86	2.83	2.79	2.75	2.71
10	2.98	2.91	2.85	2.77	2.74	2.70	2.66	2.62	2.58	2.54
11	2.85	2.79	2.72	2.65	2.61	2.57	2.53	2.49	2.45	2.40
12	2.75	2.69	2.62	2.54	2.51	2.47	2.43	2.38	2.34	2.30
13	2.67	2.60	2.53	2.46	2.42	2.38	2.34	2.30	2.25	2.21
14	2.60	2.53	2.46	2.39	2.35	2.31	2.27	2.22	2.18	2.13
15	2.54	2.48	2.40	2.33	2.29	2.25	2.20	2.16	2.11	2.07
16	2.49	2.42	2.35	2.28	2.24	2.19	2.15	2.11	2.06	2.01
17	2.45	2.38	2.31	2.23	2.19	2.15	2.10	2.06	2.01	1.96
18	2.41	2.34	2.27	2.19	2.15	2.11	2.06	2.02	1.97	1.92
19	2.38	2.31	2.23	2.16	2.11	2.07	2.03	1.98	1.93	1.88
20	2.35	2.28	2.20	2.12	2.08	2.04	1.99	1.95	1.90	1.84
21	2.32	2.25	2.18	2.10	2.05	2.01	1.96	1.92	1.87	1.81
22	2.30	2.23	2.15	2.07	2.03	1.98	1.94	1.89	1.84	1.78
23	2.27	2.20	2.13	2.05	2.01	1.96	1.91	1.86	1.81	1.76
24	2.25	2.18	2.11	2.03	1.98	1.94	1.89	1.84	1.79	1.73
25	2.24	2.16	2.09	2.01	1.96	1.92	1.87	1.82	1.77	1.71
26	2.22	2.15	2.07	1.99	1.95	1.90	1.85	1.80	1.75	1.69
27	2.20	2.13	2.06	1.97	1.93	1.88	1.84	1.79	1.73	1.67
28	2.19	2.12	2.04	1.96	1.91	1.87	1.82	1.77	1.71	1.65
29	2.18	2.10	2.03	1.94	1.90	1.85	1.81	1.75	1.70	1.64
30	2.16	2.09	2.01	1.93	1.89	1.84	1.79	1.74	1.68	1.62
40	2.08	2.00	1.92	1.84	1.79	1.74	1.69	1.64	1.58	1.51
60	1.99	1.92	1.84	1.75	1.70	1.65	1.59	1.53	1.47	1.39
120	1.91	1.83	1.75	1.66	1.61	1.55	1.50	1.43	1.35	1.25
∞	1.83	1.75	1.67	1.57	1.52	1.46	1.39	1.32	1.22	1.00

* Rounded off to three significant figures from tables of M. Merrington and C. M. Thomspon in *Biometrika*, Vol. 33, pp. 78–87, 1943. Reproduced with the kind permission of the authors and the editor.

UPPER α POINT* OF F WITH ν_1 AND ν_2 D.F.

$\alpha = 0.025$

ν_2 \ ν_1	1	2	3	4	5	6	7	8	9
1	648	800	864	900	922	937	948	957	963
2	38.5	39.0	39.2	39.2	39.3	39.3	39.4	39.4	39.4
3	17.4	16.0	15.4	15.1	14.9	14.7	14.6	14.5	14.5
4	12.2	10.6	9.98	9.60	9.36	9.20	9.07	8.98	8.90
5	10.0	8.43	7.76	7.39	7.15	6.98	6.85	6.76	6.68
6	8.81	7.26	6.60	6.23	5.99	5.82	5.70	5.60	5.52
7	8.07	6.54	5.89	5.52	5.29	5.12	4.99	4.90	4.82
8	7.57	6.06	5.42	5.05	4.82	4.65	4.53	4.43	4.36
9	7.21	5.71	5.08	4.72	4.48	4.32	4.20	4.10	4.03
10	6.94	5.46	4.83	4.47	4.24	4.07	3.95	3.85	3.78
11	6.72	5.26	4.63	4.28	4.04	3.88	3.76	3.66	3.59
12	6.55	5.10	4.47	4.12	3.89	3.73	3.61	3.51	3.44
13	6.41	4.97	4.35	4.00	3.77	3.60	3.48	3.39	3.31
14	6.30	4.86	4.24	3.89	3.66	3.50	3.38	3.29	3.21
15	6.20	4.77	4.15	3.80	3.58	3.41	3.29	3.20	3.12
16	6.12	4.69	4.08	3.73	3.50	3.34	3.22	3.12	3.05
17	6.04	4.62	4.01	3.66	3.44	3.28	3.16	3.06	2.98
18	5.98	4.56	3.95	3.61	3.38	3.22	3.10	3.01	2.93
19	5.92	4.51	3.90	3.56	3.33	3.17	3.05	2.96	2.88
20	5.87	4.46	3.86	3.51	3.29	3.13	3.01	2.91	2.84
21	5.83	4.42	3.82	3.48	3.25	3.09	2.97	2.87	2.80
22	5.79	4.38	3.78	3.44	3.22	3.05	2.93	2.84	2.76
23	5.75	4.35	3.75	3.41	3.18	3.02	2.90	2.81	2.73
24	5.72	4.32	3.72	3.38	3.15	2.99	2.87	2.78	2.70
25	5.69	4.29	3.69	3.35	3.13	2.97	2.85	2.75	2.68
26	5.66	4.27	3.67	3.33	3.10	2.94	2.82	2.73	2.65
27	5.63	4.24	3.65	3.31	3.08	2.92	2.80	2.71	2.63
28	5.61	4.22	3.63	3.29	3.06	2.90	2.78	2.69	2.61
29	5.59	4.20	3.61	3.27	3.04	2.88	2.76	2.67	2.59
30	5.57	4.18	3.59	3.25	3.03	2.87	2.75	2.65	2.57
40	5.42	4.05	3.46	3.13	2.90	2.74	2.62	2.53	2.45
60	5.29	3.93	3.34	3.01	2.79	2.63	2.51	2.41	2.33
120	5.15	3.80	3.23	2.89	2.67	2.52	2.39	2.30	2.22
∞	5.02	3.69	3.12	2.79	2.57	2.41	2.29	2.19	2.11

* Rounded off to three significant figures from tables of M. Merrington and C. M. Thompson in *Biometrika*, Vol. 33, pp. 78–87, 1943. Reproduced with the kind permission of the authors and the editor.

Upper α Point* of F with ν_1 and ν_2 d.f.

$\alpha = 0.025$

ν_2 \ ν_1	10	12	15	20	24	30	40	60	120	∞
1	969	977	985	993	997	1000	1010	1010	1010	1020
2	39.4	39.4	39.4	39.4	39.5	39.5	39.5	39.5	39.5	39.5
3	14.4	14.3	14.3	14.2	14.1	14.1	14.0	14.0	13.9	13.9
4	8.84	8.75	8.66	8.56	8.51	8.46	8.41	8.36	8.31	8.26
5	6.62	6.52	6.43	6.33	6.28	6.23	6.18	6.12	6.07	6.02
6	5.46	5.37	5.27	5.17	5.12	5.07	5.01	4.96	4.90	4.85
7	4.76	4.67	4.57	4.47	4.42	4.36	4.31	4.25	4.20	4.14
8	4.30	4.20	4.10	4.00	3.95	3.89	3.84	3.78	3.73	3.67
9	3.96	3.87	3.77	3.67	3.61	3.56	3.51	3.45	3.39	3.33
10	3.72	3.62	3.52	3.42	3.37	3.31	3.26	3.20	3.14	3.08
11	3.53	3.43	3.33	3.23	3.17	3.12	3.06	3.00	2.94	2.88
12	3.37	3.28	3.18	3.07	3.02	2.96	2.91	2.85	2.79	2.72
13	3.25	3.15	3.05	2.95	2.89	2.84	2.78	2.72	2.66	2.60
14	3.15	3.05	2.95	2.84	2.79	2.73	2.67	2.61	2.55	2.49
15	3.06	2.96	2.86	2.76	2.70	2.64	2.59	2.52	2.46	2.40
16	2.99	2.89	2.79	2.68	2.63	2.57	2.51	2.45	2.38	2.32
17	2.92	2.82	2.72	2.62	2.56	2.50	2.44	2.38	2.32	2.25
18	2.87	2.77	2.67	2.56	2.50	2.44	2.38	2.32	2.26	2.19
19	2.82	2.72	2.62	2.51	2.45	2.39	2.33	2.27	2.20	2.13
20	2.77	2.68	2.57	2.46	2.41	2.35	2.29	2.22	2.16	2.09
21	2.73	2.64	2.53	2.42	2.37	2.31	2.25	2.18	2.11	2.04
22	2.70	2.60	2.50	2.39	2.33	2.27	2.21	2.14	2.08	2.00
23	2.67	2.57	2.47	2.36	2.30	2.24	2.18	2.11	2.04	1.97
24	2.64	2.54	2.44	2.33	2.27	2.21	2.15	2.08	2.01	1.94
25	2.61	2.51	2.41	2.30	2.24	2.18	2.12	2.05	1.98	1.91
26	2.59	2.49	2.39	2.28	2.22	2.16	2.09	2.03	1.95	1.88
27	2.57	2.47	2.36	2.25	2.19	2.13	2.07	2.00	1.93	1.85
28	2.55	2.45	2.34	2.23	2.17	2.11	2.05	1.98	1.91	1.83
29	2.53	2.43	2.32	2.21	2.15	2.09	2.03	1.96	1.89	1.81
30	2.51	2.41	2.31	2.20	2.14	2.07	2.01	1.94	1.87	1.79
40	2.39	2.29	2.18	2.07	2.01	1.94	1.88	1.80	1.72	1.64
60	2.27	2.17	2.06	1.94	1.88	1.82	1.74	1.67	1.58	1.48
120	2.16	2.05	1.94	1.82	1.76	1.69	1.61	1.53	1.43	1.31
∞	2.05	1.94	1.83	1.71	1.64	1.57	1.48	1.39	1.27	1.00

* Rounded off to three significant figures from tables of M. Merrington and C. M. Thompson in *Biometrika*, Vol. 33, pp. 78–87, 1943. Reproduced with the kind permission of the authors and the editor.

UPPER α POINT* OF F WITH ν_1 AND ν_2 D.F.

$\alpha = 0.01$

ν_2 \ ν_1	1	2	3	4	5	6	7	8	9
1	4050	5000	5400	5620	5760	5860	5930	5980	6020
2	98.5	99.0	99.2	99.2	99.3	99.3	99.4	99.4	99.4
3	34.1	30.8	29.5	28.7	28.2	27.9	27.7	27.5	27.3
4	21.2	18.0	16.7	16.0	15.5	15.2	15.0	14.8	14.7
5	16.3	13.3	12.1	11.4	11.0	10.7	10.5	10.3	10.2
6	13.7	10.9	9.78	9.15	8.75	8.47	8.26	8.10	7.98
7	12.2	9.55	8.45	7.85	7.46	7.19	6.99	6.84	6.72
8	11.3	8.65	7.59	7.01	6.63	6.37	6.18	6.03	5.91
9	10.6	8.02	6.99	6.42	6.06	5.80	5.61	5.47	5.35
10	10.0	7.56	6.55	5.99	5.64	5.39	5.20	5.06	4.94
11	9.65	7.21	6.22	5.67	5.32	5.07	4.89	4.74	4.63
12	9.33	6.93	5.95	5.41	5.06	4.82	4.64	4.50	4.39
13	9.07	6.70	5.74	5.21	4.86	4.62	4.44	4.30	4.19
14	8.86	6.51	5.56	5.04	4.69	4.46	4.28	4.14	4.03
15	8.68	6.36	5.42	4.89	4.56	4.32	4.14	4.00	3.89
16	8.53	6.23	5.29	4.77	4.44	4.20	4.03	3.89	3.78
17	8.40	6.11	5.18	4.67	4.34	4.10	3.93	3.79	3.68
18	8.29	6.01	5.09	4.58	4.25	4.01	3.84	3.71	3.60
19	8.18	5.93	5.01	4.50	4.17	3.94	3.77	3.63	3.52
20	8.10	5.85	4.94	4.43	4.10	3.87	3.70	3.56	3.46
21	8.02	5.78	4.87	4.37	4.04	3.81	3.64	3.51	3.40
22	7.95	5.72	4.82	4.31	3.99	3.76	3.59	3.45	3.35
23	7.88	5.66	4.76	4.26	3.94	3.71	3.54	3.41	3.30
24	7.82	5.61	4.72	4.22	3.90	3.67	3.50	3.36	3.26
25	7.77	5.57	4.68	4.18	3.85	3.63	3.46	3.32	3.22
26	7.72	5.53	4.64	4.14	3.82	3.59	3.42	3.29	3.18
27	7.68	5.49	4.60	4.11	3.78	3.56	3.39	3.26	3.15
28	7.64	5.45	4.57	4.07	3.75	3.53	3.36	3.23	3.12
29	7.60	5.42	4.54	4.04	3.73	3.50	3.33	3.20	3.09
30	7.56	5.39	4.51	4.02	3.70	3.47	3.30	3.17	3.07
40	7.31	5.18	4.31	3.83	3.51	3.29	3.12	2.99	2.89
60	7.08	4.98	4.13	3.65	3.34	3.12	2.95	2.82	2.72
120	6.85	4.79	3.95	3.48	3.17	2.96	2.79	2.66	2.56
∞	6.63	4.61	3.78	3.32	3.02	2.80	2.64	2.51	2.41

* Rounded off to three significant figures from tables of M. Merrington and C. M. Thompson in *Biometrika*, Vol. 33, pp. 78–87, 1943. Reproduced with the kind permission of the authors and the editor.

UPPER α POINT* OF F WITH ν_1 AND ν_2 D.F.

$\alpha = 0.01$

ν_2 \ ν_1	10	12	15	20	24	30	40	60	120	∞
1	6060	6110	6160	6210	6230	6260	6290	6310	6340	6370
2	99.4	99.4	99.4	99.4	99.5	99.5	99.5	99.5	99.5	99.5
3	27.2	27.1	26.9	26.7	26.6	26.5	26.4	26.3	26.2	26.1
4	14.5	14.4	14.2	14.0	13.9	13.8	13.7	13.7	13.6	13.5
5	10.1	9.89	9.72	9.55	9.47	9.38	9.29	9.20	9.11	9.02
6	7.87	7.72	7.56	7.40	7.31	7.23	7.14	7.06	6.97	6.88
7	6.62	6.47	6.31	6.16	6.07	5.99	5.91	5.82	5.74	5.65
8	5.81	5.67	5.52	5.36	5.28	5.20	5.12	5.03	4.95	4.86
9	5.26	5.11	4.96	4.81	4.73	4.65	4.57	4.48	4.40	4.31
10	4.85	4.71	4.56	4.41	4.33	4.25	4.17	4.08	4.00	3.91
11	4.54	4.40	4.25	4.10	4.02	3.94	3.86	3.78	3.69	3.60
12	4.30	4.16	4.01	3.86	3.78	3.70	3.62	3.54	3.45	3.36
13	4.10	3.96	3.82	3.66	3.59	3.51	3.43	3.34	3.25	3.17
14	3.94	3.80	3.66	3.51	3.43	3.35	3.27	3.18	3.09	3.00
15	3.80	3.67	3.52	3.37	3.29	3.21	3.13	3.05	2.96	2.87
16	3.69	3.55	3.41	3.26	3.18	3.10	3.02	2.93	2.84	2.75
17	3.59	3.46	3.31	3.16	3.08	3.00	2.92	2.83	2.75	2.65
18	3.51	3.37	3.23	3.08	3.00	2.92	2.84	2.75	2.66	2.57
19	3.43	3.30	3.15	3.00	2.92	2.84	2.76	2.67	2.58	2.49
20	3.37	3.23	3.09	2.94	2.86	2.78	2.69	2.61	2.52	2.42
21	3.31	3.17	3.03	2.88	2.80	2.72	2.64	2.55	2.46	2.36
22	3.26	3.12	2.98	2.83	2.75	2.67	2.58	2.50	2.40	2.31
23	3.21	3.07	2.93	2.78	2.70	2.62	2.54	2.45	2.35	2.26
24	3.17	3.03	2.89	2.74	2.66	2.58	2.49	2.40	2.31	2.21
25	3.13	2.99	2.85	2.70	2.62	2.54	2.45	2.36	2.27	2.17
26	3.09	2.96	2.81	2.66	2.58	2.50	2.42	2.33	2.23	2.13
27	3.06	2.93	2.78	2.63	2.55	2.47	2.38	2.29	2.20	2.10
28	3.03	2.90	2.75	2.60	2.52	2.44	2.35	2.26	2.17	2.06
29	3.00	2.87	2.73	2.57	2.49	2.41	2.33	2.23	2.14	2.03
30	2.98	2.84	2.70	2.55	2.47	2.39	2.30	2.21	2.11	2.01
40	2.80	2.66	2.52	2.37	2.29	2.20	2.11	2.02	1.92	1.80
60	2.63	2.50	2.35	2.20	2.12	2.03	1.94	1.84	1.73	1.60
120	2.47	2.34	2.19	2.03	1.95	1.86	1.76	1.66	1.53	1.38
∞	2.32	2.18	2.04	1.88	1.79	1.70	1.59	1.47	1.32	1.00

* Rounded off to three significant figures from tables of M. Merrington and C. M. Thompson in *Biometrika*, Vol. 33, pp. 78–87, 1943. Reproduced with the kind permission of the authors and the editor.

UPPER α POINT* OF F WITH ν_1 AND ν_2 D.F.

$\alpha = 0.005$

ν_2 \ ν_1	1	2	3	4	5	6	7	8	9
1	16200	20000	21600	22500	23100	23400	23700	23900	24100
2	199	199	199	199	199	199	199	199	199
3	55.6	49.8	47.5	46.2	45.4	44.8	44.4	44.1	43.9
4	31.3	26.3	24.3	23.2	22.5	22.0	21.6	21.4	21.1
5	22.8	18.3	16.5	15.6	14.9	14.5	14.2	14.0	13.8
6	18.6	14.5	12.9	12.0	11.5	11.1	10.8	10.6	10.4
7	16.2	12.4	10.9	10.1	9.52	9.16	8.89	8.68	8.51
8	14.7	11.0	9.60	8.81	8.30	7.95	7.69	7.50	7.34
9	13.6	10.1	8.72	7.96	7.47	7.13	6.88	6.69	6.54
10	12.8	9.43	8.08	7.34	6.87	6.54	6.30	6.12	5.97
11	12.2	8.91	7.60	6.88	6.42	6.10	5.86	5.68	5.54
12	11.8	8.51	7.23	6.52	6.07	5.76	5.52	5.35	5.20
13	11.4	8.19	6.93	6.23	5.79	5.48	5.25	5.08	4.94
14	11.1	7.92	6.68	6.00	5.56	5.26	5.03	4.86	4.72
15	10.8	7.70	6.48	5.80	5.37	5.07	4.85	4.67	4.54
16	10.6	7.51	6.30	5.64	5.21	4.91	4.69	4.52	4.38
17	10.4	7.35	6.16	5.50	5.07	4.78	4.56	4.39	4.25
18	10.2	7.21	6.03	5.37	4.96	4.66	4.44	4.28	4.14
19	10.1	7.09	5.92	5.27	4.85	4.56	4.34	4.18	4.04
20	9.94	6.99	5.82	5.17	4.76	4.47	4.26	4.09	3.96
21	9.83	6.89	5.73	5.09	4.68	4.39	4.18	4.01	3.88
22	9.73	6.81	5.65	5.02	4.61	4.32	4.11	3.94	3.81
23	9.63	6.73	5.58	4.95	4.54	4.26	4.05	3.88	3.75
24	9.55	6.66	5.52	4.89	4.49	4.20	3.99	3.83	3.69
25	9.48	6.60	5.46	4.84	4.43	4.15	3.94	3.78	3.64
26	9.41	6.54	5.41	4.79	4.38	4.10	3.89	3.73	3.60
27	9.34	6.49	5.36	4.74	4.34	4.06	3.85	3.69	3.56
28	9.28	6.44	5.32	4.70	4.30	4.02	3.81	3.65	3.52
29	9.23	6.40	5.28	4.66	4.26	3.98	3.77	3.61	3.48
30	9.18	6.35	5.24	4.62	4.23	3.95	3.74	3.58	3.45
40	8.83	6.07	4.98	4.37	3.99	3.71	3.51	3.35	3.22
60	8.49	5.79	4.73	4.14	3.76	3.49	3.29	3.13	3.01
120	8.18	5.54	4.50	3.92	3.55	3.28	3.09	2.93	2.81
∞	7.88	5.30	4.28	3.72	3.35	3.09	2.90	2.74	2.62

* Rounded off to three significant figures from tables of M. Merrington and C. M. Thompson in *Biometrika*, Vol. 33, pp. 78–87, 1943. Reproduced with the kind permission of the authors and the editor.

UPPER α POINT* OF F WITH ν_1 AND ν_2 D.F.

$\alpha = 0.005$

ν_2 \ ν_1	10	12	15	20	24	30	40	60	120	∞
1	24200	24400	24600	24800	24900	25000	25100	25300	25400	25500
2	199	199	199	199	199	199	199	199	199	200
3	43.7	43.4	43.1	42.8	42.6	42.5	42.3	42.1	42.0	41.8
4	21.0	20.7	20.4	20.2	20.0	19.9	19.8	19.6	19.5	19.3
5	13.6	13.4	13.1	12.9	12.8	12.7	12.5	12.4	12.3	12.1
6	10.3	10.0	9.81	9.59	9.47	9.36	9.24	9.12	9.00	8.88
7	8.38	8.18	7.97	7.75	7.65	7.53	7.42	7.31	7.19	7.08
8	7.21	7.01	6.81	6.61	6.50	6.40	6.29	6.18	6.06	5.95
9	6.42	6.23	6.03	5.83	5.73	5.62	5.52	5.41	5.30	5.19
10	5.85	5.66	5.47	5.27	5.17	5.07	4.97	4.86	4.75	4.64
11	5.42	5.24	5.05	4.86	4.76	4.65	4.55	4.44	4.34	4.23
12	5.09	4.91	4.72	4.53	4.43	4.33	4.23	4.12	4.01	3.90
13	4.82	4.64	4.46	4.27	4.17	4.07	3.97	3.87	3.76	3.65
14	4.60	4.43	4.25	4.06	3.96	3.86	3.76	3.66	3.55	3.44
15	4.42	4.25	4.07	3.88	3.79	3.69	3.58	3.48	3.37	3.26
16	4.27	4.10	3.92	3.73	3.64	3.54	3.44	3.33	3.22	3.11
17	4.14	3.97	3.79	3.61	3.51	3.41	3.31	3.21	3.10	2.98
18	4.03	3.86	3.68	3.50	3.40	3.30	3.20	3.10	2.99	2.87
19	3.93	3.76	3.59	3.40	3.31	3.21	3.11	3.00	2.89	2.78
20	3.85	3.68	3.50	3.32	3.22	3.12	3.02	2.92	2.81	2.69
21	3.77	3.60	3.43	3.24	3.15	3.05	2.95	2.84	2.73	2.61
22	3.70	3.54	3.36	3.18	3.08	2.98	2.88	2.77	2.66	2.55
23	3.64	3.47	3.30	3.12	3.02	2.92	2.82	2.71	2.60	2.48
24	3.59	3.42	3.25	3.06	2.97	2.87	2.77	2.66	2.55	2.43
25	3.54	3.37	3.20	3.01	2.92	2.82	2.72	2.61	2.50	2.38
26	3.49	3.33	3.15	2.97	2.87	2.77	2.67	2.56	2.45	2.33
27	3.45	3.28	3.11	2.93	2.83	2.73	2.63	2.52	2.41	2.29
28	3.41	3.25	3.07	2.89	2.79	2.69	2.59	2.48	2.37	2.25
29	3.38	3.21	3.04	2.86	2.76	2.66	2.56	2.45	2.33	2.21
30	3.34	3.18	3.01	2.82	2.73	2.63	2.52	2.42	2.30	2.18
40	3.12	2.95	2.78	2.60	2.50	2.40	2.30	2.18	2.06	1.93
60	2.90	2.74	2.57	2.39	2.29	2.19	2.08	1.96	1.83	1.69
120	2.71	2.54	2.37	2.19	2.09	1.98	1.87	1.75	1.61	1.43
∞	2.52	2.36	2.19	2.00	1.90	1.79	1.67	1.53	1.36	1.00

* Rounded off to three significant figures from tables of M. Merrington and C. M. Thompson in *Biometrika*, Vol. 33, pp. 78–87, 1943. Reproduced with the kind permission of the authors and the editor.

UPPER α POINT* OF STUDENTIZED RANGE, $q_{k,\nu} = R/s$
k = SAMPLE SIZE FOR RANGE R, ν = NO. OF D.F. FOR s
$\alpha = 0.10$

ν \ k	2	3	4	5	6	7	8	9	10	11	12	13	14	15	16	17	18	19	20
1	8.93	13.4	16.4	18.5	20.2	21.5	22.6	23.6	24.5	25.2	25.9	26.5	27.1	27.6	28.1	28.5	29.0	29.3	29.7
2	4.13	5.73	6.77	7.54	8.14	8.63	9.05	9.41	9.72	10.0	10.3	10.5	10.7	10.9	11.1	11.2	11.4	11.5	11.7
3	3.33	4.47	5.20	5.74	6.16	6.51	6.81	7.06	7.29	7.49	7.67	7.83	7.98	8.12	8.25	8.37	8.48	8.58	8.68
4	3.01	3.98	4.59	5.03	5.39	5.68	5.93	6.14	6.33	6.49	6.65	6.78	6.91	7.02	7.13	7.23	7.33	7.41	7.50
5	2.85	3.72	4.26	4.66	4.98	5.24	5.46	5.65	5.82	5.97	6.10	6.22	6.34	6.44	6.54	6.63	6.71	6.79	6.86
6	2.75	3.56	4.07	4.44	4.73	4.97	5.17	5.34	5.50	5.64	5.76	5.87	5.98	6.07	6.16	6.25	6.32	6.40	6.47
7	2.68	3.45	3.93	4.28	4.55	4.78	4.97	5.14	5.28	5.41	5.53	5.64	5.74	5.83	5.91	5.99	6.06	6.13	6.19
8	2.63	3.37	3.83	4.17	4.43	4.65	4.83	4.99	5.13	5.25	5.36	5.46	5.56	5.64	5.72	5.80	5.87	5.93	6.00
9	2.59	3.32	3.76	4.08	4.34	4.54	4.72	4.87	5.01	5.13	5.23	5.33	5.42	5.51	5.58	5.66	5.72	5.79	5.85
10	2.56	3.27	3.70	4.02	4.26	4.47	4.64	4.78	4.91	5.03	5.13	5.23	5.32	5.40	5.47	5.54	5.61	5.67	5.73
11	2.54	3.23	3.66	3.96	4.20	4.40	4.57	4.71	4.84	4.95	5.05	5.15	5.23	5.31	5.38	5.45	5.51	5.57	5.63
12	2.52	3.20	3.62	3.92	4.16	4.35	4.51	4.65	4.78	4.89	4.99	5.08	5.16	5.24	5.31	5.37	5.44	5.49	5.55
13	2.50	3.18	3.59	3.88	4.12	4.30	4.46	4.60	4.72	4.83	4.93	5.02	5.10	5.18	5.25	5.31	5.37	5.43	5.48
14	2.49	3.16	3.56	3.85	4.08	4.27	4.42	4.56	4.68	4.79	4.88	4.97	5.05	5.12	5.19	5.26	5.32	5.37	5.43
15	2.48	3.14	3.54	3.83	4.05	4.23	4.39	4.52	4.64	4.75	4.84	4.93	5.01	5.08	5.15	5.21	5.27	5.32	5.38
16	2.47	3.12	3.52	3.80	4.03	4.21	4.36	4.49	4.61	4.71	4.81	4.89	4.97	5.04	5.11	5.17	5.23	5.28	5.33
17	2.46	3.11	3.50	3.78	4.00	4.18	4.33	4.46	4.58	4.68	4.77	4.86	4.93	5.01	5.07	5.13	5.19	5.24	5.30
18	2.45	3.10	3.49	3.77	3.98	4.16	4.31	4.44	4.55	4.65	4.75	4.83	4.90	4.98	5.04	5.10	5.16	5.21	5.26
19	2.45	3.09	3.47	3.75	3.97	4.14	4.29	4.42	4.53	4.63	4.72	4.80	4.88	4.95	5.01	5.07	5.13	5.18	5.23
20	2.44	3.08	3.46	3.74	3.95	4.12	4.27	4.40	4.51	4.61	4.70	4.78	4.85	4.92	4.99	5.05	5.10	5.16	5.20
24	2.42	3.05	3.42	3.69	3.90	4.07	4.21	4.34	4.44	4.54	4.63	4.71	4.78	4.85	4.91	4.97	5.02	5.07	5.12
30	2.40	3.02	3.39	3.65	3.85	4.02	4.16	4.28	4.38	4.47	4.56	4.64	4.71	4.77	4.83	4.89	4.94	4.99	5.03
40	2.38	2.99	3.35	3.60	3.80	3.96	4.10	4.21	4.32	4.41	4.49	4.56	4.63	4.69	4.75	4.81	4.86	4.90	4.95
60	2.36	2.96	3.31	3.56	3.75	3.91	4.04	4.16	4.25	4.34	4.42	4.49	4.56	4.62	4.67	4.73	4.78	4.82	4.86
120	2.34	2.93	3.28	3.52	3.71	3.86	3.99	4.10	4.19	4.28	4.35	4.42	4.48	4.54	4.60	4.65	4.69	4.74	4.78
∞	2.33	2.90	3.24	3.48	3.66	3.81	3.93	4.04	4.13	4.21	4.28	4.35	4.41	4.47	4.52	4.57	4.61	4.65	4.69

* Computed under the direction of Dr. James Pacheres of Hughes Aircraft Co., Culver City, Calif. Reproduced with his kind permission.

UPPER α POINT* OF STUDENTIZED RANGE, $q_{k,\nu} = R/s$
k = SAMPLE SIZE FOR RANGE R, ν = NO. OF D.F. FOR s
$\alpha = 0.05$

ν \ k	2	3	4	5	6	7	8	9	10	11	12	13	14	15	16	17	18	19	20
1	18.0	27.0	32.8	37.1	40.4	43.1	45.4	47.4	49.1	50.6	52.0	53.2	54.3	55.4	56.3	57.2	58.0	58.8	59.6
2	6.08	8.33	9.80	10.9	11.7	12.4	13.0	13.5	14.0	14.4	14.7	15.1	15.4	15.7	15.9	16.1	16.4	16.6	16.8
3	4.50	5.91	6.82	7.50	8.04	8.48	8.85	9.18	9.46	9.72	9.95	10.2	10.3	10.5	10.7	10.8	11.0	11.1	11.2
4	3.93	5.04	5.76	6.29	6.71	7.05	7.35	7.60	7.83	8.03	8.21	8.37	8.52	8.66	8.79	8.91	9.03	9.13	9.23
5	3.64	4.60	5.22	5.67	6.03	6.33	6.58	6.80	6.99	7.17	7.32	7.47	7.60	7.72	7.83	7.93	8.03	8.12	8.21
6	3.46	4.34	4.90	5.30	5.63	5.90	6.12	6.32	6.49	6.65	6.79	6.92	7.03	7.14	7.24	7.34	7.43	7.51	7.59
7	3.34	4.16	4.68	5.06	5.36	5.61	5.82	6.00	6.16	6.30	6.43	6.55	6.66	6.76	6.85	6.94	7.02	7.10	7.17
8	3.26	4.04	4.53	4.89	5.17	5.40	5.60	5.77	5.92	6.05	6.18	6.29	6.39	6.48	6.57	6.65	6.73	6.80	6.87
9	3.20	3.95	4.41	4.76	5.02	5.24	5.43	5.59	5.74	5.87	5.98	6.09	6.19	6.28	6.36	6.44	6.51	6.58	6.64
10	3.15	3.88	4.33	4.65	4.91	5.12	5.30	5.46	5.60	5.72	5.83	5.93	6.03	6.11	6.19	6.27	6.34	6.40	6.47
11	3.11	3.82	4.26	4.57	4.82	5.03	5.20	5.35	5.49	5.61	5.71	5.81	5.90	5.98	6.06	6.13	6.20	6.27	6.33
12	3.08	3.77	4.20	4.51	4.75	4.95	5.12	5.27	5.39	5.51	5.61	5.71	5.80	5.88	5.95	6.02	6.09	6.15	6.21
13	3.06	3.73	4.15	4.45	4.69	4.88	5.05	5.19	5.32	5.43	5.53	5.63	5.71	5.79	5.86	5.93	5.99	6.05	6.11
14	3.03	3.70	4.11	4.41	4.64	4.83	4.99	5.13	5.25	5.36	5.46	5.55	5.64	5.71	5.79	5.85	5.91	5.97	6.03
15	3.01	3.67	4.08	4.37	4.59	4.78	4.94	5.08	5.20	5.31	5.40	5.49	5.57	5.65	5.72	5.78	5.85	5.90	5.96
16	3.00	3.65	4.05	4.33	4.56	4.74	4.90	5.03	5.15	5.26	5.35	5.44	5.52	5.59	5.66	5.73	5.79	5.84	5.90
17	2.98	3.63	4.02	4.30	4.52	4.70	4.86	4.99	5.11	5.21	5.31	5.39	5.47	5.54	5.61	5.67	5.73	5.79	5.84
18	2.97	3.61	4.00	4.28	4.49	4.67	4.82	4.96	5.07	5.17	5.27	5.35	5.43	5.50	5.57	5.63	5.69	5.74	5.79
19	2.96	3.59	3.98	4.25	4.47	4.65	4.79	4.92	5.04	5.14	5.23	5.31	5.39	5.46	5.53	5.59	5.65	5.70	5.75
20	2.95	3.58	3.96	4.23	4.45	4.62	4.77	4.90	5.01	5.11	5.20	5.28	5.36	5.43	5.49	5.55	5.61	5.66	5.71
24	2.92	3.53	3.90	4.17	4.37	4.54	4.68	4.81	4.92	5.01	5.10	5.18	5.25	5.32	5.38	5.44	5.49	5.55	5.59
30	2.89	3.49	3.85	4.10	4.30	4.46	4.60	4.72	4.82	4.92	5.00	5.08	5.15	5.21	5.27	5.33	5.38	5.43	5.47
40	2.86	3.44	3.79	4.04	4.23	4.39	4.52	4.63	4.73	4.82	4.90	4.98	5.04	5.11	5.16	5.22	5.27	5.31	5.36
60	2.83	3.40	3.74	3.98	4.16	4.31	4.44	4.55	4.65	4.73	4.81	4.88	4.94	5.00	5.06	5.11	5.15	5.20	5.24
120	2.80	3.36	3.68	3.92	4.10	4.24	4.36	4.47	4.56	4.64	4.71	4.78	4.84	4.90	4.95	5.00	5.04	5.09	5.13
∞	2.77	3.31	3.63	3.86	4.03	4.17	4.29	4.39	4.47	4.55	4.62	4.68	4.74	4.80	4.85	4.89	4.93	4.97	5.01

* From pp. 176–177 of *Biometrika Tables for Statisticians*, Vol. 1, by E. S. Pearson and H. O. Hartley, published by the Biometrika Trustees, Cambridge University Press, Cambridge (1954). Reproduced with the kind permission of the authors and the publisher. Corrections of ± 1 in the last figure, supplied by Dr. James Pacheres, have been incorporated in 41 entries.

UPPER α POINT* OF STUDENTIZED RANGE, $q_{k,\nu} = R/s$
k = SAMPLE SIZE FOR RANGE R, ν = NO. OF D.F. FOR s
$\alpha = 0.01$

ν \ k	2	3	4	5	6	7	8	9	10	11	12	13	14	15	16	17	18	19	20
1	90.0	135	164	186	202	216	227	237	246	253	260	266	272	277	282	286	290	294	298
2	14.0	19.0	22.3	24.7	26.6	28.2	29.5	30.7	31.7	32.6	33.4	34.1	34.8	35.4	36.0	36.5	37.0	37.5	37.9
3	8.26	10.6	12.2	13.3	14.2	15.0	15.6	16.2	16.7	17.1	17.5	17.9	18.2	18.5	18.8	19.1	19.3	19.5	19.8
4	6.51	8.12	9.17	9.96	10.6	11.1	11.5	11.9	12.3	12.6	12.8	13.1	13.3	13.5	13.7	13.9	14.1	14.2	14.4
5	5.70	6.97	7.80	8.42	8.91	9.32	9.67	9.97	10.2	10.5	10.7	10.9	11.1	11.2	11.4	11.6	11.7	11.8	11.9
6	5.24	6.33	7.03	7.56	7.97	8.32	8.61	8.87	9.10	9.30	9.49	9.65	9.81	9.95	10.1	10.2	10.3	10.4	10.5
7	4.95	5.92	6.54	7.01	7.37	7.68	7.94	8.17	8.37	8.55	8.71	8.86	9.00	9.12	9.24	9.35	9.46	9.55	9.65
8	4.74	5.63	6.20	6.63	6.96	7.24	7.47	7.68	7.87	8.03	8.18	8.31	8.44	8.55	8.66	8.76	8.85	8.94	9.03
9	4.60	5.43	5.96	6.35	6.66	6.91	7.13	7.32	7.49	7.65	7.78	7.91	8.03	8.13	8.23	8.32	8.41	8.49	8.57
10	4.48	5.27	5.77	6.14	6.43	6.67	6.87	7.05	7.21	7.36	7.48	7.60	7.71	7.81	7.91	7.99	8.07	8.15	8.22
11	4.39	5.14	5.62	5.97	6.25	6.48	6.67	6.84	6.99	7.13	7.25	7.36	7.46	7.56	7.65	7.73	7.81	7.88	7.95
12	4.32	5.04	5.50	5.84	6.10	6.32	6.51	6.67	6.81	6.94	7.06	7.17	7.26	7.36	7.44	7.52	7.59	7.66	7.73
13	4.26	4.96	5.40	5.73	5.98	6.19	6.37	6.53	6.67	6.79	6.90	7.01	7.10	7.19	7.27	7.34	7.42	7.48	7.55
14	4.21	4.89	5.32	5.63	5.88	6.08	6.26	6.41	6.54	6.66	6.77	6.87	6.96	7.05	7.12	7.20	7.27	7.33	7.39
15	4.17	4.83	5.25	5.56	5.80	5.99	6.16	6.31	6.44	6.55	6.66	6.76	6.84	6.93	7.00	7.07	7.14	7.20	7.26
16	4.13	4.78	5.19	5.49	5.72	5.92	6.08	6.22	6.35	6.46	6.56	6.66	6.74	6.82	6.90	6.97	7.03	7.09	7.15
17	4.10	4.74	5.14	5.43	5.66	5.85	6.01	6.15	6.27	6.38	6.48	6.57	6.66	6.73	6.80	6.87	6.94	7.00	7.05
18	4.07	4.70	5.09	5.38	5.60	5.79	5.94	6.08	6.20	6.31	6.41	6.50	6.58	6.65	6.72	6.79	6.85	6.91	6.96
19	4.05	4.67	5.05	5.33	5.55	5.73	5.89	6.02	6.14	6.25	6.34	6.43	6.51	6.58	6.65	6.72	6.78	6.84	6.89
20	4.02	4.64	5.02	5.29	5.51	5.69	5.84	5.97	6.09	6.19	6.29	6.37	6.45	6.52	6.59	6.65	6.71	6.76	6.82
24	3.96	4.54	4.91	5.17	5.37	5.54	5.69	5.81	5.92	6.02	6.11	6.19	6.26	6.33	6.39	6.45	6.51	6.56	6.61
30	3.89	4.45	4.80	5.05	5.24	5.40	5.54	5.65	5.76	5.85	5.93	6.01	6.08	6.14	6.20	6.26	6.31	6.36	6.41
40	3.82	4.37	4.70	4.93	5.11	5.27	5.39	5.50	5.60	5.69	5.77	5.84	5.90	5.96	6.02	6.07	6.12	6.17	6.21
60	3.76	4.28	4.60	4.82	4.99	5.13	5.25	5.36	5.45	5.53	5.60	5.67	5.73	5.79	5.84	5.89	5.93	5.98	6.02
120	3.70	4.20	4.50	4.71	4.87	5.01	5.12	5.21	5.30	5.38	5.44	5.51	5.56	5.61	5.66	5.71	5.75	5.79	5.83
∞	3.64	4.12	4.40	4.60	4.76	4.88	4.99	5.08	5.16	5.23	5.29	5.35	5.40	5.45	5.49	5.54	5.57	5.61	5.65

* From pp. 176–177 of *Biometrika Tables for Statisticians*, Vol. I, by E. S. Pearson and H. O. Hartley, published by the Biometrika Trustees, Cambridge University Press, Cambridge (1954). Reproduced with the kind permission of the authors and the publisher.

Pearson and Hartley Charts and Fox Charts for the Power of the F-Test

Pearson and Hartley Charts* for the Power of the *F*-Test

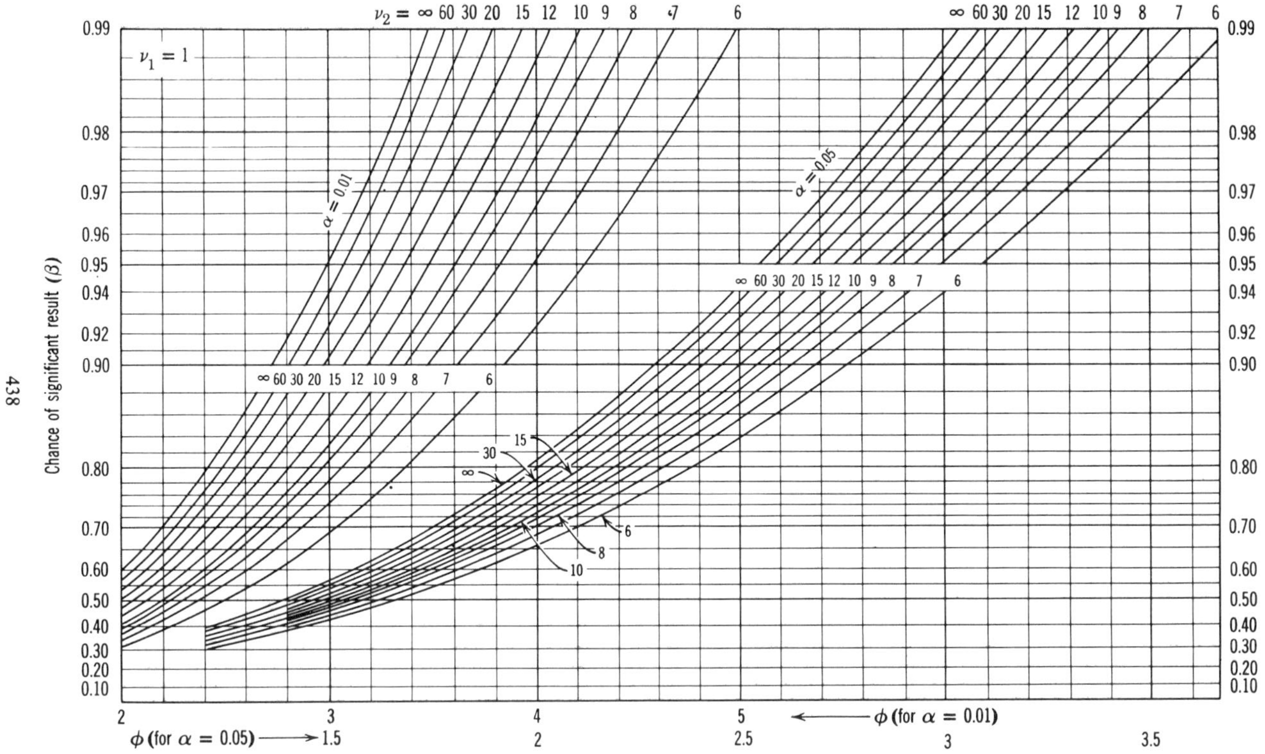

* By E. S. Pearson and H. O. Hartley in *Biometrika*, Vol. 38, pp. 115–122 (1951). Reproduced with the kind permission of the authors and the editor.

Pearson and Hartley Charts* for the Power of the F-Test

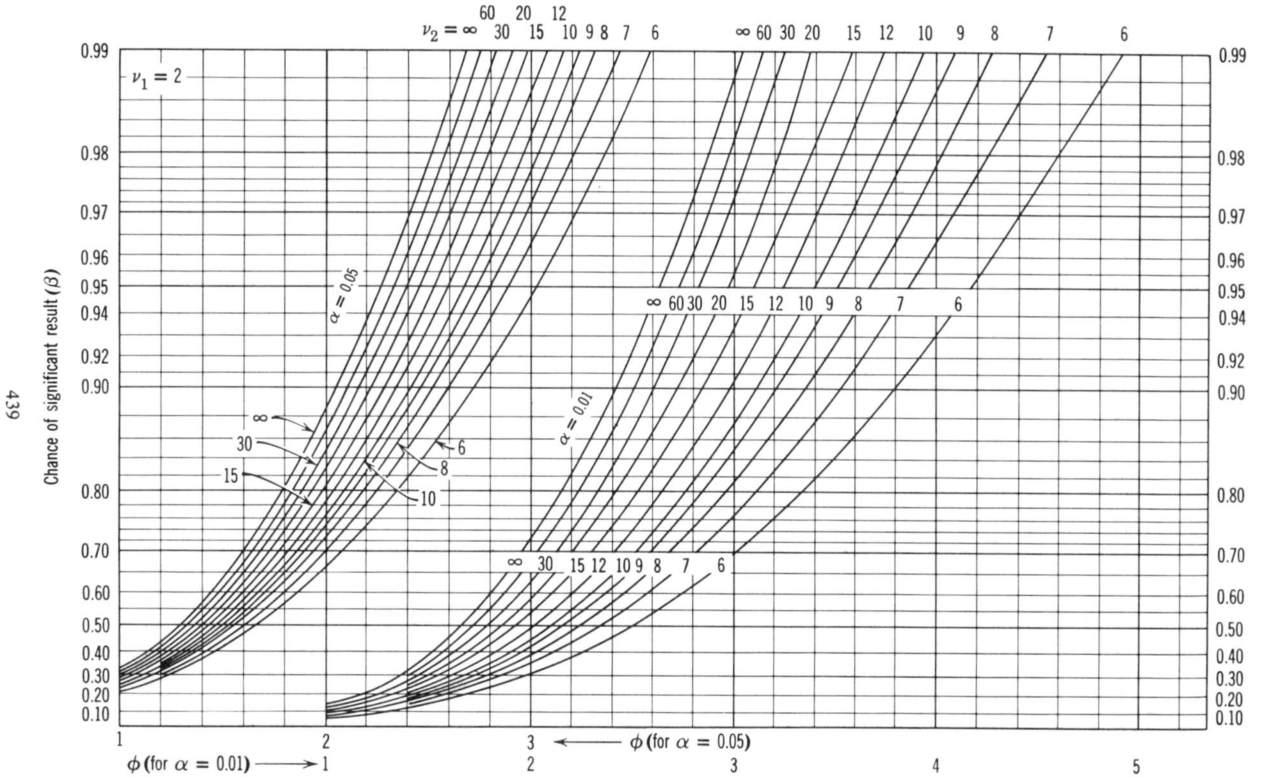

* By E. S. Pearson and H. O. Hartley in *Biometrika*, Vol. 38, pp. 115–122 (1951). Reproduced with the kind permission of the authors and the editor.

Pearson and Hartley Charts* for the Power of the F-Test

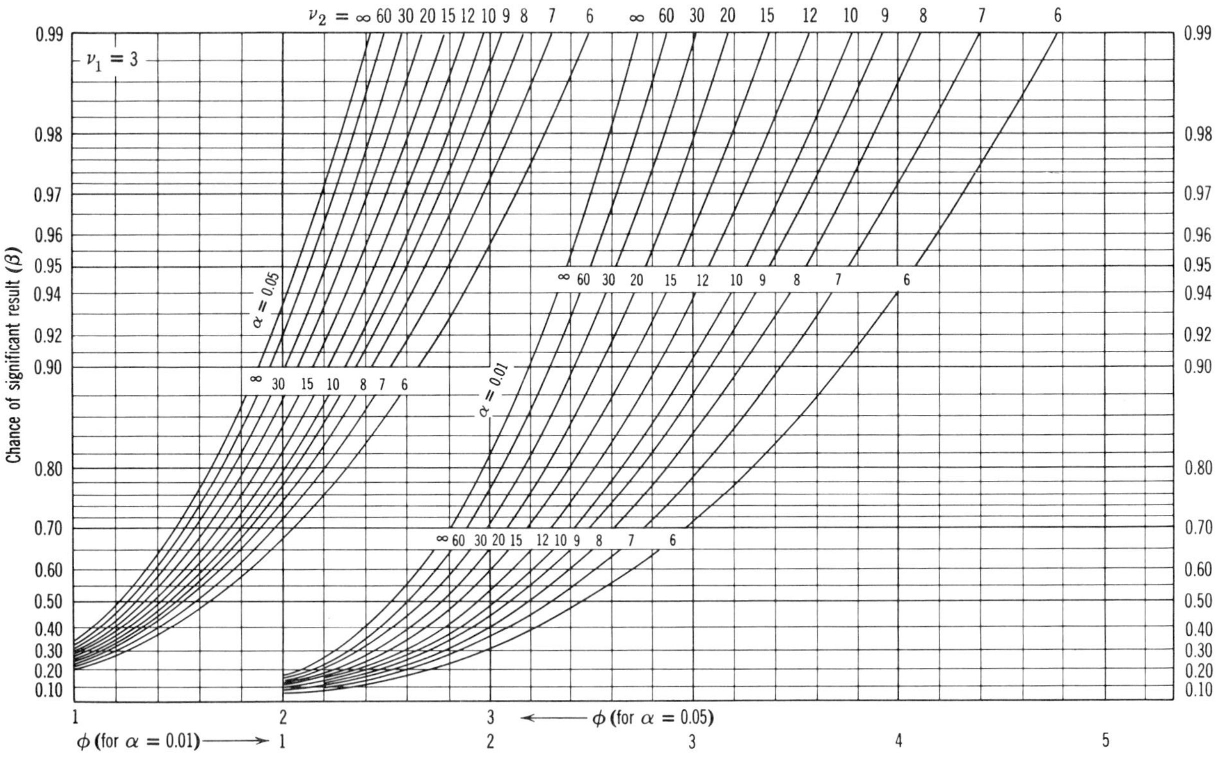

* By E. S. Pearson and H. O. Hartley in *Biometrika*, Vol. 38, pp. 115–122 (1951). Reproduced with the kind permission of the authors and the editor.

Pearson and Hartley Charts* for the Power of the F-Test

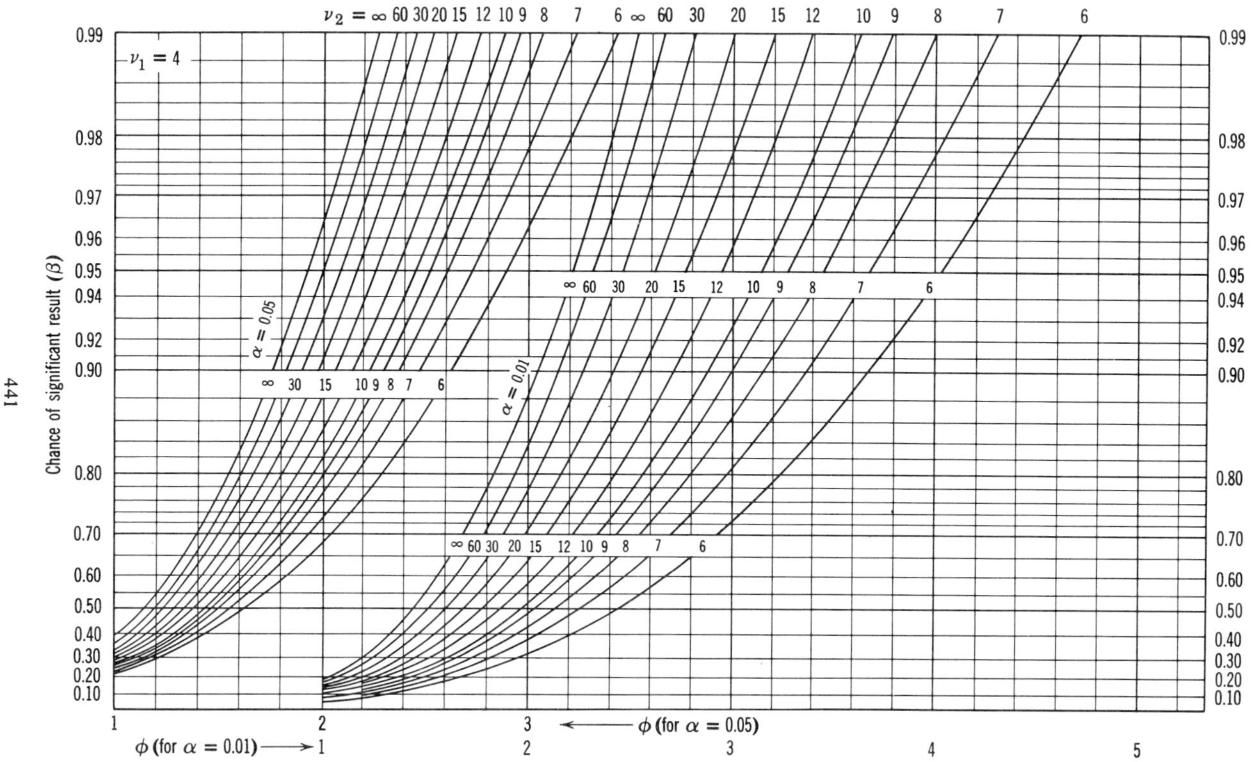

* By E. S. Pearson and H. O. Hartley in *Biometrika*, Vol. 38, pp. 115–122 (1951). Reproduced with the kind permission of the authors and the editor.

Pearson and Hartley Charts* for the Power of the F-Test

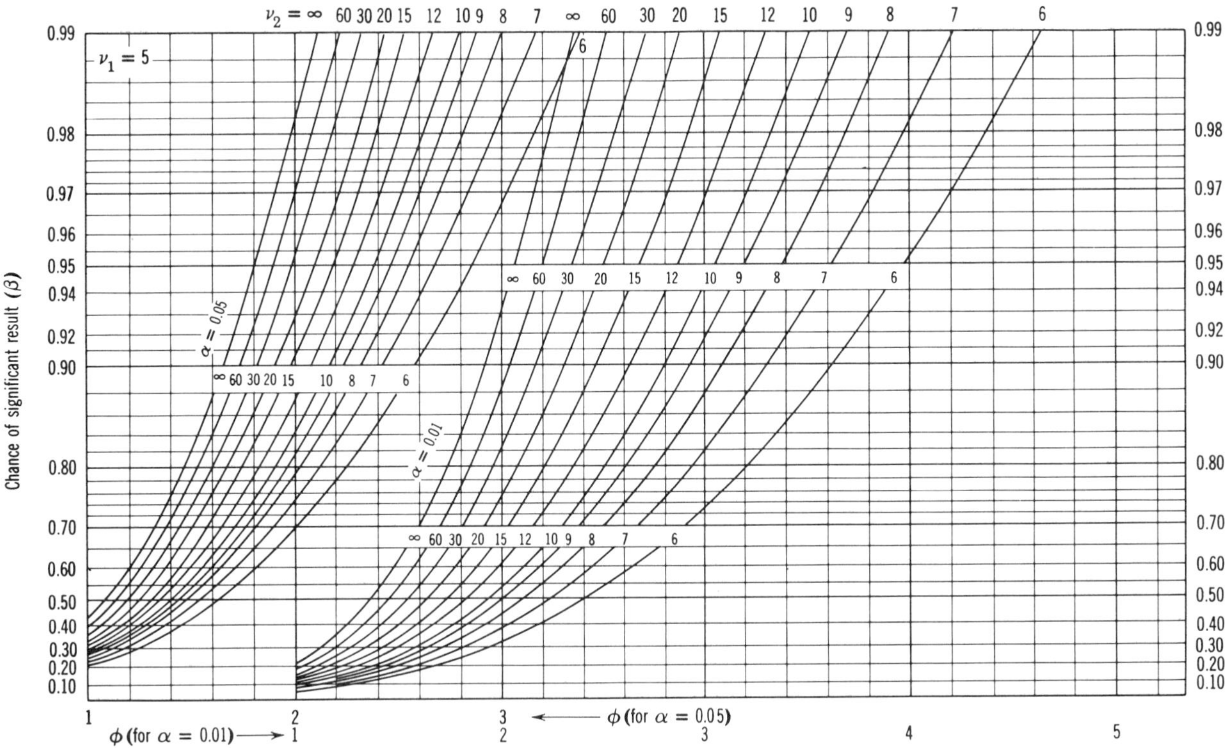

* By E. S. Pearson and H. O. Hartley in *Biometrika*, Vol. 38, pp. 115–122 (1951). Reproduced with the kind permission of the authors and the editor.

Pearson and Hartley Charts* for the Power of the F-Test

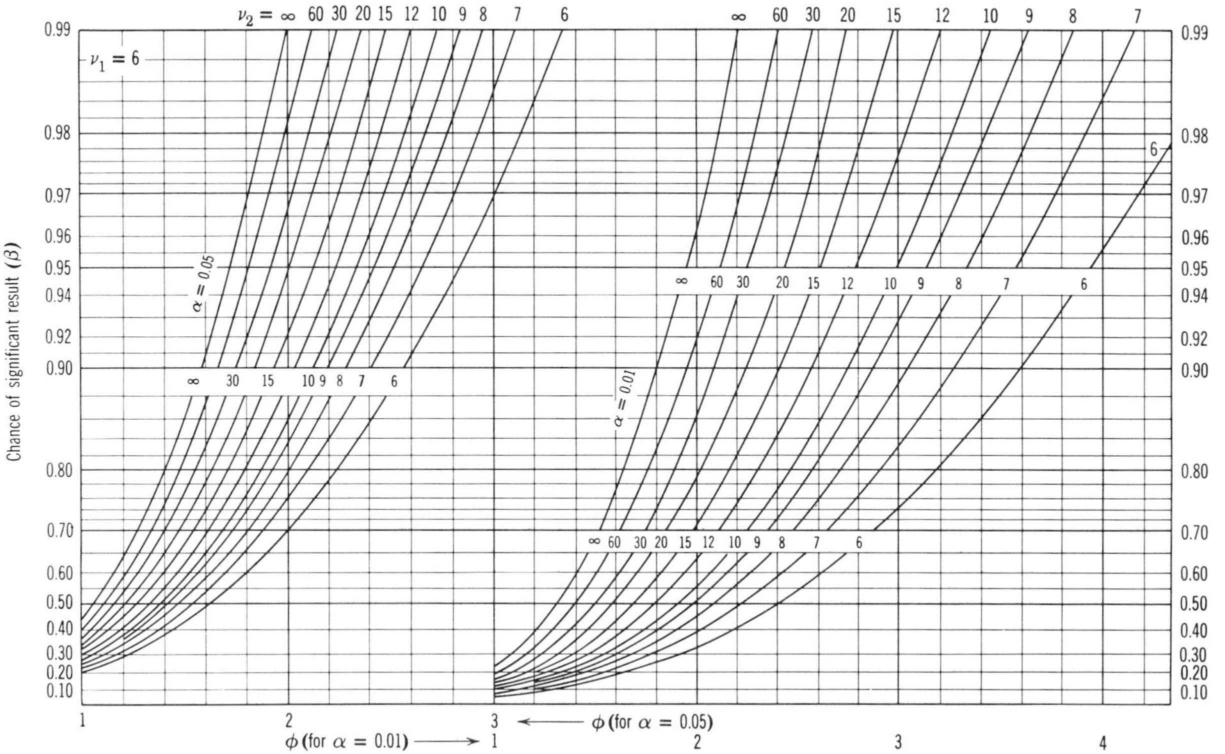

* By E. S. Pearson and H. O. Hartley in *Biometrika*, Vol. 38, pp. 115–122 (1951). Reproduced with the kind permission of the authors and the editor.

Pearson and Hartley Charts* for the Power of the *F*-Test

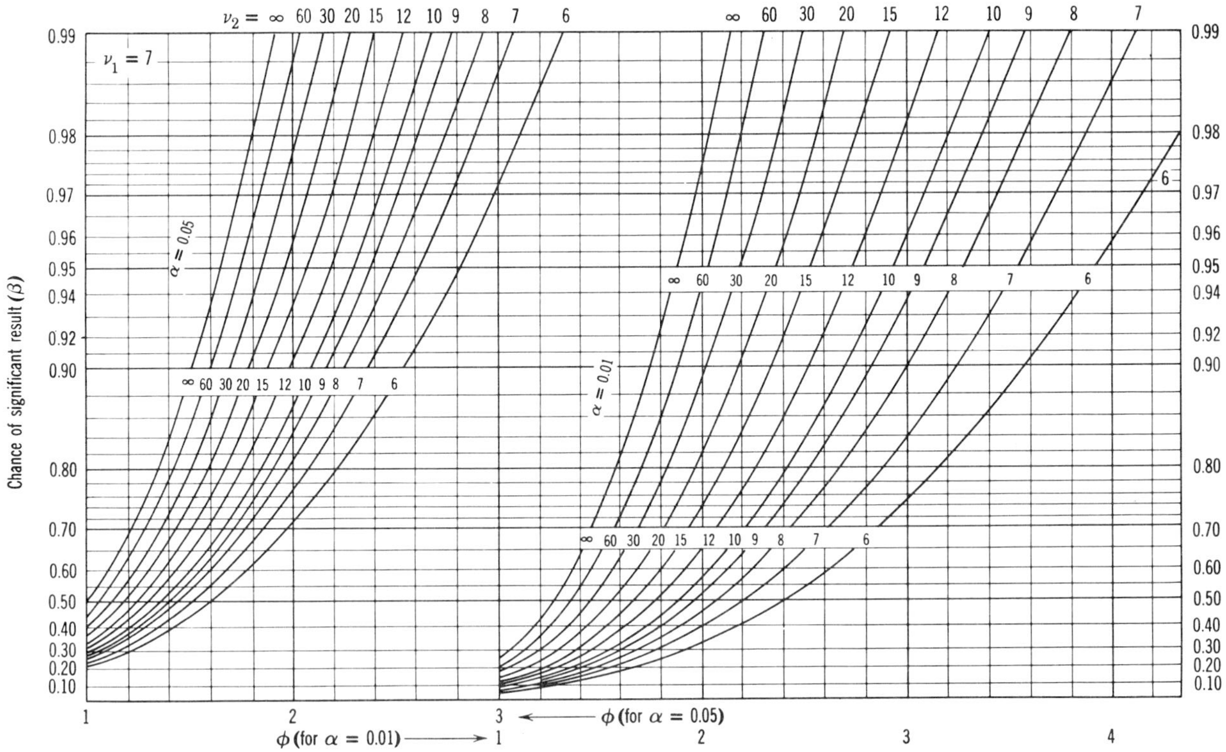

* By E. S. Pearson and H. O. Hartley in *Biometrika*, Vol. 38, pp. 115–122 (1951). Reproduced with the kind permission of the authors and the editor.

Pearson and Hartley Charts* for the Power of the F-Test

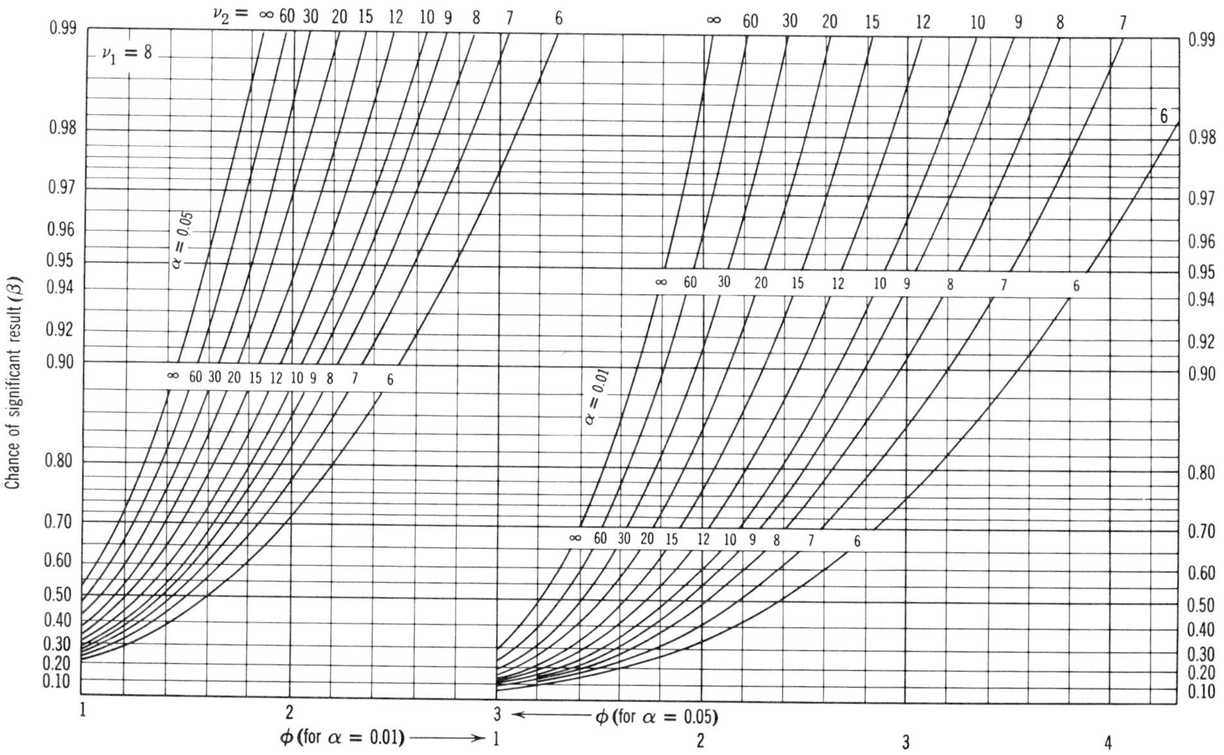

* By E. S. Pearson and H. O. Hartley in *Biometrika*, Vol. 38, pp. 115–122 (1951). Reproduced with the kind permission of the authors and the editor.

Fox Charts* for the Power of the F-Test

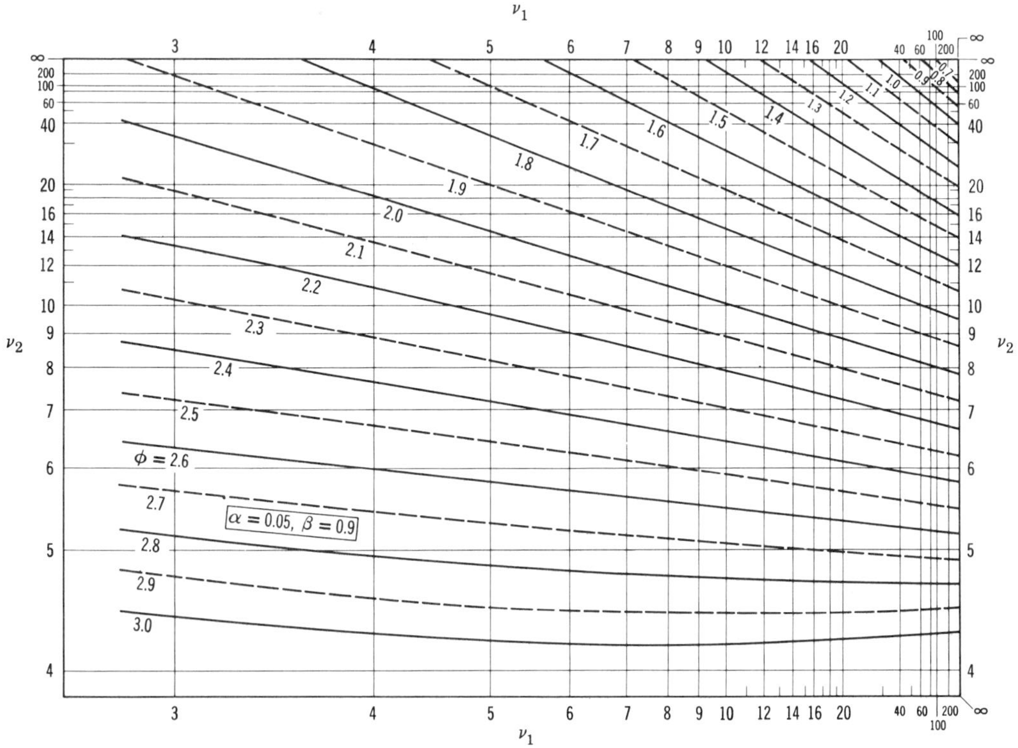

* By M. Fox in *Annals Math. Stat.*, Vol. 27, pp. 485–494 (1956). Reproduced with the kind permission of the author and the editor.

Fox Charts* for the Power of the *F*-Test

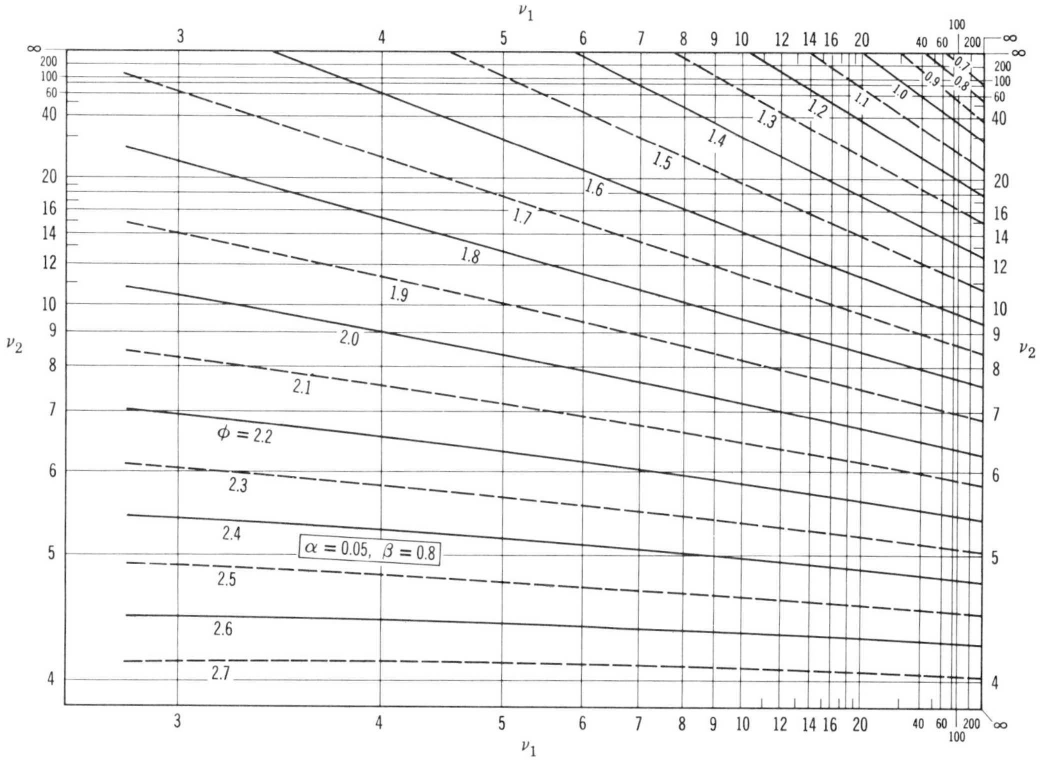

* By M. Fox in *Annals Math. Stat.*, Vol. 27, pp. 485–494 (1956). Reproduced with the kind permission of the author and the editor.

Fox Charts* for the Power of the F-Test

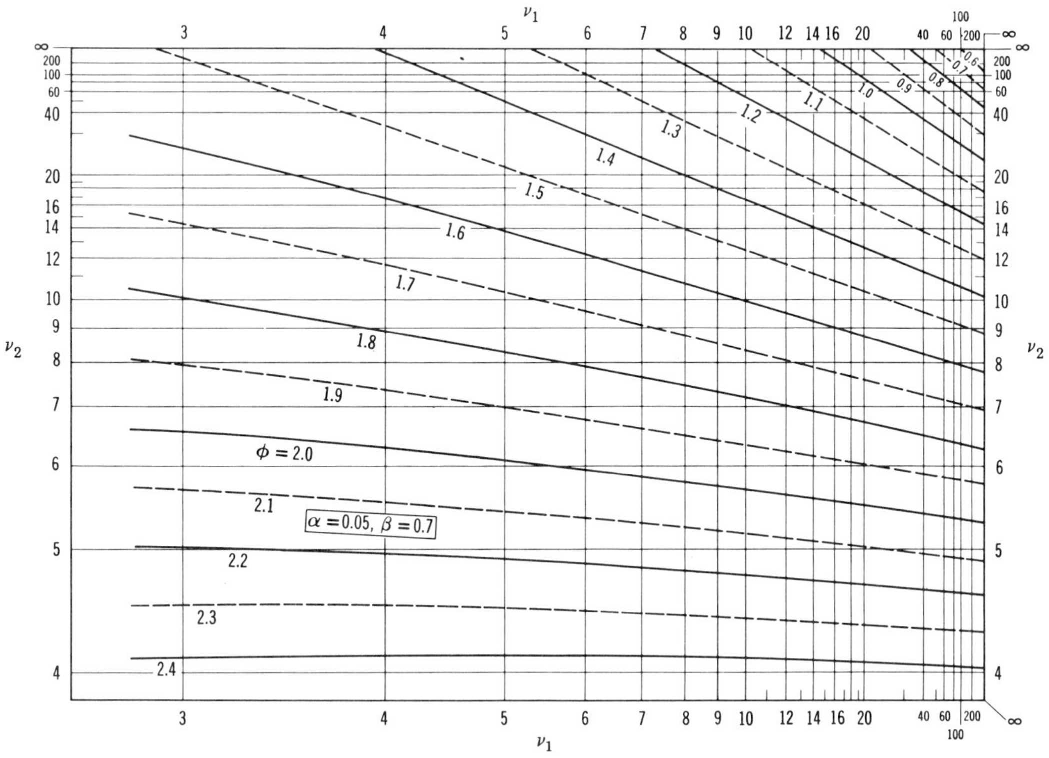

* By M. Fox in *Annals Math. Stat.*, Vol. 27, pp. 485–494 (1956). Reproduced with the kind permission of the author and the editor.

Fox Charts* for the Power of the F-Test

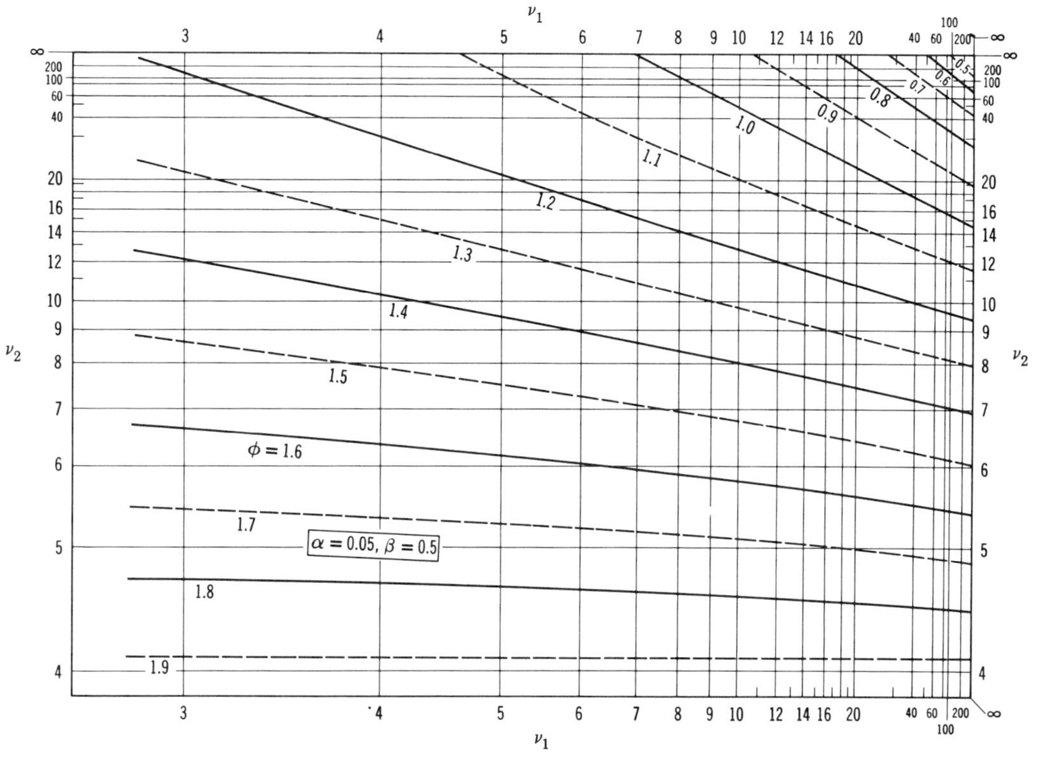

* By M. Fox in *Annals Math. Stat.*, Vol. 27, pp. 485–494 (1956). Reproduced with the kind permission of the author and the editor.

FOX CHARTS* FOR THE POWER OF THE F-TEST

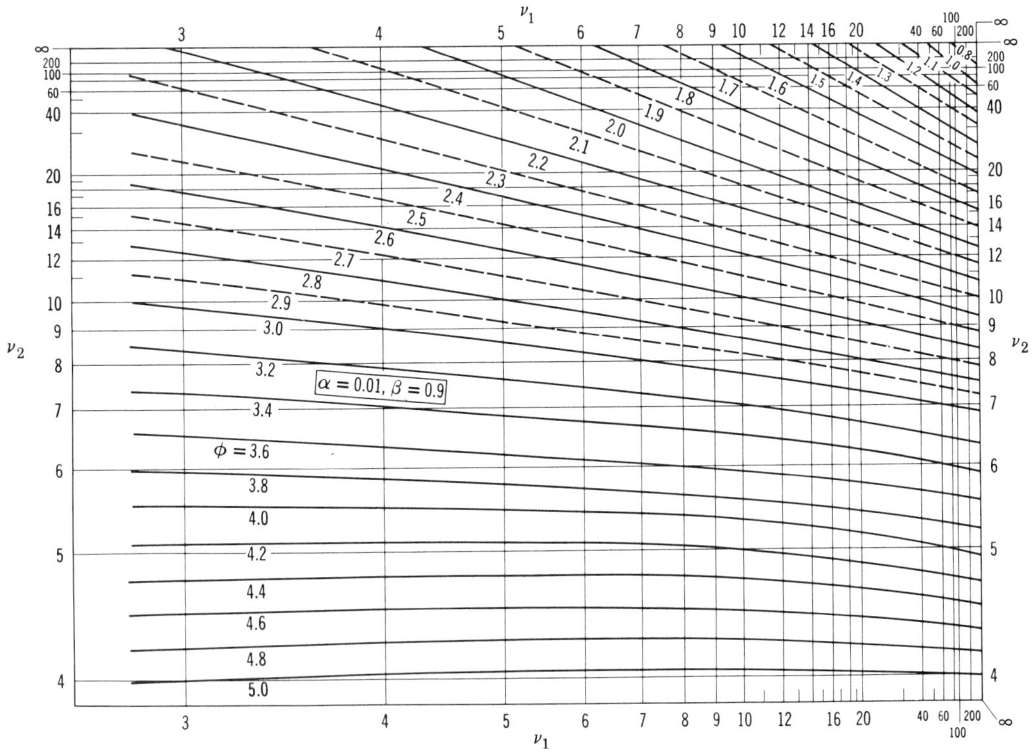

* By M. Fox in *Annals Math. Stat.*, Vol. 27, pp. 485–494 (1956). Reproduced with the kind permission of the author and the editor.

Fox Charts* for the Power of the F-Test

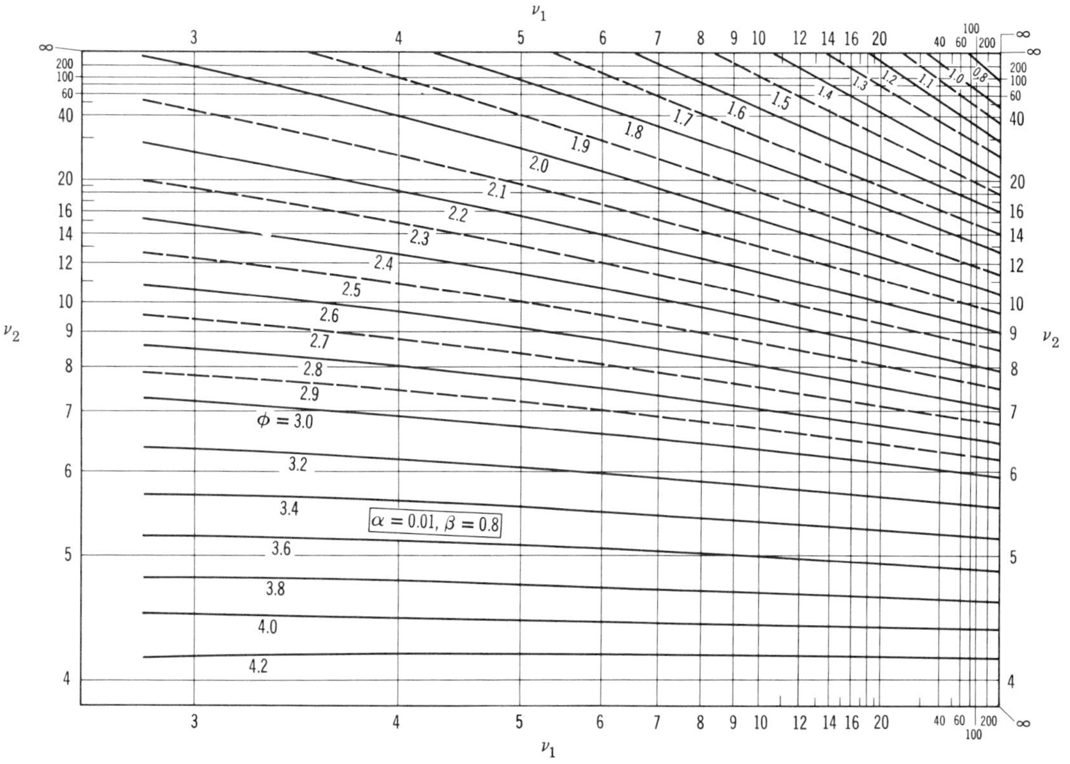

* By M. Fox in *Annals Math. Stat.*, Vol. 27, pp. 485–494 (1956). Reproduced with the kind permission of the author and the editor.

Fox Charts* for the Power of the F-Test

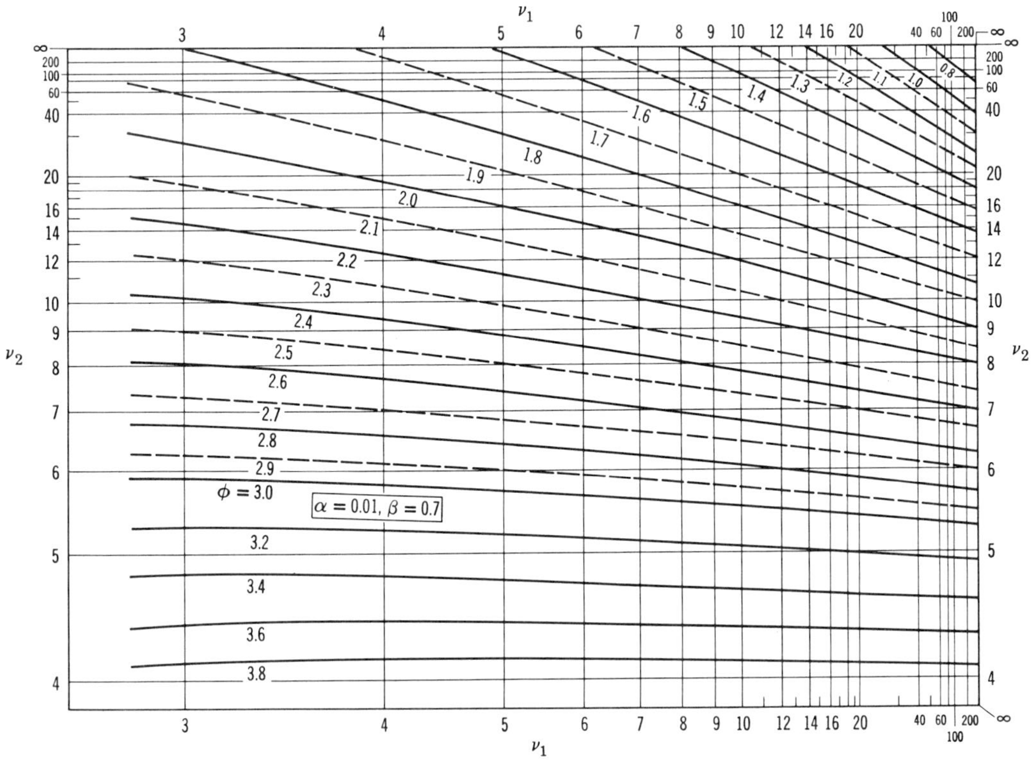

* By M. Fox in *Annals Math. Stat.*, Vol. 27, pp. 485–494 (1956). Reproduced with the kind permission of the author and the editor.

Fox Charts* for the Power of the F-Test

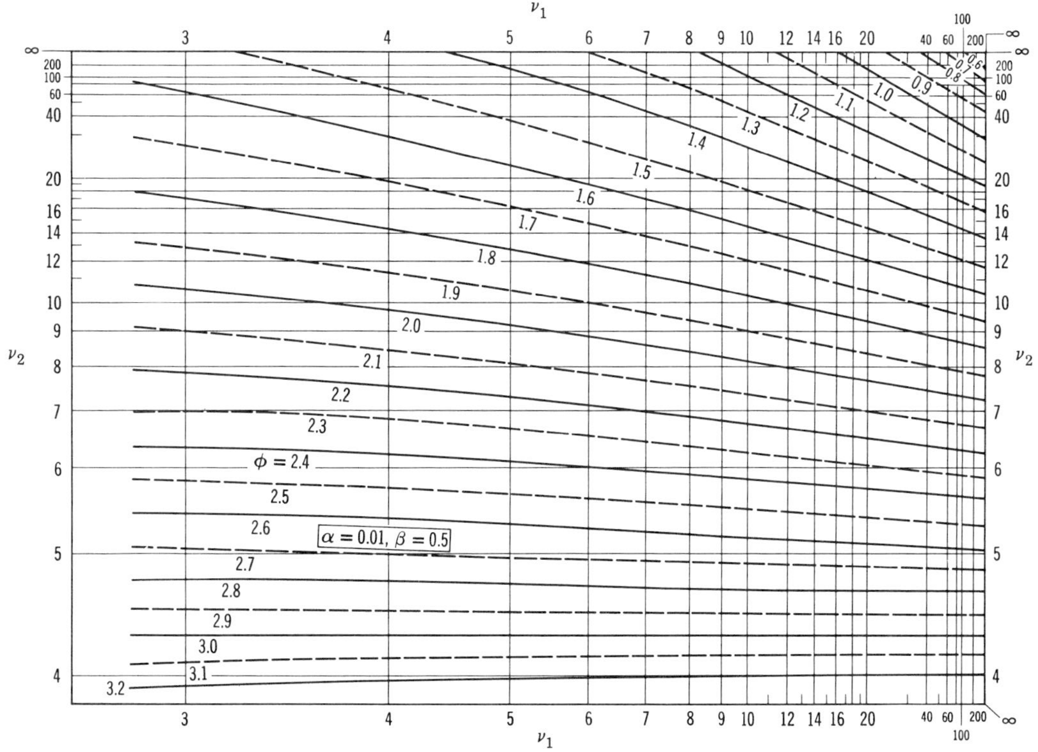

* By M. Fox in *Annals Math. Stat.*, Vol. 27, pp. 485–494 (1956). Reproduced with the kind permission of the author and the editor.

Fox Charts* for the Power of the F-Test

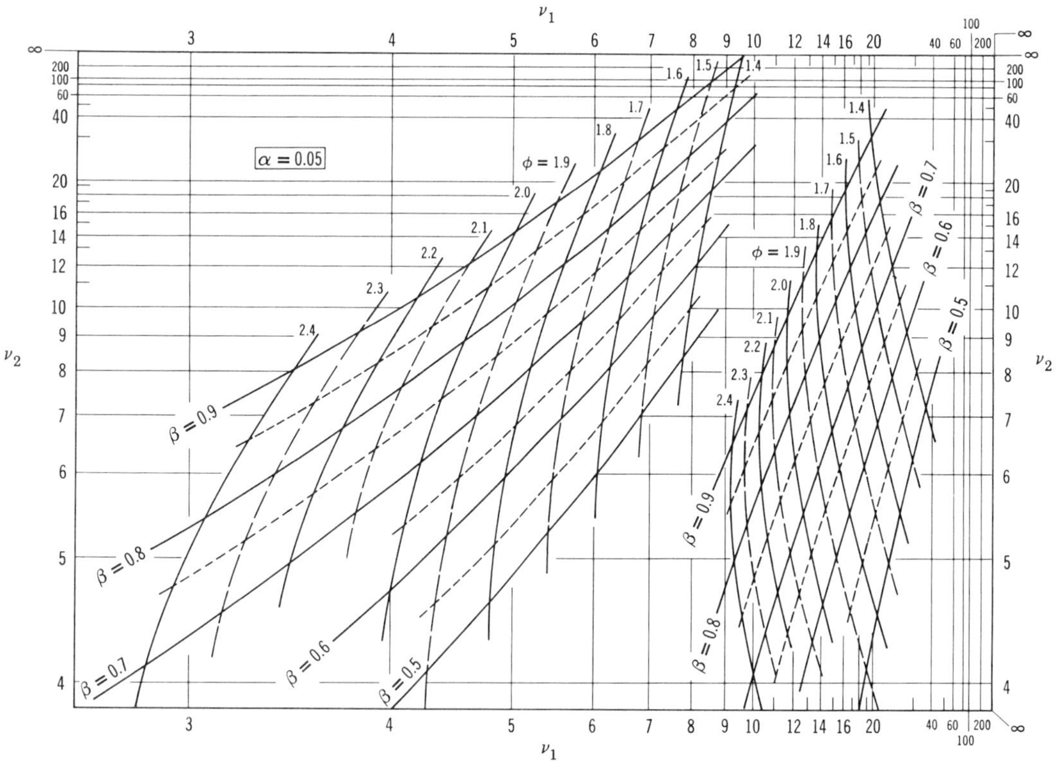

* By M. Fox in *Annals Math. Stat.*, Vol. 27, pp. 485–494 (1956). Reproduced with the kind permission of the author and the editor.

Fox Charts* for the Power of the *F*-Test

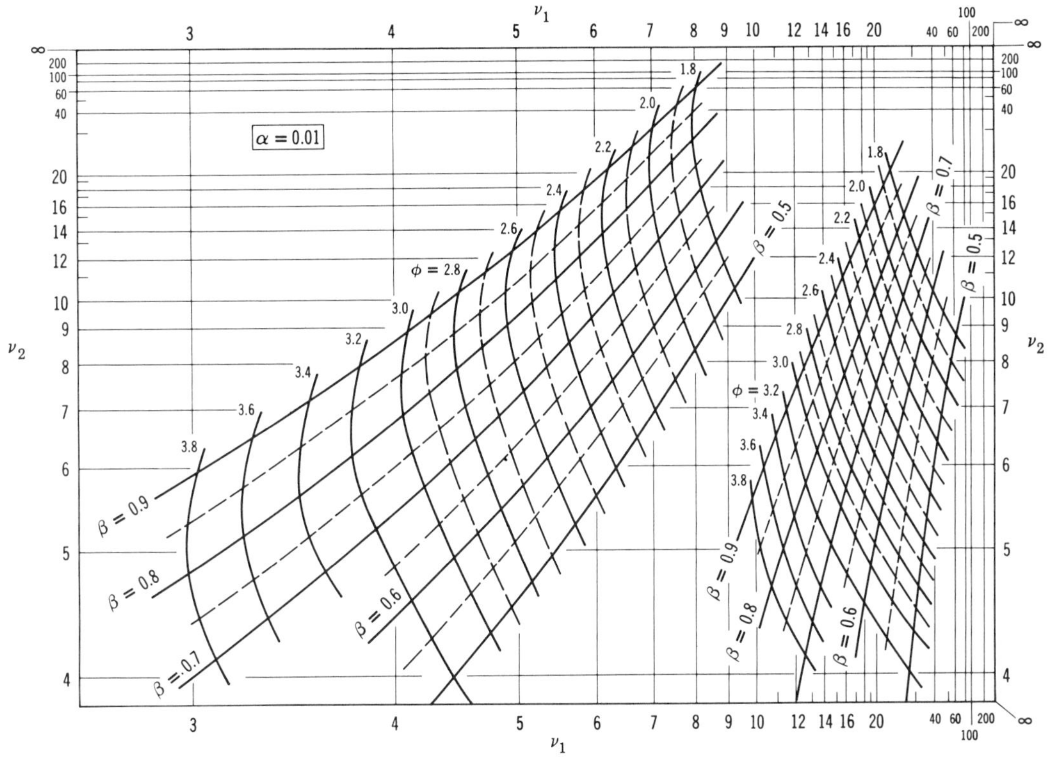

* By M. Fox in *Annals Math. Stat.*, Vol. 27, pp. 485–494 (1956). Reproduced with the kind permission of the author and the editor.

Author Index and Bibliography

[Pages on which author is cited are given in brackets.]

S. L. Andersen, *see* G. E. P. Box and . . . (1955).

R. L. Anderson and T. A. Bancroft (1952), *Statistical Theory in Research*, McGraw-Hill, New York. [129]

T. W. Anderson (1958), *An Introduction to Multivariate Statistical Analysis*, John Wiley, New York. [196]

Anonymous [325]

T. A. Bancroft, *see* R. L. Anderson and . . . (1952).

T. A. Bancroft, *see also* H. Bozivich, . . . , and H. O. Hartley (1956).

M. S. Bartlett (1934), "The vector representation of a sample," *Proc. Cambridge Philos. Soc.*, Vol. 30, pp. 327–340. [42]

M. S. Bartlett (1935), "The effect of non-normality on the *t*-distribution," *Proc. Cambridge Philos. Soc.*, Vol. 31, pp. 223–231. [346]

M. S. Bartlett (1937), "Properties of sufficiency and statistical tests," *Proc. Roy. Soc. London*, series A, Vol. 160, pp. 268–282. [83, 362]

M. S. Bartlett and D. G. Kendall (1946), "The statistical analysis of variance–heterogeneity and the logarithmic transformation," *J. Royal. Stat. Soc.*, Series B, Vol. 8, pp. 128–138. [84]

R. E. Bechhofer (1951), *The Effect of Preliminary Tests of Significance on the Size and Power of Certain Tests of Univariate Linear Hypotheses*, unpublished Ph.D. thesis, Columbia Univ. [126]

R. E. Bechhofer (1954), "A single-sample multiple-decision procedure for ranking means of normal populations with known variances," *Annals Math. Stat.*, Vol. 25, pp. 16–39. [81].

R. E. Bechhofer (1958), "A sequential multiple decision procedure for selecting the best one of several normal populations with a common unknown variance, and its use with various experimental designs," *Biometrics*, Vol. 14, pp. 408–429. [81]

C. A. Bennett and N. L. Franklin (1954), *Statistical Analysis in Chemistry and the Chemical Industry*, John Wiley, New York. [126, 274, 284]

J. Berkson (1950), "Are there two regressions?" *J. Amer. Stat. Assoc.*, Vol. 45, pp. 164–180. [214]

G. Birkhoff and S. MacLane (1953), *A Survey of Modern Algebra*, Macmillan, New York. [371, 380, 387]

R. C. Bose (1944), "The fundamental theorem of linear estimation," *Proc. 31st Indian Sci. Congress*, pp. 2–3 (abstract). [13, 23]

R. C. Bose (1949*a*), *Least Squares Aspects of Analysis of Variance*, Inst. Stat. Mimeo, series 9, Chapel Hill, N.C. [127]

R. C. Bose (1949b), "A note on Fisher's inequality for balanced incomplete block designs," *Annals Math. Stat.*, Vol. 20, pp. 619–620. [162]

R. C. Bose, W. H. Clatworthy, and S. S. Shrikhande (1954), *Tables of Partially Balanced Designs with Two Associate Classes*, Inst. Stat. Univ. N. C., reprint series 50, Raleigh. [163]

G. E. P. Box (1953), "Non-normality and tests on variances," *Biometrika*, Vol. 40, pp. 318–335. [87, 337, 360, 362]

G. E. P. Box (1954*a*), "Some theorems on quadratic forms applied in the study of analysis of variance problems: I. Effect of inequality of variance in the one-way classification," *Annals Math. Stat.*, Vol. 25, pp. 290–302. [247, 343, 354, 355]

G. E. P. Box (1954*b*), "Some theorems on quadratic forms applied in the study of analysis of variance problems: II. Effect of inequality of variance and of correlation of errors in the two-way classification," *Annals Math. Stat.*, Vol. 25, pp. 484–498. [355, 356, 359, 360, 364]

G. E. P. Box and S. L. Andersen (1955), "Permutation theory in the derivation of robust criteria and the study of departures from assumption," *J. Roy. Stat. Soc.*, series B, Vol. 17, pp. 1–34. [327, 347, 348, 349, 350]

G. E. P. Box, personal communication. [83]

H. Bozivich, T. A. Bancroft, and H. O. Hartley (1956), "Power of analysis of variance test procedures for certain incompletely specified models," *Annals Math. Stat.*, Vol. 27, pp. 1017–1043. [126]

I. Bross (1950), "Fiducial intervals for variance components," *Biometrics*, Vol. 6, pp. 136–144. [231]

R. H. Bruck and H. J. Ryser (1949), "The nonexistence of certain finite projective geometries," *Canadian J. Math.*, Vol. 1, pp. 88–93. [159]

M. G. Bulmer (1957), "Approximate confidence limits for components of variance," *Biometrika*, Vol. 44, pp. 159–167. [231, 235]

W. H. Clatworthy, *see* R. C. Bose, . . . , and S. S. Shrikhande (1954).

W. G. Cochran (1934), "The distribution of quadratic forms in a normal system, with applications to the analysis of covariance," *Proc. Cambridge Philos. Soc.*, Vol. 30, pp. 178–191. [419, 420]

W. G. Cochran and G. Cox (1957), *Experimental Designs*, second edition, John Wiley, New York. [105, 129, 149, 163, 170]

J. Cornfield and J. W. Tukey (1956), "Average values of mean squares in factorials," *Annals Math. Stat.*, Vol. 27, pp. 907–949. [284]

R. Courant (1950), *Differential and Integral Calculus*, Interscience, New York. [118]

G. Cox, *see* W. G. Cochran and . . . (1957).

H. Cramér (1946), *Mathematical Methods of Statistics*, Princeton Univ. Press. [236, 335, 336, 349, 364, 369, 416]

C. Daniel, personal communication. [63]

F. N. David and N. L. Johnson (1951), "The effects of non-normality on the power function of the F-test in the analysis of variance," *Biometrika*, Vol. 38, pp. 43–57. [356]

O. L. Davies, editor (1956), *The Design and Analysis of Industrial Experiments*, second edition, Oliver & Boyd, Edinburgh, and Hafner Publishing Co., New York. [133, 159]

D. B. Duncan (1952), "On the properties of the multiple comparison test," *Virginia J. of Sci.*, Vol. 3, pp. 49–67. [78]

D. B. Duncan (1955), "Multiple range and multiple F-tests," *Biometrics*, Vol. 11, pp. 1–42. [78]

D. B. Duncan (1957), "Multiple range tests for correlated and heteroscedastic means," *Biometrics*, Vol. 13, pp. 164–176. [78]

D. B. Duncan [67]

P. S. Dwyer (1951), *Linear Computations*, John Wiley, New York. [205]

T. Eden and F. Yates (1933), "On the validity of Fisher's z-test when applied to an actual sample of non-normal data," *J. Agricultural Sci.*, Vol. 23, pp. 6–16. [324]

C. Eisenhart (1947), "The assumptions underlying the analysis of variance," *Biometrics*, Vol. 3, pp. 1–21. [6]

R. A. Fisher (1918), "The correlation between relatives on the supposition of Mendelian inheritance," *Trans. Roy. Soc. Edinburgh*, Vol. 52, pp. 399–433. [3]

R. A. Fisher (1925), *Statistical Methods for Research Workers*, first edition, Oliver & Boyd, Edinburgh. [3, 36, 90, 223, 313, 359]

R. A. Fisher (1926), "The arrangement of field experiments," *J. Ministry Agriculture*, Vol. 33, pp. 503–513. Included as paper 17 in *Contributions to Mathematical Statistics* by R. A. Fisher, John Wiley, New York, 1950. [147]

R. A. Fisher (1932), *Statistical Methods for Research Workers*, fourth edition, Oliver & Boyd, Edinburgh. [195]

R. A. Fisher (1935), *The Design of Experiments*, Oliver & Boyd, Edinburgh. [3, 9C, 313, 324]

R. A. Fisher (1940), "An examination of the different possible solutions of a problem in incomplete blocks," *Annals Eugenics*, Vol. 10, pp. 52–75. [162]

R. A. Fisher and F. Yates (1943), *Statistical Tables for Biological, Agricultural, and Medical Research*, Oliver & Boyd, Edinburgh. [76, 129, 150, 159, 193, 364]

R. A. Fisher [364]

E. Fix (1949), "Tables of noncentral χ^2," *Univ. Calif. Publications in Statistics*, Vol. 1, no. 2, pp. 15–19. The headings $\alpha = .01$ and $\alpha = .05$ on pp. 17 and 19 should be interchanged. [42]

M. Fox (1956), "Charts of the power of the F-test," *Annals Math. Stat.*, Vol. 27, pp. 484–497. [42]

N. L. Franklin, *see* C. A. Bennett and . . . (1954).

H. A. Freeman (1942), *Industrial Statistics*, John Wiley, New York. [193]

M. F. Freeman and J. W. Tukey (1950), "Transformations related to the angular and the square root transformations," *Annals Math. Stat.*, Vol. 21, pp. 607–611. [365]

K. F. Gauss (1809), *Theoria motus corporum coelestium in sectionibus conicis solem ambientium*, Perthes and Besser, Hamburg. [9]

W. Gautschi, personal communication. [14]

A. K. Gayen (1950), "The distribution of the variance ratio in random samples of any size drawn from non-normal universes," *Biometrika*, Vol. 37, pp. 236–255. [350]

R. C. Geary (1936), "The distribution of 'Student's' ratio for non-normal samples," *J. Roy. Stat. Soc., Suppl.*, Vol. 3, pp. 178–184. [346]

R. C. Geary (1947), "Testing for normality," *Biometrika*, Vol. 34, pp. 209–242. [346]

F. A. Graybill (1954), "On quadratic estimation of variance components," *Annals Math. Stat.*, Vol. 25, pp. 367–372. [221]

U. Grenander and M. Rosenblatt (1956), *Statistical Analysis of Stationary Time Series*, John Wiley, New York. [334]

J. M. Hammersley (1949), "The unbiased estimate and standard error of the interclass variance," *Metron*, Vol. 15, pp. 173–188. [237, 346]

H. O. Hartley, *see* E. S. Pearson and . . . (1951).

H. O. Hartley, S. S. Shrikhande, and W. B. Taylor (1953), "A note on incomplete block designs with row balance," *Annals Math. Stat.*, Vol. 24, pp. 123–126. [169]

H. O. Hartley, *see also* E. S. Pearson and . . . (1954).

H. O. Hartley, *see also* H. Bozivich, T. A. Bancroft, and . . . (1956).

W. C. Healy Jr. (1956), "Two-sample procedures in simultaneous estimation," *Annals Math. Stat.*, Vol. 27, pp. 687–702. [65]

L. H. Herbach (1957), *Optimum Properties of Analysis of Variance Tests Based on Model II and Some Generalizations of Model II*," Scientific Paper 6, Engineering Statistics Laboratory, New York Univ. College of Engineering, Research Division, New York, 170 pp. lithographed. [221]

J. L. Hodges Jr., and E. L. Lehmann (1956), "The efficiency of some nonparametric competitors of the *t*-test," *Annals Math. Stat.*, Vol. 27, pp. 324–335. [361]

J. L. Hodges Jr., personal communication. [65, 230]

W. Hoeffding (1952), "The large sample power of tests based on permutations of observations," *Annals Math. Stat.*, Vol. 23, pp. 169–192. [317, 323]

R. Hooke (1956), "Some applications of bipolykays to the estimation of variance components and their moments," *Annals Math. Stat.*, Vol. 27, pp. 80–98. [346]

G. Horsnell (1953), "The effect of unequal group variances on the *F*-test for the homogeneity of group means," *Biometrika*, Vol. 40, pp. 128–136. [356, 357, 358]

H. Hotelling, *see* H. Working and . . . (1929).

H. Hotelling (1931), "The generalization of 'Student's' ratio," *Annals Math. Stat.*, Vol. 2, pp. 360–378. [29, 270]

P. L. Hsu (1938*a*), "Contribution to the theory of 'Student's' *t*-test as applied to the problem of two samples," *Stat. Research Memoirs*, Vol. 2, pp. 1–24. [353, 355, 356, 364]

P. L. Hsu (1938*b*), "On the best unbiased quadratic estimate of the variance," *Stat. Research Memoirs*, Vol. 2, pp. 91–104. [23]

P. L. Hsu (1938*c*), "Notes on Hotelling's generalized *T*," *Annals Math. Stat.*, Vol. 9, pp. 231–243. [272, 418]

P. L. Hsu (1941), "Analysis of variance from the power function standpoint," *Biometrika*, Vol. 32, pp. 62–69. [48]

G. A. Hunt and C. M. Stein, unpublished work. [50]

J. S. Hunter, *see* W. J. Youden and . . . (1955).

J. P. Imhof (1958), *Contributions to the Theory of Mixed Models in the Analysis of Variance*, Ph.D. thesis, Department of Statistics, Univ. Calif., Berkeley. [270, 289]

G. S. James (1951), "The comparison of several groups of observations when the ratios of the population variances are unknown," *Biometrika*, Vol. 38, pp. 324–329. [364]

N. L. Johnson, *see* F. N. David and . . . (1951).

O. Kempthorne (1952), *The Design and Analysis of Experiments*, John Wiley, New York. [5, 291, 293, 297, 321, 324]

O. Kempthorne (1955), "The randomization theory of experimental inference," *J. Amer. Stat. Assoc.*, Vol. 50, pp. 946–967. [291]

O. Kempthorne, *see also* M. B. Wilk and . . . (1955).

O. Kempthorne, *see also* M. B. Wilk and . . . (1957).

D. G. Kendall, *see* M. S. Bartlett and . . . (1946).

M. G. Kendall (1946), *The Advanced Theory of Statistics*, Vol. II, Charles Griffin, London. [199]

J. Kiefer and J. Wolfowitz (1958), *Optimum Designs in Regression Problems*, mimeographed report, 35 pp., Cornell Univ. [ix]

S. Kolodziejczyk (1935), "On an important class of statistical hypotheses," *Biometrika*, Vol. 27, pp. 161–190. [32]

C. H. Kraft, personal communication. [230]

C. Y. Kramer (1956), "Extension of multiple range tests to group means with unequal numbers of replications," *Biometrics*, Vol. 12, pp. 307–310. [78]

W. H. Kruskal, personal communication. [5, 98, 136, 366]

A. M. Legendre (1806), *Nouvelles Méthodes pour la Détermination des Orbites des Comètes; avec un Supplément Contenant Divers Perfectionnements de ces Méthodes et Leur Application aux Deux Comètes de 1805*, Courcier, Paris. [9]

E. L. Lehmann and H. Scheffé (1955), "Completeness, similar regions, and unbiased estimation: Part II," *Sankhyā*, Vol. 15, pp. 219–236. [47]

E. L. Lehmann, *see* J. L. Hodges Jr., and . . . (1956).

E. L. Lehmann (1959*a*), *Testing Statistical Hypotheses*, John Wiley, New York. [33, 47, 50, 317]

E. L. Lehmann (1959*b*), "Optimum properties of the analysis of variance test," submitted to *Annals Math. Stat.* [47]

E. L. Lehmann, personal communication. [223]

E. Lehmer (1944), "Inverse tables of probabilities of errors of the second kind," *Annals Math. Stat.*, Vol. 15, pp. 388–398. [42]

G. J. Lieberman, *see* G. J. Resnikoff and . . . (1957).

P. K. Loraine (1952), "On a useful set of orthogonal comparisons," *J. Royal Stat. Soc.*, series B, Vol. 14, pp. 234–237. [129]

S. MacLane, *see* G. Birkhoff and . . . (1953).

W. G. Madow (1940), "The distribution of quadratic forms in noncentral normal random variables," *Annals Math. Stat.*, Vol. 11, pp. 100–103. [420]

W. G. Madow (1948), "On a source of downward bias in the analysis of variance and covariance," *Annals Math. Stat.*, Vol. 19, pp. 351–359. [135]

H. B. Mann (1949), *Analysis and Design of Experiments*, Dover Publications, New York. [44, 159]

S. Moriguti (1954), "Confidence limits for a variance component," *Reports of Stat. Applications in Research, Japanese Union of Scientists and Engineers*, Vol. 3, no. 2, pp. 29–41. [231]

M. E. Muller, personal communication. [5]

D. C. Murdoch (1957), *Linear Algebra for Undergraduates*, John Wiley, New York. [371, 387]

J. Neyman (1923), "Sur les applications de la théorie des probabilités aux expériences agricoles: Essay des principes," *Roczniki Nauk Rolniczch*, Vol. 10, pp. 1–51. Polish with German summary. [291]

J. Neyman and E. S. Pearson (1928), "On the use and interpretation of certain test criteria for purposes of statistical inference," *Biometrika*, Vol. 20A, pp. 175–240, 263–294. [32]

J. Neyman and E. S. Pearson (1931), "On the problem of k samples," *Bull. Acad. Polonaise Sci. Lett.*, series A, pp. 460–481. [83]

J. Neyman and E. S. Pearson (1933), "On the problem of the most efficient tests of statistical hypotheses," *Philos. Trans. Roy. Soc. London*, series A, Vol. 231, pp. 289–337. [46]

J. Neyman (1935), with cooperation of K. Iwaszkiewicz and S. Kolodziejczyk, "Statistical problems in agricultural experimentation," *J. Royal Stat. Soc., Suppl.*, Vol. 2, pp. 107–154. [291]

J. Neyman (1937), "Outline of a theory of statistical estimation based on the classical theory of probability," *Philos. Trans. Roy. Soc. London*, series A, Vol. 236, pp. 333–380. [29]

P. B. Patnaik (1949), "The noncentral χ^2 and F-distributions and their approximations," *Biometrika*, Vol. 36, pp. 202–232. [413, 414]

A. E. Paull (1950), "On a preliminary test for pooling mean squares in the analysis of variance," *Annals Math. Stat.*, Vol. 21, pp. 539–556. [126]
E. S. Pearson, *see* J. Neyman and . . . (1928).
E. S. Pearson (1929), "The distribution of frequency constants in small samples from non-normal symmetrical and skew populations," *Biometrika*, Vol. 21, pp. 259–286. [346, 350]
E. S. Pearson (1931), "The analysis of variance in cases of non-normal variation," *Biometrika*, Vol. 23, pp. 114–133. [333, 337, 346, 347]
E. S. Pearson, *see also* J. Neyman and . . . (1931).
E. S. Pearson, *see also* J. Neyman and . . . (1933).
E. S. Pearson and H. O. Hartley (1951), "Charts of the power function of the analysis of variance tests, derived from the non-central *F*-distribution," *Biometrika*, Vol. 38, pp. 112–130. [41]
E. S. Pearson and H. O. Hartley (1954), *Biometrika Tables for Statisticians*, Vol. I, Cambridge Univ. Press, Cambridge. [27, 83]
K. Pearson (1934), *Tables of the Incomplete Beta-Function*, Cambridge Univ. Press, Cambridge. [28, 135]
A. M. Peiser (1943), "Asymptotic formulas for significance levels of certain distributions," *Annals Math. Stat.*, Vol. 14, pp. 56–62. Correction, Vol. 20 (1949), pp. 128–129. [80]
K. C. S. Pillai and K. V. Ramachandran (1954), "On the distribution of the ratio of the i-th observation in an ordered sample from a normal population to an independent estimate of the standard deviation," *Annals Math. Stat.*, Vol. 25, pp. 565–572. [78]
E. J. G. Pitman (1937), "Significance tests which may be applied to samples from any populations: III. The analysis of variance test," *Biometrika*, Vol. 29, pp. 322–335. [291, 321, 322, 324, 325]
S. J. Pretorius (1930), "Skew bivariate frequency surfaces, examined in the light of numerical applications," *Biometrika*, Vol. 22, pp. 109–223. [333]
K. V. Ramachandran, *see* K. C. S. Pillai and . . . (1954).
C. R. Rao (1946), "On the linear combination of observations and the general theory of least squares," *Sankhyā*, Vol. 7, pp. 237–256. [200]
C. R. Rao (1952), *Advanced Statistical Methods in Biometric Research*, John Wiley, New York. [196, 205]
G. J. Resnikoff and G. J. Lieberman (1957), *Tables of the Non-central t-distribution*, Stanford Univ. Press, Stanford, Calif. [42]
M. Rosenblatt, *see* U. Grenander and . . . (1957).
H. J. Ryser, *see* R. H. Bruck and . . . (1949).
F. E. Satterthwaite (1946), "An approximate distribution of estimates of variance components," *Biometrics Bulletin*, Vol. 2, pp. 110–114. [247]
H. Scheffé (1942), "On the ratio of the variances of two normal populations," *Annals Math. Stat.*, Vol. 13, pp. 371–388. [360]
H. Scheffé (1943), "Statistical inference in the non-parametric case," *Annals Math. Stat.*, Vol. 14, pp. 305–332. [317]
H. Scheffé (1952), "An analysis of variance for paired comparisons," *J. Amer. Stat. Assoc.*, Vol. 47, pp. 381–400. [5]
H. Scheffé (1953), "A method for judging all contrasts in the analysis of variance," *Biometrika*, Vol. 40, pp. 87–104. [67, 69, 71, 77]
H. Scheffé, *see also* E. L. Lehmann and . . . (1955).

H. Scheffé (1956*a*), "A 'mixed model' for the analysis of variance," *Annals Math. Stat.*, Vol. 27, pp. 23–36. [261, 265, 270]

H. Scheffé (1956*b*), "Alternative models for the analysis of variance," *Annals Math. Stat.*, Vol. 27, pp. 251–271. [3, 221]

H. Scheffé (1958), "Fitting straight lines when one variable is controlled," *J. Amer. Stat. Assoc.*, Vol. 53, pp. 106–117. [214]

C. J. Seelye (1958), "Conditions for a positive definite quadratic form established by induction," *Amer. Math. Monthly*, Vol. 65, pp. 355–356. [398]

W. A. Shewhart [333]

S. S. Shrikhande (1951), "Designs for two-way elimination of heterogeneity," *Annals Math. Stat.*, Vol. 22, pp. 235–247. [169]

S. S. Shrikhande, *see also* H. O. Hartley, . . . , and W. B. Taylor (1953).

S. S. Shrikhande, *see also* R. C. Bose, W. H. Clatworthy, and . . . (1954).

S. D. Silvey (1954), "The asymptotic distributions of statistics arising in certain non-parametric tests," *Proc. Glasgow Math. Assoc.*, Vol. 2, pp. 47–51. [324]

H. F. Smith [247]

G. W. Snedecor (1934), *Analysis of Variance and Covariance*, Collegiate Press, Inc., Ames. [36]

D. A. Sprott (1956), "A note on combined interblock and intrablock estimation in incomplete block designs," *Annals Math. Stat.*, Vol. 27, pp. 633–641. Correction, Vol. 28 (1957), p. 269. [176]

C. M. Stein (1945), "A two-sample test for a linear hypothesis whose power is independent of the variance," *Annals Math. Stat.*, Vol. 16, pp. 243–258. [41, 65]

C. M. Stein (1948), "The selection of the largest of a number of means," Abstract 5, *Annals Math. Stat.*, Vol. 19, p. 429. In the displayed formula the denominator of the last fraction should be $2t_j$ instead of t_j. Five lines from bottom, $\ln\alpha$ means $\log_e \alpha$. [81]

C. M. Stein, *see also* G. A. Hunt and . . . .

Student (1927), "Errors of routine analysis," *Biometrika*, Vol. 19, pp. 151–164. [332, 333, 334]

P. C. Tang (1938), "The power function of the analysis of variance tests with tables and illustrations of their use," *Stat. Research Memoirs*, Vol. 2, pp. 126–149. [41]

G. Tarry (1900), "Le problème de 36 officiers," *Compte Rendu de l'Association Française pour l'Avancement de Science Naturel*, Vol. 1 (1900), pp. 122–123; Vol. 2 (1901), pp. 170–203. [159]

W. B. Taylor, *see* H. O. Hartley, S. S. Shrikhande, and . . . (1953).

W. A. Thompson Jr. (1955), "The ratio of variances in a variance components model," *Annals Math. Stat.*, Vol. 26, pp. 325–329. [221]

J. W. Tukey (1949*a*), "One degree of freedom for non-additivity," *Biometrics*, Vol. 5, pp. 232–242. [130, 134]

J. W. Tukey (1949*b*), "Dyadic anova, an analysis of variance for vectors," *Human Biology*, Vol. 21, pp. 65–110. [196]

J. W. Tukey (1949*c*), *Interaction in a Row-by-Column Design*, Memo. Report 18, Stat. Research Group, Princeton Univ., 14 pp., dittoed. This material is included in Cornfield and Tukey (1956). [241]

J. W. Tukey, *see also* M. F. Freeman and . . . (1950).

J. W. Tukey (1951), "Components in regression," *Biometrics*, Vol. 7, pp. 33–69. [231]

J. W. Tukey (1953), *The Problem of Multiple Comparisons*, dittoed MS of 396 pages, Princeton Univ. [78]

J. W. Tukey (1955), Answer to query no. 113, *Biometrics*, Vol. 11, pp. 111–113. [134, 144]

J. W. Tukey (1956), "Variances of variance components: I. Balanced designs." *Annals Math. Stat.*, Vol. 27, pp. 722–736. [229, 346]

J. W. Tukey, *see also* J. Cornfield and . . . (1956).

J. W. Tukey (1957*a*), "Variances of variance components: II. The unbalanced single classification," *Annals Math. Stat.*, Vol. 28, pp. 43–56. [229, 346]

J. W. Tukey (1957*b*), "Approximations to the upper 5% point of Fisher's B distribution and noncentral χ^2," *Biometrika*, Vol. 44, pp. 528–530. [414]

J. W. Tukey, personal communication. [83, 89, 273]

J. W. Tukey [67]

A. Wald (1940), "A note on the analysis of variance with unequal class frequencies," *Annals Math. Stat.*, Vol. 11, pp. 96–100. [224]

A. Wald (1942*a*), "On the power function of the analysis of variance test," *Annals Math. Stat.*, Vol. 13, pp. 434–439. [48]

A. Wald (1942*b*), *On the Principles of Statistical Inference*, Notre Dame Math. Lectures, no. 1, Edwards Brothers, Ann Arbor, Mich. [50]

A. Wald and J. Wolfowitz (1944), "Statistical tests based on permutations of the observations," *Annals Math. Stat.*, Vol. 15, pp. 358–372. [324]

G. S. Watson (1955), "Serial correlation in regression analysis," *Biometrika*, Vol. 42, pp. 327–341. [21]

B. L. Welch (1937), "On the z-test in randomized blocks and Latin Squares," *Biometrika*, Vol. 29, pp. 21–52. [291, 321, 324, 327, 328]

B. L. Welch (1951), "On the comparison of several mean values: an alternative approach," *Biometrika*, Vol. 38, pp. 330–336. [364]

R. A. Wijsman (1957), "Random orthogonal transformations and their use in some classical distribution problems in multivariate analysis," *Annals Math. Stat.*, Vol. 28, pp. 415–423. [418]

R. A. Wijsman, personal communication. [202, 421]

M. B. Wilk (1954), *The Logical Derivation of Linear Models for Experimental Situations and Their Use in Selecting the Appropriate Error Term in the Analysis of Variance. VI. Interactions with Experimental Units* (AV-10), Analysis of Variance Project, Stat. Laboratory, Iowa State College, Ames (dittoed report). [297]

M. B. Wilk (1955), *Linear Models and Randomized Experiments*, Ph.D. thesis, Iowa State College. [291]

M. B. Wilk and O. Kempthorne (1955), "Fixed, mixed, and random models in the analysis of variance," *J. Amer. Stat. Assoc.*, Vol. 50, pp. 1144–1167. [284]

M. B. Wilk and O. Kempthorne (1957), "Non-additivities in a Latin Square design," *J. Amer. Stat. Assoc.*, Vol. 52, pp. 218–236. [304, 309, 311]

S. S. Wilks (1932), "Certain generalizations of the analysis of variance," *Biometrika*, Vol. 24, pp. 471–494. [417]

J. Wolfowitz, *see* A. Wald and . . . (1944).

J. Wolfowitz, *see also* J. Kiefer and . . . (1958)

H. Working and H. Hotelling (1929), "Application of the theory of error to the interpretation of trends," *J. Amer. Stat. Assoc.*, Mar. Suppl., pp. 73–85. [29, 68]

F. Yates, *see* T. Eden and . . . (1933).

F. Yates (1935), Discussion of Neyman's 1935 paper, *J. Roy. Stat. Soc., Suppl.*, Vol. 2, pp. 161–166. [324]

F. Yates (1936), "Incomplete randomized blocks," *Annals Eugenics*, Vol. 7, pp. 121–140. [161]

F. Yates (1940), "The recovery of inter-block information in balanced incomplete block designs," *Annals Eugenics*, Vol. 10, pp. 317–325. [170]

F. Yates, *see also* R. A. Fisher and . . . (1943).

W. J. Youden and J. S. Hunter (1955), "Partially replicated Latin Squares," *Biometrics*, Vol. 11, pp. 399–405. (Dr. Hunter kindly informed me of the following misprints: In the formula at the bottom of p. 409 for the adjusted treatment mean the 6 should be replaced by G, and in the formula at the top of p. 410 for the estimated variance of a difference of adjusted treatment means the $(k + 1)$ should be replaced by $(k + 3)$.) [160]

Subject Index